VEGETATION ECOLOGY

Vegetation Ecology

Edited by
Eddy van der Maarel

Blackwell
Publishing

BLACKWELL PUBLISHING
350 Main Street, Malden, MA 02148-5020, USA
108 Cowley Road, Oxford OX4 1JF, UK
550 Swanston Street, Carlton, Victoria 3053, Australia

First published 2005 by Blackwell Science Ltd

Library of Congress Cataloging-in-Publication Data

Vegetation ecology / edited by Eddy van der Maarel.
p. cm.
Includes bibliographical references and index.
ISBN 0-632-05761-0 (pbk. : alk. paper)
1. Plant ecology. 2. Plant communities. I. van der Maarel, E.

QK901.V35 2005
581.7—dc22

2004006037

A catalogue record for this title is available from the British Library.

Set in 9.5/12pt Caslon
by Graphicraft Limited, Hong Kong
Printed and bound in the United Kingdom
by TJ International, Padstow, Cornwall

The publisher's policy is to use permanent paper from mills that operate a sustainable forestry policy, and which has been manufactured from pulp processed using acid-free and elementary chlorine-free practices. Furthermore, the publisher ensures that the text paper and cover board used have met acceptable environmental accreditation standards.

For further information on
Blackwell Publishing, visit our website:
www.blackwellpublishing.com

Contents

Contributors

Dr Mike P. Austin, *CSIRO Sustainable Ecosystems, G.P.O. 284, Canberra ACT 2601, Australia*
Tel. +61 2 62421758; Fax +61 2 62421705; e-mail mike.austin@csiro.au

Prof. Dr Jan P. Bakker, *Community and Conservation Ecology Group, University of Groningen, P.B. 14, NL-9750 AA Haren, The Netherlands*
Tel. +31 50 3632221; Fax +31 50 3632273; e-mail j.p.bakker@biol.rug.nl

Dr Robert Baxter, *Environmental Research Centre, School of Biological and Biomedical Sciences, University of Durham, South Road, Durham DH1 3LE, United Kingdom*
Tel. +44 191 3341261; Fax +44 191 3341201; e-mail robert.baxter@durham.ac.uk

Susanne Bonn, *Institute of Botany, Faculty of Biology, University of Regensburg, D-93040 Regensburg, Germany*
Tel. +49 711 6016715; Fax +49 941 9433106; e-mail susanne.bonn@biologie.uni-regensburg.de

Prof. Elgene O. Box, *Department of Geography, University of Georgia, Athens, GA 30602-2502, USA*
Tel. +1 706 5422369; Fax +1 706 5422388; e-mail boxeo@uga.edu

Dr Mary L. Cadenasso, *Institute of Ecosystem Studies, P.B. AB, Millbrook, NY 12545-0129, USA*
Tel. +1 845 6775343; Fax +1 845 6775976; e-mail cadenassom@ecostudies.org

Dr Bengt Å. Carlsson, *Department of Ecology and Evolution, Uppsala University, Villavägen 14, SE-752 36 Uppsala, Sweden*
Tel. +46 18 4712891; Fax +46 18 553419; e-mail bengt.carlsson@ebc.uu.se

Prof. Kazue Fujiwara, *Vegetation Science, Department of Natural Environment & Information, Yokohama National University, Tokiwadai 79-7, Hodogaya-ku, Yokohama 240-8501, Japan*
Tel. +81 45 3394355; Fax +81 45 3394370; e-mail kazue@ynu.ac.jp

Dr Ron G.M. de Goede, *Department of Soil Quality, Wageningen University, P.O. Box 8005, 6700 EC Wageningen, The Netherlands*
Tel. +31 317 485048; Fax +31 317 483766; e-mail ron.degoede@wur.nl

Prof. Brian Huntley, *Environmental Research Centre, School of Biological and Biomedical Sciences, University of Durham, South Road, Durham DH1 3LE, United Kingdom*
Tel. +44 191 3341282; Fax +44 191 3341201; e-mail brian.huntley@durham.ac.uk

Dr Thomas W. Kuyper, *Department of Soil Quality, Wageningen University, P.O. Box 8005, 6700 EC Wageningen, The Netherlands*
Tel. +31 317 482352; Fax +31 317 483766; e-mail thom.kuyper@wur.nl

Prof. Jan Lepš, *Faculty of Biological Sciences, University of South Bohemia, and Institute of Entomology, Czech Academy of Sciences, Branišovská 31, CZ-370 05 české Budějovice, Czech Republic*
Tel. +420 38 7772374; Fax +420 38 7772345; e-mail suspa@bf.jcu.cz

Prof. Dr Christoph Leuschner, *Pflanzenökologie, Abt. Ökologie und Ökosystemforschung, Albrecht-von-Haller-Institut für Pflanzenwissenschaften, Universität Göttingen, Untere Karspüle 2, D-37073 Göttingen, Germany*
Tel. +49 551 395718; Fax +49 551 395701; e-mail cleusch@gwdg.de

Prof. Samuel J. McNaughton, *Biological Research Labs, Syracuse University, 130 College Place, Syracuse, NY 13244-1220, USA*
Tel. +1 315 4433507; Fax +1 315 4432012; e-mail sjmcnaug@mailbox.syr.edu

Dr Steward T.A. Pickett, *Institute of Ecosystem Studies, P.B. AB, Millbrook, NY 12545-0129, USA*
Tel. +1 845 6775343; Fax +1 845 6775976; e-mail picketts@ecostudies.org

Prof. Dr Peter Poschlod, *Institute of Botany, Faculty of Biology, University of Regensburg, D-93040 Regensburg, Germany*
Tel. +49 941 9433108; Fax +499419433106; e-mail peter.poschlod@biologie.uni-regensburg.de

Dr Petr Pyšek, *Institute of Botany, Academy of Sciences of the Czech Republic, CZ-252 43 Průhonice, Czech Republic*
Tel. +420 27 1015266; Fax +420 27 1015266; e-mail pysek@ibot.cas.cz

Prof. Marcel Rejmánek, *Department of Evolution & Ecology, University of California, Davis, CA 95616, USA*
Tel. +1 530 7521092; Fax +1 530 7521449; e-mail mrejmanek@ucdavis.edu

Dr David M. Richardson, *Institute for Plant Conservation, Botany Department, University of Cape Town, Rondebosch 7700, South Africa*
Tel. +27 21 6502440; Fax +27 21 6503726; e-mail rich@botzoo.uct.ac.za

Prof. Håkan Rydin, *Department of Ecology and Evolution, Uppsala University, Villavägen 14, SE-752 36 Uppsala, Sweden*
Tel. +46 18 4712854; Fax +46 18 553419; e-mail hakan.rydin@ebc.uu.se

Dr Mahesh Sankaran, *Centre for Population Biology, Imperial College at Silwood Park, Ascot, Berkshire, SL57PY, UK; also Natural Resource Ecology Laboratory, Colorado State University, Fort Collins, CO 80523-1499, USA*
Tel. +1 970 4911964; Fax +1 970 4911965; e-mail mahesh@nrel.colostate.edu

Dr Brita M. Svensson, *Department of Ecology and Evolution, Uppsala University, Villavägen 14, SE-752 36 Uppsala, Sweden*
Tel. +46 18 4712879; Fax +46 18 553419; e-mail brita.svensson@ebc.uu.se

Dr Oliver Tackenberg, *Institute of Botany, Faculty of Biology, University of Regensburg, D-93040 Regensburg, Germany*
Tel. +49 941 9433105; Fax +499419433106; e-mail oliver.tackenberg@biologie.uni-regensburg.de

Prof. Dr Jelte van Andel, *Community and Conservation Ecology Group, CEES, University of Groningen P.B. 14, NL-9750 AA Haren, The Netherlands*
Tel. +31 50 3632224; Fax +31 50 3632273; e-mail j.van.andel@biol.rug.nl

Prof. Dr Eddy van der Maarel, *Community and Conservation Ecology Group, CEES, University of Groningen, P.B. 14, NL-9750 AA Haren, The Netherlands*
Tel. +31 561 430760; 06 63421030; Fax +31 561 430941; e-mail eddy.arteco@planet.nl

Preface

This book is a multi-authored account of the many-sided topic of vegetation ecology. It is usual nowadays to involve many authors in a textbook since it is almost impossible for one, or a few, authors to cover the full breadth of a modern field of science. As editor I have certainly had some influence on the choice and contents of the various chapters, but nevertheless the chapters are independent essays on important aspects of vegetation ecology.

During the development of the first, introductory chapter the idea grew not only to introduce some concepts, but also to present some more personal views in a historical perspective on developments in various parts of the world. In this connection I have taken the chance to comment on the use (and abuse) of terms borrowed from other sciences and from society which are less appropriate for vegetation ecology or even ecology at large. I have also paid some attention to topics either not, or only partly, covered by the following chapters and discuss various important new publications which could not all be elaborated in the individual chapters.

Vegetation ecology is linked to a series of other bio- and geoscience disciplines such as animal ecology, climatology, nature conservation, palaeo-ecology, plant geography, plant physiology and population ecology. Some of these transitional fields are, at least partly, treated in separate chapters; others could only be mentioned in passing. In retrospect, two actual topics which were not treated in full in any of the chapters, 'pattern and process in the plant community' and 'plant functional types' deserve special attention, and are treated in sections 1.15 and 1.16 of the introductory chapter, respectively. In order to keep the book to its contracted length, we have cited original research publications only if they were considered to be crucial and were not covered in any generally available textbook or review article. However, in some expanding fields, the literature from the 1990s and even from this century is overwhelming and more citations were necessary here.

I have enjoyed the lively cooperation of the chapter authors and hope that their chapters will be appreciated both as essays in their own right and as intrinsic parts of this book. I should like to thank first of all my wife Marijke van der Maarel for her continuous support and the careful copy-editing of all chapters before they were submitted to the publishers. I am also grateful to the following people – listed in chronological order – who have helped in organizational and technical matters on behalf of Blackwell Publishing: Ian Sherman, Sarah Shannon, William Maddox, Rosie Hayden, Cee Pike and Janey Fisher.

Finally I hope that this book will find its way across the world of vegetation scientists and plant ecologists.

Eddy van der Maarel

We are grateful to all of the authors and publishers who have granted permission for us to reproduce material in this book, and would be pleased to be notified of any omissions.

1

Vegetation ecology – an overview

Eddy van der Maarel

1.1 Vegetation ecology, vegetation and plant community

1.1.1 Vegetation, phytocoenose, plant community

Vegetation ecology, the study of the plant cover and its relationships with the environment, also called synecology, is a complex scientific undertaking, both regarding the overwhelming variation of its object of study in space and time, and its intricate interactions with abiotic and biotic factors. It is also a very modern science with important applications in well-known social activities, notably nature management, in particular the preservation of biodiversity, sustainable use of natural resources, and detecting 'global change' in the plant cover of the Earth.

Vegetation, the central object of study in vegetation ecology, can be loosely defined as a system of largely spontaneously growing plants. Not all growing plants form vegetation; for instance, a sown corn field or a flowerbed in a garden do not. But the weeds surrounding such plants do form vegetation. A pine plantation will become vegetation after some years of spontaneous growth of the pine trees and the subsequent development of an understorey.

From the early 19th century on, vegetation scientists have studied pieces of vegetation which they considered samples of a plant community (see Mueller-Dombois & Ellenberg 1974; Allen & Hoekstra 1992). Intuitively, and later on explicitly, such stands were selected on the basis of uniformity and discreteness. The vegetation included in the sample should look uniform and should be discernable from surrounding vegetation. From early on, plant communities have been discussed as possibly or certainly integrated units which can be studied as such and classified.

In order to elucidate these points a distinction is necessary between concrete stands of vegetation and the abstract concept of the plant community, a distinction not explicitly found in the works of most early European and American vegetation scientists, but characteristic of the Braun-Blanquet approach, i.e. phytosociology as it was developed in Central Europe, notably by J. Braun-Blanquet, with a strong emphasis on typology, the establishment of plant community types based on descriptions of stands (called *relevés*).

Westhoff & van der Maarel (1978) proposed to reserve the term 'phytocoenose' for the concrete stand of vegetation; their definition may be reformulated as 'a piece of vegetation in a uniform environment with a relatively uniform floristic composition and structure that is distinct from the surrounding vegetation'. The corresponding definition of the abstract plant community or phytocoenon is a type of phytocoenose derived from the characterization of a group of phytocoenoses corresponding with each other in all characters that are considered typologically relevant. A plant community can be conveniently studied while separated from its abiotic and biotic environment with which it forms an ecosystem, even if this separation is artificial. Conceptually this can be solved by extrapolating a phytocoenose to a biocoenose by including all other organisms interacting with the above-ground and below-ground parts of the phytocoenose. For a community of birds, insects, molluscs or any other taxonomic group under study (see Barkman 1978) we adopt the term taxocoenosis and for all organisms included in a part of the ecosystem, e.g. all on a certain trophic level or in one layer of vegetation, the term merocoenose can be used (Westhoff & van der Maarel 1978).

As follows from the definition, the delimitation in the field of phytocoenoses is based on distinctiveness and uniformity. Distinctiveness of a stand has been much discussed and interpreted. Distinctiveness implies discontinuity towards the surrounding vegetation. This can be environmentally obvious, for example in the case of a depression in a dry area, or, in a man-made landscape, roadside vegetation between the road and a ditch. However, more usually the distribution of the local plant populations should be checked. This has been actual since H.A. Gleason (e.g. 1926) observed that species are 'individualistically' distributed along omni-present environmental gradients and thus cannot form bounded communities (Nicholson & McIntosh 2002). Note that this observation referred to stands of vegetation, even if the word 'community' was used! The wealth of literature on ordination (see Chapter 2) offered ample proof of the 'continuum concept of vegetation' (McIntosh, see Nicholson & McIntosh 2002).

Gleason and many of his adherers were of the opinion that plant species could not form integrated communities because of their individualistic behaviour and criticized the community concept of F.E. Clements (e.g. 1916; see also Mueller-Dombois & Ellenberg 1974), the pioneer in succession theory, who compared the community with an organism and, apparently, recognized plant community units in the field. However, this 'holistic approach' to the plant community had little to do with the recognition of phytocoenoses in the field.

Shipley & Keddy (1987) simplified the controversy by reducing it to the occurrence of different boundary patterns in the field. They devised a field method to test the 'individualistic and community-unit concepts as falsifiable hypotheses'. They detected the concentration of species distribution boundaries at certain points along environmental gradients. In their study – as in other studies – boundary clusters are found in some cases and not in others. Coincidence of distribution boundaries occur at a steep part of an environmental gradient, and at places with a sharp spatial boundary or strong fluctuations in environmental conditions.

The occurrence of different boundary situations as such is of theoretical importance. They can be linked to the two types of boundary distinguished by van C.G. van Leeuwen and put in a vegetation ecological framework by van der Maarel (e.g. 1990). The first type is the limes convergens which can be identified with ecotones s.s. or tension zones. Here species boundaries can be determined strictly by abiotic conditions, but interference between species may play a part (e.g. Shipley & Keddy 1987), and the ecotone may also be caused or sharpened by plants, the so-called vegetation switch (Wilson & Agnew 1992). The opposite type of boundary, limes divergens or ecocline, is typically what we now call a gradient situation where species reach local distribution boundaries in an 'individualistic' way (van der Maarel 1990).

Despite the general appreciation of the individualistic character of species distributions, Gleason himself has never doubted the reality of plant communities. He even used the term association – albeit in a different way from European phytosociologists. Later 'Gleasonians' also recognized that 'there is a certain pattern to the vegetation with more or less similar groups of species recurring from place to place' (Curtis 1959). This was further elucidated by R.H. Whittaker (e.g. 1978; see also Mueller-Dombois & Ellenberg 1974). It is thus quite possible to arrive at plant community types by comparing phytocoenoses which may lack sharp boundaries – or/and floristic uniformity – but which appear to be sufficiently similar. The same situation occurs in soil science.

Unlike distinctiveness, the aspect of uniformity has been more generally accepted as a prerequisite for the plant community. Most vegetation ecologists agree on a certain uniformity a phytocoenose should show in order to be included in any plant community study. A check on the uniformity would include that, at least over a certain area, the vegetation has the same appearance, physiognomy, i.e. the same height and the same plant species in a dominant position. In addition the floristic composition should not vary too much. In very few cases quantitative checks have been developed, such as in the school of Curtis (e.g. 1959) where quadrants of a stand-to-be were compared with a $\chi 2$ test for homogeneity (note that statistical homogeneity would imply a random distribution of plants over the stand and this restriction is too rigorous; therefore the term uniformity is preferred, even if its test is not quantitative). Together with the vegetation the environment is checked for obvious variation, for instance water level.

Even Gleason, although convinced of the continuity in vegetational variation (see Nicholson & McIntosh 2002), considered uniformity as essential. However, Gleason and others have also recognized that stands of vegetation are seldom uniform. Apparently a certain degree of variation within a phytocoenose is generally accepted. See further Chapter 2.

1.1.2 Plant communities: integrated, discrete units?

Within the neutral definitions of plant community and phytocoenose, quite different ideas and opinions on the essence of the plant community have been expressed since the early 20th century and the discussion is still going on. The above-mentioned controversy between Clements and Gleason has been an important element in this discussion and also a confusing one because it has not always been clear whether statements referred to the phytocoenose or the phytocoenon level. As a concluding remark on this controversy and an introduction to this section, an original interpretation of the difference between the two masters by Allen & Hoekstra (1992) and a comment on this interpretation will follow. The interpretation is based on the differences in the landscapes the two grew up in: Clements was brought up in the prairie landscape of Nebraska and viewed plant communities as units from horseback, while Gleason walked through the forest, from tree to tree, aware of the small-scale differences within the community. Thus, the different environments may have had a decisive influence on their 'perspective'.

However, two outstanding European contemporaries of Clements and Gleason do not fit this interpretation at all! The Russian plant ecologist G.I. Ramenskiy, who is generally considered the father of ordination and who was a Gleasonian *avant la lettre*, demonstrated the individuality of species distributions along gradients with meadow vegetation. On the other hand, the Finnish forest ecologist A.K. Cajander developed an authoritative typology of Finnish forests (e.g. Trass & Malmer 1978). Apparently, emphasizing continuities or rather discontinuities can be done in any plant community type and this has to do with intellectual attitude rather than upbringing and field experience.

The ideas of Clements and Gleason can be seen as different concepts of the plant community. Westhoff & van der Maarel (1978) distinguished them from each other as 'organismal concept' and 'individualistic concept', respectively, and from two other concepts: the 'social structure' concept and the 'population structure' concept. The concept of social structure goes back to J.K. Paczoski (see Braun-Blanquet 1932) and early Russian authors, who recognized the multitude of interactions between species, such as competition, symbiosis and saprophytism. The population structure was highlighted by R.H. Whittaker (e.g. Whittaker 1975; White 1985). It added to the community concept that the properties of each participating plant species are not only determined by interactions between species but also within species.

More recently, ecologists became interested in the possibility (for others the necessity) that a plant community has (must have) 'emergent properties', causing the whole to be more than the sum of its parts. One of the few properties investigated with this idea in mind was the development of dominance and diversity. Wilson *et al.* (1998) considered that relative abundance distributions are 'an important feature of community structure', and Wilson (1999a) discussed different types of assembly rule for plant communities, based on (i) the performance of individual species, (ii) species numbers, (iii) plant characters (traits) and (iv) species abundance relations, while emphasizing the need to compare actual community properties with a null model assuming the absence of such rules. Weiher & Keddy (1999) highlighted the relevance of constraints in the representation of traits in relation to environmental variation and suggested that traits related to the availability of mineral resources, such as maximum biomass and leaf shape, are more tightly constrained as soil fertility increases. They proposed the term 'assembly rules' as a key concept. They suggest that the plant community needs more structure, 'order', if more energy flows through the system. Finally, Grime (2001) paid attention to the mechanisms of plant community assembly. He concentrated on dominance-diversity relations while dividing the participating species into dominant, subordinate and transient, and taking into account the different plant functional types which play a part. When we look at the plant community as a component of an ecosystem, overall community characteristics such as a certain level of evapotranspiration and biomass production are considered emergent properties as well (Chapter 3).

One or more of these different plant community concepts are reflected in the many plant community definitions available. The definition by Westhoff & van der Maarel (1978) is representative of phytosociology as it was developed in Central Europe, notably by J. Braun-Blanquet, and in Northern Europe by G.E. Du Rietz. However, it also reflects ideas from early Anglo-American plant ecology, both in Great Britain (A.G. Tansley) and the USA (F.E. Clements), notably the emphasis on the interrelations between community and environment and on species interactions. The definition is in fact a double one because of the separation between concrete and abstract units (see above). The concrete phytocoenose is defined as 'a part of a vegetation consisting of interacting populations growing in a uniform environment and showing a floristic composition and structure that is relatively uniform and distinct from the surrounding vegetation'.

Several later definitions of the plant community reflected the outcome of the more recent debates on the holistic and individualistic concepts, and on the reality of emergent properties. Looijen & van Andel (1999) reviewed some community definitions (but not the above-mentioned one) and focused on problems of ambiguity and distinctiveness. They tried to cope with this problem by redefining the community as 'a set of individuals that occur in the intersection of the areas occupied by populations of these species'. The attempt to spatially delimit a community in this way is theoretically possible but in practice it will fail; also, the delimitation of population boundaries is dependent on the scale of observation (Parker 2001). Parker developed a 'focal-individual model' of a community, which is based on a web of interactions between individuals. According to this model the plant community has no fixed boundaries, neither in space nor in time. The 'phenomenological' definition by Grootjans *et al.* (1996) tries to avoid any element of discussion but maintains the need for a spatial limit. They add a time dimension because of the changes over time of a community which is not necessarily undergoing succession: 'largely spontaneously growing plant individuals which are present within a distinguishable space-time unit'.

In view of these comments, the 1978 definition should be adapted. First, in extreme environments, the representatives of one taxonomical group under study, for instance vascular plants, may belong to only one species, for instance *Salicornia europaea* in European low salt marshes or *Ipomoea pes-caprae* on tropical beaches. So we had better delete the implicit assumption that at least two species are involved. Second, it may be realistic to avoid a discussion on whether the boundary of a phytocoenosis coincides with the formal or actual boundary of the populations of participating species. Third, it may not be necessary to include interactions. For most of us these are obvious and universal, but theoretical cases of plants growing together without interactions should not be excluded a priori. In conclusion the following definition may be presented:

> a phytocoenosis is a piece of vegetation in a uniform environment with a relatively uniform floristic composition and structure that is distinct from the surrounding vegetation.

1.1.3 Plant communities and plant community types

From the phytocoenosis to the plant community is not only moving from analysis to typology and from the concrete to the abstract, it is also changing the perspective from a neutral description of the local plant cover to a certain conception of the community.

Plant community types must be based on characteristics analysed in phytocoenoses. Naturally, these are derived from the plant species present. Originally, the decisive characteristic was the physiognomy, i.e. the dominance of certain growth forms such as trees, shrubs and grasses. The different physiognomic types were called formations and were usually described for large areas by plant geographers, such as E. Warming (see Mueller-Dombois & Ellenberg 1974). Later on the combination of plant species, the floristic composition, became decisive, albeit in combination with physiognomy (implicitly or explicitly). For this community type the term association became standard under the definition adopted at the 1910 Botanical Congress (see section 1.1.1): 'A plant community of definitive floristic composition, presenting a uniform physiognomy, and growing in uniform habitat conditions. The association is the fundamental unit of synecology' (see Westhoff & van der Maarel 1978).

The idea of a community type can be conceived irrespective of boundaries, either between phytocoenoses in the field or between phytocoena in environmental or ordination space. R. Tüxen considered a type as an ideal concept – in line with German philosophers – which could empirically be recognized as a 'correlation concentrate'. Tüxen's idea was elaborated by H. von Glahn who distinguished three steps in classification: (i) identification, through reconnaissance and comparison; (ii) elaboration of a maximal correlative concentration, i.e. first of vegetation, second of environmental characteristics, through tabular treatment (and nowadays multivariate methods); and (iii) systematic categorization, i.e. arranging the type in a system of plant communities (see Westhoff & van der Maarel 1978). The concise description of the classification process by Whittaker (e.g. 1978) came very close to this European approach. See further Chapter 4.

1.1.4 Above-ground and below-ground components of the plant community

Vegetation studies are usually restricted to the above-ground components, even if it is known long since (see, e.g. Braun-Blanquet 1932) that the below-ground components are of decisive importance for the anchoring of plants, the uptake of water and nutrients, and the storage of photosynthates. Most of the large biomass is made up by roots (including storage organs such as corms, bulbs, tubers and crowns, showing stratification and occurring in various growth forms) (e.g. McMichael & Persson 1991), rhizomes (horizontal stems sending out shoots upward and roots downward) and seeds. It would be logical, although technically complicated and inevitably semi-destructive, to include below-ground components in the definition and description of plant communities. Nevertheless this has not become common practice, not even for the soil seed bank (see, however, Chapter 6).

Also, root-related phenomena such as nitrogen-fixing and and mycorrhizal symbioses are well known as important but they were only recently included in vegetation studies (see Chapter 11). Evidently, the dense contacts between roots, biological turnover (through biomass consumption and decomposition, humus formation and partial re-use of mineralized components) and nutrient cycling are convincing contributions to the notion of the integrated plant community.

1.1.5 Vegetation observed at different spatial and temporal scales and levels of integration

The plant community as defined above is a realistic concept only at a certain scale of observation, i.e. the scale at which it is possible to judge the relative uniformity and distinctness. This 'community scale' will vary with the structure of the community, from some m^2 for short grassland to several thousand m^2 in giant forest. Within the relative uniformity many plant communities show differentiation, both vertically and horizontally. The vertical differentiation is most pronounced in woody vegetation where the different layers – tree, shrub, herb and moss layers – are usually described separately. Particularly in Fennoscandia such layers were found to occur independently of each other, for instance a moss layer of a certain composition could occur both in heathland and pine forest. Horizontal differentiation arises through animal excrements, bark formation on trees, fallen trees, dead shrubs and grasses. For these partial phytocoenoses and other microcommunities of other organisms than vascular plants, mosses and lichens (as usually included in vegetation studies), the term *synusia* (plural *synusiae*) has come into existence, largely through the work of H. Gams (see Barkman 1978). Such microcommunities are dependent on the phytocoenosis they form part of – and are therefore also indicated as dependent communities – but are units in their own right. An example is the work of J.J. Barkman on microcommunities within juniper scrub, formally described as *Dicrano-Juniperetum*, which differ in floristic composition and particularly in microclimate (Stoutjesdijk & Barkman 1992).

Many of these synusiae have been subjected to formal classification and environmental characterization (Westhoff & van der Maarel 1978). Many microcommunities result from internal changes in the community, including the natural death of plants, animal activities and other disturbances.

Plant communities are also part of larger units. In the usual hierarchy the next higher unit above the community is the ecosystem, which in its turn is part of a **biome**, a formation together with its fauna and environment. In vegetation ecology communities are rather considered as components of landscape units (**ecotopes**). Mueller-Dombois & Ellenberg (1974) distinguish four types of community complex:

1 *Mosaic complex*, such as the hummock-hollow complex in bogs;
2 *Zonation complex* along a local gradient, e.g. a lake shore;
3 *Vegetation region*, roughly equivalent to a formation;
4 *Vegetation belt*, a zonation complex along an elevational gradient, *inter alia* a mountain.

In the Braun-Blanquet approach a formal typology of such units has been developed by R. Tüxen, based on the analysis of the occurrence of associations and other syntaxa and distinguishing characteristic syntaxon combinations, so-called **sigma-syntaxa** or **sigmeta**. Such vegetation complexes are united into **geosigmeta**, being the phytosociological characterization of larger landscape types (see Schwabe 1989; Dierschke 1994).

Botanical object	Ecological object	Plant phys	Plant ecol	Plant geo	Pop ecol	Veg ecol	Micro clim	Soil sci	Land ecol	Clim sci	Geo chem
1. Cell											
2. Organism	Organism–environment system										
3. Population	Population–environment system										
4. Microcommunity	Microcommunity–environment ecosystem										
5. Plant community	Phytocoenosis–environment ecosystem										
6. Plant community complex	Regional ecosystem complex - landscape										
7. Formation	Biome										
	Biosphere (ecosphere)										

Fig. 1.1 Objects of botanical, ecological and environmental studies – plant physiology; plant ecology; plant geography; population ecology; microclimate studies; vegetation ecology; soil science; landscape ecology; climatology; geochemistry – at different levels of organization and different scales of observations.

Textbooks usually present 'levels of organization': for instance Allen & Hoekstra (1992): 1. Cell ς, 2. Organism ς, 3. Population ς, 4. Community ς, 5. Ecosystem ς, 6. Landscape ς, 7. Biome ς, 8. Biosphere. However, the objects of study arranged in such sequences are not all comparable. Even if plant organisms and populations appear as community constituents in many community definitions, they are not organizational subunits of ecosystems.

We should in fact distinguish two sequences (Fig. 1.1). Moreover, as Allen & Hoekstra (1992) made clear, there is some confusion between level of organization and scale of observation which run parallel, but only to a certain extent. They give the (slightly exaggerated) example of moss populations (level 3) on a dead tree (level 2) in a forest (4) which occur in a moss community complex (6) while the organisms in the rotting log form an ecosystem (5). What they mean is that at one scale of observation quite different levels of organization are studied. Conversely, different scales of observation can be included at one level of organization. A vegetation ecologist may study quadrats of 1 m^2 in short grassland or moss communities and plots of 10^4 m^2 in rain forest. Detailed vegetation maps have been made of only 10^{-2} km^2 while the phytosociological vegetation map of Europe covers *c.* 10^7 km^2.

Finally, as also indicated in Fig. 1.1, each discipline or approach involved in the study of plants and ecosystems, respectively, usually extends beyond its 'central' level of organization. The intricate relations between organization and scale are extended by including temporal scales (section 1.7). A summary of these considerations is presented in Fig. 1.2, which combines a scheme relating levels of organization to temporal scales of vegetation dynamics with a scheme relating spatial to temporal research scales. Essential elements in the hierarchical approach of organization levels and scales are the recognition of (i) mosaic structures, with elements of a mosaic of a smaller grain size being mosaics of their own at a larger grain size; (ii) different processes governing patterns at different scales; and (iii) different degrees of correlation between vegetational and environmental variables at different grain sizes.

	Fluctuation	Gap, patch dynamics	Cyclic succession	Secondary succession	Primary succession	Secular succession
Organism–environment	10^{-1}–1 yr 10^{-2}–10 m	1–10 yr 10^{-2}–10 m				
Population–environment	1 yr 1–10 m	1–10 yr 10–10^{2} m	1–10^{2} yr 10–10^{2} m	10–10^{2} yr 10–10^{2} m	10–10^{3} yr 10–10^{2} m	
Microcommunity–environment	1 yr 1–10 m	1–10 yr 10–10^{2} m				
Phytocoenosis–environment	1 yr 1–10 m	1–10 yr 10–10^{2} m	1–10^{2} yr 10–10^{2} m	10–10^{2} yr 10–10^{2} m	10–10^{3} yr 10–10^{2} m	
Regional landscape				10^{0}–10^{2} yr 10^{2}–10^{4} m	10^{2}–10^{3} yr 10^{2}–10^{4} m	10^{2}–10^{4} yr 10^{2}–10^{4} m
Biome						10^{3}–10^{6} yr 10^{4}–10^{6} m
Biosphere						10^{6}–10^{7} yr

Fig. 1.2 Spatial scales (m) and temporal scales (yr) of studies of ecological objects and their dynamics, based on adapted schemes of van der Maarel (1988) and Gurevitch *et al.* (2002).

1.1.6 Vegetation survey and sampling

Whatever our aim, approach and scale of observation, vegetation, whether loosely defined or approached as a phytocoenosis or a unit on a higher level of integration, should be described and measured. Vegetation characteristics are either derived from plant morphological characters, usually called **structure**, or from the plant species recognized, the **floristic composition**. Our description or analysis will only include a relatively small piece of vegetation which is considered representative of a larger unit. This leads to the issue of **sampling**.

In statistics the members of the universe can usually be identified without problems, e.g. the individuals of a crop and the trees in a plantation. In vegetation ecology this is much less simple. Moreover, the variables we measure and compare are partly composite and difficult to measure, notably the species composition, or only measurable through destructive sampling, notably above-ground biomass (and most below-ground characteristics as well).

Our universe to be sampled is the total area occupied by a certain type of vegetation. However, (i) it is difficult to identify this type, even if we have some previous knowledge about it; and (ii) this total area can be too large to be encompassed in one sampling. The first problem can be approached as follows – partly following the classical textbook by Cain & Castro (1959): sampling is preceded by **reconnaissance,** an inspection of a local area where the vegetation pattern with its dominant species and species combinations is recognized and related to topography and other apparent environmental conditions. The second problem can be solved by restricting ourselves to a 'local universe'.

The reconnaissance is followed by a **primary survey,** including a brief description of the dominant vegetation types. Of course the areal extent and amount of detail will depend on the specific objective of the study. Amongst the many different objectives four common ones are:

1 Phytosociological, with the intention to analyse phytocoenoses for a subsequent community classification, either of one particular type, or, preferably, of the all the types locally recognized;

2 Ecological, with the intention to correlate the local variation in vegetation composition with variation in environmental factors; ideally ecological and phytosociological sampling are integrated;

3 Dynamical, with the intention to establish or revisit pieces of vegetation for describing vegetation changes;

4 Applied, for instance to investigate the effect of a management measure.

The next phase in the analysis is an **intensive survey**, usually including a more complete description and measurement of the structure and species composition of the vegetation, and analysis of soil and microclimate characteristics.

In any case a unit of investigation has to be located and delimited. Since this is usually a well-delineated piece of vegetation the indication plot or **sample plot** is obvious. Other terms in use include stand, site and **quadrat**, a sample area within a frame, usually a square. For time series of observations the terms **permanent plot** or **permanent quadrat** are in use. Often terms for the analysis of the plot – analysis itself, sample, record, relevé – are used as equivalent to the sample plot. We better stick to the terms **plot** and **sample**, while adding the term **relevé** (French for record) for the special case it was originally meant for: a phytosociological record of a phytocoenosis for classification purposes.

As to the phytosociological sampling objective, the Braun-Blanquet approach has often been criticized (e.g. by Mueller-Dombois & Ellenberg 1974) for a neglect of the primary survey and a 'subjective' selection of sample plots which are recognized as representative stands of a plant community type from which relevés have been taken elsewhere. The problem with selective sampling is not so much that the sample is not representative but rather that other, related stands of vegetation are not sampled. However, the personal bias of phytosociologists and the lack of representative samples will be compensated for if larger numbers of samples of a certain type are available. This is the case in most European countries. For example, the survey of British plant communities (Rodwell 1991–2000) was based on 35,000 samples. In the survey of plant communities of The Netherlands (Schaminée *et al.* 1995–1999) lower units are documented by up to several hundreds of relevés, and higher units by several thousands. The database behind this survey, which is still growing, contained 350,000 samples (relevés) in 2001, and software to handle such huge data sets has been developed (Hennekens & Schaminée 2001).

If vegetation analysis proceeds on the basis of sample plots the plot is usually analysed completely, at least regarding species composition ('**single-plot analysis**'). In certain cases and in certain traditions **multiple-plot analysis** is preferred on the basis of systematically or randomly located small squares.

In cases where the delineation of a sample plot is not possible – or not desired – so-called **plotless sampling** can be applied (see Mueller-Dombois & Ellenberg 1974). This proceeds along lines where contacts with vegetation are recorded at regular distances, or with networks of points, or with small quadrats. Lines, points or quadrats can be laid out at random, in a systematical way, or in a combination (stratified random sampling). The formerly practised vegetation analysis in Northern Europe (e.g. by Du Rietz 1931; see also Trass & Malmer 1978) can be considered as a transition between multiple-plot analysis s.s. and plotless sampling.

1.1.7 Size of the sample plot; minimal area

When an intensive survey is carried out in a sample plot the size of this plot has to be determined. Usually the entire local phytocoenosis (according to its definition) is inspected as to uniform environment, floristic composition and structure, as distinct from the surrounding vegetation. Depending on the type of vegetation the area covered may vary from a few m^2 to several ha. If species composition is one of the descriptors, the sample plot should not be too small because only a few species would then be included. This leads to a discussion of the concept of minimal area, defined (Westhoff & van der Maarel 1978) as a 'representative area, as an adequate sample of species of regular occurrence', which is related to the total number of species in the stand. A definition such as that of Mueller-Dombois & Ellenberg (1974): 'the smallest area on which the species composition of the community in question is adequately represented' is what Westhoff & van der Maarel called a 'synthetic minimal area'. Such an area cannot be determined without previous knowledge of the community sampled; this can have been acquainted during the successive approximation.

To this end the determination of a species-area relationship has been recommended, both in classical phytosociology (Braun-Blanquet 1932) and in Anglo-American textbooks of vegetation analysis (Cain & Castro 1959). The usual way of determining this relation is to start with a very small quadrat, count the number of species, enlarge the quadrat, usually with a factor 2, count the number of additional species, etc. until the boundaries of the local stand are reached. Instead of such a series of nested plots randomly located plots of increasing size are theoretically preferred and, still better, several such series should be analysed (see Dietvorst *et al.* 1982 for references). The species-area relation is usually plotted as number of species against area. The resulting curve is quasi-asymptotic and the suggestion was to consider as minimal area an area beyond which the curve levels off.

Although this procedure has been long criticized as it is subjective, i.e. dependent on the choice of the ratio of the y-axis to the x-axis, it has remained standard practice in phytosociology, partly under the influence of Tüxen (1970) who collected many species-area curves from the literature and spoke of a '**saturated community**' if the minimal area had been reached. It is curious that the decisive arguments against this approach had already been provided by the Swedes O. Arrhenius, L.G. Romell and H. Kylin before Braun-Blanquet had published the first edition of his textbook. They had developed three models of species-area relationships, in graphical terms the log-log, the linear-log and the linear-linear relation, respectively. The third relation, the real saturation curve was only found in species-poor communities, while the first was fully developed by Preston (1962) and the second is well-known as the basis for the α-diversity of Fisher, as elaborated by Williams (1964). Braun-Blanquet (1932) mentioned these models but he was only vaguely aware of the repercussions for the minimal area approach by confirming that Kylin's model of a linear-linear species-area relation was appropriate for the minimal area approach but by not realizing that the other two models, which are much more commonly applicable to natural communities, question the validity of the species-area based determination of the minimal area (see Westhoff & van der Maarel 1978).

While these considerations on minimal area refer to numbers of species represented, the area to be sampled should also be large enough to represent the abundance relations of the participating species. This idea was expressed for the first time by E. Meijer Drees in 1954, who distinguished between qualitative (species-area-based) and quantitative minimal area. In his case the latter concept referred to the area where most of the timber species in tropical rain forest stands were represented with trees of more than 100 cm circumference.

Another quantitative approach had already been proposed by G.E. Du Rietz in the 1920s: the frequency of species in series of quadrats of increasing size is determined and the number of 'constant' species (with frequency of at least 90%) is plotted against area. M. Gounot, C. Roux and other French investigators calculated the floristic similarity between quadrats of increasing size. However, none of these more sophisticated methods produced saturation curves in most cases. Dietvorst *et al.* (1982; also for references) elaborated the similarity approach by comparing values with the maximum similarity values obtained in models with 5000 cells with varying numbers of species and mean cover. Critical quantitative similarity levels varied from more than 90% in salt marsh to 50–80% in open sand dune vegetation; qualitative levels from 50% in *Calluna* heath to 80% in salt marsh. The highest of the two corresponding minimal area levels was chosen as minimal area. These values were within the range indicated by Westhoff & van der Maarel (1978). It was also shown that the sizes of the two minimal areas are related to species richness and amount of dominance (Table 1.1).

Barkman (1989) advocated an additional method by plotting the increase in species number against log area (based on large numbers of replicates). If the increase is zero over short trajectories this would be an indication that the size of some within-community pattern is exceeded. Following E. Meijer Drees and others, he also emphasized the concept of **'biological minimal area'**, the area needed for a local phytocoenosis to maintain itself, including patch dynamics. For forests this area

Table 1.1 Relation between the size of the qualitative (MA_{qual}) and quantitative (MA_{quan}) minimal area in relation to species richness and mean species dominance. Based on Dietvorst *et al.* (1982). Reproduced by permission of Kluwer Academic Publishers.

		Mean species dominance	
		High	Low
Species richness	High	MA_{quan} small $MA_{qual} >> MA_{quan}$	MA_{quan} large MA_{qual} large
	Low	MA_{quan} small MA_{qual} small	MA_{quan} large $MA_{qual} < MA_{quan}$

could be several ha. The species richness of the total vegetation stand is the same as what nowadays is called the community species pool (Zobel *et al.* 1998).

In conclusion, a 'minimal area' to be sampled, related to species richness, canopy height and species dominance relations, remains difficult to determine. Instead a 'representative' sampling area should be selected the size of which can be chosen on the basis of field experience with different vegetation types as represented in various textbooks. Table 1.2 presents size intervals for representative sample plots.

Table 1.2 Minimal area values (m^2) for various plant communities, largely according to Westhoff & van der Maarel (1978) with additions, partly following Knapp (1984) and Dierschke (1994). Nomenclature largely according to Rodwell *et al.* (2002). In case of mosaic complexes, e.g. in heathland, bog and forest, intervals refer to the elements of the complex and not to the complex as a whole.

Epiphytic moss and lichen communities		0.1–0.4
Terrestrial moss and lichen communities		1–2
Free-floating aquatic communities (*Lemnetea*)		2–5
Hygrophilous pioneer communitie (*Isoeto-Nanojuncetea*)		2–5
Vegetation of trampled habitats (*Polygono-Poetea annuae*)		2–5
Lower salt marshes (*Thero-Salicornietea*)		4–10
Open dune and sand grasslands (*Koelerio-Corynephoretea*)		4–10
(Sub-)Mediterranean therophyte communities (*Helianthemetea guttati*)		4–10
Heavily managed grasslands (*Cynosurion cristati*)		4–10
Upper salt marshes (*Juncetea maritimi*)		10–25
Rooted floating aquatic communities (*Potametea*)		10–25
Temperate pastures and meadows (*Molinio-Arrhenatheretea*)		10–25
Basiphilous grasslands (*Festuco-Brometea*)		10–25
Ombrotrophic bog vegetation (*Oxycocco-Sphagnetea*)		10–25
Bog-pool and mire vegetation (*Scheuchzerio-Caricetea fuscae*)		10–25
Steppes (*Festuco-Brometea*)		20–50
(Sub-)Alpine calcareous grasslands (*Elyno-Seslerietea*)		20–50
Coastal yellow dune communities (*Ammophiletea*)		20–50
Tall swamp vegetation (*Phragmito-Magnocaricetea*)		20–50
Heathlands (*Calluno-Ulicetea*)		20–50
Weed communities (*Stellarietea mediae*)		40–100
Woodland fringe and gap vegetation		40–100
Perennial ruderal vegetation (*Artemisietea vulgaris*)		40–100
Temperate scrub (*Rhamno-Prunetea*)		40–100
Mediterranean maquis (*Quercetea ilicis*), chaparral		40–100
Mediterranean low scrub (*Cisto-Lavanduletea*)		40–100
Willow and poplar scrub and woodland (*Salicetea purpureae*)		100–250
Fynbos		100–250
Deciduous forest on rich soils in Europe (*Querco-Fagetea*):	herb layer	100–250
Swamp woodland (*Alnetea glutinosae*)		100–250
Coniferous forest (*Vaccinio-Piceetea*)		200–500
Managed deciduous forest on rich soils in Europe (*Querco-Fagetea*):	canopy	200–500
Mature deciduous forest on rich soils in Europe (*Querco-Fagetea*):	canopy	400–1000
Ibid. in North America:	canopy	400–1000
Desert vegetation		400–1000
(Sub-)Tropical dry forest	canopy	400–1000
Tropical secondary forest	canopy	2000–5000
Tropical rain forest	canopy	4000–10,000

1.1.8 Sampling of vegetation characteristics

Vegetation structure and floristic composition are usually measured or estimated on a plant community basis. Barkman (1979) distinguished between **texture**, the composition of morphological elements, and **structure s.s.**, the spatial arrangement of these elements – the temporal arrangement, including **phenology** (see, *inter alia* Mueller-Dombois & Ellenberg 1974; Dierschke 1994), can be included here. However, most ecologists still use structure as a general term.

Four overall measurements, some of them more widely used than others, may be mentioned:

1 **Stratification**, the arrangement of phytomass in layers. Usually a tall tree, low tree, tall shrub, low shrub, dwarf-shrub, tall herb, low herb and moss layer are distinguished if separated from each other. See further, *inter alia*, Mueller-Dombois & Ellenberg (1974).

2 **Cover.** Percentage cover is the relative area occupied by the vertical projection of all aerial parts of plants, as a percentage of the surface area of the sample plot. This can be determined for the vegetation as a whole or for separate layers. Cover is usually estimated by eye, but can also be determined more accurately through the line-intercept method – in sparse vegetation – where contacts between the line and plant parts are counted, or the point-intercept method – in dense short vegetation – where contacts with a cross-wire grid are counted, or the cover pin frame – in dense taller vegetation – where pins are moved vertically downwards and contacts with plant parts are counted (because pins can hit plants at several heights total cover can exceed 100%).

3 **Phytomass.** Total phytomass (= plant biomass) in the plant community, is expressed as dry-weight $g{\cdot}m^2$, $kg{\cdot}m^2$ or $t{\cdot}ha$ ($t{\cdot}ha^{-1}$ = 10 $kg{\cdot}m^2$). Phytomass is usually determined by removing the **standing crop**, the above-ground phytomass during the period of maximal development. The standing crop is related to, but by no means identical to, what is produced during the growing season – which varies from weeks in arctic to 12 months in moist tropical environments. Plant production, i.e. production by **autotrophic plants**, also called **primary production** – to distinguish it from secondary production, which is the transformation of phytomass by heterotrophic organisms, animals and saprobes – is usually expressed in terms of **productivity**, production per time unit, usually $g{\cdot}m^2{\cdot}yr$ (see Chapter 3). The destructive sampling necessary for phytomass measurements usually requires an adapted sampling scheme so that a sufficient area of the same vegetation remains undisturbed.

Phytomass can be determined per layer so that a vertical phytomass profile can be obtained and interpreted in terms of species interactions and light climate (e.g. Fliervoet 1985). Barkman (1988) developed a method and apparatus to determine **phytomass denseness**, and its horizontal and vertical distribution. This method is also destructive, but only small sections of plant mass are cut. Such profiles can be fruitfully linked to measurements of microclimate (Stoutjesdijk & Barkman 1992).

4 **Leaf area index.** The total area of leaf surface (actually photosynthetic surface) expressed in m^2 per m^2 surface area is known as **leaf area index**, LAI; it can be determined per layer and can thus also be used for a refined description of the architecture of vegetation. A derivate characteristic is **specific leaf area**, SLA = leaf (lamina) area

per unit leaf (lamina) dry mass. LAI and cover are related, but no studies of the correlation between the two characteristics for individual species are known to the author.

Next, structural-physiognomic characteristics can be determined. Typical textural characters, as mentioned by Barkman (1979), are leaf size, leaf consistency, leaf orientation, leaf longevity and plant growth form. The consistent analysis (rather the detailed description) of such characters as developed by P. Dansereau, F.R. Fosberg and A.W. Küchler, and life-form categories based on, or elaborated from, C. Raunkiær's system – and summarized by Mueller-Dombois & Ellenberg (1974), Kent & Coker (1992) and Dierschke (1994) – is usually related to the respective classification systems developed. The description of the characteristics and spatial position of organs, as in textural descriptions, including drawings of vegetation profiles, has not become a standard procedure. Structural research rather proceeds via the species composition combined with the allocation of species to life form or other categories.

Structural analysis of above-ground plant parts should be (but is seldom) completed with an analysis of the below-ground parts. For instance, Dierschke (1994) presented examples of root stratification.

1.1.9 Sampling of species characteristics

The species composition of a plant community, the key element in its definition, is described in its simplest form by a list of species occurring in the sample plot. The list is mostly restricted to vascular plants, and almost always to their above-ground parts; often easily recognizable mosses, liverworts and lichens are included. The quantity a species attains can be called its **performance**, but often the term abundance is used, even if this is only one of the following quantitative measures:

1 **Abundance**, the number of individuals on the sample plot. Because individuality in many (clonal) plant species is difficult to determine (see Chapter 5), the concept of **plant unit**, a plant or part of a plant (notably a shoot) behaving like an individual, is needed, if only for a quantitative approach of species diversity based on the distribution of plant units over species (Williams 1964). **Density** is a derivate variable, being the abundance per unit area.

2 **Frequency** is the number of times a species occurs in subplots within the sample plot – or within an undelimited phytocoenosis (formally plottless sampling).

3 **Cover** can be measured species-wise (see section 1.1.8); it is usually estimated along a cover scale. Many scales have been proposed (van der Maarel 1979), some of which more or less linear (e.g. with 10% intervals), some geometrical, e.g. the still-used five-point geometrical Hult–Sernander–Du Rietz scale (after R. Hult, R. Sernander and G.E. Du Rietz) developed during the 1910s by the so-called Uppsala school (see Trass & Malmer 1978).

4 **Cover-abundance** is a combined parameter of cover – in case the cover exceeds a certain level, e.g. 5% – and abundance. This 'total estimate' (Braun-Blanquet 1932) has been both criticized as a wrong combination of two independently varying parameters and praised as a brilliant integrative approach. It reminds us of the **importance value** developed by Curtis (1959), the product of density, frequency and cover, which has been popular in the US for some decades. Several proponents of a combined

cover-abundance estimation have nevertheless found it realistic to convert the abundance categories in the combined scale into approximate cover values. The two combined scales still in use are the Domin or Domin–Krajina scale (after the Czech ecologists K. Domin and V.J. Krajina; see Mueller-Dombois & Ellenberg 1974) and the Braun-Blanquet scale which, in several variants, has been in use since the 1920s. Van der Maarel (1979) suggested an 'ordinal transform' (OTV) scale replacing the modern nine-point Braun-Blanquet scale by the values 1–9, which could be used, if not as arithmetic at least as ordinal values. This scale was also included in Westhoff & van der Maarel (1978) and has found wide acceptance.

Van der Maarel (in Fresco *et al.* 2001) suggested a cover-based interpretation of this scale by replacing the original abundance categories 1–4 by mean cover % values and by interpreting the high ordinal transform values (OTVs) 8 and 9 as corresponding to in reality much higher cover % values than those indicated by the Braun-Blanquet scale, in fact > 100% – because in dense vegetation the dominant species develop phytomass in several vegetation layers (as is in fact measured with the vertical pin method mentioned in section 1.1.8). We can then draw a regression line (as was in fact suggested by Jongman *et al.* 1995) and obtain the equation

$$\text{OTV} = 1.415 \ln C + 2 \quad (1.1)$$

where C the cover value in %. Herewith we would have a fair approximation to a ratio scale, where the means of the cover classes form a geometrical (× 2) series. See Table 1.3.

5 Basal area, the area outline of a plant near the surface, is of particular interest for trees and can be used for tree volume estimations (see Mueller-Dombois & Ellenberg 1974). A related measure is tree diameter at breast height (DBH; at 1.30 m), which is more often used in standard forest descriptions.

6 Phytomass can be measured per species, even if this is a very tedious work. These data can be used to accurately relate species performances to each other and to follow species performances in time series of observations and experiments.

Table 1.3 Extended Braun-Blanquet cover-abundance scale and ordinal transform values (OTV) according to van der Maarel (1979) with interpreted cover value intervals for low cover values. From Fresco *et al.* (2001). Reproduced by permission of the publisher.

Braun-Blanquet	Abundance category	Cover: interpreted interval	OTV cover interval	OTV
r	1–3 Individuals	$c \leq 5\%$		1
+	Few individuals	$c \leq 5\%$	$0.5 < c \leq 1.5\%$	2
1	Abundant	$c \leq 5\%$	$1.5 < c \leq 3\%$	3
2m	Very abundant	$c \leq 5\%$	$3 < c \leq 5\%$	4
2a	Irrelevant	$5 < c \leq 12.5\%$		5
2b	"	$12.5 < c \leq 25\%$		6
3	"	$25 < c \leq 50\%$		7
4	"	$50 < c \leq 75\%$		8
5	"	$c > 75\%$		9

7 **Sociability**, the gregariousness of plant units of a species, has been a standard parameter included in phytosociological relevés (Braun-Blanquet 1932). Five degrees are distinguished, varying from 1 = plant units growing singly to 5 = growing in great crowds over most of the sample plot. However, this parameter has seldom been used in the comparison of relevés, mainly because sociability is species-specific for many species and also because there is no numerical way to treat the data. See also section 1.1.10.

Species data should not only be collected above-ground but also below-ground. Dierschke (1994) presented examples of root:shoot ratio differentiation within a plant community. Titlyanova *et al.* (1999) showed how in steppes the below-ground phytomass (which can store 70% of the net primary production) is more homogeneously distributed, both over the area and over the species. The dominance-diversity curves of 19 species in steppe vegetation based on percentage dry weight contributions of species to green phytomass and below-ground organs are quite different. Where in both cases the top species are *Stipa krylovii* and *Potentilla acaulis*, the other species have different sequences and the below-ground curve is much less steep.

The main use of data on species characteristics is in the classification and ecology of communities, but these data also form the basis for the analysis of vegetation dynamics. For this purpose permanent sample plots can be established which are regularly, preferably annually, investigated. In order to interpret changes in species characteristics the data should be more accurate than in a spatial context. In relevés of permanent plots and in the analysis of chronosequences (section 1.7) a more detailed cover scale can be used. However, to reduce the effects of subjectivity more exact data, notably on phytomass, are preferred.

1.1.10 Standardized phytosociological relevé

In the framework of the European Vegetation Survey a template for common data standards for phytosociological relevés has been proposed (Mucina *et al.* 2000), including area and form of plot, exact location, distinction of vegetation layers, and use of any scale for species quantities which can be converted to the ordinal transform scale.

1.1.11 Spatial pattern analysis

The notion of **pattern** in vegetation has become standard since Watt's (1947) paper on pattern and process in the plant community. The basic idea is that in many communities many plant species occur in **patches** which occur in **mosaics** and that these patches 'are dynamically related to each other'. The occurrence of patches is one of three ways plant units of a species can be located: the clumped or underdispersed distribution. In the opposite case the dispersion is regular (overdispersion). The intermediate situation is a random dispersion – note that in a perfectly homogeneous community all species should have such a dispersion.

There exists a wide variety of methods of pattern detection. Dale (1999) treated most of them.

In one group of methods, presence-absence data of one (or more) plant species are collected in small quadrats, usually located in transects – a line-intercept approach is also possible. The selected size of the quadrats is dependent on the structure of vegetation. In each quadrat the presence, or a quantitative measure, e.g. cover degree, is recorded and the variance of the distribution calculated. Next quadrats are lumped in blocks (pairs, quadruples, etc.) and each time variance is calculated again. The graph of variance against block size may reveal patterns. This method was developed by P. Greig-Smith and K.A. Kershaw in the 1950s. Several statistically elaborated variants have been developed and of which three-term local quadrat variance is recommended by Dale (1999). Here the mean variance between the variance of a block and the sum of two adjacent blocks on either side is calculated.

In another group of methods dating from the 1970s, plants are replaced by a point and the distances of plant points to each other, or to nearest neighbours, or to fixed reference points are determined. The frequency distribution of the distances can reveal patterns. One of the popular statistics is Ripley's *K*, called after the statistician B.D. Ripley (e.g. Haase 1995). This type of method is effective when plant units (notably trees) or patches (in desert vegetation) can be easily individualized; data may be obtained from remote sensing and then large-scale patterns can be detected.

New variants and new methods are regularly being published. As an example of a new method, Dai & van der Maarel (1997) suggested patch-size frequency analysis as an easy, straightforward – but statistically complicated – approach based on presence (or cover) records in transects of small quadrats. Patterns may occur in a hierarchical way: a species may occur in patches of a given size, where the patches may have a clumped dispersion. Also, the size of gaps between patches may show a pattern. Patterns of correlated species dispersions may also be detected. Finally, the pattern of abiotic or biotic parameters possibly show interference with the plant pattern. For example, Dai & van der Maarel found that the plant species *Filipendula vulgaris* showed patches of occurrence with ≥ 5% cover with the same size (*c.* 50 cm diameter) as patches of cow dung in the grazed limestone grassland this species is characteristic of.

1.2 Vegetation and environment: discontinuities and continuities

1.2.1 European and Anglo-American development of vegetation ecology

In Chapter 2 Austin treats vegetation and environment in a coherent way, indeed as vegetation ecology. This term was coined by Mueller-Dombois & Ellenberg (1974) who both were educated in Germany in the ecological tradition of continental-European phytosociology. Anglo-American vegetation ecology has its roots in plant ecology – and is usually called so. Despite this and earlier textbooks, including that of Braun-Blanquet (1932), Anglo-American plant ecology has for some time identified European phytosociology with the mere description of plant communities. This was in the 1940s and 1950s when the 'Braun-Blanquet system' of plant communities became fully developed. Extensive surveys were published in the newly established journal *Vegetatio* and nomenclature rules were discussed. Possibly Anglo-American readers may have concluded that phytosociology was synonymous with rigid community typology. The confusion has since long been resolved. Continental-European vegetation ecology has developed in many directions (e.g. Ellenberg 1988).

Community typology is still important, but nowadays particularly as a pragmatic tool in communication, both between ecologists and between ecology and society. Simultaneously, plant community description has become an appreciated branch of ecology in Anglo-American ecology. In Europe the European Vegetation Survey project, initiated by S. Pignatti, is well under way. It was the obvious next step in 'phytosociological systematics' developed by R. Tüxen and stimulated during many symposia led by him (see Dierschke 1994). The importance of such surveys for society has always been pointed out by Tüxen, who was the father of applied phytosociology. And society is increasingly aware of this importance. For instance, the European project has recently resulted in a very useful and sponsored survey of plant communities each with an English name referring to the type of habitat and distribution area (Rodwell *et al.* 2002).

1.2.2 Discontinuities and continuities; classification and ordination

Austin makes clear that both vegetation and environment are characterized by discontinuities and continuities and that their interrelationships should be described by multivariate methods of ordination and classification. It shows how mainly three paradigms have emerged during the history of vegetation ecology, which we can conveniently label 'association', 'indirect gradient' and 'direct gradient'; the differences between the paradigms are smaller than is often believed and vegetation ecology can further develop when a synthesis of the three paradigms is developed.

The concept of environmental gradient plays a central part here. Where vegetation varies continuously in relation to environmental variation a series of communities would form a **coenocline** (see Westhoff & van der Maarel 1978). The description of clinal variation and particularly the choice of proper ordination methods can be optimized by finding a realistic model of plant species behaviour along the underlying environmental gradients. Where vegetation shows more abrupt transitions numerical classification is appropriate, but again the application of methods can be optimized by adopting a proper species model. In this respect, the concept of **compositional turnover**, i.e. the coming and going of plant species along a gradient, may be mentioned. Through a measure of this turnover the length of a gradient can be estimated. Økland (1990) presented a summary and recommended SD, the number of mean standard deviations of species occurring along the gradient. Formally this measure can only be used if all species are distributed according to a Gaussian response curve, which, as Austin shows, they are not.

As to numerical ordination, Austin treats two methods as representative for the plethora of methods, Correspondence Analysis (CA) and its derivate Canonical Correspondence Analysis (CCA), and Non-metric Multidimensional Scaling (NMDS). CA-CCA has become the standard ordination approach, even if it is based on assumptions which are not often met. It has certainly become so popular because of the effective computer program available. The scaling methods are computationally more complicated, but theoretically to be preferred.

As to numerical classification, Austin treats again two methods, the divisive method TWINSPAN and the agglomerative method UPGMA. Basically both approaches achieve a hierarchical structure, graphically shown as a dendrogram. TWINSPAN is by far the most popular method and its popularity has only grown since it was incorporated in the program TURBOVEG for phytosociological classification of very large data sets (Hennekens & Schaminée 2001). Attractions of these programs are the capacity and speed and the relatively low number of options one has to consider, but this has distracted the attention from their weaknesses: the strictly hierarchical approach and the fact that it is based on Correspondence Analysis, with its problems.

UPGMA and some related methods have been used both for obtaining groups of similar species and groups of similar sample plots. Program TABORD by E. van der Maarel *et al.* and its elaborated version FLEXCLUS by O. van Tongeren (see Jongman *et al.* 1995) are in a way a combination of clustering and ordination with the result presented in an ordered phytosociological table. Essential here is that a cluster structure is searched for on an optimal level of similarity and then ordered, so that the approach is reticulate rather than hierarchical. This is considered a more realistic phytosociological approach and at the same time a step towards the paradigm synthesis advocated by Austin.

1.2.3 Ideas of environment

Under this heading Austin emphasizes the importance of a framework of 'broad environmental factors' which should be developed for any study of vegetation and environment. The special attention paid to climatic and derivate microclimatic factors leads to the notion of the hierarchy of spheres influencing vegetation in this order of impact (see also Chapter 12):

Atmosphere	> Lithosphere	> Hydrosphere	> Pedosphere	> Biosphere
Climate temperature, precipitation	Bedrock	Groundwater	Soil factors	Fauna

A useful distinction within the environmental factors is between (i) indirect factors, notably altitude, topography and landform; (ii) direct factors such as temperature, groundwater level and pH – which are determined by indirect factors; and (iii) resource factors such as water availability and nutrients, which are determined by indirect and direct factors. Another distinction is between distal and proximal factors, where the proximal factors are operating directly on the plant. Generally, vegetation ecology is more meaningful if the environmental factors available for vegetation – environment studies are more physiologically relevant.

1.2.4 Ecological characterization of communities by means of indicator values

An additional way of characterizing the environment of a plant community is to use indicator values assigned to the participating plant species. The best known system of values is that of H. Ellenberg (Ellenberg *et al.* 1992), with indicator values for most of the Central European vascular plant species regarding moisture, soil nitrogen status, soil reaction (acidity/lime content), soil chloride concentration, light regime, temperature and continentality. The values generally follow a, typically ordinal, nine-point scale, based on field experience and some measurements. They reflect, as Austin describes it, the ecological behaviour of species, their realized niche. Even if these values are typically a 'distant' approach of the environment, they have been used abundantly, also in north-west Europe. Using them to calculate (weighted) mean values for plots and communities is a calibration problem, discussed by ter Braak (see Jongman *et al.* 1995).

1.3 Vegetation and ecosystems

Leuschner, in Chapter 3 on ecosystems, concentrates his essay on trophic levels between which matter and energy is exchanged. Elaborations of this theme are found particularly in Chapters 10 on plant–herbivore relations and 11 on interactions between plants and soil-dwelling organisms. It is also worth repeating the message on the importance of the below-ground parts of the plant which are parts of the plant community but also the physical link between the biotic and abiotic parts of the ecosystems.

An important part of the primary production ends up in the below-ground plant parts. Here both decomposition and humus formation take place. In an ecosystem in steady state there is a balance between net primary production and organic matter decomposition. This balance is reached in later stages of succession. As Leuschner states, after perturbation an ecosystem can often rapidly regain certain structural properties. As an example, Titlyanova & Mironycheva-Tokareva (1990) described the building up of the below-ground structure during secondary succession in just a few years. On the other hand, the re-development of a steady state in steppe grassland may take 200 yr. This also relates to the actual discussion on the relation between diversity and ecosystem function (Chapter 8).

It is interesting that ecosystem ecologists have no doubt about the reality of emergent properties. It is as if these properties appear clearer, the higher the level of integration is at which we are looking at ecosystems. Ultimately we are facing clear aspects of regulation at the 'Gaia' level of the global ecosystem.

Leuschner finishes his chapter with a treatment of four biogeochemical cycles, of carbon, nitrogen, phosphorus and water. These cycles are studied on the global level and these processes on this level return in Chapter 14.

1.4 Vegetation types and their broad-scale distribution

1.4.1 Physiognomic-ecological classification; formations

In Chapter 4, Box & Fujiwara treat vegetation typology mainly in relation to the broad-scale distribution of vegetation types. On a world scale types have largely been

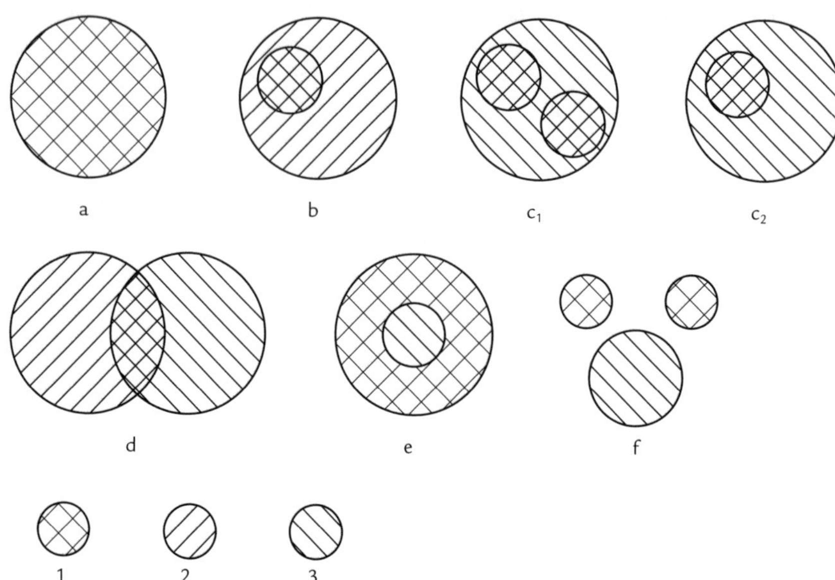

Fig. 1.3 Different types of character taxa (CT) based on the relation between the distribution area of a taxon (A_t) and that of a syntaxon (A_s) it should be characteristic of. **a.** General CT: $A_t \approx A_s$; **b.** Local CT: $A_t < A_s$ and included in A_s; **c.** Regional (superregional) CT: $A_t > A_s$; also CT in other syntaxa (c_1) or diffuse behaviour outside syntaxon and then rather CT of a higher-rank syntaxon (c_2); **d.** Local CT: A_t and A_s different but with overlap; **e.** Regional CT: A_t and A_s overlap but taxon is genetically and ecologically more variable in the centre of A_t and only CT in the periphery of A_s; **f.** Differential CT: $A_t >> A_s$ while locally present in some syntaxa and absent in others. Figure from Dierschke (1994), text based on Westhoff & van der Maarel (1978). Reproduced by permission of the publisher.

defined physiognomically, in the beginning (early 19th century) by plant geographers, including A. Grisebach, who, as early as 1838, coined the term formation. Some readers may share the author's memory of the famous world map of formations by H. Brockmann-Jerosch & E. Rübel decorating the main lecture hall of many botanical institutes. Box & Fujiwora emphasize the ecological context in which these physiognomic systems were developed. In fact, the English term plant ecology was coined in the translation of the book on ecological plant geography by Warming (1909).

1.4.2 Plant community classification; the Braun-Blanquet approach

Plant community classifications have now been developed to the extent that they can be used in broad-scale distribution studies and mapping projects. Some further essentials may be added. This undoubtedly most influential community classification system, called **syntaxonomical system**, is hierarchical and resembles the taxonomy of plants (and animals). The association – although originally given a wider definition (section 1.1.2) – is the central community type; similar associations are united in **alliances**, alliances in **orders**, and orders in **classes.** As in the case of species with subspecies, associations may be differentiated into **subassociations**. During the development of syntaxonomy various intermediate ranks have been introduced, but the four levels mentioned here remain the most important ones. Each syntaxon is defined by a characteristic species combination, a group of **diagnostic taxa** which may include **character taxa, differential taxa** and **companions**. The degree of differentiation within the taxon is usually the species, but in regions with detailed floras subspecies and varieties are also used, while in certain cases (see below) genera may appear. Character species are ideally confined to one syntaxon **S** – but need not occur there in all or most of the relevés assigned to **S**; differential species occur in **S** but not in syntaxa within the next higher rank, while again in other syntaxa; companions occur in most relevés of **S** but also in other syntaxa. As a further parallel to the taxonomy of plants a syntaxonomical nomenclature has been formalized (Weber *et al.* 2000).

The confinement of taxa to syntaxa is seldom absolute and degrees of **fidelity** have been recognized. The classical fidelity degrees (Braun-Blanquet 1932) are seldom seriously applied. Szafer & Pawlowsky (see Dierschke 1994) proposed a refined scale – which does not seem to be applied either. The main problem is that the distribution area of characteristic species seldom coincide with that of their syntaxon: they can be much wider, but also smaller, or overlap only partly. Westhoff & van der Maarel (1978) and particularly Dierschke (1994) have discussed this (including some ideas of J. Barkman) and distinguished local, regional, superregional and absolute character species (Fig. 1.3). In practice a syntaxon is often quite loosely defined by a species combination which is more or less characteristic.

Equally little discussion has been published on the decision whether syntaxa of a given lower rank shall be united with a syntaxon of the next highest rank – or maybe the next highest rank still. Now syntaxonomy is carried out numerically (Mucina & Dale 1989) it is obvious to link syntaxonomical ranks to levels of similarity, as has already been initiated by Westhoff & van der Maarel (1978), but which was surprisingly not developed any further.

Another major problem, related to the former two problems – which is hardly investigated – is the imperfectness of the hierarchy. Syntaxa of a lower rank often show floristic similarities to syntaxa from different classes, and many character species of a given syntaxon show a low degree of fidelity. Obviously problems of fidelity and hierarchy will only get worse when phytosociological surveys are extended over large areas. The crucial question is not whether a syntaxonomical hierarchy can be maintained and extended, but whether it will work.

1.4.3 Plant community classification; large-scale approaches

Extension of phytosociological surveys over large areas is both a great perspective and a formidable challenge, not only scientifically, also organisationally. Box & Fujiwara mention the examples of Japan, Europe and the USA. The European Vegetation Survey has recently resulted in a survey of 928 phytosociological alliances in 80 classes, each with an English name referring to the physiognomy, type of habitat and distribution area (Rodwell *et al.* 2002). The classes are grouped into 15 formations. A similar project was initiated in North America. Notably, the Federal Geographic

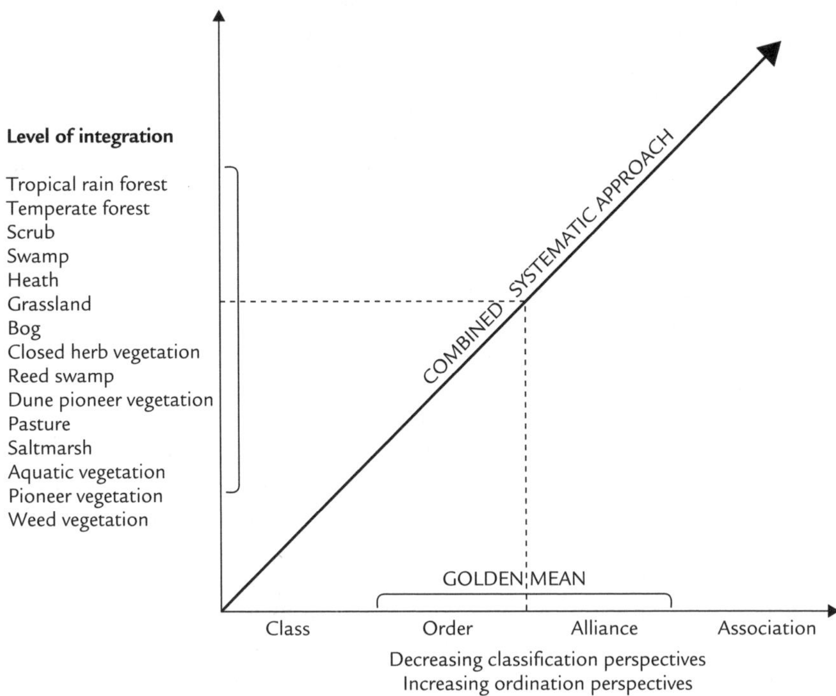

Fig. 1.4 Relation between the level of integration in vegetation and the relative success of classification versus ordination in a 'combined systematic approach'. (Slightly changed after van der Maarel 1979, following an idea of E. van der Maarel, V. Westhoff & C.G. van Leeuwen.)

Data Committee of the US Geological Survey developed a classification with a hierarchy of higher-level physiognomic units and two lower-level floristic units, associations and superimposed alliances. US vegetation ecologists have been linked to this project under the auspices of the Ecological Society of America and have developed standards for the description of the floristic units (Jennings *et al.* 2002). In this project, the association is defined as a physiognomically uniform group of vegetation stands that share one or more diagnostic (dominant, differential, indicator or character) species. The alliance is a physiognomically uniform group of associations sharing one or more diagnostic (dominant, differential, indicator or character) species which, as a rule, are found in the uppermost stratum of the vegetation. These units should occur as repeatable patterns of assemblages across the landscape, and are generally found under similar habitat conditions (Grossman *et al.* 1998). Note that uniform physiognomy is emphasized more than in the European Vegetation Survey. In the preliminary survey 4657 associations have been listed, nested within 1522 alliances. The number of alliances compares well with the above-mentioned number for Europe.

It is clear from this short description that there is a growing interest in subordinating floristic units to physiognomic ones. This is also directly relevant for vegetation mapping. The integrated physiognomic-floristic approach has indeed been proven to be effective since its apparently first attempt by van der Maarel & Westhoff in the 1960s (see van Dorp *et al.* 1985).

1.4.4 Potential natural vegetation, biomes and ecosystems

Box & Fujiwara pay some attention to the problem met with in the mapping of vegetation of larger areas which have lost most of their original vegetation due to human land use and to the development of the concept of **potential natural vegetation** for large-scale vegetation mapping. Reconstruction of vegetation types developing as new climax after human impact would have stopped is of course difficult and can proceed in different ways (see Bredenkamp *et al.* 1998 for different possibilities and an example map of Ireland by J.R. Cross with only 19 units).

Formations can be extended to biomes by including the characteristic fauna of the regions involved and to ecosystems by including global climate and soil characteristics. Distribution patterns can be better understood by means of a plant functional approach – a development in which Box has been directly involved.

1.4.5 Classification and ordination as complementary approaches in phytosociology

The broad-scale vegetation surveys in Europe and the USA consider the alliance as a sort of central unit – doubtless mainly for practical reasons. There was a similar focus on the alliance level from a theoretical viewpoint A cross-section of NW European vegetation types was arranged according to their environment, ranging from coarse-grained relatively dynamic and homogeneous to fine-grained relative constant and divergent environments (Fig. 1.4), or ecotone versus ecocline environments (van der Maarel 1990). The hypothesis was that with increasing environmental complexity classification becomes less and ordination becomes more effective. As a 'golden mean' it was recommended to apply both techniques while approaching the syntaxonomy on the alliance/order level (van der Maarel 1979). The confusing complication is that the most integrated communities are the most difficult to discern and classify. Anyway, the combined approach suggested 40 years ago is in the same spirit as the conclusion by Austin in Chapter 2.

1.5 Clonal growth of plants in the community

Svensson, Rydin & Carlsson give a clear account on the ways and ecological significances of vegetative spread by clonal plants, and there is little to add to that from a vegetation ecology viewpoint. They make clear that clonal spread is a form of dispersal – even if (diaspore) dispersal as discussed in Chapter 6 will be seen as dispersal proper. An important source of variation is in the length of stolons and runners and the speed with which these are formed. The distinction between 'phalanx' and 'guerilla' behaviour of species, as endpoints along this line of variation, is real but it is curious that other terms seem never to have been proposed. Apart from the mental hesitation one should have to consider plants as warriors, the two strategies can hardly be compared in terms of history and number of 'genets' involved. The description of the types in Chapter 5 makes clear how little the behaviour of clonal plants resembles military strategies, particularly 'guerilla' (even the spelling of the word causes confusion; one should properly stick to the Spanish 'guerrilla'). The form of extension with a dense network of short, slowly growing clones may simply be called 'frontal', while plants with rapidly growing loose, long internodes have an 'errant' extension. The latter terms characterize clonal extension more realistically.

Of special interest for vegetation ecology is the characterization of vegetation types regarding the relative importance of clonal species and their role in patch dynamics (section 1.15).

1.6 Dispersal

As Chapter 5, which deals with clonal dispersal, Chapter 6 on diaspore dispersal by Poschlod, Tackenberg & Bonn is a well-defined contribution to which little needs to be added from a vegetation ecology viewpoint. Nevertheless there are important links to other chapters, first of all the next two chapters 7 and 8.

As to vegetation succession, the availability of diaspores is one of the major characteristics of secondary (post-agricultural and post-disturbance) succession, versus the lack of diaspores on the virginal substrates of a primary succession. On a smaller temporal and spatial scale the mobility of plants through clonal and diaspore dispersal is a driving force in 'pattern and process' in the plant community. Fine-scale mobility of plants as described in the carousel model and similar contexts is very much a matter of dispersal to open space becoming available.

Poschlod *et al.* make clear that dispersal is one of the essential factors which determine the composition of the species pool of a plant community (Zobel *et al.* 1998, who consider species reservoir a better, i.e. a more appropriate term). The community reservoir is supplied through dispersion from the local reservoir around the community, which in its turn is supplied by the regional reservoir through migration and speciation (for a different species pool approach see Chapter 9).

This chapter is also a natural place to treat the soil seed bank – which, as Poschlod *et al.* also state, would better be called diaspore bank. Zobel *et al.* (1998) suggested inclusion of the diaspore bank in the community pool, thus including the so-called persistent diaspores. It is a matter of discussion whether to also include species that never germinate in the target community – because they do not fit the environment. However, there are many examples of species apparently not fitting an environment and nevertheless occurring there, if only ephemerally. This is usually a matter of 'mass effect', the availability of numerous diaspores meeting favourable conditions for germination. This has long been recognized in phytosociology as vicinism (van der Maarel 1995).

1.7 Vegetation succession

1.7.1 Analysis of vegetation dynamics

In Chapter 7, Pickett & Cadenasso present results from analyses of both **chronosequences** ('space-for-time substitutions') and **permanent plots** and they indicate how the long-term, ideally yearly, analysis of permanently marked plots is gaining importance. There is a long tradition of permanent plot studies in Europe, starting in 1856 with the Park Grass Experiment at Rothamsted near London (mentioned in Bakker *et al.* 1996). Phytosociological studies of permanent plots started in the 1930s and there are now thousands of such plots under regular survey, many of them in the first place to help in solving management problems (Chapter 12). Bakker *et al.* (1996), who summarized this development, mention as perspectives of permanent plot studies the monitoring of changes, separating fluctuations from successional trends, and enabling extrapolations both in time and space.

Two parallel time series of great use in vegetation succession are first of all the census of individual plants. Mapping and counting plants in permanent plots started in the early 1900s (White 1985). A more recent time series is that of remote sensing data. If such data are accurate enough and can be phytosociologically interpreted after field work both short-term and long-term changes on the landscape level can be followed. Van Dorp *et al.* (1985) described the succession (largely a primary one) of a young dune landscape over 50 yr through interpretation of five aerial photos and two phytosociological maps of the area.

1.7.2 Causes of vegetation dynamics

Pickett & Cadenasso start by drawing a parallel to the theory of natural selection and a theory of vegetation dynamics and making an 'if then' statement: if a site becomes available, species are differentially available at that site, and/or species perform differentially at that site, then the composition and/or structure of vegetation will change. Succession is a special form of vegetation dynamics with a discrete starting point, a directional trajectory, and an unambiguous end. As Pickett *et al.* (1987; see also Glenn-Lewin & van der Maarel 1992) explained, site availability is largely the result of a disturbance, differential species availability is a matter of dispersal (see Chapter 6), and differential species performance is based on the differences in ecophysiology and life history, the outcome of species interactions (see Chapters 9 and 11) and herbivory (see Chapter 10).

1.7.3 Types of disturbance and types of vegetation dynamics

One of the interesting consequences of this approach is that the classical distinction between primary and secondary succession disappears. The two decisive factors, disturbance and dispersal, vary gradually. At one extreme the disturbance is intensive and/or coarse-scale and the newly available site is really virginal, e.g. the island

Surtsey which arose after a volcanic eruption. In the beginning there was no community pool and no local species pool and dispersal was a limiting factor. Here we have a primary succession s.s. At the other extreme a disturbance may leave part of the substrate and the local species pool in tact and a regeneration succession will result which is hardly a secondary succession in the usual sense. This is an example of the 'fine-scale dynamics' Pickett & Cadenasso describe. The chapter gives several other examples of within-community processes.

Since disturbance is the major trigger of vegetation dynamics, some disturbance typology is useful. Following Grubb (1985), White & Pickett (1985) and Glenn-Lewin & van der Maarel (1992) we can distinguish between extent, time (frequency) and magnitude (intensity) of a disturbance. Pickett & Cadenasso give examples of how these aspects can vary – while presenting details on the relatively unknown dynamics of riparian wetlands. Vegetation change after a disturbance will vary in the time period needed to reach a new stable state. Fig. 1.2 indicates how we can distinguish between fluctuation (on the population level), patch dynamics, secondary succession, primary succession, and secular succession, long-term vegetation change in response to (global) changes in climate (see Chapter 14), and how the time scale varies from less than a year to thousands of years. Dynamic studies of plant populations, expecially clonal plants, may vary from 10 to 10^3 yr (examples in White 1985). Cyclic successions may take only some years in grasslands rich in short-lived species (e.g. van der Maarel & Sykes 1993), 30–40 yr in heathlands (e.g. Gimingham 1988) and 50–500 yr in forests (e.g. Veblen 1992). The duration of successional stages at the plant community level ranges from less than a year in early secondary stages in the tropics to up to 1000 yr in late temperate forest stages. Finally, long-term succession in relation to global climate change may take a hundred to a million years (e.g. Prentice 1992).

1.7.4 Development of vegetation and soil

Pickett & Cadenasso point to the fact that in between disturbances biomass will accumulate. More generally, succession is a process of building up biomass and structure, both above-ground in the form of vegetation development, and below-ground in the form of soil building. Odum (1969) in his classical paper on ecosystem development was one of the first to present an overall scheme of gradual asymptotic biomass accumulation and a peak in gross production in the 'building phase' of a succession. This scheme was refined and extended to soil development by Peet (1992) and others.

The contribution to these developments by individual species varies with the type of succession and the successional phase. The old phytosociological literature paid attention to different types of species while emphasizing the 'constructive species', i.e. the species with a high biomass production which build up the vegetation (Braun-Blanquet 1932). Russian ecologists have used the term edificator for this type of species (see, e.g. J. White 1985). Usually these species are dominants. Grime (2001) summarized the conditions for the development of dominance and mentioned maximum plant height, plant morphology, relative growth rate and accumulation of litter as important traits for dominants.

1.8 Diversity and ecosystem function

1.8.1 Different aspects of diversity

Chapter 8 by Lepš is on diversity or biodiversity as it is called nowadays. It starts with a brief treatment of some diversity indices. These are all concerned with species diversity, or rather taxon diversity, the variation in taxa. In addition to α-diversity within-taxon or genetic diversity is mentioned and β- (between-community) diversity is briefly treated. The β-index presented, i.e. Whittaker's, has also been recommended by Magurran (1988) and Fresco *et al.* (2001). It is related to measure of compositional turnover along gradients discussed in section 1.2.2.

At the landscape level, γ-diversity, some combination of α- and β-diversity, can be used, but no special measure has been proposed according to Fresco *et al.* (2001) in their survey of diversity indices. Its simplest form should be $\alpha \times \beta$. In practice it is the total number of species in a landscape unit (community complex).

In a review of biodiversity aspects by van der Maarel (1996) some forms of distinctiveness are also considered aspects of biodiversity: **phylogenetic distinctiveness**, based on taxonomic distinctiveness, **numerical distinctiveness**, based on the rarity of occurrence, and **distributional distinctiveness**, i.e. endemism of taxa. Lepš makes the point that the diversity of a community is largely a function of the species pool and the forms of distinctiveness can indeed be determined in the species pool.

1.8.2 Measures of α-diversity

Two different approaches for the determination of species diversity are distinguished, both based on the relationships between the quantities with which the different species occur. One approach describes the way in which species number increases with increasing sample size; the other calculates some index of diversity based on relative quantities of species. In the first approach diversity indices are derived from models about the distribution of plant units (section 1.1.9), with α the parameter of the log series distribution (see Williams 1964) being the best known index. The second basic approach makes use of the distribution of species quantities over the species without assuming any particular distribution model. Indices of this type, the Shannon index being the best known, are, as it were, parameter-free diversity measurements (van der Maarel 1996).

As Lepš makes clear, diversity has both an aspect of species **richness**, i.e. the number of species, and of **evenness**, the way species quantities are distributed. These two aspects are more related than is generally recognized by users of diversity indices. According to the relation between the various diversity indices described by M.O. Hill (equation 8.3, p. 201) the well-known indices of Simpson and Shannon are similar – which almost all studies using both indices have demonstrated – in that the most abundant species determines, to some extent, the diversity, but Simpson does this more than Shannon. If species proportions are relative biomass or cover values, even the Shannon index would indicate the non-evenness rather than the richness. Since the ordinal transform scale may be considered an approximative geometric cover scale (section 1.1.9) ordinal transformation of cover-abundance values – or logarithmic transformation of biomass values – before applying the Shannon index would give more realistic results.

1.8.3 Diversity and function

Chapter 8 was planned as a contribution to diversity but developed into an essay on the relation beween diversity and ecosystem function. Much research – and speculation – has been triggered by the symposium volume by that name edited by Schulze & Mooney (1994). As Lepš elucidates, biotic diversity can be better understood if it can be divided into functional components. If we manage to distinguish such types and allocate each species to a type, diversity – i.e. species richness – can then be approached as the number of functional types multiplied by the mean number of species per type.

1.9 Species interactions structuring plant communities

The concise chapter on species interactions by van Andel gives a survey of the different types of interaction and then pays attention to the following types of interaction: competition, allelopathy, parasitism, facilitation and mutualism. Interestingly the attention for competition, the classical main type of interaction, is no longer predominant. Competition as a mechanism to arrange species packing along gradients (see Chapter 2) remains important, particularly for vegetation ecologists (e.g. Mueller-Dombois & Ellenberg 1974). Still, the typically community-structuring force of facilitation is a more fascinating topic in vegetation ecology – compare, for instance, its place in Chapter 9 with that in the recent but more autecologically oriented textbook by Crawley (1997). Facilitation, in particular, nursery effects has been described for almost all plant community types where one or more environmental factors (nutrient shortage, drought, temperature) could be critical for some plants but not for others. On the other hand, succession may be hampered by processes of inhibition.

Another important community-structuring interaction type with a rapidly growing body of literature devoted to it is mycorrhiza. Van Andel treats it as an important aspect of mutualism, but it forms also part of the topic 'interactions between higher plants and soil-dwelling organisms', elaborated in Chapter 11.

Van Andel's chapter is one of the few where bryophytes are treated in some detail. In addition to his description of bryophyte pattern formation the review paper by Rydin (1997) can be mentioned, which shows how, despite the generally low resource levels bryophytes have available, they can compete intensively. Zamfir & Goldberg (2000) described experiments showing how competition between mosses on the community level can be different from competition on the individual level (and also described the difference between the two approaches).

1.10 Plant–herbivore interactions

In Chapter 10 Sankaran & McNaughton present an integrative account on herbivory, linking up with Chapters 3, 7, 8, 9 and 12. In view of the broad spectrum of plant types and plant parts being eaten and the equally broad spectrum of eating animals, as well as the often intricate mutual adaptations between plants and animals in each case of interaction, the idea of co-evolution comes into mind (e.g. Howe & Westley 1988).

Plants deal with herbivory by avoidance or tolerance, i.e. compensation for damage; a range of compensatory responses is discussed. Another interesting range is discussed between symbiotic and parasitic aspects of grazing. Finally an actual twofold theme is treated: effects of herbivores on plant diversity and effects of herbivore diversity.

Chapter 7 discusses another aspect of herbivory, its contribution to pattern formation. This aspect has been looked upon from another angle by Olff & Ritchie (mentioned in Chapter 10) who relate the impact of herbivores to species diversity in relation to productivity and palatability along gradients of soil fertility and precipitation. These authors continued to elaborate on the relation between plant species diversity and environmental conditions (Olff *et al.* 2002) and particularly the significance of variation in herbivore size (Haskell *et al.* 2002).

1.11 Interaction between higher plants and soil-dwelling organisms

Kuyper & de Goede concentrate in Chapter 11 on the interactions between plants and soil organisms occurring around and in the roots. Naturally, their topic shows overlaps with the two preceding chapters. The three major processes described are nitrogen fixation by bacteria, mycorrhizae with fungi and root-feeding by invertebrates. As in the previous chapter the gradual transition and alteration between symbiotic and antagonistic aspects is emphasized. It is in the context of the environmental conditions how costs and benefits of symbioses affect plants. There is also a large variation in the degree of specificity regarding the interaction. Both plants and mutualistic and antagonistic soil organisms show different degrees of specificity or selectivity. Both private and shared associations could amplify or reduce the differences in competitive abilities between plant species. As a consequence, there is no one-to-one relationship between the below-ground mechanisms and processes, and the effects on plant species richness in vegetation.

There is an interesting link to Chapter 13 on plant invasion in that exotic plants, introduced in another region where some of the accompanying soil-dwelling organisms do not occur, may behave quite differently. A link with Chapter 7 follows from the elucidation of the two hypotheses as to the driving force of succession. If mycorrhizal fungi are causes of plant dynamics (driver hypothesis), the presence of specific mycorrhizal fungi is required for the growth of specific plants. If soil organisms are merely passive followers of plant species dynamics (passenger hypothesis), specific plants are required to stimulate the growth of specific mycorrhizal fungi.

1.12 Vegetation conservation, management and restoration

The chapter on management by Bakker is ample proof of the profit made by conservation, management and restoration ecology of the development of vegetation ecology, at least in Europe and Japan (Chapter 4). Phytosociological classification facilitates communication over national boundaries on target plant communities; vegetation mapping can be used for land use planning. Still more importantly, ecological theory regarding the behaviour of plant species along gradients, ecological indicator values, diaspore dispersal, species pool, seed bank dynamics and succession has been developed (see Chapters 2, 6, 7 and 8). The development of ecohydrology as a basis for the restoration of nutrient-poor wetlands is particularly impressive.

Most of Bakker's examples of successful management projects are from Western Europe where indeed both theory and practice have been developed constantly. For a world perspective, see Perrow & Davy (2002).

1.13 Invasive species and invasibility of plant communities

Chapter 13 by M. Rejmánek, D.M. Richardson & P. Pyšek deals with the main characteristics of invasive species; main pathways of migration of invasive species and characteristics of environments and plant communities open to invasions. Of special interest are the relations between invasive and local native species and the often different behaviour of invasive species in their new, alien environment. An interesting suggestion is that invasibility of plant communities by exotics is mainly caused by fluctuations in resource availability (cf. Grime 2001).

Only few invasive species become dominant in new environments and act as a 'transformer species'. They have major effects on the biodiversity of the local native community. They all transform the environment and different ways of transformation are treated. Useful information is provided on the perspectives of eradication of invasive species. As a rule of thumb, species which have invaded an alien area for more than 1 ha can hardly be eradicated.

A very interesting and important conclusion which is emerging is that stable environments with little anthropogenic disturbance tend to be less open to invasive species.

1.14 Vegetation ecology and global change

Chapter 14 by B. Huntley & R. Baxter deals with global pollution problems, notably eutrophication, increasing CO_2 concentration and sea-level rise, but particularly global warming. Of interest in this connection are models to help understand and predict future changes of some main ecosystem types, and problems of species to cope with changes and of dispersing to newly available suitable environments.

Studies on effects of global changes rely heavily on palaeo-ecological studies. In a way these studies are extrapolations into the future of the processes of secular succession, already mentioned in section 1.7. Secular succession, also called vegetation history (Huntley & Webb 1988), was already recognized in early phytosociology, e.g. by Braun-Blanquet (1932) under the name synchronology, as the ultimate vegetation succession.

A major problem in global change studies is that communities subjected to the impact of global changes can be in the process of long-term responses to disturbances which happened long ago. As an example the case of the Fiby forest near Uppsala can be mentioned. Palaeo-ecological research by R. Bradshaw & G. Hannon and gap-dynamical studies by I.C. Prentice and H. Hytteborn *et al.* (review in Engelmark & Hytteborn 1999) showed how more than 400 yr ago *Picea abies* entered the forest upon the termination of the domestic grazing of the mixed *Quercus robur* forest, and how 200 yr ago the *P. abies* stage was largely destroyed by a gale, upon which *Populus tremula* took over. This is still present with trees originating from that period in the now regenerated *Picea* forest, which still has some isolated broad-leaved *Quercus* and *Tilia* trees. It will be very difficult here to separate possible trends following global warming from further changes resulting from the two major historical 'disturbances'.

Models – which are continuously developed but do not play a major part in Chapter 14 – suffer from the uncertainty regarding estimations of crucial parameters, leading to an often broad range of the parameter predicted. Moreover, it may appear that essential parameters have been overlooked. Nevertheless, the further development of predictive models must be encouraged.

1.15 Pattern and process in the plant community

1.15.1 Patch dynamics, cyclic succession and mosaic cycle

In various chapters elements of pattern and elements of process have been put forward. In this section spatial and temporal aspects of vegetation dynamics within the plant community will be discussed. The word combination pattern and process has become a standard feature of community ecology since A.S. Watt published his seminal paper (Watt 1947). The basic idea is that within a phytocoenosis which is in a steady state on the community level, changes may occur patchwise as a result of local disturbance (exogenous factors) or plant senescence (endogenous factors); in the gaps formed regeneration will occur which will in the first instance lead to a patch of vegetation which is different from its surroundings. It should be noted that 'pattern' as an object of pattern analysis (section 1.1.11) may have different causes. With Kershaw (see Kershaw & Looney 1985) we may distinguish between morphological, environmental and sociological pattern. Morphological pattern arises from the growth form of plants, in particular clonal plants (Chapter 5). Environmental pattern is related to spatial variation in environmental factors, for instance soil depth. Sociological pattern results from species interactions and temporal changes in the behaviour of plants and should be the 'pattern' in Watt's pattern and process.

The phenomenon of patch dynamics has been discovered and described in woodland and forest by Watt since the 1920s, and by R. Sernander in Sweden some years earlier (van der Maarel 1996). They observed that gaps arise in the forest through disturbance of some kind and that vegetation regenerates via herb, shrub and low tree stages. If such gaps are large enough, the various stages of this regeneration succession (see Chapter 7) can be described as own plant communities. In the European syntaxonomical system (section 1.4.2) the various regeneration succession stages and the mature forest parts are placed in different classes (e.g. Rodwell *et al.* 2002), but only in ecological surveys (e.g. Ellenberg 1988) is the successional coherence clear. If the gaps are large, as with hurricanes and fire, the following succession proceeds as it were on the community level and is called secondary succession (Chapter 7).

Watt described similar patch dynamics in bogs (where he had studied the work of H. Osvald from 1923), heathlands and grasslands. In bogs the well-known mosaic of hollows and hummocks appeared to be dynamically related and was described as a 'regeneration complex'. As Watt described it: 'each patch in this space-time mosaic is dependent on its neighbours and develops under conditions partly imposed by them'. He adds that the different stages were considered seral and also as separate communities. That is why the term cyclic succession was coined for the complex. However, for Watt it was just one community with a homogeneous substrate.

The case of heathland, which was elaborated by Ph. Stoutjesdijk in The Netherlands (see Stoutjesdijk & Barkman 1992) and particularly in Scotland by C.H. Gimingham (e.g. 1988), concentrated on the population dynamics of the dominant species *Calluna vulgaris*. *Calluna* heath is a model community where four stages – in this case at the same time growth phases of the species – can be distinguished: pioneer (establishment of *Calluna*, up to 6 yr), building (development of hemispherical bushes up to 50 cm height, cover up to 100%, up to 15 yr), mature (lichens on old stems, mosses under the canopy, up to 25 yr) and degenerate (central branches dying, establishment of grasses, up to 40 yr). During the first two, 'upgrade' phases there is a net gain in biomass, during the second two, 'downgrade' there is a net loss. In the case of *Calluna* it is largely the growth and later senescence of the dominant species which determines the cycle. Because the stages are related to individual *Calluna* bushes and the dominant species is always present, they are not recognized as separate communities.

Whether or not to call these cyclical processes succession is a matter of definition and of scale. When the stages have a relatively large spatial extent and are recognized as communities the term cyclic succession is appropriate. See, e.g. Glenn-Lewin & van der Maarel (1992) for exceptions (e.g. *Calluna* heath does not always show this cyclic process) and other examples. An alternative term *Mosaik-Zyklus* has been proposed by the German animal ecologist H. Remmert; this term was introduced in the English literature as mosaic-cycle (Remmert 1991). A mosaic-cycle is a special case of patch dynamics where the changes are triggered largely by endogenous factors, in particular plant senescence. The bogs and heathlands of Watt may fulfil this criterion, although exogenous factors may play some part here as well (Burrows 1990).

1.15.2 Patch dynamics in forest

Although gap regeneration in temperate (beech) forest was Watt's first example of pattern and process, and Sernander had studied regeneration in Swedish boreal forest, the first impact of Watt's 1947 paper was on heathland and grassland ecologists – partly because he was working and publishing on these systems after 1947, and, for a different reason, that forest ecologists, particularly in North America were preoccupied with secondary, old-field succession, which took place over large areas in a spectacular way. The scheme of 'ecosystem development' (section 1.7.4) by Odum

(1969), which had a considerable impact on succession theory, was based on old-field succession. During the 1970s interest in forest dynamics became broadened. The review by White (1979) and the book by White & Pickett (1985) put disturbance and patch dynamics in the focus of forest dynamics (see Chapter 7). As to gap dynamics, points of special interest are the influence of the size, form and orientation of gaps on regeneration and the differences in gap light climate dependent on latitude, tree replacement series, the release of suppressed shade-tolerant mature tree species when light becomes available in the gap (see, e.g. Burrows 1990; Veblen 1992). In temperate and particularly boreal forest not only regeneration after windthrow but also after fire is of particular concern (e.g. Engelmark *et al.* 2000).

1.15.3 Patch dynamics in grassland

The work by Watt on grasslands inspired P.J. Grubb, one of his pupils, to elaborate the concept of regeneration niche in a paper as influential as Watt's (Grubb 1977). The essence of this concept is that gaps arise everywhere, every time through the death or partial destruction of plant units through natural death of short-lived species and all sorts of animal activities, and the open space can be occupied by a germinating seed or by a runner of a clonal plant. The concept was found especially useful in grasslands where many such gaps are so small that they are hardly noticed. In grazed grasslands local removal of plant parts, trampling and deposition of dung are additional causes of gaps, often larger ones. An example of an intricate relation between the availability of microsites and the activities of voles (*Microtus arvalis*) in grassland was given by Gigon & Leutert (1996). See Fig. 1.5.

Grasslands are expecially suited for the study of patch dynamics because the gaps – microsites – are small, can be easily manipulated and the appearance and disappearance of plant units, particularly seedlings, can be accurately followed. An example of manipulation with fire and soil disturbance is the work of Collins (e.g. 1989), who found that spatial heterogeneity is created under such circumstances.

1.15.4 The carousel model

Where gaps arise more or less continuously in grasslands and plant species both become locally extinct because of disturbances and/or death, but also get continuous opportunities to re-establish, species may show a high fine-scale mobility. At the same time, as in the example of Fig. 1.6, patch dynamics can contribute considerably to the co-existence – albeit co-occurrence in many cases – of many plant species on small areas of grassland. During the careful observation over several years of hundreds of very small subplots in plots of limestone grassland on the alvars of southern Öland the impression arose that the grassland community as a whole, described as *Veronica spicata-Avenula pratensis* association was remarkably constant in floristic composition, while the species lists on subplots of 10 cm^2 and 100 cm^2 seemed to change from year to year. Van der Maarel & Sykes (1993) quantified this mobility in two ways, species-wise and subplot-wise: (1) cumulative frequency, i.e. the cumulative number of subplots a species is observed in over the years; (2) cumulative species richness, i.e. the mean number of species that is observed in a subplot over the

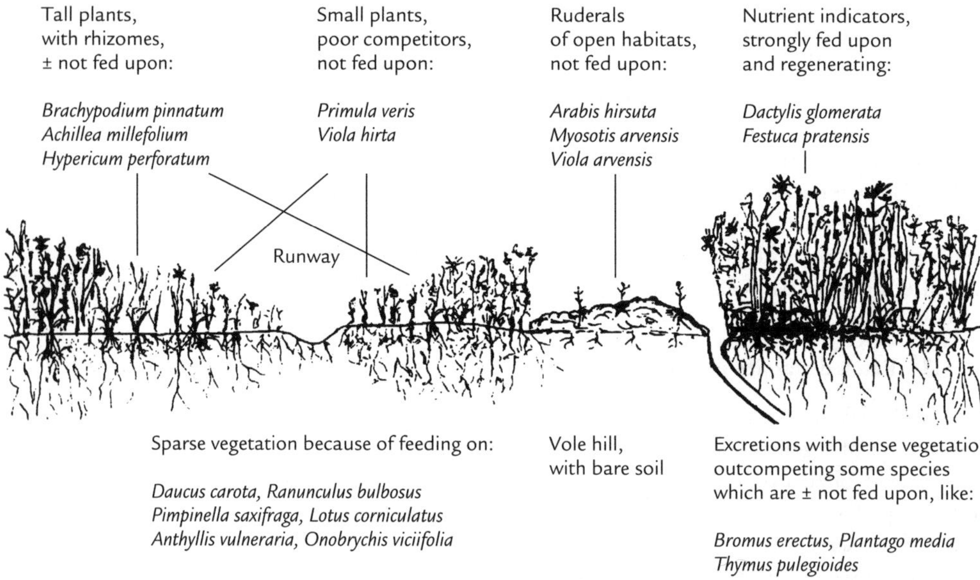

Fig. 1.5 Fine-scale vegetation pattern related to a pattern of microsites in limestone grassland created by *Microtus arvalis*. Reproduced by permission of the publisher. Gigon & Leutert (1996). From Scale not indicated; width of the figure represents *c*. 13 cm.

years. Indeed, many species appeared in many new subplots, which led to the suggestion of a carousel with a 'merry-go-round' of most species. In this short, open grassland on summer-dry soil many short-lived species are involved and germination is a main process in (re-)establishment of species. Simultaneously Herben *et al.* (1993) described fine-scale mobility in a mountain grassland and used persistence as their measure, i.e. the tendency of a species to remain in the subplot where it occurs, as the complement of mobility. Here the mobility of clonal plants (see Chapter 5) was in the focus of interest.

Amongst the comments on the carousel model were the lack of a null model of mobility. Van der Maarel & Sykes (1997) showed how observed mobility can be compared with a model of random allocation of species over subplots taking their frequencies into account, upon a comment by J.B. Wilson in line with his null model for species richness variance (Wilson *et al.* 1995). Also a minimum mobility model was suggested where species occurrences were allocated as far as possible to the same subplots where they occurred earlier – so persistence was maximized. Both cumulative species richness and cumulative frequency were compared with values obtained with the random and the minimum model. Fig. 1.7 shows an example of an increase in cumulative richness – considerable in itself – which is approximately intermediate between the 'minima' and the 'random' accumulation. As to the cumulative frequency, a species mobility index was proposed:

$$MR = (cf_{obs} - f_{min})/cf_R \qquad (1.2)$$

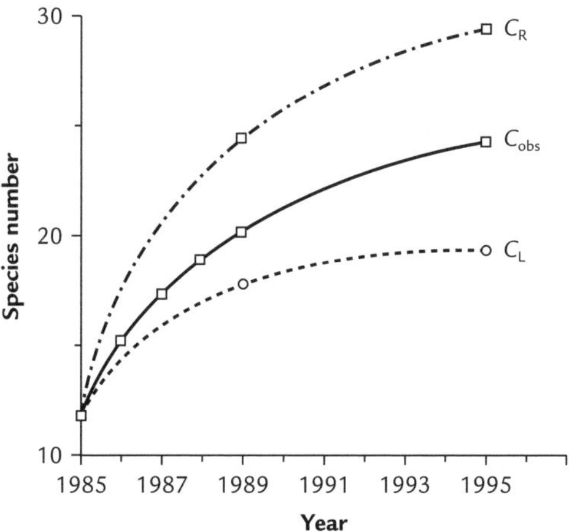

Fig. 1.6 Average species accumulation over 10 yr in 10 cm × 10 cm subplots in a limestone grassland permanent plot. C_{obs} is the observed number; C_R is the accumulation according to a random mobility model; C_L is the accumulation according to a minimum mobility model. Note that in 10 yr mean accumulating number of species increased from *c.* 12 to *c.* 24. After van der Maarel & Sykes (1997).

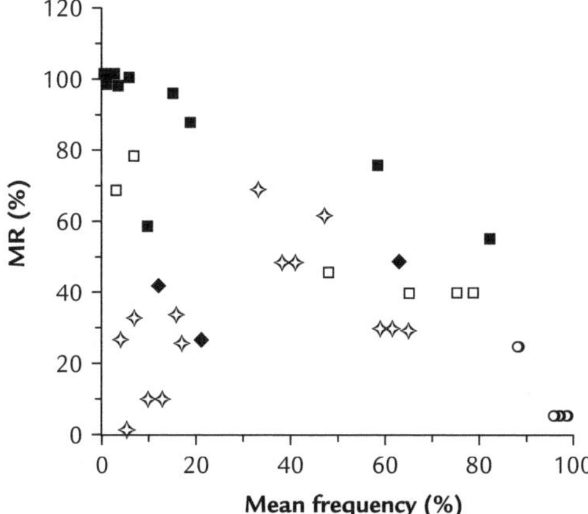

Fig. 1.7 Mobility rate (MR) of plant species in limestone grassland plots in relation to their mean frequency in 10 cm × 10 cm subplots. ❍ = constant species; ☐ = species cumulative frequency not significantly different from random (black: annuals); ✧ = species cumulative frequency significantly lower than random (black: short-lived perennials). After van der Maarel & Sykes (1997). Reproduced by permission of the publisher.

where cf_{obs} is the observed cumulative frequency, in this case over a 10-yr period in the same grassland as mentioned above, f_{min} is the lowest annual frequency observed during the period of observation, and cf_R is the expected frequency according to the random model. Fig. 1.7 shows how this index is related to the mean frequency of

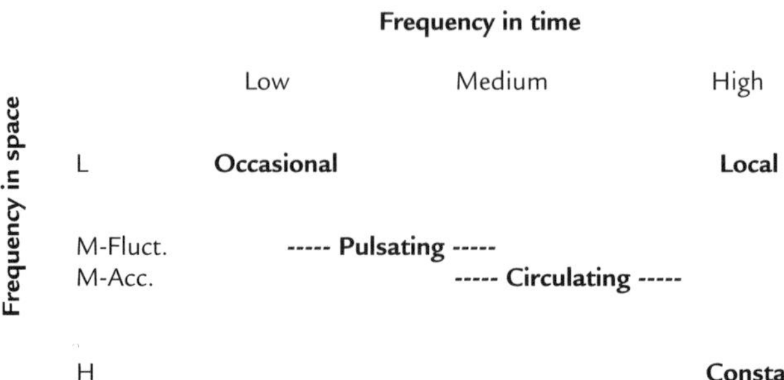

Fig. 1.8 Types of within-community plant species mobility based on frequency in space and time in 10 cm × 10 cm subplots in limestone grassland during 1986–94. Mean spatial frequency values divided into high, > 75% (H), medium, 35–75% (M; M-Fluct. = with large between-year differences; M-Acc = accumulating frequency) and low, < 25% (L). Temporal frequency values divided into H (occurring in > 66% of the years), M (33–66%) and L (< 33%). After van der Maarel (1996). Reproduced by permission of the publisher.

species, in such a way that, naturally, a high mean frequency prevents a species from being mobile. It also indicates that most species with a natural mobility near to the random model are annuals or short-lived perennials. This picture was elaborated as in Fig. 1.8 (van der Maarel 1996) into a division into five categories (van der Maarel & Sykes 1997) with 'local' species (with low mean and low cumulative frequency) at one end, followed by 'occasional species' (with low mean frequency and relatively high cumulative frequency), 'pulsating' species (with medium mean frequency, but low frequency in at least one year, and high cumulative frequency), 'circulating species' (with relatively constant medium mean frequency and high cumulative frequency) and 'constant species' (with high mean frequency).

The carousel model has helped to discover species mobility in several other grasslands and also in woodlands, where, as was expected, mobility is low for most species and different forest patches may have own carousels (Maslov & van der Maarel 2000). This was also concluded by Palmer & Rusch (2001), who at the same time tried to find solutions for some shortcomings in the carousel model. They focused on species mobility and stated first that this was not only a matter of appearance in new microsites, 'immigration', with immigration rate i being the probability of occupying an empty plot per year, but also of disappearance in old microsites, 'extinction', with extinction rate e being the probability of an occupied plot becoming empty per year. Palmer & Rusch also introduced an equilibrium frequency p where immigration and extinction are in balance, and where the three parameters are related as follows:

$$p = i/(i + e) \tag{1.3}$$

They made clear that turnover rate can be defined as the reciprocal of residence time and this is equal to the extinction rate. They also showed that persistence V (e.g. Herben *et al.* 1995) is related to i and e in a simple way:

$$V = 1 - e - i \qquad (1.4)$$

The turn-around time of a species, the carousel time, CT, follows from

$$CT = (-\log N - \log(1 - P_0))/\log(1 - i) \qquad (1.5)$$

where N is the number of plots and P_0 is the initial frequency of the species. Note that carousel time is dependent on the initial frequency of the species and on the number of plots, the number of 'seats in the carousel'.

Palmer & Rusch also make some remarks on the reality of random models and show that for the random model of van der Maarel & Sykes (considered less realistic) RCT can be related to CT under the random model and an index 'relative carousel time' can be derived which is simply:

$$RCT = \log(1 - p)/\log(1 - i) \qquad (1.6)$$

They then present some results based on (only) four years of observation and find carousel times for the largest plots (1 m^2) from 13 to > 4000 yr. They wonder whether this parameter is useful for explaining species co-existence on a short time scale. One may agree with this, but one can also comment that part of the high mobility in the grassland is caused by pulsating and occasional species and it may need a longer series of observations to include 'top years' for certain species. Finally, Palmer & Rusch comment that the number of plots and their size are decisive for the value of CT. Therefore it may be better to use RCT as defined above. Indeed, results presented by Palmer & Rusch show that RCT values are more constant across subplot size and differences between species are large enough to compare species and hypothesize about the differences.

1.16 Plant functional types, life-form types and plant strategy types

1.16.1 Plant functional types and guilds

Plant functional type (PFT) is a group of plant species sharing certain morphological-functional characteristics. PFTs are in the first place used in ecological studies, but they are also important in plant geography (see Chapter 4). The use of plant function seems to go back to Knight & Loucks (1969) who related plant function and morphology to environmental gradients, and Box (1981) who correlated 'ecophysiognomic' plant types with climatic factors, and used climatic envelopes for selected sites to predict the combination of forms (Chapter 4) – Peters (1991) mentioned this study with its validated global model as a good example (one of the few) of predictive ecology. Smith & Huston (1989) stated that the PFT concept is analogous to the (zoological) concept of guild' and described it as a group of species that 'use the same type of resources in more or less the same way'. Indeed botanical guild systems have been developed (see Semenova & van der Maarel 2000). Wilson (1999b) considered

Table 1.4 Some classical life-form systems of vascular plants. **A.** Main life-form groups according to Du Rietz (1931). **B.** Growth forms according to Warming (1909); only main groups distinguished. **C.** Main terrestrial life forms according to Raunkiær (1934), largely following Braun-Blanquet (1964). **D.** Hydrotype groups according to Iversen (1936).

A.	
Physiognomic forms	Based on general appearance at full development
Growth forms	Largely based on shoot formation (*sensu* Warming)
Periodicity-based life forms	Based on seasonal physiognomic differences
Bud height-based life forms	Based on height of buds in the unfavourabler season (*sensu* Raunkiær)
Bud type-based life forms	Based on differences in type and structure of buds
Leave-based life forms	Based on form, size, duration of the leaves
B.	
Hapaxanthic (monocarpic) plants	Plants which reproduce only once and then die; including annuals, biennials and certain perennials, e.g. *Agave*
Pollakanthic (polycarpic) plants	Plants which reproduce repeatedly
Sedentary generative	Primary root or corm long-lived, with only generative reproduction
Sedentary vegetative	Primary root short-lived, with both generative and some vegetative reproduction
Mobile stoloniferous	Creeping above-ground with stolons which develop rootlets
Mobile rhizomatous	Extending below-ground with rhizomes
Mobile aquatic	Free-floating aquatic plants
C.	
Phanerophytes (P)	Perennial plants with perennating organs (buds) at heights > 50 cm Tree P; Shrub P; Tall herb P; Tall stem succulent P.
Chamaephytes (Ch)	Perennial plants with perennating organs at heights < 50 cm Woody (frutescent) dwarf-shrub Ch; Semi-woody (suffrutescent) dwarf-shrub Ch; herbaceous Ch, low succulent Ch, pulvinate Ch
Hemicryptophytes (H)	Perennial plants with periodically dying shoots and perennating organs near the ground Rosette H; Caespitose H; Reptant H
Geophytes (Cryptophytes) (G)	Perennials loosing above-ground parts and surviving below-ground during the unfavourable period Root-budding G; Bulbous G; Rhizome G; Helophyte G
Therophytes (T)	Annuals, completing their life cycle within one favourable growing period, surviving during the unfavourable period as seed or young plant near the ground Ephemeral T (completing cycle several times per growing period); Spring-green T; Summer-green T; Rain-green T; Hibernating green T (green almost all year).
D.	
Terriphytes	Terrestrial plants without aerenchyma
Seasonal xerophytes	
Euxerophytes	
Hemixerophytes	
Mesophytes	
Hygrophytes	
Telmatophytes	Paludal plants (growing in swamps and marshes) with aerenchyma
Amphiphytes	Aquatic plants with both aquatic and terrestrial growth forms
Limnophytes	Aquatic plants in a strict sense

guilds and PFTs as synonyms and preferred the older term guild. Nevertheless, PFT became the dominant term and was redefined as a grouping 'of species which perform similarly in an ecosystem based on a set of common biological attributes' (Gitay & Noble 1997).

During the 1990s PFT systems were particularly used in relation to climate (hence the interest in the typology of Box 1981), and more particularly climate change (e.g. Woodward & Cramer 1996), and to disturbance (Lavorel & Cramer 1999).

In a way, the abundant use of PFTs is a revival of the attention paid to **life forms** during the period 1900–1930. Life forms were seen as types of adaptation to environmental conditions, first of all by E. Warming who spoke of epharmonic convergence after the term **epharmony** = 'the state of the adapted plant', coined as early as 1882 by J. Vesque. Life-form systems from this early period include those of E. Warming from 1895, C. Raunkiær from 1907, G.E. Du Rietz from 1931 and J. Iversen from 1936 (Table 1.4). Environmental adaptation is most obvious in the life form system of Raunkiær. Modern, extended versions of this system, e.g. by Braun-Blanquet (1964), Dierschke (1994) and particularly Mueller-Dombois & Ellenberg (1974) give subdivisions of the main Raunkiær types according to differentiation into growth form and plant height.

1.16.2 Plant strategy types

A concept more recent than life form which is also closely related to PFT is **plant strategy.** The best known system of plant strategies is that by Grime (2001; earlier publications cited there). A special introduction in this book is devoted to PFTs, where he treats the two concepts as equivalent, if not synonymous, but maintains the term strategy in the rest of the book. However, there is a simple difference: strategies, 'groupings of similar or analogous genetic characteristics which recur widely among species or populations and cause them to exhibit similarities in ecology' (Grime) are the combined characteristics of a PFT. These characteristics have also been called **attributes**, e.g. the '**vital attributes**' of Noble & Slatyer (1980), used in relation to community changes caused by disturbances. However, nowadays the term **trait** (probably borrowed from genetics) is used, while attribute is now also used for the different expressions of a trait, which should rather be called states (see further Semenova & van der Maarel 2000).

The three strategy types proposed by Grime have been maintained virtually unchanged, even if the system has been regularly criticized. They enable plants to cope with environmental constraints of two kinds, **stress**, 'external constraints which limit the rate of dry matter production', and **disturbance**, 'destruction of plant biomass arising from outside abiotic, biotic and human impact'. The constraints leading to stress can be both shortages and excesses in the supply of resources, but in practice the focus is on shortages.

Plants in the adult stage have developed three types of strategy. **Competitors** (C) are adapted to environments with low levels of stress and disturbance; **stress-tolerators** (S) to high stress and low disturbance, and **ruderals** (R) to low stress and high disturbance. No plants have developed a 'viable strategy' for the combination of high stress and high disturbance. By distinguishing intermediate levels of stress

Table 1.5 Some adaptations of stress-tolerant, competitive and ruderal plants, selected from the list presented by Grime (2001).

	Stress-tolerant	Competitive	Ruderal
Life form	Herbs, shrubs, trees	Herbs, shrubs, trees	Herbs
Shoots	Wide range of growth forms	Dense leaf canopy Wide lateral spread	Small stature Little lateral spread
Leaves	Small, leathery or needle-like	Robust, mesomorphic	Various, often mesomorphic
Maximum potential growth	Slow	Rapid	Rapid
Response to stress	Responses slow and minor	Maximizing vegetative growth	Less vegetative growth, flowering
Acclimation of photosynthesis and mineral nutrition to seasonal changes in resources	Strongly developed	Weakly developed	Weakly developed
Storage of photosynthates and mineral nutrients	In leaves, stems and/or roots	Rapidly incorporated into vegetative structures, partly stored	In seeds
Longevity of established phase	Long to very long	Long or relatively short	Very short
Longevity of leaves and roots	Long	Relatively short	Short
Leaf phenology	Evergreen; various patterns of leaf production	Peaks of leaf production in periods of maximum	Short phase of leaf production
Perennation	Stress-tolerant leaves and roots	Dormant buds and seeds	Dormant seeds

and disturbance intermediate (so-called secondary) strategy types are distinguished: **competitive ruderals** (C-R) adapted to low stress and moderate disturbance; **stress-tolerant ruderals** (S-R) adapted to high stress and moderate disturbance; **stress-tolerant competitors** (S-C) adapted to moderate stress and low disturbance; and **C-S-R strategists** adapted to moderate stress and moderate disturbance. Table 1.5 presents some traits in which the three main strategy types differ. Obviously, competitors and ruderals are quite similar, apparently because of their link to 'low stress', which means soil fertility (see also Chapter 5).

CSR theory has some predecessors, mentioned by Grime (2001). The most interesting is the theory of L.G. Ramenskiy, who distinguished three types of life-history strategies which are astonishingly similar to the CSR types, as concluded by Grime (2001). The first introduction to Ramenskiy's work in English (mediated by D. Mueller-Dombois) was by Rabotnov (1975), who spoke of 'phytocoenotypes'. His

Table 1.6 Some adaptations of persistent ('P'), vigorous ('V') and exploiting ('E') plant populations based on Onipchenko *et al.* (1998).

Type	Persistent (P)	Vigorous (V)	Exploiting (E)
Synonym Rabotnov	Patient	Violent	Explerent
Synonym Onipchenko *et al.*	Endurant	Dominant	Explorative
Equivalent Grime	Stress-tolerating	Competitive	Ruderal
Resource capture	Low requirement	High when resources available	High when resources available
Stage with high tolerance	Juvenile	Adult	Dormant
Conditions of occurrence	Abiotic and biotic stress	Productive conditions - large biomass production	Productive conditions - moderate biomass production
Response to biomass removal	Negative	Negative	Weak
Response to disturbance s.l.	Negative	Negative, dependent on intensity	Positive
Response to resource availability	Slow	Positive	Positive and rapid
Seed production	High; small seeds	High; large seeds	Intermediate; small seeds
Persistent seed bank	Small	No	Large
Flexibility of leaf longevity	Low	Low (seedling); high (adult)	High
Change in rate of gas exchange per unit leaf area	Low	Low	High

pupil V.G. Onipchenko used Ramenskiy's ideas in combination with ideas by Yu.E. Romanovskiy on two ways a population can succeed in the competition for limiting resources, i.e. reducing the equilibrium resource requirement R* (Tilman 1982) and developing a high resource capture capacity and a high population growth rate when the resource is available. Onipchenko *et al.* (1998) elucidated the 'RRR', Ramenskiy/Rabotnov/Romanovskiy, approach in a functional study of alpine plant communities. Table 1.6 presents some main differences and similarities between the three strategies, which are here called **persistent**, **vigorous** and **exploiting** strategies. Onipchenko *et al.* (1998) measured nine characteristics, both vegetative and generative, of 42 alpine species of environments strongly varying in fertility, stress and disturbance, ordinated the species on the basis of these characteristics and concluded that differences in seed and seed bank characteristics were more informative than the two parameters according to which the Grime triangle is filled, i.e. the morphology index M (biomass development) and relative growth rate, RGR. It was also concluded that there is a continuous variation between extreme combinations of characteristics – as Grime in fact also demonstrates – which leads to a preference for indicating degrees of persistence, vigour and exploitation rather than speak of persistent, vigorous and exploiting species and intermediate types.

1.16.3 On disturbance and stress

A more direct criticism of the Grime triangle was presented by Grubb (1985): 'stress' is a complex phenomenon which, in addition to shortage of nutrients, water and light, includes seasonal drought, high salt concentration, waterlogging, frost and heat. Grubb also differentiated 'disturbance' according to frequency, extent and intensity, and considered the division between continual and periodic disturbance as basic, connected to entirely different adaptations. Based on work by S.D. Fretwell and L. Oksanen, Oksanen & Ranta (1992) emphasized the role of grazing, the most important continual disturbance, in many plant communities and made clear that grazing has no proper place in the CSR system; they suggested to differentiate between frequency and intensity of disturbance; grazing is a form of disturbance with high frequency but low intensity, which indeed occurs in many stressful (nutrient-poor) environments. Their alternative triangle includes the well-known *r*- and *K*-selection mechanisms and '*g*-selection', which operates under high-frequency but low-intensity disturbance.

Grubb (1998) distinguished three adaptations to resource shortage instead of one: (i) 'low flexibility' strategy, with long-lived leaves, low maximum relative growth rate and no changes in form and gas exchange rate when the shortage is relieved; (ii) 'switching' strategy, with low flexibility in the juvenile stage but changing form, particularly specific leaf area (SLA) and growth rate when resources become available in the adult stage; (iii) 'gearing down' strategy, involving strong reduction in respiration rate, both in the juvenile and adult stages, when a resource becomes scarce (Table 1.7).

As remarked above, stress resulting from resource excess has so far hardly been considered, although Grubb (1985) mentioned several forms of excess stress. Yet,

Table 1.7 Some characteristics of stress-tolerant species with low flexibility, switching and gearing-down specialization, according to Grubb (1998).

Type	Switching	Gearing down	Flexibility
RGR_{max} of seedling relative to seed mass	Low	Low	Low to high
RGR_{max} of adult	Low	High	Low to high
Flexibility of SLA in response to relief of shortage	Low	Low (seedling); high (adult)	High
Flexibility of leaf longevity	Low	Low (seedling); high (adult)	High
Change in rate of gas exchange per unit leaf area	Low	Low	High

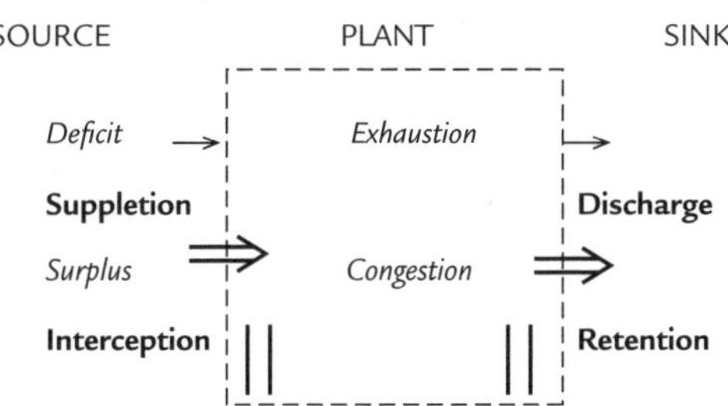

Fig. 1.9 Stress situations (italics) and adaptation mechanisms (bold) of plants, based on a model for ecosystems by G. van Wirdum & C.G. van Leeuwen as rephrased by van der Maarel (1980).

Shelford's 'law of tolerance' from 1913 (e.g. Kent & Coker 1992; Fresco *et al.* 2001) is based on a species-response curve showing two zones of stress for any relevant environmental factor: a minimum level needed and a maximum level tolerated. This double stress situation has been elaborated in a more strategic way by using a model for optimal matter and energy flow in an ecosystem where the plant is seen as a link in a source-sink system (Fig. 1.9).

1 There can be a deficit of a necessary resource – Grime's stress situation – leading to exhaustion which can be coped with in two ways: suppletion and retention. Suppletion can be achieved, for instance by extension of the root system in dry soil, foraging for water; retention for instance by storing nutrients in plants in a nutrient-poor environment.

2 There can also be a surplus of a resource, leading to congestion which can be compensated for by interception, as in the case of a coastal cliff plant under salt spray adopting a prostrate growth form and increased leaf hairiness, or by discharge, for instance salt exudation by a salt-marsh species.

Finally, a pragmatic alternative was proposed by Westoby (1998), avoiding the controversies around the concepts in Grime's system and easily applicable outside the flora of north-west Europe. Three traits are used as axes in a three-dimensional scheme: specific leaf area, plant height at maturity, and seed mass. These traits are said to vary largely independently. SLA and plant height are related to Grime's RGR and morphology index M, respectively.

1.17 Epilogue

Vegetation ecology has grown tremendously since its first textbook appeared (Mueller-Dombois & Ellenberg 1974). and since then, many thousands of papers have been published in international journals. Although only a small minority of them have been cited in this book, it is hoped that the growth of the science, both in depth and in breadth, will become clear from the 13 chapters that follow. The growing breadth is also expressed in the involvement of scientists from other disciplines in vegetation ecology, notably population ecology, ecophysiology, microbiology, soil biology, entomology, animal ecology, physical geography, geology and climatology.

It is encouraging that the international cooperation between plant ecologists all over the world has grown impressively as well. The authorship of this book includes colleagues from Africa, Asia, Australia, Europe and the USA.

Several chapters conclude with a summary of achievements, others offer perspectives for the future of our science. Let us hope that the book will indeed contribute to the further development of vegetation ecology.

Acknowledgements

I thank Mike P. Austin, Jan P. Bakker, Thom Kuyper, Marcel Rejmánek, Mahesh Sankaran, Brita Svensson and Jelte van Andel for comments on earlier versions of this chapter; Dick Visser at the Biological Centre of the University of Groningen and Joost van der Maarel of Opulus Press for drawing some figures; and last but not least Marijke van der Maarel for both reviewing and copy-editing this chapter.

References

Allen, T.F.H. & Hoekstra, T.W. (1992) *Toward a unified ecology*. Columbia University Press, New York.

Bakker, J.P., Olff, H., Willems, J.H. & Zobel, M. (1996) Why do we need permanent plots in the study of long-term vegetation dynamics? *Journal of Vegetation Science* 7, 147–156.

Barkman, J.J. (1978) Synusial Approaches to Classification. In: *Classification of Plant Communities* (ed. R.H. Whittaker), 2nd ed., pp. 111–165. Junk, The Hague.

Barkman, J.J. (1979) The investigation of vegetation texture and structure. In: *The Study of Vegetation* (ed. M.J.A. Werger), pp. 123–160. Junk, The Hague.

Barkman, J.J. (1988) A new method to determine some characters of vegetation structure. *Vegetatio* **78**, 81–90.

Barkman, J.J. (1989) A critical evaluation of minimum area concepts. *Vegetatio* **85**, 89–104.

Box, E.O. (1981) *Macroclimate and Plant Forms: an Introduction to Predictive Modeling in Phytogeography*. Junk, The Hague.

Box, E.O. (1996) Plant functional types and climate at the global scale. *Journal of Vegetation Science* 7, 309–320.

Braun-Blanquet, J. (1932) *Plant Sociology. The Study of Plant Communities*. Authorized English translation of 'Pflanzensoziologie' by G.D. Fuller & H.S. Conard. McGraw-Hill Book Company, New York.

Braun-Blanquet, J. (1964) *Pflanzensoziologie*. 3. Auflage. Springer-Verlag, Wien.

Bredenkamp, G., Chytrý, M. Fischer, H.S., Neuhäuslová, Z. & van der Maarel, E. (eds) (1998) Vegetation mapping: Theory, methods and case studies. *Applied Vegetation Science* **1**, 161–266.

Burrows, C.J. (1990) *Processes of Vegetation Change*. Unwin Hyman, London.

Cain, S.A. & Castro, G.M. de Oliveira (1959) *Manual of Vegetation Analysis*. Harper & Brothers, New York.

Clements, F.E. (1916) *Plant Succession. An Analysis of the Development of Vegetation*. Carnegie Institution, Washington.

Collins, S.L. (1989) Experimental analysis of patch dynamics and community heterogeneity in tallgrass prairie. *Vegetatio* **85**, 57–66.

Crawley, M.J. (ed.) (1997) *Plant Ecology*. 2nd ed. Blackwell Science, Oxford.

Curtis, J.T. (1959) *The vegetation of Wisconsin*. University of Wisconsin Press, Madison.

Dai, X. (2000) Impact of cattle dung deposition on the distribution pattern of plant species in an alvar limestone grassland. *Journal of Vegetation Science* **11**, 715–724.

Dai, X. & van der Maarel, E. (1997) Transect-based patch size frequency analysis. *Journal of Vegetation Science* **8**, 865–872.

Dale, M.T.R. (1999) Spatial *Pattern Analysis in Plant Ecology*. Cambridge University Press, Cambridge.

Dierschke, H. (1994) *Pflanzensoziologie*. Verlag Eugen Ulmer, Stuttgart.

Dietvorst, P., van der Maarel, E. & van der Putten, H. (1982) A new approach to the minimal area of a plant community. *Vegetatio* **50**, 77–91.

Du Rietz, G.E. (1931) Life-forms of terrestrial flowering plants. *Acta Phytogeographica Suecica* **3**, 1–95.

Ellenberg, H. (1988) *Vegetation Ecology of Central Europe*. Cambridge University Press, Cambridge – translation of 'Vegetation Mitteleuropas mit den Alpen', the 5th ed. of which appeared in 1996 at Eugen Ulmer Verlag, Stuttgart.

Ellenberg, H., Weber, H.E., Düll, R., Wirth, V., Werner, W. & Paulißen, D. (1992) Zeigerwerte von Pflanzen in Mitteleuropa, 2nd ed. *Scripta Geobotanica* **18**, 1–258.

Engelmark, O. & Hytteborn, H. (1999) Coniferous forests. In: Swedish plant geography – Dedicated to Eddy van der Maarel on his 65th birthday (eds H. Rydin, P. Snoeijs & M. Diekmann). *Acta Phytogeographica Suecica* **84**, 55–74.

Engelmark, O., Gauthier, S. & van der Maarel, E. (eds) (2000) Disturbance dynamics in boreal and temperate forests. *Journal of Vegetation Science* **11**, 777–880.

Fliervoet, L.M. (1985) Characterization of the canopy structure of Dutch grasslands. *Vegetatio* **70**, 105–117.

Fresco, L.F.M., van der Maarel, E. & Kaźmierczak, E. (2001) VEGRON v. 7.0. *Numerical analysis in vegetation ecology: Program package with Introduction and User manual.* Opulus Press, Uppsala.

Gigon, A. & Leutert, A. (1996) The Dynamic keyhole-key model of coexistence to explain diversity of plants in limestone and other grasslands. *Journal of Vegetation Science* **7**, 29–40.

Gimingham, C.H. (1988) A reappraisal of cyclical processes in Calluna heath. *Vegetatio* **77**, 61–64.

Gitay, H. & Noble, I.R. (1997) What are functional types and how should we seek them? In: *Plant functional types: their relevance to ecosystem properties and global change* (eds T.M. Smith, H.H. Shugart & F.I. Woodward), pp. 3–19. Cambridge University Press, Cambridge.

Glenn-Lewin, D.C. & van der Maarel, E. (1992) Patterns and processes of vegetation dynamics. In: *Plant succession – theory and prediction* (eds D.C. Glenn-Lewin, R.K. Peet & T.T. Veblen), pp. 11–59. Chapman & Hall, London.

Grime, J.P. (2001) *Plant Strategies, Vegetation Processes, and Ecosystem Properties*. 2nd ed. John Wiley & Sons, Chichester.

Grootjans, A.P., Fresco, L.F.M., de Leeuw, C.C. & Schipper, P.C. (1996) Degeneration of species-rich Calthion palustris hay meadows; some considerations on the community concept. *Journal of Vegetation Science* **7**, 185–194.

Grossman, D.H., Faber-Langendoen, D., Weakley, A.S. *et al.* (1998) *International Classification of Ecological Communities*. Vol. 1. *The National Vegetation Classification system.* The Nature Conservancy, Arlington.

Grubb, P.J. (1977) The maintenance of species-richness in plant communities: the importance of the regeneration niche. *Biological Reviews of the Cambridge Philosophical Society* **52**, 107–145.

Grubb, P.J. (1985) Plant populations and vegetation in relation to habitat, disturbance and competition: problems of generalization. In: *The Population Structure of Vegetation* (ed. J. White), pp. 595–621. Junk, The Hague.

Grubb, P.J. (1998) A reassessment of the strategies of plants which cope with shortages of resources. *Perspectives in Plant Ecology, Evolution and Systematics* **1**, 3–31.

Gurevitch, J., Scheiner, S.M. & Fox, G.A. (2002) *The Ecology of Plants*. Sinauer Associates, Sunderland.

Haase, P. (1995) Spatial pattern analysis in ecology based on Ripley's K-function: Introduction and methods of edge correction. *Journal of Vegetation Science* **6**, 583–592.

Haskell, J.P., Ritchie, M.E. & Olff, H. (2002) Fractal geometry predicts varying body size scaling relationships for mammal and bird home ranges. *Nature* **418**, 527–430.

Hennekens, S.M. & Schaminée, J.H.J. (2001) TURBOVEG, a comprehensive data base management system for vegetation data. *Journal of Vegetation Science* **12**, 589–591.

Herben, T., Krahulec, F., Hadincová, V. & Skálová, H. (1993) Small-scale variability as a mechanism for large-scale stability in mountain grasslands. *Journal of Vegetation Science* **4**, 163–170.

Herben, T., During, H.J. & Krahulec, F. (1995) Spatiotemporal dynamics in mountain grasslands – species autocorrelations in space and time. *Folia Geobotanica & Phytotaxomia Phytotax.* **30**, 185–196.

Hill, M.O. (1973) Diversity and evenness: a unifying notation and its consequences. *Ecology* **54**, 427–432.

Howe, H.F. & Westley, L.C. (1988) *Ecological Relationships of Plants and Animals*. Oxford University Press, New York.

Huntley, B. & Webb III, T. (1988) *Handbook of Vegetation Science, part 7, Vegetation history*. Kluwer Academic Publishers, Dordrecht.

Iversen, J. (1936) Biologische Pflanzentypen als Hilfsmittel in der Vegetationsforschung. Ein Beitrag zur ökologischen Charakterisierung und Anordnung der Pflanzengesellschaften. *Meddelelser fra Skalling-Laboratoriet* **4**, 1–224.

Jennings, M., Loucks, O., Glenn-Lewin, D. *et al.* (2002) *Standards for Associations and Alliances of the U.S. National Vegetation Classification. Version 1.0.* The Ecological Society of America Vegetation Classification Panel. www.esa.org/vegweb/vegstds_v1.htm

Jongman, R.H.G., ter Braak, C.J.F. & van Tongeren, O.F.R. (1995) *Data analysis in community and landscape ecology*. Cambridge University Press, Cambridge.

Kent, M. & Coker, P. (1992) *Vegetation Description and Analysis – A Practical Approach*. Belhaven Press, London.

Kershaw, K.A. & Looney, J.H.H. (1985) *Quantitative and Dynamic Plant Ecology*. 3rd ed. Edward Arnold, London.

Knapp, R. (1984) Sample (relevé) areas (Distribution, homogeneity, size, shape) and plot-less sampling. In: *Sampling Methods and Taxon Analysis in Vegetation Science* (ed. R. Knapp), pp. 191–119. Junk, The Hague.

Knight, D.H. & Loucks, O.L. (1969) A quantitative analysis of Wisconsin forest vegetation on the basis of plant function and gross morphology. *Ecology* **50**, 219–234.

Lavorel, S. & Cramer, W. (eds) (1999) Plant functional types and disturbance dynamics. *Journal of Vegetation Science* **10**, 603–730.

Looijen, R.C. & van Andel, J. (1999) Ecological communities: conceptual problems and definitions. *Perspectives in Plant Ecology, Evolution and Systematics* **2**, 210–222.

Magurran, A.E. (1988) *Ecological Diversity and Its Measurement*. Princeton University Press, Princeton.

Maslov, A.A. & van der Maarel, E. (2000) Limitations and extensions of the carousel model in boreal forest communities. In: *Vegetation Science in Retrospect and Perspective. Proceedings*

41th IAVS Symposium (eds P.J. White, L. Mucina & J. Lepš), pp. 362–365. Opulus Press, Uppsala.

McMichael, B.L. & Persson, H. (eds) (1991) *Plant Roots and their Environment.* Elsevier Science, Amsterdam.

Meijer Drees, E. (1954) The minimum area in tropical rain forest with special reference to some types in Bangka (Indonesia). *Vegetatio* **5/6**, 517–523.

Mucina, L. & Dale, M.B. (1989) Numerical taxonomy. *Vegetatio* **81**, 1–215.

Mucina, L., Schaminée, J.H.J. & Rodwell, J.S. (2000) Common data standards for recording relevés in field survey for vegetation classification. *Journal of Vegetation Science* **11**, 769–772.

Mueller-Dombois, D. & Ellenberg, H. (1974) *Aims and Methods of Vegetation Ecology.* John Wiley and Sons, New York.

Nicholson, M. & McIntosh, R.P. (2002) H.A. Gleason and the Individualistic Hypothesis Revisited. *Bulletin of the Ecological Society of America* **83**, 133–142.

Noble, I.R. & Slayter, R.O. (1980) The use of vital attributes to predict successional changes in plant communities subject to recurrent disturbances. *Vegetatio* **43**, 5–21.

Odum, E.P. (1969) The strategy of ecosystem development. *Science* **164**, 262–270.

Økland, R.H. (1990) Vegetation ecology: theory, methods and applications with reference to Fennoscandia. *Sommerfeltia Supplements* **1**, 1–233.

Oksanen, L. & Ranta, E. 1992. Plant strategies along mountain vegetation gradients: a test of two theories. *Journal of Vegetation Science* **3**: 175–186.

Oksanen, L., Fretwell, S.D., Arruda, J. & Niemel, P. (1981) Exploitation ecosystems in gradients of primary productivity. *American Naturalist* **118**, 240–261.

Olff, H. & Ritchie, M.E. (1998) Effects of herbivores on grassland plant diversity. *Trends in Ecology and Evolution* **13**, 261–265.

Olff, H., Ritchie, M.E. & Prins, H.H.T. (2002) Global environmental determinants of diversity in large herbivores. *Nature* **415**, 901–905.

Onipchenko, V.G., Semenova, G.V. & van der Maarel, E. (1998) Population strategies in severe environments: alpine plants in the northwestern Caucasus. *Journal of Vegetation Science* **12**, 305–318.

Palmer, M.W. & Rusch, G.M. (2001) How fast is the carousel? Direct indices of species mobility with examples from an Oklahoma grassland. *Journal of Vegetation Science* **12**, 305–318.

Parker, V.T. (2001) Conceptual problems and scale limitations of defining ecological communities: a critique of the CI concept (Community of Individuals). *Perspectives in Plant Ecology, Evolution and Systematics* **4**, 80–96.

Peet, R.K. (1992) Community structure and ecosystem function. In: *Plant succession – theory and prediction* (eds D.C. Glenn-Lewin, R.K. Peet & T.T. Veblen), pp. 103–151. Chapman & Hall, London.

Perrow, M.R. & Davy, A.J. (2002) *Handbook of Ecological Restoration. Vol. 1 Principles of Restoration. Vol. 2. Restoration in Practice.* Cambridge University Press, Cambridge.

Peters, R.H. (1991) *A Critique for Ecology.* Cambridge University Press, Cambridge.

Pickett, S.T.A., Collins, S.L. & Armesto, J.J. (1987) Models, mechanisms and pathways of succession. *Botanical Review* **53**, 335–371.

Prentice, I.C. (1992) Climate change and long-term vegetation dynamics. In: *Plant succession – theory and prediction*, (eds D.C. Glenn-Lewin, R.K. Peet & T.T. Veblen), pp. 293–339. Chapman & Hall, London.

Preston, F.W. (1962) The canonical distribution of commonness and rarity. *Ecology* **43**, 185–215, 410–432.

Rabotnov, T.A. (1975) On phytocoenotypes. *Phytocoenologia* **2**, 66–72.

Raunkiær, C. (1934) *The Life Forms of Plants and Statistical Plant Geography*. Clarendon Press, Oxford.

Remmert, H. (1991) The Mosaic Cycle Concept of Ecosystems. *Ecological Studies* **85**. Springer-Verlag, Berlin.

Rodwell, J.S. (1991–2000) *British Plant Communities. Volumes I-V.* Cambridge University Press, Cambridge.

Rodwell, J.S., Schaminée, J.H.J., Mucina, L., Pignatti, S., Dring, J. & Moss, D. (2002) The Diversity of European Vegetation. An overview of phytosociological alliances and their relationships to EUNIS habitats. *Report EC-LNV 2002/054*, Wageningen.

Rydin, H. (1997) Competition among bryophytes. *Advances in Bryology* **6**, 135–168.

Schaminée, J.H.J., Hommel, P.W.F.M., Stortelder, A.H.F., Weeda, E.J. & Westhoff, V. (1995–1999). *De Vegetatie van Nederland, Vols. 1–5*. Opulus Press, Uppsala/Leiden.

Schulze, E.-D. & Mooney, H.A. (eds) (1994) *Biodiversity and ecosystem function*. Springer-Verlag, Berlin.

Schwabe, A. (1989) Vegetation complexes of flowing-water habitats and their importance for the differentiation of landscape units. *Landscape Ecology* **2**, 237–253.

Semenova, G.V. & van der Maarel, E. (2000) Plant functional types – a strategic perspective. *Journal of Vegetation Science* **11**, 917–922.

Shipley, B. & Keddy, P.A. (1987) The individualistic and community-unit concepts as falsifiable hypotheses. *Vegetatio* **69**, 47–55.

Smith, T. & Huston, M. (1989) A theory of spatial and temporal dynamics of plant communities. *Vegetatio* **83**, 49–69.

Stoutjesdijk, Ph. & Barkman, J.J. (1992) *Microclimate, Vegetation and Fauna*. Opulus Press, Uppsala.

Tilman, D. (1982) *Resource Competition and Community Structure*. Princeton University Press, Princeton.

Titlyanova, A.A. & Mironycheva-Tokareva, N.P. (1990) Vegetation succession and biological turnover on coal-mining spoils. *Journal of Vegetation Science* **1**, 643–652.

Titlyanova, A.A., Romanova, I.P., Kosykh, N.P. & Mironycheva-Tokareva, N.P. (1999) Pattern and process in above-ground and below-ground components of grassland ecosystems. *Journal of Vegetation Science* **10**, 307–320.

Trass, H. & Malmer, N. (1978) North European Approaches to Classification. In: *Classification of Plant Communities* (ed. R.H. Whittaker), 2nd ed., pp. 201–245. Junk, The Hague.

Tüxen, R. (1970) Einige Bestandes- und Typenmerkmale in der Struktur der Pflanzengesellschaften. In: *Gesellschaftsmorphologie* (ed. R. Tüxen), pp. 76–98. Junk, The Hague.

van der Maarel, E. (1979) Transformation of cover-abundance values in phytosociology and its effects on community similarity. *Vegetatio* **39**, 97–114.

van der Maarel, E. (1980) Towards an ecological theory of nature management. *Verhandlungen Gesellschaft für Ökologie* **8**, 13–24.

van der Maarel, E. (1988) Vegetation dynamics: patterns in time and space. *Vegetatio* **77**, 7–19.

van der Maarel, E. (1990) Ecotones and ecoclines are different. *Journal of Vegetation Science* **1**, 135–138.

van der Maarel, E. (1995) Vicinism and mass effect in a historical perspective. *Journal of Vegetation Science* **6**, 445–446.

van der Maarel, E. (1996) Pattern and process in the plant community: Fifty years after A.S. Watt. *Journal of Vegetation Science* **7**, 19–28.

van der Maarel, E. & Sykes, M.T. (1993) Small-scale plant species turnover in a limestone grassland: the carousel model and some comments on the niche concept. *Journal of Vegetation Science* **4**, 179–188.

van der Maarel, E. & Sykes, M.T. (1997) Rates of small-scale species mobility in alvar limestone grassland. *Journal of Vegetation Science* **8**, 199–208.

van Dorp, D., Boot, R. & van der Maarel, E. (1985) Vegetation succession in the dunes near Oostvoorne, The Netherlands, since 1934, interpreted from air photographs and vegetation maps. *Vegetatio* **58**, 123–136.

Veblen, T.T. (1992) Regeneration dynamics. In: *Plant succession – theory and prediction* (eds D.C. Glenn-Lewin, R.K. Peet & T.T. Veblen), pp. 152–187. Chapman & Hall, London.

Warming, E. (1909) *Oecology of Plants: An Introduction to the Study of Plant Communities. English edition of the Danish textbook Plantesamfund. Grundtræk af den økologiske Plantegeografi.* (1895) by M. Vahl, P. Groom & B. Balfour. Oxford University Press, Oxford.

Watt, A.S. (1947) Pattern and process in the plant community. *Journal of Ecology* **35**, 1–22.

Weber, H.E., Moravec, J. & Theurillat, J.-P. (2000) International Code of Phytosociological Nomenclature. 3rd ed. *Journal of Vegetation Science* **11**, 739–768.

Weiher, E. & Keddy, P. (1999) Assembly rules as general constraints on community composition. In: *Ecological Assembly Rules. Perspectives, advances, retreats* (eds E. Weiher & P. Keddy), pp. 251–271. Cambridge University Press, Cambridge.

Westhoff, V. & van der Maarel, E. (1978) The Braun-Blanquet approach. In: *Classification of Plant Communities* (ed. R.H. Whittaker), 2nd ed., pp. 287–297. Junk, The Hague.

Westoby, M. (1998) A leaf-height-seed (LHS) plant ecology strategy scheme. *Plant and Soil* **199**, 213–227.

White, J. (ed.) (1985) *Handbook of Vegetation Science, Part 3, The Population Structure of Vegetation.* Junk, Dordrecht.

White, P.S. (1979) Pattern, process, and natural disturbance in vegetation. *The Botanical Review* **45**, 229–299.

White, P.S. & Pickett, S.T.A. (eds) (1985) *The Ecology of Natural Disturbance and Patch Dynamics.* Academic Press, Orlando.

Whittaker, R.H. (1975) *Communities and Ecosystems.* 2nd ed. Macmillan & Co., New York.

Whittaker, R.H. (1978) Approaches to Classifying Vegetation. In: *Classification of Plant Communities* (ed. R.H. Whittaker), 2nd ed., pp. 1–31. Junk, The Hague.

Williams, C.B. (1964) *Patterns in the Balance of Nature and Related Problems in Quantitative Ecology.* Academic Press, London. New York.

Wilson, J.B. (1999a) Assembly rules in plant communities. In: *Ecological Assembly Rules. Perspectives, advances, retreats* (eds E. Weiher & P. Keddy), pp. 130–164. Cambridge University Press, Cambridge.

Wilson, J.B. (1999b) Guilds, functional types and ecological groups. *Oikos* **86**, 507–522.

Wilson, J.B. & Agnew, A.D.Q. (1992) Positive-feedback switches in plant communities. *Advances in Ecological Research* **23**, 263–336.

Wilson, J.B., Sykes, M.T. & Peet, R.K. (1995) Time and space in the community structure of a species-rich limestone grassland. *Journal of Vegetation Science* **6**, 729–740.

Wilson, J.B., Gitay, H., Steel, J.B. & King, W. McG. (1998) Relative abundance distributions in plant communities: effects of species richness and of spatial scale. *Journal of Vegetation Science* **9**, 213–220.

Woodward, T.I. & Cramer, W. (eds) (1996) Plant functional types and climate change. *Journal of Vegetation Science* **7**, 305–430.

Zamfir, M. & Goldberg, D.E. (2000) The effect of density on interactions between bryophytes at the individual and community levels. *Journal of Ecology* **88**, 243–255.

Zobel, M., van der Maarel, E. & Dupré, C. (1998) Species pool: the concept, its determination and significance for community restoration. *Applied Vegetation Science* **1**, 55–66.

2

Vegetation and environment: discontinuities and continuities

M.P. Austin

2.1 Introduction

> The pattern of variation shown by the distribution of species among quadrats of the earth's surface chosen at random hovers in a tantalizing manner between the continuous and the discontinuous.
>
> (Webb 1954)

The issue can be expressed as: is vegetation organized into discrete recognizable communities or as a continuum of gradually changing composition? Answering this question has a history of confused debate between conflicting schools of research, tedious descriptive accounts and a lack of hypothesis testing. McIntosh (1985) demonstrated the important role it has played in all ecological disciplines, not just plant community ecology. Lack of resolution of the issue has led some ecologists to conclude that it is irrelevant to the advancement of ecological science. However, a study of a forest, grassland or the population of a species can have little practical value without an adequate description of the associated vegetation and environment.

2.1.1 Vegetation concepts

Two terms are used extensively in vegetation ecology, community and continuum. Definitions of the term community vary (see Chapter 1). A plant community can be broadly defined as (i) having a consistent floristic composition; (ii) having uniform physiognomy; (iii) occurring in a particular environment; and (iv) usually occurring at several locations. Implicit in the definition is the assumption that the consistent composition and uniform physiognomy is the result of biotic interactions between the species, particularly competition. The individualistic continuum concept differs in that each species is considered to have an individualistic response to both abiotic and biotic factors such that when vegetation is viewed in relation to an environmental variable, variation in floristic composition and structure is continuous.

2.1.2 Relationship between vegetation and environment

Detailed analysis of the relationship between vegetation and environment requires a detailed understanding of the environmental processes that influence vegetation, for example, knowledge of the processes that link rainfall to the availability of water to plants and of the physiological processes that govern its use by different species is essential.

In ecology there is often a dichotomy between experimental and observational studies. Very few studies combine rigorous observational analysis with detailed manipulative experiments on vegetation composition. Grime (2001) provides an exception with experiments based on extensive surveys of grasslands in a local area together with examples from other regions. The focus in this chapter is on what observational studies can tell us about vegetation/environment relationships. There is an intimate dependence between developments in vegetation concepts, mathematical methods of analysis and knowledge of environmental processes. An account of these aspects is presented in the context of three questions:

1 Is vegetation pattern continuous or discontinuous and how is this pattern related to environment?

2 What theory and methods are most appropriate for investigating such patterns?

3 What is the relative importance of environment and factors intrinsic to the vegetation in determining the observed patterns?

2.2 Early history

In order to evaluate the relative merits of different approaches to vegetation/environment patterns it is important to know the history that has led to the current research paradigms (Kuhn 1970) being adopted. A Kuhnian paradigm consists of an agreed collection of facts, a conceptual framework concerning those facts, a restricted set of problems selected from within the framework and studied with an accepted array of methods (Austin 1999b). Over the past 50 years, developments in three areas have contributed to the development of current paradigms in observational plant community ecology. These were recognition of (i) alternative theoretical frameworks; (ii) the need for rigorous quantitative methods of sampling and analysing vegetation; and (iii) the need to measure more precisely the potential causal environmental variables.

2.2.1 Continuum versus community controversy

Two conceptual approaches dominated plant community ecology prior to the 1950s: the climax community as a super-organism associated with its proponent F.E. Clements, and the association as a vegetation unit that could be classified in a similar way to a species often associated with the name of J. Braun-Blanquet. The first paradigm predominated in North America and Britain, the second in continental Europe (see section 1.1.1). Both 'schools' accepted that vegetation could be classified into units (communities) though their assumptions and methods differed (Whittaker 1962; Mueller-Dombois & Ellenberg 1974; Westhoff & van der Maarel 1978; McIntosh 1985). Gleason (1926) had advanced an alternative conceptual framework: 'the individualistic concept of vegetation'. This framework attracted intense opposition and subsequent neglect until the 1950s. Whittaker (1967) restated Gleason's ideas as two principles:

> (1) The principle of species individuality – each species is distributed in relation to the total range of environmental factors (including effects of other species) it encounters according to its own genetic structure, physiological characteristics and population dynamics.

No two species are alike in these characteristics, consequently, with few exceptions, no two species have the same distributions.
(2) The principle of community continuity – communities which occur along continuous environmental gradients usually intergrade continuously, with gradual changes in population levels of species along the gradient.

Subsequently, it was found that other European ecologists, particularly Ramensky in Russia, had put forward similar ideas and received similar negative responses (Whittaker 1967).

In the late 1940s and 1950s, Whittaker (1956), identified with gradient analysis, and Curtis (1959), identified with the continuum concept, began to examine patterns of vegetation composition using explicit though different numerical methods. They and their students concluded that vegetation patterns were better explained by the continuum concept of continuous variation in relation to environmental gradients. Their studies generated further controversy. Three issues were often confused in the debate: (i) the discrete community versus continuum issue; (ii) the use of objective numerical methods to analyse vegetation data on composition as opposed to subjective methods; and (iii) whether disturbed or heterogeneous stands of vegetation had been included in the sampling.

By 1970 it was recognized that quantitative methods could be applied to either continuum or community approaches and that the two approaches were not necessarily incompatible (see Mueller-Dombois & Ellenberg 1974 and Whittaker 1978a, b, for further commentary).

2.3 Development of quantitative methods

2.3.1 Direct ordination

Direct ordination, originally termed direct gradient analysis by Whittaker (1956), is the analysis of species distributions (presence/absence or abundance data) and collective properties (e.g. species richness) in relation to environmental variables conventionally referred to as environmental gradients. Initially, the methods used were graphical and the environmental measures were crude, often simply subjective estimates of moisture (Fig. 2.1). Relatively independent developments of this graphic

Fig. 2.1 (*opposite*) Two early examples of direct gradient analyses from Britain and Central Europe. **a.** Distribution of tree species along gradients of acidity and moisture in Central European forests (Ellenberg 1988). Thick black border encloses zone where species is dominant. Broken border encloses zone where species is co-dominant. These borders define zones where species have ecological optima as opposed to hatched zones which indicate species' estimated physiological optima. Reproduced by permission of Cambridge University Press. **b.** Distribution of five chalk grassland species in relation to slope and aspect as represented by a diagrammatic hemispherical hill in four regions of north-west Europe. The spokes represent the eight cardinal points of the compass and slope increases in steps of five degrees from the centre (Perring 1960). Note for species I and II: within the contour species is present while hatched area has > 5% cover. For other species: within the contour species is present, simple hatching > 10% cover, cross hatch > 20%; black > 25%.

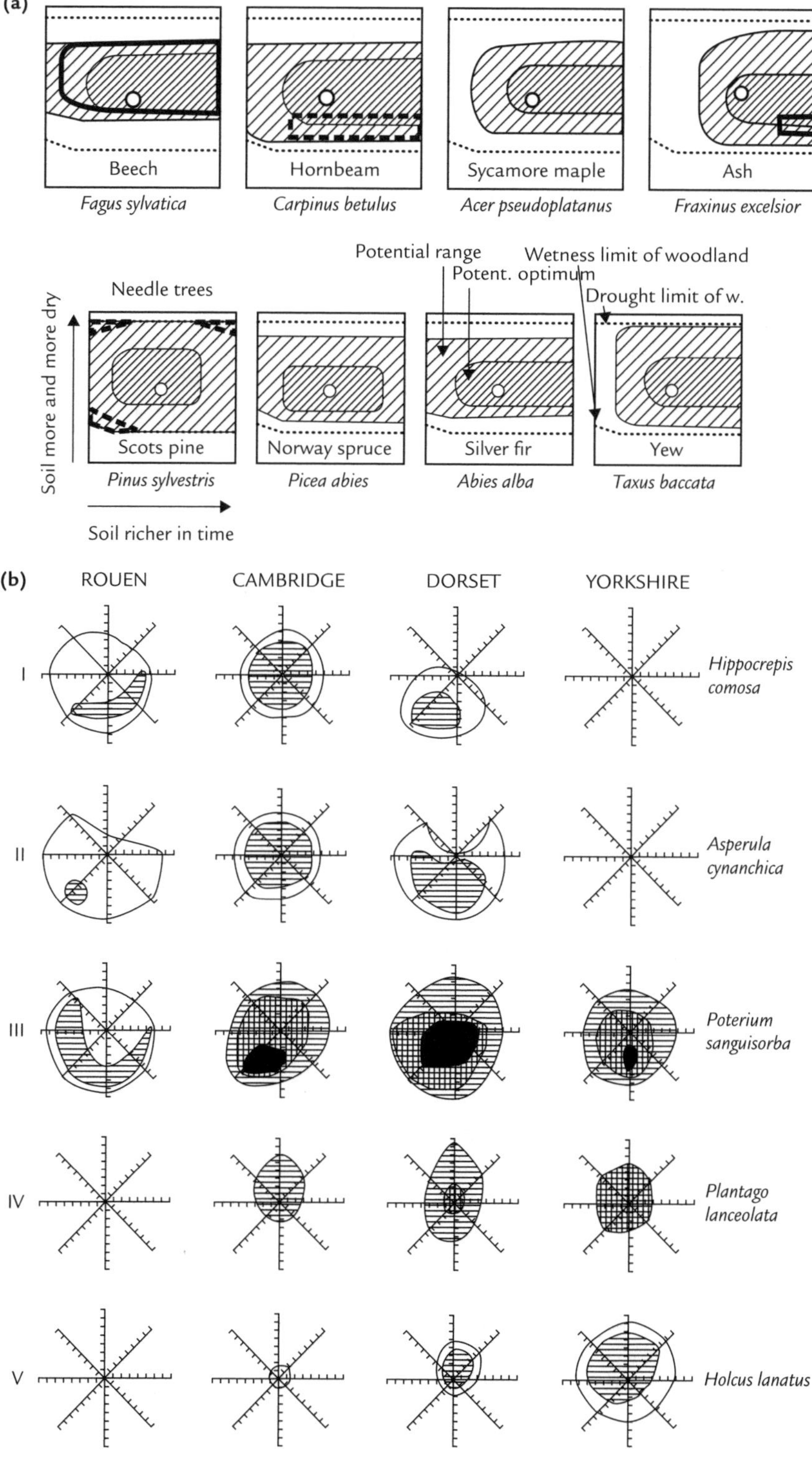
(a)
Beech
Fagus sylvatica
Hornbeam
Carpinus betulus
Sycamore maple
Acer pseudoplatanus
Ash
Fraxinus excelsior
Potential range
Wetness limit of woodland
Potent. optimum
Drought limit of w.
Needle trees
Soil more and more dry
Scots pine
Pinus sylvestris
Norway spruce
Picea abies
Silver fir
Abies alba
Yew
Taxus baccata
Soil richer in time
(b)
ROUEN
CAMBRIDGE
DORSET
YORKSHIRE
I
Hippocrepis comosa
II
Asperula cynanchica
III
Poterium sanguisorba
IV
Plantago lanceolata
V
Holcus lanatus

analysis appear to have occurred in America (Whittaker 1956), England (Perring 1959) and Europe (Ellenberg 1988, first German edition 1963). Fig. 2.1 shows two examples. No species were found to have similar patterns of distribution. The evidence is not presented, only the interpretation which would not satisfy modern standards of statistical rigor. The evidence of dissimilar patterns of distribution among species as opposed to the long-held assumption of coincident distributions of species was however clear. There has been a progressive improvement in the statistical methods used since this early work.

2.3.2 Indirect ordination

This numerical approach determines the major gradients of variation to be found in the vegetation data itself. A graphical representation of the variation in vegetation across all sites can be constructed by measuring the similarity between each site based on the species composition. Such an ordination diagram summarizes the major axes of variation in the vegetation data matrix. The earliest method, the continuum (or compositional) index, took account of only a single dimension. Methods were quickly recognized or developed which would allow several dimensions to be estimated and displayed, pioneered by the Wisconsin school (Fig. 2.2; Bray & Curtis 1957; Curtis 1959; see Greig-Smith 1983; Kent & Coker 1992; Jongman *et al.* 1995 for details). The gradients estimated in this way need not represent environmental gradients but may represent successional changes or differences in grazing regimes. The indirect methods do not make the assumption that all major variations are due to environment as direct methods often do.

An early example investigated the variation in a small limestone grassland area in Wales (Gittins 1965). The example shows how the method could display patterns of variation in an individual species and detect discontinuities in vegetation composition where they existed (Fig. 2.2). The two plots in the bottom left of

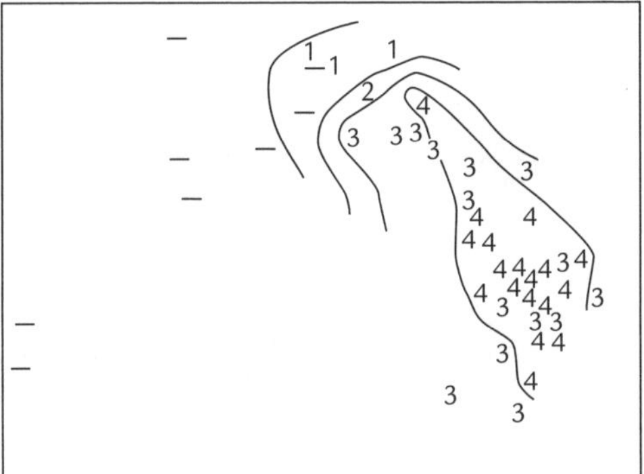

Fig. 2.2 An early example of indirect ordination analysis with two axes from a Welsh limestone grassland showing the distribution of *Helianthemum chamaecistus* with four levels of abundance plus absence. Note the two outliers in the bottom left-hand corner, which were from a sheep camp. After Gittins (1965).

the figure are very different in composition from the rest and hence disjunct. These plots were from a sheep night camp, which had become enriched with nutrients and therefore supported a flora distinct from the surrounding nutrient-poor limestone grassland.

2.3.3 Numerical classification

Early classification of vegetation was subjective. Numerical methods were developed to provide objective procedures. They were based on use of a similarity or association measure between plots of vegetation, grouping together those plots, which were most similar. The results are usually presented in the form of a dendrogram where progressively less similar groups are combined (Fig. 2.3). Numerous methods of classification were developed with various similarity measures and different strategies for grouping plots together. See Kent & Coker (1992), Greig-Smith (1983) and Jongman *et al.* (1995).

Ross (1986) surveyed seven roadside verges 10 years after they had been sown with a seed mixture of five grass species. Numerical classification with centroid sorting on presence/absence data was used (Fig. 2.3). The dendrogram shows the average similarity at which quadrats and groups of quadrats are combined. Four broad groups are readily recognized in the classification:

1 Flat mown verges close to the road with *Lolium perenne*, *Holcus lanatus* and *Trifolium repens*;

2 Verges invaded by trees (*Fagus sylvatica* and *Acer pseudoplatanus*);

3 Rock/scree cuttings with *Festuca rubra* and *Chamaenerion angustifolium*;

4 Grassed embankments subdivided into two types.

The dendrogram summarizes the impact of different environments on the successional history of the sown verges' floristic composition of the quadrats.

Initially ordination and numerical classification were contrasted as supporting the different concepts of vegetation organization, continuum or community respectively. However these methods can be applied without regard for the different concepts. The two methods provide complementary information about the composition of the vegetation and its relationship to the environment. Ordination displays the major axes of variation while classification identifies clusters of sites and outliers. The objectivity of the methods was also seen to be illusory. Each method was explicit, consistent and repeatable but the choice of method was a highly subjective decision.

Central to the use of indirect methods of ordination and classification is the use of similarity measures (section 2.6.2).

2.4 Ideas of environment

Early ecology texts emphasized the multitude of environmental factors and the complexity of their effects on different species. In contrast, Jenny (1941) had presented a simple conceptual framework for soils. The equation for soil formation put forward by Jenny is a list of factors that should be taken into account when examining how soils form. As modified for vegetation it is:

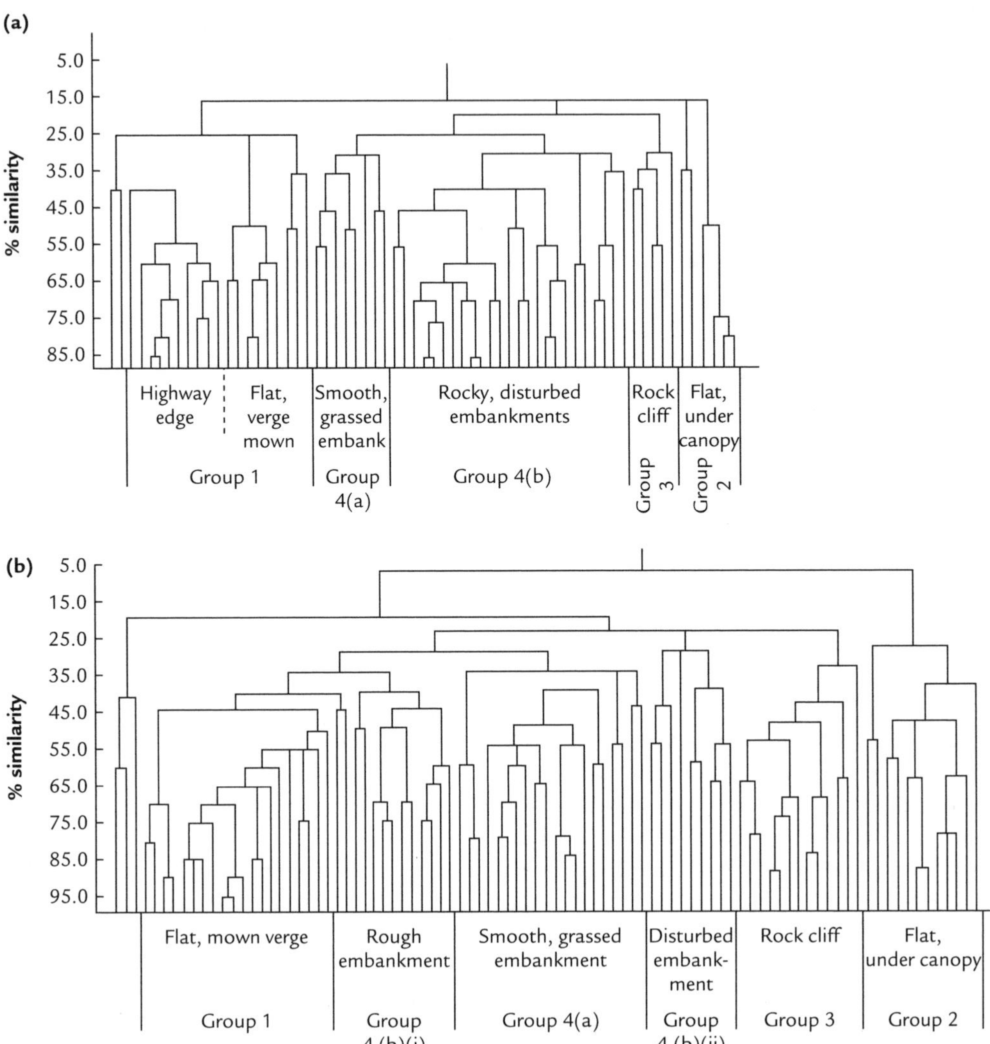

Fig. 2.3 Example of numerical classification dendrograms. **a.** Plot data from road verges 10 yr after establishment; **b.** Same plots after 20 yr. After Ross (1986).

$$V = f(cl, p, r, o, t,) \tag{2.1}$$

where V is some property of vegetation, which is a function (f) of cl = climate, p = parent material, r = topography, o = organisms and t = time. Each of these factors may influence plants in numerous complex ways. No mathematical expression can summarize the processes involved. As a minimum it provides a checklist of broad environmental factors to be considered (see discussion in Mueller-Dombois & Ellenberg 1974). Maximally, it can provide a conceptual framework for both survey design and environmental analysis. Comprehensive use of this framework was made by Perring (1958, 1959, 1960) to design a survey of chalk grasslands in England and northern France analysing the vegetation data graphically (Fig. 2.1b). He restricted the study by parent material (p), and then stratified sampling by climate and topography. Topography was idealized as a hemispherical hill (an inverted pudding basin) and a stratified sample taken by slope and aspect. The results show individualistic responses by species to climate, slope and aspect and complex interactions between these variables.

Few studies since have used such an explicit approach to the analysis of vegetation/ environment relationships. One contributing factor is the variety of ways environmental variables can be expressed and measured. Different types of environmental variable can be recognized, e.g. abiotic and biotic. Abiotic variables such as rainfall and soil nitrogen content directly determine plant growth and success. Biotic variables such as the competition from other plants (see Chapter 9), pathogens, herbivores (see Chapter 10) and mycorrhizae, may destroy plants (pathogens), enhance growth (mycorrhizae) or have complex effects contingent on abiotic variables (see Chapter 11). Environmental variables may be considered to be either distal or proximal. Distal variables like rainfall influence plant growth through various intermediate variables (e.g. soil permeability and soil water-holding capacity) while the equivalent proximal variable is water availability at the root hair.

Another alternative classification of environmental variables or gradients is into indirect, direct and resource gradients (Austin & Smith 1989). Indirect variables or gradients are those that have no direct influence on plant growth. An example is altitude, a variable often correlated with vegetation composition. Altitude can only have an influence via some correlated variable, which has a direct influence (e.g. temperature or rainfall). However these variables have correlations with altitude that are specific to a locality. Correlations based on indirect variables can be used for local prediction but cannot provide explanation in terms of ecological process. Direct environmental gradients are those where the variable has a direct influence on plant growth. Examples are pH and temperature. Resource variables are those which are consumed by plants in the course of growth, e.g. phosphorus or nitrogen. There is no absolute set of categories for these variables. Water is both a consumable resource and a direct variable when excess creates anaerobic conditions.

Loucks (1962) developed environmental scalars as a means of estimating proximal environmental variables that might show greater explanatory power than the arbitrary choice of variables based on ease of measurement that often occurs even today. The scalars combined distal variables based on environmental process knowledge to give estimates of proximal, direct variables which may establish relationships with vegetation that are more robust and less dependent on location-specific correlations.

2.5 Theory: Continuum and community

2.5.1 Need for consensus

Progress in vegetation science depends on the development of explicit theory and numerical methods capable of discriminating between rival theories. At the present

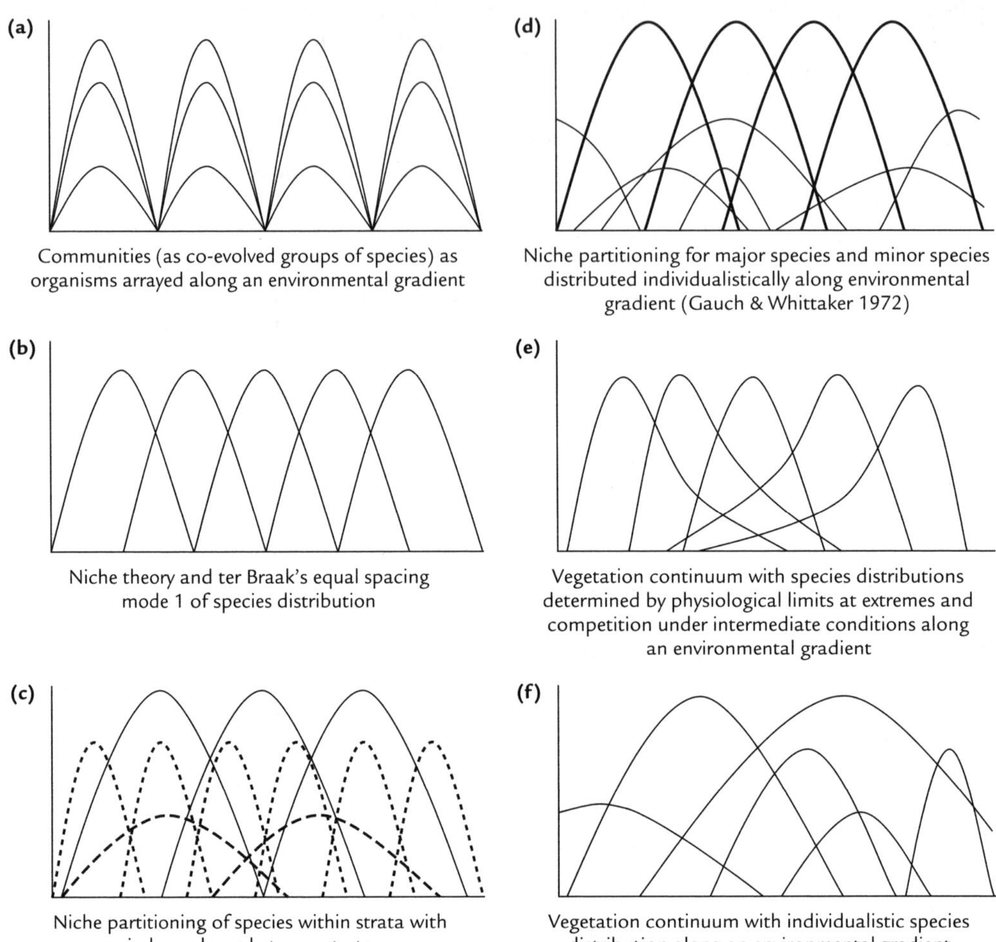

Fig. 2.4 Six hypothetical patterns of vegetation composition along an environmental gradient corresponding to different theories. See text for details.

time there is little consensus on even rival theories and little agreement as to what constitute suitable methods for discrimination (cf. Faith *et al.* 1987; Austin & Smith 1989; Jongman *et al.* 1995).

2.5.2 Continuum

Alternative realizations. Fig. 2.4 shows a spectrum of possibilities from the super-organism concept of a community (a) to a totally individualistic organization (f). The second realization (b) is based on the niche concept of species partitioning a resource gradient. Species have equal ranges and amplitudes and are equally spaced along the

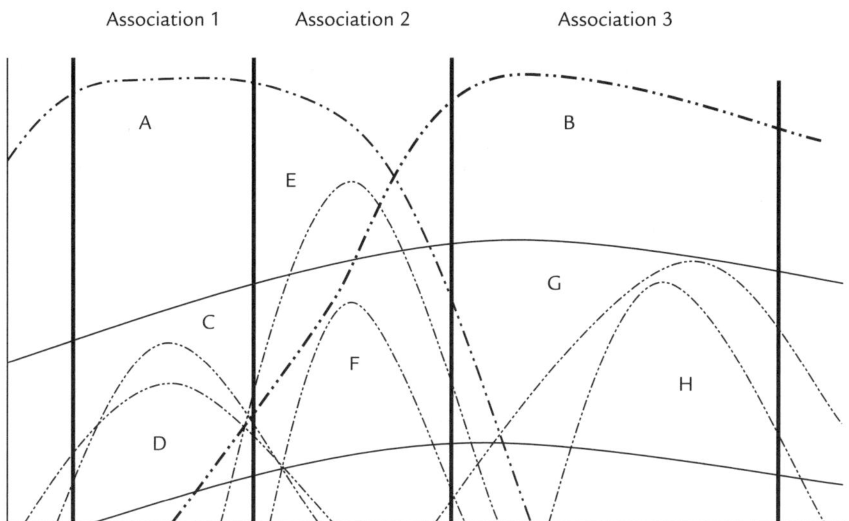

Fig. 2.5 A possible representation of phytosociological associations along an environmental gradient showing constant species (heavy broken line), differential species (light broken line C-H) and indifferent species (light solid line). Associations are distinguished by different combinations of constant and differential species.

gradient. This is the model explicitly underlying correspondence analysis (CA) and canonical correspondence analysis (CCA) (ter Braak 1986; Jongman *et al.* 1995; see section 2.6.3). This niche representation can be combined with the idea that each stratum (trees, shrubs etc.) partitions the gradient independently of the other strata (c). The result is a continuum with each species showing a response partially determined by growth form.

Gauch & Whittaker (1972) put forward a detailed set of hypotheses about the patterns of species response observed along a gradient. These included equal spacing of the dominants (trees) equivalent to resource partitioning and individualistic patterns for understorey species (d). Austin (1999a) summarized results for eucalypt species along a mean annual temperature gradient. These suggest that patterns of species response change depending on position on the gradient (e). It was hypothesized that the physiology of individual species determined limits towards the extremes of the gradient while competition determined species occurrence and shape of response in the central mesic portion of the gradient. This observation applies only to the tree stratum and there is no independent confirmation of these ideas. The individualistic continuum (f) shows no patterns of species behaviour along the gradient.

It is possible to represent phytosociological associations, as they might exist along an environmental gradient (Fig. 2.5). Identification of each association depends on recognition of the constant species with a wide environmental range, and differential species with narrower ranges that distinguish each association. For example, association 1 is characterized by constant species A and differential species C and D with association 2 having constant species A and B and differential species E and F.

The presence of indifferent and rare species would result in a diagram that would not easily be distinguished from the continuum models presented in Fig. 2.4.

This series of hypothetical vegetation patterns demonstrates that (i) the differences between phytosociological concepts and continuum concepts may be smaller than sometimes imagined; and (ii) discriminating between these hypotheses will require detailed data and rigorous statistical methods. No comprehensive tests have been published. There is a complication. The species responses shown in Fig. 2.4 are, with one exception, presented as symmetric bell-shaped curves. If species responses are not bell-shaped and symmetric, what implications does this have for theories of vegetation composition?

Niche theory and continuum concepts. Niche theory assumes each species has a fundamental niche (in the absence of competitors) in relation to some resource gradient. Species niche response is usually assumed to be a symmetric bell-shaped curve (Fig. 2.4b). Each species is usually shown as having the same response with equal width and amplitude. In the presence of competitors the species is restricted to a realized niche. The optima for both the fundamental and realized niches are co-incident as in example 1a of Fig. 2.6. This is a special case of a more general theory advanced by Ellenberg (see Mueller-Dombois & Ellenberg 1974). Species realized niche (ecological response) may be displaced from its physiological response (fundamental niche) by a superior competitor. This can result in bimodal curves. Each species may have different shaped responses to different environmental gradients (Fig. 2.6). Ellenberg's ideas of plant species niche shapes have received little recognition in the general ecological literature (Austin 1999b). Neither niche theorists nor plant community ecologists have considered in detail the patterns suggested in Figs. 2.4–2.6.

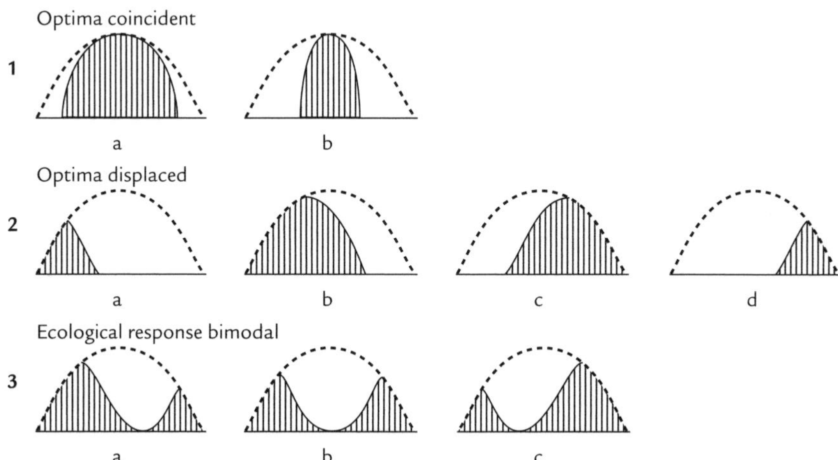

Fig. 2.6 Ellenberg's theory of species response patterns. Example 1a corresponds to classical niche theory. Other examples show interpretations of the possible responses of various species in relation to different environmental gradients. Competition from a superior competitor results in different shapes of ecological response for the species displaced from their physiological optima. Reproduced by permission of the authors.

Table 2.1 Frequency distributions of different shapes of ecological response surfaces for 100 common species in Tasmanian montane vegetation. From Minchin (1989). Data for the 100 species occurring in at least 20 quadrats. A Monte-Carlo test showed no difference between structural groups in the relative frequencies of the shape categories. Reproduced by permission of the author.

Structural group	**Response surface shape**		
	Symmetric	Skewed	Complex
Trees	3	4	1
Shrubs	24	13	7
Herbs	8	11	7
Graminoids	8	5	6
Pteridophytes	2	0	4
All species	45	33	22

The various continuum and community concepts are basically phenomenological; they are descriptive without an explicit mechanistic basis. Ellenberg's hypothesis introduces species-specific physiological limits and competition as organizing processes to produce the observed patterns. Some numerical methods of vegetation analysis are explicitly based on symmetric bell-shaped curves and equal partitioning of the gradient without considering the ecological processes involved (Jongman *et al.* 1995). They have a restrictive theoretical basis that needs to be tested.

Evidence from ordinations. Ordinations can only provide evidence that a particular pattern of vegetation exists along a gradient. Indirect ordinations may display sharp discontinuities in the ordination space (Fig. 2.2). These have usually been ascribed to major differences in environmental conditions. Indirect ordinations cannot be used as proof of the existence of a continuum. The mathematical methods employed have implicit ecological assumptions about how species respond to gradients, which may determine the outcome. Most early analyses using direct ordination indicated the existence of varied species response shapes. No obvious coincidences of species limits were observed which would support the community concept. Symmetric bell-shaped curves were no more abundant than other shapes. Minchin (1989) found only 45% of species had response surfaces which appeared unimodal and symmetric, the so-called Gaussian responses (Table 2.1).

Relatively few studies have directly tested continuum concepts or attempted to discriminate between the two concepts of community or continuum. Austin (1987) examined the continuum propositions put forward by Gauch & Whittaker (1972) using tree species in south-east Australia. A marked preponderance of skewed curves was found for the major species along a mean annual temperature gradient. Recent statistical modelling has supported this conclusion (Austin 1999a). A test of Gauch & Whittaker's second proposition 'the modes of major species are evenly distributed along environmental gradients while those of minor species tend to be randomly distributed' rejected the proposition for major species. Other propositions could not be tested due to confounding with species richness, which increased steadily with temperature from one species at the tree line.

Minchin (1989) undertook a fuller analysis with 100 species in relation to two environmental gradients, altitude and soil drainage, in a montane environment in Tasmania. A test for the even distribution of modes indicated that species modes were clumped (all species), random (major species) or varied with structural group (growth form). Herbs had clumped species modes while other growth forms were random; alpha diversity (species number per unit area) was examined. Unimodal species richness patterns were evident for the different growth forms. The modes of richness for each growth form occupied different positions in the environmental space.

Shipley & Keddy (1987) attempted to distinguish between the continuum and community concepts using species limits. They examined species limits along transects following a water table gradient. If the community concept holds then there should be more limits in some intervals than others along the gradient and species limits both upper and lower should coincide i.e. cluster (Fig. 2.4a). If the individualistic continuum concept holds then the average number of limits per interval along the gradient should be equal apart from random effects (Fig. 2.4f). In addition, the number of upper limits of species should be independent of the number of lower limits in each interval for the continuum concept to hold. Both upper and lower limits were found to be clustered. The individualistic continuum is rejected. No correlation between the number of upper and lower limits per interval was found. The community concept is also rejected. The results are equivocal and address only two of the possibilities represented in Figs. 2.4–2.6. The transects ran from the edge of an *Acer saccharinum* forest into a marsh as far as the edge of the zone of aquatic species with floating leaves. This is a steep gradient from a terrestrial to an aquatic environment. Only six of the 43 species have both upper and lower limits recorded within the gradient. Most species have either an upper limit (aquatics) or a lower limit (terrestrial plants). A gradient length with less extreme moisture conditions or a less steep gradient might yield a different result.

Shipley & Keddy (1987) pointed out that they used an indirect gradient or factor-complex gradient in Whittaker's terms, namely water table depth. A clustering of limits in one interval might then indicate a discontinuity in one of the many environmental variables correlated with the factor-complex represented by water table depth. For example, anaerobic soil conditions may occur as a step function at a particular water depth in the marsh. The sharp increase in anaerobic conditions might appear to limit species at the same water level when in fact they are actually limited by different degrees of anaerobic conditions. Distance along a transect cannot be equated directly with changes in an environmental variable. The correlation of species patterns with an indirect distal variable may yield very different results from those using a direct proximal variable.

Plant community ecologists have yet to specify the properties of either the community or continuum concepts in sufficient detail for any variant to be statistically distinguished from another.

2.5.3 Community

The term **community** is used with various meanings in the ecological literature (see McIntosh 1985 and section 1.1.1). Drake (1991) applied it to an experimental foodweb involving algae, bacteria, protozoans and cladocerans, plants and animals. He explored the mechanics of community assembly. The composition of invasion-resistant communities was found to depend on the order of invasion by species, as was the foodweb structure. The results demonstrate a number of important features involving the invasibility of some communities and the predictability of the outcome of competition among the primary producers in multiple trophic level experiments. The conclusions are relevant to vegetation ecology.

The definition of an 'ecological community' used by Drake (1991) is 'an ensemble of individuals representing numerous species which coexist and interact in an area or habitat'. This and other definitions could apply to almost any combination of species under any circumstances. They are non-operational for any form of comparative analysis of observations or experiments. To compare the results of foodweb or other experiments we need to know how different each 'ecological community' is from another. This leads naturally to the use of multivariate methods such as ordination and classification to measure the difference.

2.5.4 Possible synthesis

The controversy between the community-unit or association concept and the continuum concept arises because the former is an abstraction based on geographical space and the latter an abstraction based on environmental space (Austin & Smith 1989). Fig. 2.7a represents a hypothetical transect up a mountain in an area with four species showing an altitudinal zonation. Five communities can be distinguished – A, AB, B, C and D – if species associations are recognized based on their frequency of occurrence with the combinations BC and CD as ecotones (Fig. 2.7b). The communities are a result of the frequency of different altitudes along the transect, particularly the community AB on the bench at 200 m and community B on the bench at 400 m. The distribution of the four species in relation to altitude however is a continuum (Fig. 2.7a). Each species is spaced along the gradient approximating Gauch & Whittaker's (1972) conception of the continuum, yet communities are clearly recognizable along the transect.

Another hypothetical transect in an adjacent area where the two benches were at altitudes of 170 m and 430 m instead of 200 m and 400 m gives a different result. Here the communities would be A, B, BC, C, D, with ecotones AB and CD (Fig. 2.7c). The frequency of altitudinal classes has changed and hence the most frequent combinations of species (communities) are different. So, communities are a function of the frequency of different environments in the landscape examined. However, the altitudinal continuum would be unchanged: continua are a function of the environmental space measured. Note that the environmental gradient used here is an indirect (factor-complex) gradient. The continuum pattern observed only applies where the correlations between altitude and direct or resource gradients remain constant.

Mueller-Dombois & Ellenberg (1974, p. 205) discuss the definition of the phytosociological association and the choice of characteristic species. They point out that when an investigator is concerned with a small geographical area, many characteristic and differential species can be identified for each association. If the geographic range

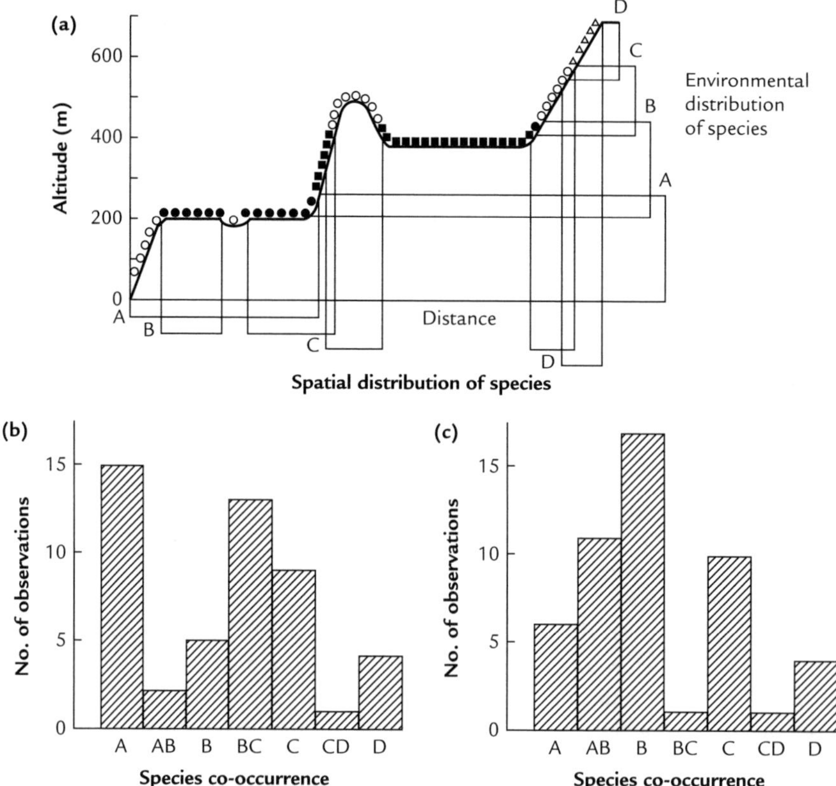

Fig. 2.7 **a.** A hypothetical transect up an altitudinal gradient showing the spatial extent of the possible combinations of species. Each species has a distinct but overlapping niche with respect to the indirect environmental gradient of altitude (Austin & Smith 1989). Reproduced by permission of Kluwer Academic Publishers. **b.** A histogram of the frequency of species combinations from the transect shown in Fig. 2.7a (Austin 1991). **c.** A histogram of the frequency of the same species combinations but from a different transect where the benches occur at different altitudes (Austin 1991). Reproduced by permission of CSIRO.

is increased, more and more species, which locally had a strong correlation with one association, are now found in other associations. Enlarging a study region will result in the inclusion of entirely new environments. The difficulties of identifying diagnostic species are consequences that follow naturally from the ideas presented in Fig. 2.7. However, where a single gradient is studied the phytosociological model and the continuum model may appear very similar (Figs. 2.4 and 2.5). Austin & Smith (1989) concluded:

1 The continuum concept applies to an abstract environmental space, not necessarily to any geographical distance on the ground or to any indirect environmental gradient;

2 The abstract concept of a community of co-occurring species can only be relevant to a particular landscape and its pattern of environmental variables; community is a property of the landscape.

Such a community concept is compatible with the different concepts of a continuum (Fig. 2.4). For communication and ecological management the community will be the preferred concept to use, provided the applicable region is clearly defined. For investigation of vegetation/environmental relationships the continuum concept is preferable.

2.6 Multivariate methods

2.6.1 Introduction

Multivariate analysis has been an integral part of vegetation studies since the advent of computers. There is now an extensive literature on the subject (see Greig-Smith 1983; Kent & Coker 1992; Jongman *et al.* 1995; Legendre & Legendre 1998).

Multivariate analysis has also been subject to sweeping criticism (e.g. by May 1985, p. 112), echoed by Crawley (1997, p. 510). Van der Maarel (1989) provided a spirited defence; see also comments by Kent & Coker (1992). There is some truth in the criticism, but in much the same way as there is truth in the comment that decades of reductionist analysis of pair-wise species competition has not led to great advances in our understanding of the vegetation patterns in nature.

Recent use of multivariate methods has been directed at correlating vegetation with environment. Attention has been focused on the particular correlations found for vegetation in specific habitats. Few generalizations about vegetation/environment relationships have emerged.

The process of multivariate pattern analysis has several steps, explained by Kent & Coker (1992), Jongman *et al.* (1995), Legendre & Legendre (1998) and Guisan & Zimmerman (2000). This chapter concentrates on the data matrix and correlation methods.

2.6.2 Data matrix

Central to multivariate analysis is the matrix of species measurements for each of a number of plots, plus a series of environmental measurements for each plot. These two data matrices can be combined in different ways, depending on the nature of the question. If the vegetation pattern *per se* is explored an indirect approach may be called for. When variation in vegetation is to be related to environmental variation a direct approach may be required. Where data are available on species abundances and attributes measured at different times together with environmental data, elaborate procedures can be applied (see Legendre & Legendre 1998).

2.6.3 Species measurements

Measurement of species and environmental variables is a critical component of any analysis. See Greig-Smith (1983), Kent & Coker (1992) and Chapter 1 for possible species measures. The species measures most frequently used at the present time are a cover-abundance scale with unequal class intervals or the presence or absence of a species. Cover/abundance is an ordinal variable that only provides information about the ranking of sites in terms of the abundance of that species. This restricts greatly the statistical procedures, which can be used. Presence/absence data is a binary variable, which has implications for any statistical analysis used; the correct binomial error function must be used.

2.6.4 Environmental measurements

Traditionally the environmental data collected consist of variables such as altitude, slope and aspect. These variables are typically indirect variables and little thought is given to the processes that may result in the variable being correlated with vegetation.

Slope and aspect are examples of variables where much is known about the environmental processes that are likely to be responsible for any correlation. Re-expression of these indirect variables as direct or resource variables should improve and clarify any observed correlation. Aspect is the compass direction in which a quadrat may be facing. A compass bearing is a circular measure where 2° and 358° are closer to each other than either is to 340°. Various data transformations have been used to correct this problem without reference to the environmental processes involved.

A major difference between north- and south-facing slopes is the amount of solar radiation each receives. The potential solar radiation at a point can be calculated as a complex trigonometric function of the aspect and slope of the site, depending on the position of the sun, which varies with the time of year and latitude. No simple data transformation of aspect will capture this information about radiation, a variable that has numerous direct effects on the physiology of plants. Many different combinations of aspect and slope have equivalent radiation climates.

Ecologists have long recognized the influence of aspect. Whittaker (1956) used subjective estimates to take account of high hills cutting off the direct rays of the sun. Radiation on protected north-facing valleys can be 10% of that on exposed south-facing slopes in winter at northern temperate latitudes. Today, algorithms exist for calculating radiation, which may include horizon effects, direct and diffuse radiation components, sunshine hours and atmospheric properties. Dubayah & Rich (1995) provide a modern account of the equations involved. Radiation integrates many features of the plant's environment in terms of explicit physical processes, and hence is a more relevant variable than aspect.

This process of deriving more physiologically relevant environmental variables can be taken further. Actual evapotranspiration, the amount of water transpired in a given time, is a crucial indicator of the drought stress suffered by a plant. A simple model of the physical process can estimate actual evapotranspiration. Rainfall is a measure of water supply in a given period. Current storage is estimated from the available soil water capacity. Potential evapotranspiration is a measure of demand and can be derived from weather records. Water available for transpiration is given by the sum of rainfall and the amount in the soil store. If the total of these is greater than the demand, actual transpiration is equal to potential. If the total is less, then actual transpiration is equal to the total available. A measure of moisture stress is the ratio of actual to potential evapotranspiration. This measure estimates the extent to which supply satisfies demand. At the local scale, radiation is the dominant term in the equation estimating potential evapotranspiration (Ep). Differences in moisture relations of plants on different aspects depend on the relative amounts of radiation they receive. The Ep on different aspects will vary proportional to the radiation received relative to the radiation received on a flat surface.

These processes can be incorporated into a simple water balance model to estimate moisture stress (MI) scalar for different aspects:

$$\begin{aligned}
&\mathrm{Ea}_t = \mathrm{Ep}_t \text{ if } (R_t + S_{t-1}) > \mathrm{Ep}_t \text{ where } \mathrm{Ep}_t = \mathrm{Ep.RI} \\
&\mathrm{Ea}_t = (R_t + S_{t-1}) \text{ if } (R_t + S_{t-1}) < \mathrm{Ep}_t \\
&\mathrm{MI} = \mathrm{Ea}_t/\mathrm{Ep}_t \\
&S_t = R_t + S_{t-1} - \mathrm{Ea}_t \quad S_t = 0.0 \text{ if } S_t < 0.0 \\
&S_t = \mathrm{Smax} \text{ if } S_t > \mathrm{Smax}
\end{aligned} \tag{2.2}$$

where Ea_t is the actual evapotranspiration for time step t; Ep_t is the potential evapotranspiration for time step t adjusted by the relative radiation index RI; R_t is rainfall for time step t; Smax is maximum soil moiture storage; S_{t-1} is the soil moisture remaining from the last time step $t - 1$.

The water balance model can use monthly average values for Ep and rainfall or actual data. When the average values reached equilibrium, the moisture stress index would then be estimated for a particular season, annually or for a specific weather sequence.

Environmental scalars such as that described here for moisture stress can be made more elaborate. They may then appear more precise than those based on field observations. *Poterium sanguisorba*, which shows a distribution centred on south-west slopes in Fig. 2.1b provides an example of the problems of scalars. The species is typical of chalk grassland species that show this distribution in England (Perring 1959). It is well known to field ecologists that south-west aspects with slopes above 15° are the most drought-prone and characterized by species with more southern distributions. This is inconsistent with the moisture stress scalar and radiation model described earlier. The radiation model predicts that radiation is greatest on slopes facing directly south i.e. aspect 180°. Radiation is symmetric about south with south-east and south-west slopes receiving the same amount of radiation. The model is inadequate. Differences in potential evapotranspiration between aspects arise from differences in air temperatures. Temperatures in the afternoon are higher when radiation is falling on south-west aspects than when the same amount of radiation falls on south-east aspects in the morning. The physical model is incomplete; a significant component has been omitted. Analysis of vegetation/environment correlation is an iterative process requiring constant testing of the model with field observations. Development of environmental scalars for use in environmental modelling with geographic information systems (GISs) is a very active area of research and improvements are constantly being made. Environmental scalars integrating our physical and physiological knowledge can be generated from many environmental factors. Guisan & Zimmermann (2000) provide a useful review of current ideas. Fig. 2.8 shows the many different connections that can exist between distal variables and the more proximal direct variables. To understand vegetation patterns, we need to understand the environmental processes that are responsible.

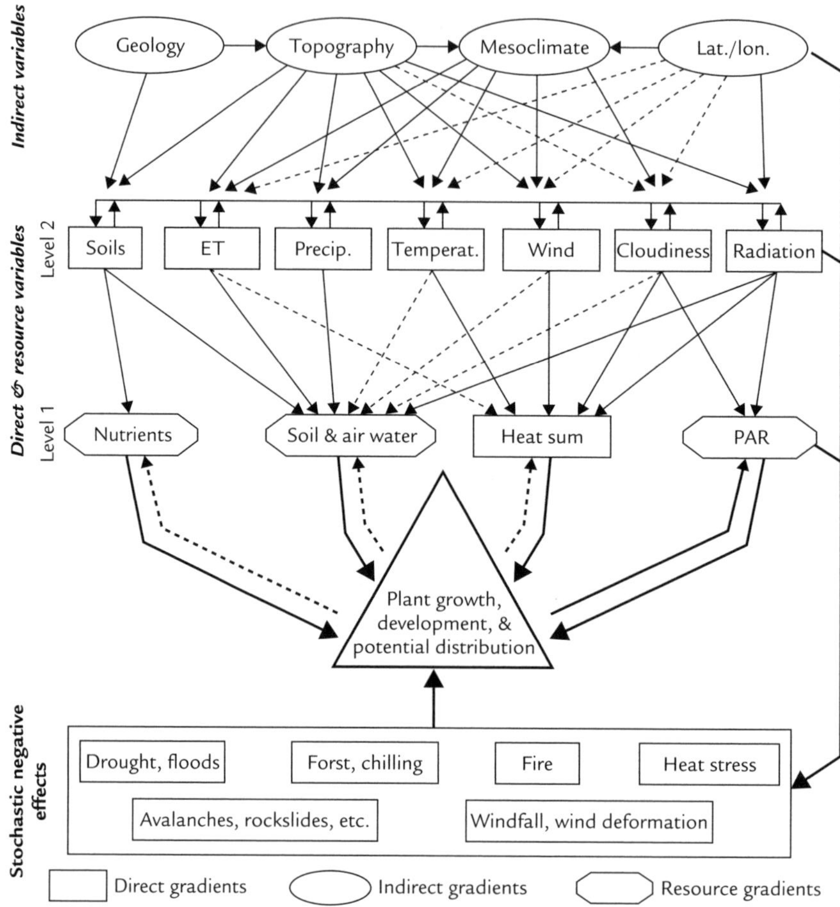

Fig. 2.8 Relationship of indirect and direct environmental variables and their possible combination into scalars. ET = evapotranspiration; PAR = photosynthetically active radiation. From Guisan & Zimmermann (2000). Reproduced by permission of Elsevier.

2.6.5 Similarity measures

In contrast to direct methods, indirect methods use a similarity matrix to summarize the species relationships between quadrats. A similarity measure (S) is calculated between every quadrat and every other quadrat. The resulting similarity matrix can be used to produce either a classification or an ordination. See Kent & Coker (1992) and Legendre & Legendre (1998) for details. The results are critically dependent on the similarity measure chosen and either the sorting strategy for numerical classification or the ordination method. Numerous similarity measures have been proposed; see Greig-Smith (1983), Faith *et al.* (1987) and Legendre & Legendre (1998) for examples. Choice depends on the assumptions the researcher is prepared to make.

The common assumption is that species' responses to an environmental gradient take the form of a bell-shaped curve (Fig. 2.4). The choice of similarity measure is often incompatible with this ecological concept (Faith *et al.* 1987). This has important consequences for the performance of similarity measures in their ability to recover information about the underlying environmental gradients.

When comparing two plots from different positions along an environmental gradient, the number of species that are absent from both plots ('double-zero matches') is critical. If two plots have no species in common it implies they are so far away from each other in environmental space that no single species can tolerate both environments. No single measure of similarity between the two plots can measure how far apart the two plots are. When plots are closer together in environmental space then some species will occur in both plots. These species contribute information about the distance beween the plots in environmental space. The number of zeros in common provides no additional information except that they are distant. Similarity measures, which incorporate double-zero matching information, distort the ecological relationships.

Similarity measures summarize information in species space, which is intended to be used to construct species patterns in environmental space (Fig. 2.9). If we assume each species has a linear or unimodal response shape along an environmental gradient CD (Fig. 2.9a), then the information available in species space depends on the shape of the species response and degree of overlap between species. When the quadrats along the gradient CD are plotted in species space, i.e. with species abundance as axes, then the gradient becomes twisted in a complex fashion. Only in the top row is there a simple relationship between gradient CD and the equivalent line CD in species space (Fig. 2.9b). It is only linear under the circumstances represented by the first row. The relationship between quadrat composition in species space and the environmental gradient is rarely linear. Similarity measures estimate distance in species space (Fig. 2.9b) not in environmental space (Fig. 2.9a). If the similarity measure is based on simple linear or Euclidean concepts of distance and a principal components analysis (PCA) ordination is applied, then severe distortions of the ordinations may result, including the 'horseshoe effect'. When a series of plots from a sequence of unimodal species along an environmental gradient (e.g. as in Fig. 2.4) are ordinated, most ordination techniques represent the plots as a horseshoe in two dimensions (e.g. CA). There are incompatibilities between the data analysis models used by different authors and theories of species responses. If skewed and bimodal species responses occur (Fig. 2.6), the problem becomes even more complicated.

Faith *et al.* (1987) examined the behaviour of 29 similarity measures and standardizations using artificial data sets. A large number of different data sets were constructed based on different assumptions about the nature of species response curves to environmental gradients. The true ecological distance between plots along the gradient was known as a consequence of using artificial data. This could be compared with the dissimilarity estimated from the compositional data. The results for three similarity measures (Fig. 2.10) show how different outcomes can occur if the data model is not equivalent to the theoretical model of species response. The Manhattan measure (a) shows the impact of total quadrat abundances on compositional distance

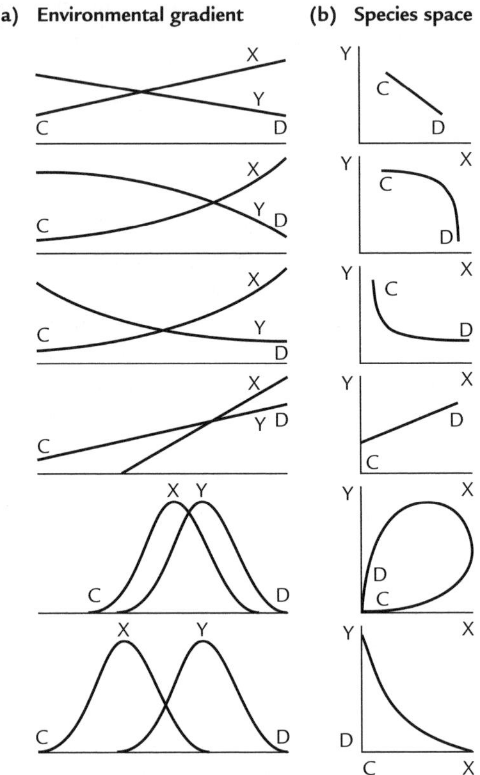

Fig. 2.9 Possible relationships between species when plotted in environmental and species space. **a.** Different performances of two species X and Y along a single environmental gradient. **b.** Relationships between X and Y when the performance of Y is plotted against X. From Greig-Smith (1983). It is the information in **b.** which contributes to the similarity measures which ordination methods use to recreate the patterns of **a.**

when the ecological distance is such that there are no species in common. Plots with no species in common appear similar if they both have low total abundances. The Kendall measure (b) reaches a limiting dissimilarity when there are no species in common but is sensitive to plot total abundances when there are many species in common. The symmetric quantitative Kulczynski (c) provides a more balanced representation of the ecological distance. The χ^2-measure of distance used in correspondence analysis was found to perform badly relative to the Kulczynski measure, which performed best when used with data standardized to species maxima.

Numerous standardizations and transformations of the data have been suggested (Greig-Smith 1983), but no general agreement has been reached on those most appropriate for vegetation data (Faith *et al.* 1987; Legendre & Legendre 1998). The relationship between data analysis models and theoretical models of the composition of vegetation remains an area of research with many unanswered technical questions.

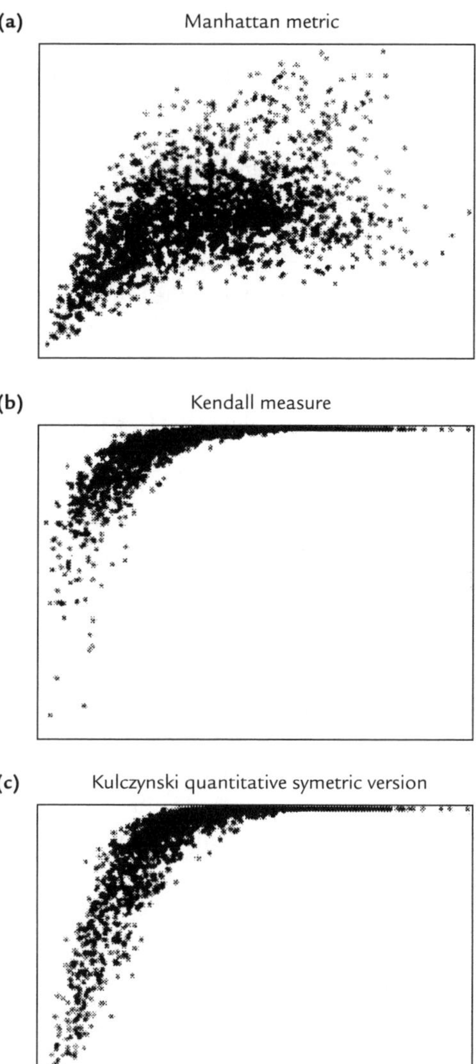

Fig. 2.10 The relationship between compositional dissimilarity (y-axis) and the 'true' ecological distance (x-axis) for artificial data calculated from known species response shapes for three different measures: **a.** Manhattan metric; **b.** Kendall measure; **c.** Kulczynski quantitative symmetric version. After Faith *et al.* (1987). Reproduced by permission of the author, and Kluwer Academic Publishers.

2.6.6 Indirect ordination

Numerous indirect ordination techniques have been proposed using these similarity measures. Many have now been shown to be effective only with certain limited types of data sets. For example, PCA ordination, although still widely used, will give distorted results when any species shows a unimodal response to the underlying gradient.

Two methods deserve special attention, correspondence analysis (CA) and non-metric multidimensional scaling (NMDS). CA is a method that uses a χ^2-distance as the similarity measure. NMDS constructs an ordination where the distance between plots has maximum rank order agreement with the similarity measures between plots. In theory, NMDS can accommodate any similarity measure provided the resulting relationship between similarity and distance in ordination space remains monotonic (Minchin 1987). These are less restrictive assumptions than apply to CA. Belbin (1991) described semi-strong hybrid scaling (SSH), a modified form of NMDS for use with vegetation data.

The most used method is CA. However, it is often used in the form of canonical correspondence analysis (CCA) where the axes of the ordination are constrained to maximize their relationship with a nominated set of environmental variables. CCA is a hybrid ordination method that combines features of direct and indirect ordination.

The choice of an ordination method requires a suitable evaluation method. One cannot use real data for evaluation. The true gradients underlying an observed vegetation pattern can never be unequivocally known. A comparison of two methods on real data may give two different answers. Artificial data where the true gradients are known is necessary to evaluate methods. However, this requires that the model used to generate the data, reflects a realistic theory of how vegetation varies in relation to environment.

Ter Braak (1986) in developing his CCA approach is very explicit in the mathematical assumptions implicit in the method:

1 The species' tolerances (niche widths) are equal;
2 The species' maxima are equal;
3 The species' optima are homogeneously distributed over a length of the gradient (A) that is large compared to individual species tolerances;
4 The site scores are distributed over a length of the gradient that is large but contained within A.

The words 'homogeneously distributed' mean either that the optima or scores are equally spaced along the gradient or that they are randomly distributed according to a uniform distribution. Assumption 4 assumes a particular sampling strategy for vegetation sampling. Assumptions 1 to 3 assume a particular species-packing model for the environmental gradient. The methods attempt, to estimate the one represented in Fig. 2.4b. Ter Braak (1986) acknowledges that assumptions 1 and 2 'are not likely to hold in most natural communities' (one wonders why 3 is not included). He then claims that the usefulness of the method 'in practice relies on its robustness against violations of these conditions'. The robustness of the method has been examined with artificial data generated with different assumptions from those above (ter Braak *et al.* 1993) and considered to be satisfactory. No comparison with the performance of alternative ordination methods was made (Austin 2002).

The claim of robustness does not accord with the work of Faith *et al.* (1987). They showed with artificial data that χ^2-distance as used by CA is unsatisfactory for estimating the true ecological distance. Minchin (1987) using similar artificial data sets showed fairly conclusively that local non-metric multidimensional scaling (LNMDS) outperformed detrended correspondence analysis (DCA) a form of CA, in recovering two-dimensional gradients. Økland (1999) examined the impact of horseshoe distortions and noise on the performance of various ordination methods including CA, DCA and CCA using artificial data. He showed that there are significant problems in distinguishing ecological signal from random noise and distortion due to the inappropriate choice of theoretical model. In many cases the importance of the ecological signal is underestimated.

These methods cannot tell us what models of vegetation organization actually exist as they are never assumption-free regarding vegetation organization. The current position is that best practice would be to use more than one method, preferably with well-known assumptions and robustness.

2.6.7 Classification

A wide variety of numerical classification techniques have been developed including monothetic divisive and polythetic agglomerative methods plus hybrid techniques such as minimum spanning trees (Greig-Smith 1983). The monothetic divisive methods are less used today. Classification methods provide a complementary analysis to ordination. An ordination summarizes the major axes of variation in a two or three-dimensional diagram. A classification dendrogram summarizes as a linear array the vegetation variations present in multiple dimensions.

Two methods are frequently used at the present time, TWINSPAN (Hill *et al.* 1975; see also Kent & Coker 1992) and UPGMA. TWINSPAN is based on the progressive division of plots based on the relevant CA axis values; technically it is a polythetic divisive method. UPGMA (unweighted pair-group method using arithmetic averages) is an agglomerative polythetic method combining plots and grouping combinations based on the highest average similarity (e.g. Legendre & Legendre 1998). The two strategies differ fundamentally. Detailed evaluation of classification methods using artificial data with appropriate ecological properties is needed. Vegetation classification using explicit, consistent and repeatable methods is essential for communication and conservation. Vegetation management requires that identifiable units be available to which the same management can be applied. Field experiments need to be applied to identifiable plant communities, so that results from similar or different communities can be compared.

2.7 Direct gradient analysis

2.7.1 Introduction

There has been a major expansion in use of statistical models to undertake direct gradient analysis in the last decade (Franklin 1995; Guisan & Zimmermann 2000;

Austin 2002) because of the development of new regression methods and associated software packages. The framework for any analysis of vegetation/environment relationships has three components. The first component is an ecological model incorporating the ecological theory to be used or tested, e.g. the likely shape of the species response to an environmental variable. The second component is a data model. This concerns how the data were collected and measured. Were the data collected using a statistically designed survey procedure or is it an *ad hoc* compilation of published data? If species abundance is estimated as a cover/abundance scale, should these values be transformed in some way? The third component of the framework is a statistical model. The choice of statistical method and of the error function to be used for the species data is part of the statistical model. Assumptions made in one of the model components can confound those of another component (Austin 2002).

2.7.2 Methods

Currently three methods are being used and evaluated by plant ecologists (Franklin 1995; Guisan & Zimmermann 2000).

1 Generalized linear modelling (GLM). This method is a generalization of normal least-squares analysis using maximum likelihood. It allows the analysis of various types of data and error functions in particular presence/absence data with its binomial error function. See McCullough & Nelder (1989).

2 Generalized additive modelling (GAM). This method is a non-parametric extension of GLM. It uses a data-smoothing procedure, which has the great advantage that the possible shape of the species response does not have to be specified by a mathematical function prior to the analysis. Yee & Mitchell (1991) introduced GAM into plant ecology. See Hastie & Tibshirani (1990).

3 Classification and regression trees (CART). The method seeks to progressively split the data set into two subsets on the basis of an environmental variable such that the difference between the subsets in abundance of a vegetation variable is maximized (Breiman *et al.* 1984). The result is a dendrogram with the data split successively into groups on the basis of different environmental predictors. Each group is defined by a set of rules (combinations of environmental predictors which predict the level of abundance of the species or other dependent variable). See Vayssieres *et al.* (2000).

Numerous other methods exist including machine-learning techniques e.g. neural nets (Franklin 1995), but no consensus yet exists on which method or combination of methods to use.

CCA (ter Braak 1986; Jongman *et al.* 1995), a combination of indirect ordination with environmental regression is the most frequently used method in vegetation science at the present time. A key step in the method constrains the ordination axes to be maximally correlated with the environmental variables included in the analysis. The assumption is that the major variation in vegetation composition is environmental and not due to succession or other historical influences, an assumption that should always be tested. In practice, most applications assume that the environmental variables are linearly correlated with the ordination axes regardless of whether they are indirect or direct variables (Austin 2002). CCA does not require this assumption.

2.7.3 Applications

CCA is usually used to determine the environmental correlates of the variation in vegetation composition while accepting the assumptions of the underlying ecological data and statistical models. The three other methods (GLM, GAM and CART) are frequently used to predict species distributions in a region from survey data in conjunction with a geographic information system (GIS). The work of Leathwick in New Zealand exemplifies the use of GLM and GAM to investigate species/environment relationships. The forests of New Zealand are composed of three groups of tree species; broad-leaved evergreen species, Gondwanan conifers and *Nothofagus* species. Composition varies across a wide range of climatic conditions (mean annual temperature from *c.* 5.0 °C to 16.0 °C, mean annual rainfall from 400 to > 10,000 mm) and in response to historical disturbances (volcanic eruptions and earthquakes). Existing extensive plot survey data on species tree density for stems > 30 cm diameter have been coupled with a GIS for New Zealand, which provides information on climate, and biophysical variables for use as environmental predictors for each plot.

Leathwick & Mitchell (1992) examined data from the central North Island of New Zealand. They modelled 11 tree species using presence/absence data and GLM with the predictors mean annual temperature, solar radiation difference, mean annual rainfall, and depth of Taupo pumice as continuous variables. Topography and drainage were treated as categorical variables. Quadratic terms for the continuous predictors were used to test for curvilinear responses. Mean annual temperature was a predictor in all models with a quadratic term significant for 10 species confirming the importance of unimodal species responses. It was the most important predictor for nine species. The solar radiation variable was the second most important predictor overall. Depth of pumice included in six models is a surrogate variable representing both a physical substrate predictor and the distance from a major historical disturbance, the Taupo volcanic eruption of 130 A.D. Both environment and succession since the volcanic eruption influence current species distribution. The statistical models demonstrated the relative importance of climatic and volcanic variables in determining forest composition. Leathwick (1995) extended the analysis to the whole of New Zealand to examine the climatic relationships of 33 tree species using GAM. Many species responses to environmental variables were shown to depart from the symmetric unimodal curves often assumed.

Leathwick (1998) explored whether the distribution of the *Nothofagus* species was due to environmental variables or due to slow dispersal after postglacial climate changes and volcanic catastrophes. A proximity factor (presence of *Nothofagus* on other plots within 5 km of the plot) was used after fitting the environmental models to demonstrate significant spatial autocorrelation in the distribution of the *Nothofagus* species. This supports the non-equilibrium explanation of slow dispersal for their observed distribution. Regression models can test hypotheses about the non-equilibrium nature of forest composition.

Nothofagus species are frequently the dominant species in the communities where they occur. Many species that occur in association with *Nothofagus* also occur in identical environments without *Nothofagus* as a dominant. It is possible therefore

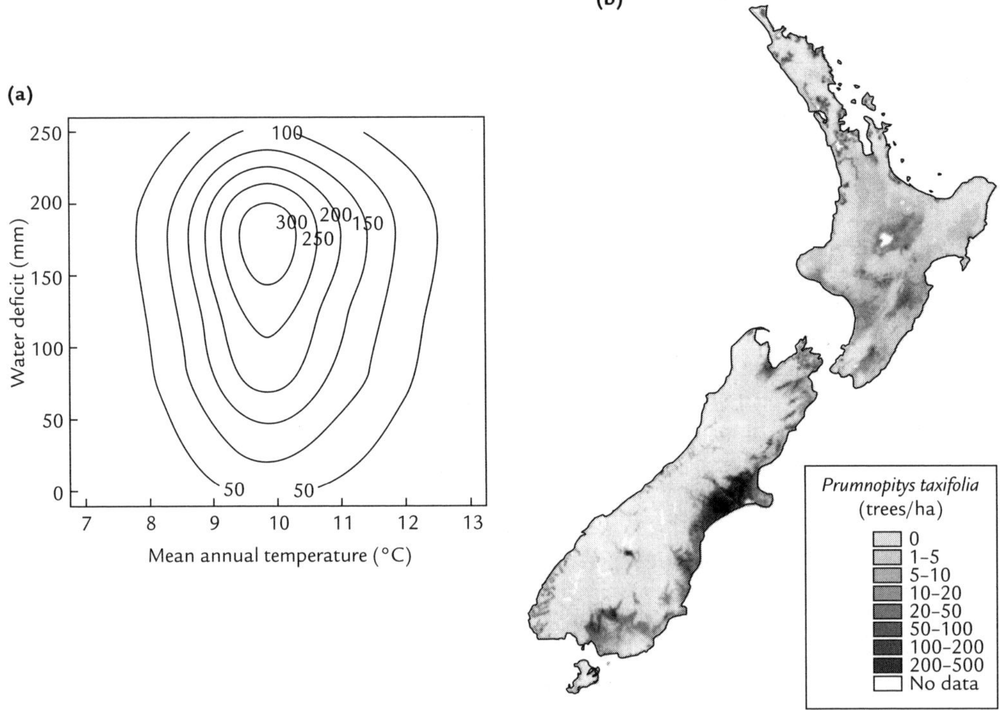

Fig. 2.11 Example of the environmental GAM model and the geographical distribution map generated from it. **a.** Part of GAM model predicting density response of *Prumnopitys taxifolia* to annual water deficit and mean annual temperature. Note skewed response to annual water deficit. **b.** Predicted geographical distribution of *P. taxifolia* density for New Zealand. Note abundance on the dry east coasts. Figure kindly supplied by J.L. Leathwick, Landcare New Zealand.

to model the impact of competition from the dominant *Nothofagus* species on the co-occurring species (Leathwick & Austin 2001). The results show that density of *Nothofagus* species has significant effects on the species composition of forests. These effects vary with the position of the plots on the environmental gradients of temperature and moisture. Competitive effects of dominant species conditional on environment have been detected from broad-scale survey data. The development of these regression models including environment, competition and historical limitations on the dispersal of dominant species has allowed the prediction of New Zealand's potential forest composition and pattern across the whole country (Leathwick 2001).

An example of a skewed response surface found using GAM and the geographical prediction possible when a GIS with suitable layers is available is shown in Fig. 2.11. *Prumnopitys taxifolia* is a Gondwanan conifer with a distinct dry east coast distribution in New Zealand. The methods used were based on those in Leathwick (2001).

The progressive development of realistic environmental processes is well demonstrated by these New Zealand studies in the case of the moisture stress indices discussed earlier. The initial environmental predictor for the moisture component was mean annual rainfall (Leathwick & Mitchell 1992), then the ratio of summer rainfall to summer potential evapotranspiration was used (Leathwick 1995). Leathwick *et al.* (1996) developed a soil water balance model to estimate an annual integral of water deficit based on a 1 in 10 yr drought rainfall. Monthly relative humidity was found to be a significant predictor for many species (Leathwick 1998). The three important points to note are:

1 The progressive incorporation of better statistical methods more consistent with current ecological concepts;

2 The increasing realism of the ecological concepts incorporated into the models;

3 The improved representation of environmental processes.

2.7.4 Limitations

Observational analysis of the kind described above is often denigrated as 'mere correlation' and not causation. Shipley (2000) reminded us that correlation is merely 'unresolved' causation. Resolving an observed correlation may result in a causal explanation of no ecological interest, or it may yield a detailed set of relevant hypotheses. The New Zealand studies have provided a set of hypotheses and estimates of the relative importance of different environmental variables. This needs to be contrasted with the intuitive correlations on which many ecological hypotheses and subsequent experiments are based. One frequent limitation of species distributional modelling is the lack of a dynamic component. Vegetation is often assumed to be in equilibrium with the environment. Repeated measurements on the same plots can be used to study the successional changes in vegetation by means of a trajectory analysis using ordination (e.g. Greig-Smith 1983, p. 309). The impact of historical events can be incorporated into regression studies, e.g. time since fire. Leathwick (2001) takes an alternative approach where competition from dominants and spatial autocorrelation are introduced into the predictors to account for non-equilibrium effects due to historical events and slow dispersal of certain species.

A common limitation is the miss-match between ecological assumptions and the statistical methods used. Studies with both CCA and GLM often assume that vegetation variables have a straight-line relationship with environmental variables. There is no ecological or statistical reason to impose this limitation. Theory suggests a curvilinear unimodal response with a maximum occurring between upper and lower limits beyond which the species does not occur (Austin 2002).

2.8 Synthesis

2.8.1 Theory

Currently three research paradigms (Kuhn 1970) can be recognized in vegetation science as it concerns questions of whether vegetation is continuous or discontinuous and how it relates to environment. In Kuhnian paradigms, confirmatory studies, those providing supporting evidence, are more usual than tests of the basic assumptions of the paradigm whether the assumptions concern facts, theory or methodology. A willingness to recognize the strengths and weaknesses of each paradigm is needed to achieve a synthesis.

Traditional phytosociology constitutes one paradigm. The conceptual framework concerns the recognition and definition of the association and the hierarchical classification of associations; vegetation is assumed to be discontinuous. The two other recognizable paradigms tacitly accept the continuum concept; variation in vegetation composition is continuous and largely determined by environment. The two paradigms differ in the assumptions they make about species responses to environment and the methods selected to study vegetation variation. The multivariate analysis paradigm emphasizes the use of ordination and classification techniques, e.g. CCA. A recently emerging paradigm has been the use of statistical regression methods (e.g. GAM) to study the continuum, which make fewer assumptions about the species response to environment. The differences between these paradigms are less than many plant ecologists assume. The phytosociological association with its characteristic and differential species can be said to define a region in environmental space. This region of environmental space is more frequent in the landscape than others. The combinations of species characteristic of that environment are therefore frequent in the landscape and hence more recognizable (Figs. 2.4–2.6). This hypothesis relates the association with a region in the multi-dimensional continuum and requires testing. Differences between the other identified paradigms depend more on details of their analytical approaches.

2.8.2 Analytical approaches

The same methods of vegetation analysis are often shared by the three paradigms. Phytosociology can and frequently does use multivariate analysis methods to identify associations, differential and constant species. Recognition that analysis needs to be explicit, consistent and repeatable will increase this usage. The popular CCA method integrates multivariate analysis and regression. Its current weakness is that often only straight-line regressions are used. This is not a fundamental part of the method, for example, a GAM procedure can be used instead (W. Venables, pers. comm.). A synthesis of the two methods is possible. Information from all species would contribute to defining the principal gradients in vegetation composition but the correlation between these gradients and the environment would not be constrained to any particular shape.

2.8.3 Possibilities and problems

A synthesis of the three paradigms would help plant ecologists to focus on the unresolved issues of intrinsic causes of vegetation variation and the influence of environment on such variation. A possible framework is outlined here – the intention is not to suggest that this is the solution but to provide a topic for discussion.

Among the early pioneers of the direct gradient analysis of vegetation, Perring (1958, 1959) provided the most explicit conceptual framework. He proposed that vegetation properties were a function of various groups of environmental variables (factors), e.g. climate, or topography.

On this basis, vegetation studies should be undertaken with a stratified survey design based on such an explicit model of the possible processes involved.

The variable groups were all indirect variables. These could be expressed as direct or resource variables (Fig. 2.8) using our increased knowledge of environmental processes. Accepting the interpretation presented in Fig. 2.1b, there is an interaction between regional climate and the microclimate as represented by aspect and slope. Each species shows an individualistic response shifting its topographic distribution depending on climate. The use of direct gradients would simplify the figure to a single gradient of moisture stress. GAM models as used by Leathwick would provide rigorous quantitative descriptions of the relationships.

Questions of the existence of plant communities or of the relative importance of different direct gradients could be examined with appropriate stratified designs. A combination of CCA and other ordination techniques with statistical models could investigate the assumptions of different paradigms concerning the shape of species' responses. There are problems of both theory and methodology that need to be addressed. The importance of history and geographical barriers in determining current vegetation composition needs to be examined. The methodology needed will have to incorporate spatial autocorrelation into statistical models or ordination techniques. Current methods often ignore interactions, while current theory tends to focus on a single dimension or continuum of variation. The pattern of variation in multi-dimensional space is a key issue; almost nothing is known about the shapes and orientation of species distributions in multi-dimensional environmental space. Species packing and distribution of species richness are other unknown patterns in this space.

Whether vegetation is discontinuous or continuous depends on the perspective of the viewer. Viewed from a landscape perspective it is often discontinuous. In environmental space it is usually thought to be continuous. Rigorous testing of vegetation patterns in this space has yet to be achieved. The descriptive patterns resulting from multivariate pattern analysis and statistical modelling take us only so far. Understanding these patterns is an essential ingredient in sustainable vegetation management. At the present time there are many unanswered questions in vegetation science but there are also too many unquestioned answers.

Acknowledgements

I thank P. Gibbons, C.J. Krebs, R.P. McIntosh, J. Reid, B. Wellington and the editor for comments on the draft chapter.

References

Austin, M.P. (1987) Models for the analysis of species response to environmental gradients. *Vegetatio* **69**, 35–45.

Austin, M.P. (1991) Vegetation theory in relation to cost-efficient surveys. In: *Nature Conservation: Cost-effective Biological Surveys and data analysis. Proceedings of a CONCOM Workshop* (eds C.R. Margules & M.P. Austin), pp. 17–22. Canberra.

Austin, M.P. (1999a) The potential contribution of vegetation ecology to biodiversity research. *Ecography* **22**, 465–484.
Austin, M.P. (1999b) A silent clash of paradigms: some inconsistencies in community ecology. *Oikos* **86**, 170–178.
Austin, M.P. (2002) Spatial prediction of species distribution: an interface between ecological theory and statistical modelling. *Ecological Modelling* **157**, 101–118.
Austin, M.P. & Smith, T.M. (1989) A new model for the continuum concept. *Vegetatio* **83**, 35–47.
Belbin, L. (1991) Semi-strong hybrid scaling, a new ordination algorithm. *Journal of Vegetation Science* **2**, 491–496.
Bray, J.R. & Curtis, J.T. (1957) An ordination of the upland forest communities of southern Wisconsin. *Ecological Monographs* **27**, 325–349.
Breiman, L., Friedman, J.H., Olshen, R.A. & Stone, C.J. (1984) *Classification and regression trees*. Wadsworth International Group, Belmont.
Crawley, M.J. (1997) The structure of plant communities. In: *Plant Ecology*, (ed. M.J. Crawley), 2nd ed., pp. 475–531. Blackwell Scientific, Oxford.
Curtis, J.T. (1959) *The vegetation of Wisconsin: an ordination of plant communities*. University of Wisconsin Press, Madison.
Drake, J.A. (1991) Community-assembly mechanics and the structure of an experimental species ensemble. *American Naturalist* **137**, 1–25.
Dubayah, R. & Rich, P.M. (1995) Topographic solar radiation models for GIS. *International Journal of Geographical Information Systems* **9**, 405–419.
Ellenberg, H. (1988) *Vegetation Ecology of Central Europe*. 4th ed. Cambridge University Press, Cambridge.
Faith, D.P., Minchin, P.R. & Belbin, L. (1987) Compositional dissimilarity as a robust measure of ecological distance. *Vegetatio* **69**, 57–68.
Franklin, J. (1995) Predictive vegetation mapping: geographic modelling of biospatial patterns in relation to environmental gradients. *Progress in Physical Geography* **19**, 474–499.
Gauch, H.G. & Whittaker, R.H. (1972) Coencline simulation. *Ecology* **53**, 446–451.
Gittins, R. (1965) Multivariate approaches to a limestone grassland community. 1. A stand ordination. *Journal of Ecology* **53**, 385–401.
Gleason, H.A. (1926) The individualistic concept of the plant association. *Bulletin of the Torrey Botanical Club* **53**, 1–20.
Greig-Smith, P. (1983) *Quantitative Plant Ecology*. 3rd. ed. Blackwell Scientific Publications, Oxford.
Grime, J.P. (2001) *Plant strategies, vegetation processes, and ecosystem properties*. 2nd ed. Wiley & Sons, Chichester England.
Guisan, A. & Zimmermann, N.E. (2000) Predictive habitat distribution models in ecology. *Ecological Modeling* **135**, 147–186.
Hastie, T. & Tibshirani, R. (1990) *Generalised Additive Models*. Chapman and Hall, London.
Hill, M.O., Bunce, R.G.H. & Shaw, M.W. (1975) Indicator species analysis, a divisive polythetic method of classification, and its application to a survey of native pinewoods in Scotland. *Journal of Ecology* **63**, 597–613.
Jenny, H. (1941) *Factors of soil formation*. McGraw-Hill, New York.
Jongman, R.G.H., ter Braak, C.J.F. & van Tongeren, O.F. (1995) *Data Analysis in Community and Landscape Ecology*. Cambridge University Press, Cambridge.
Kent, M. & Coker, P. (1992) *Vegetation Description and Analysis: a Practical Approach*. Belhaven Press, London.
Kuhn, T.S. (1970) *The Structure of Scientific Revolutions*. 2nd ed. The University of Chicago Press, Chicago.

Leathwick, J.R. (1995) Climatic relationships of some New Zealand forest tree species. *Journal of Vegetation Science* **6**, 237–248.

Leathwick, J.R. (1998) Are New Zealand's *Nothofagus* species in equilibrium with their environment? *Journal of Vegetation Science* **9**, 719–732.

Leathwick, J.R. (2001) New Zealand's potential forest pattern as predicted from current species-environment relationships. *New Zealand Journal of Botany* **39**, 447–464.

Leathwick, J.R. & Austin, M.P. (2001) Competitive interactions between tree species in New Zealand old-growth indigenous forests. *Ecology* **82**, 2560–2573.

Leathwick, J.R. & Mitchell, N.D. (1992) Forest pattern, climate and vulcanism in central North Island, New Zealand. *Journal of Vegetation Science* **3**, 603–616.

Leathwick, J.R., Whitehead, D. & McLeod, M. (1996) Predicting changes in the composition of New Zealand's indigenous forests in response to global warming: a modelling approach. *Environmental Software* **11**, 81–90.

Legendre, P. & Legendre, L. (1998) *Numerical Ecology*. 2nd. English ed. Elsevier Science, Amsterdam.

Loucks, O. (1962) Ordinating forest communities by means of environmental scalars and phytosociological indices. *Ecological Monographs* **32**, 137–166.

May, R.M. (1985) Evolutionary ecology and John Maynard Smith. In: *Essays in Honour of John Maynard Smith*. (eds P.J. Greenwood, P.H. Harvey & M. Slatkin), pp. 107–116. Cambridge University Press, Cambridge.

McCullagh, P. & Nelder, J.A. (1989) *Generalized Linear Models*. 2nd. ed. Chapman and Hall, London.

McIntosh, R.P. (1967) The continuum concept of vegetation. *Botanical Review* **33**, 130–187.

McIntosh, R.P. (1985) *The Background of Ecology*. Cambridge University Press, Cambridge.

Minchin, P.R. (1987) An evaluation of the relative robustness of techniques for ecological ordination. *Vegetatio* **69**, 89–107.

Minchin, P.R. (1989) Montane vegetation of the Mt. Field Massif, Tasmania: a test of some hypotheses about properties of community patterns. *Vegetatio* **83**, 97–110.

Mueller-Dombois, D. & Ellenberg, H. (1974) *Aims and Methods of Vegetation Ecology*. Wiley, New York.

Økland, R.H. (1999) On the variation explained by ordination and constrained ordination axes. *Journal of Vegetation Science* **10**, 131–136.

Perring, F. (1958) A theoretical approach to a study of chalk grassland. *Journal of Ecology* **46**, 665–679.

Perring, F. (1959) Topographical gradients of chalk grassland. *Journal of Ecology* **47**, 447–481.

Perring, F. (1960) Climatic gradients of chalk grassland. *Journal of Ecology* **48**, 415–442.

Ross, S.M. (1986) Vegetation change on highway verges in south-east Scotland. *Journal of Biogeography* **13**, 109–117.

Shipley, B. (2000) *Cause and Correlation in Biology a user's guide to path analysis, structural equations and causal inference*. Cambridge University Press, Cambridge.

Shipley, B. & Keddy, P.A. (1987) The individualistic and community-unit concepts as falsifiable hypotheses. *Vegetatio* **69**, 47–55.

ter Braak, C.J.F. (1986) Canonical correspondence analysis: a new eigenvector technique for multivariate direct gradient analysis. *Ecology* **67**, 1167–1179.

ter Braak, C.J.F., Juggins, S., Birks, H.J.B. & van der Voet, H. (1993) Weighted averaging partial least squares regression (WA-PLS): definition and comparison with other methods for species-environment calibration. In: *Multivariate Environmental Statistics* (eds G.P. Patil & C.R. Rao), Vol. 6, pp. 525–560. North-Holland Publishing Company, Amsterdam.

van der Maarel, E. (1989) Theoretical vegetation science on the way. *Vegetatio* **83**, 1–6.

Vayssieres, M.P., Plant, R.E. & Allen-Diaz, B.H. (2000) Classification trees: An alternative non-parametric approach for predicting species distributions. *Journal of Vegetation Science* **11**, 679–694.

Webb, D.A. (1954) Is the classification of plant communities either possible or desirable? *Botanisk Tidsskrift* **51**, 362–370.

Westhoff, V. & van der Maarel, E. (1978) The Braun-Blanquet approach. In: *Classification of plant communities.* (ed. R.H. Whittaker), 2nd. ed., pp. 287–399. Junk, The Hague.

Whittaker, R.H. (1956) Vegetation of the Great Smoky Mountains. *Ecological Monographs* **26**, 1–80.

Whittaker, R.H. (1962) Classification of natural communities. *Botanical Review* **28**, 1–239.

Whittaker, R.H. (1967) Gradient analysis of vegetation. *Biological Review* **42**, 207–264.

Whittaker, R.H. (ed.) (1978a) *Ordination of Plant Communities.* Junk, The Hague.

Whittaker, R.H. (ed.) (1978b) *Classification of Plant Communities.* Junk, The Hague.

Yee, T.W. & Mitchell, N.D. (1991) Generalised additive models in plant ecology. *Journal of Vegetation Science* **2**, 587–602.

3

Vegetation and ecosystems

Christoph Leuschner

3.1 The ecosystem concept

The ecosystem concept was introduced by Tansley (1935) who stated that organisms cannot be separated from their environment if their ecology is to be understood. Nowadays, the ecosystem concept is one of the most influential ideas in contemporary ecology (Waring 1989). Modern definitions view the ecosystem as an energy-driven complex of the biological community (plants, animals, fungi and prokaryotes) and its physical environment which has a limited capacity for self-regulation.

Organisms and their environment form complex biophysical systems with a system being defined as a set of elements (e.g. plants and climate factors) to which a relationship of cause and effect exists. Our understanding of complex systems is hindered by the fact that organisms and atmosphere, lithosphere (soil) and hydrosphere are linked by a multitude of interactions that can rarely be disentangled completely. Moreover, many interactions are non-linear, include feedback loops, or occur accidentally which makes prediction often difficult (Pomeroy *et al.* in Pomeroy & Alberts 1988).

The adoption of the ecosystem concept starts from the notion that we cannot understand important processes on community or landscape levels from a knowledge of the ecology and interaction of the community members alone. For example, we are not able to precisely predict the consequences for the ecosystem carbon balance of a doubling of atmospheric carbon dioxide if we refer to data on plant ecophysiology and population biology only (Körner 1996). Moreover, agriculture, forestry and water resources management often deal with ecosystem-level processes such as nitrogen loss or groundwater recharge and their prediction. These goals require a shift in view from the organism and community levels to larger spatial and temporal scales.

The ecosystem approach adopts a system's perception of the living world. A hierarchy of biological organization levels can be identified with a sequence from the biomolecule through the cell, organism, population, community and ecosystem to the landscape. Ecosystems are found at the top end of this gradient in biological complexity. To study complex environmental systems, 'hierarchy theory' has been developed which recognizes ecosystems as multi-scale phenomena, ranging from the biochemical or organismic levels to the ecosystem level, and covering processes from seconds to thousands of years, with the different levels of organization being connected by asymmetric relationships (Allen & Starr 1982; O'Neill *et al.* 1986). Ecosystem studies based on hierarchy theory often use both 'bottom-up' and 'top-down' approaches. The former approaches attempt to assemble large-scale phenomena from smaller-scale components. The top-down approaches proceed in a reductionistic way by attempting to identify processes at lower levels that might cause observed ecosystem patterns.

The theory of complex systems states that the different organizational levels reveal 'emergent properties' (see Chapter 1) that cannot be predicted simply by adding up the properties of the next-lower level. The existence of emergent properties has been rejected by some ecologists (e.g. Harper 1982), but has been found by physicists to hold even at the subatomic level, and the notion of emergent properties has proved to be of practical use in ecosystem analysis. Higher levels often respond more slowly to disturbance and can buffer faster system dynamics at lower levels with the consequence that ecosystems exhibit a higher stability than do species populations because interactions and replacements do occur among organisms that dampen the rate of change at the ecosystem level. Moreover, a given property often behaves in a manner that is totally different from that of lower-level properties (Lenz *et al.* 2001). For example, negative effects of environmental factors on growth and fitness of organisms or population size may have no effect or even a positive influence on productivity at the ecosystem level because different populations interact.

Evapotranspiration or primary productivity in a patch of vegetation are characteristic emergent properties at the ecosystem level that not only depend on the ecophysiological controls of water loss and carbon gain at the leaf and plant levels. At the ecosystem level, additional factors such as canopy structure ('roughness'), atmospheric turbulence or competition with neighbours are increasingly important in regulating evapotranspiration and primary productivity while ecophysiological factors often appear to be lost during the scaling-up process. Biodiversity (the number of species per plot) is another emergent property of ecosystems that can influence ecosystem functions, and may be indispensable for ecosystem persistence in a variable environment (Naeem 2002). However, the role of biodiversity in ecosystems cannot be inferred from the species level (see also Chapter 8).

During the past 30 years, ecosystem level studies have helped to understand the regulation of energy and matter turnover in the biosphere, to comprehend the biological basis of productivity, and to secure human food production. In this period of attention for global change, ecosystem analysis has moved to a global perspective. The recent International Geosphere-Biosphere Programme (IGBP) has been designed in an attempt to predict how the biosphere will respond to an ever-growing human pressure (Walker & Steffen 1996).

3.2 The nature of ecosystems

Ecosystems are found all over the Earth's surface in terrestrial, limnic and marine environments. They stretch from the highest mountain top to the deep sea. A classification of ecosystems is often based on vegetation types because plants are the main primary producers of organic matter, and they often define the spatial structure of ecosystems.

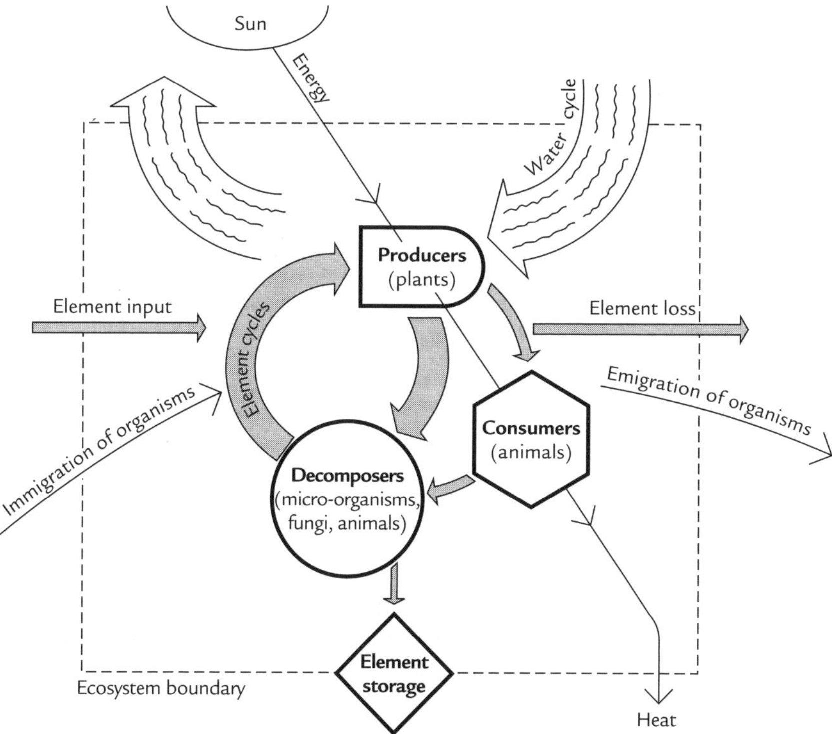

Fig. 3.1 The coupling of energy flow, and water and element cycling in ecosystems. Element cycles, which include those of carbon and nutrients, channel the bulk of matter from the producers (plants) to reducing organisms which decompose plant material and release carbon dioxide and nutrient ions to the environment.

All ecosystems are open systems with respect to the exchange of energy, matter and organisms with their surroundings (Fig. 3.1). Indeed, ecosystems typically have no clearly defined boundaries, except for (for example) small atoll islands, ponds or forest fragments that are isolated in the landscape. Instead, boundaries of most terrestrial ecosystems may be defined by the purpose of study and, thus, are somewhat arbitrary. A useful criterion for defining ecosystem boundaries would be the homogeneous distribution in space of key processes such as water and nutrient fluxes.

All ecosystems change in their structure and function over time. Change can be driven by external influences such as increased or lowered temperatures, or altered water supply. In many cases, however, change is the consequence of the activity of the community itself. Individuals change with season and age, populations increase and decrease, the number of species present may vary, and fluxes of energy and matter change with season and community age. The most fundamental change occurs during succession that often alters soil and microclimate of an ecosystem (see Chapter 8). Different ecosystem properties can change at different rates. This has the consequence that static (time-independent) perceptions of communities and ecosystems are oversimplifications which can lead to wrong conclusions. Modern ecosystem science thus concentrates on the understanding of the dynamic properties of ecosystems.

Ecosystems have a limited capacity for self-regulation. For example, gales may episodically destroy large patches of temperate forest in North America and Europe. After century-long forest succession, stand structure, primary productivity and nutrient flux rates will eventually regain a state which is close to the pre-disturbance situation. Indeed, many structural and functional properties of ecosystems are restored rapidly during the process of succession whereas others including species composition can differ substantially in communities prior to and after perturbation.

How an ecosystem responds to disturbance is a crucial characteristic in an epoch with rapidly growing human impact on nature. A widely used term is **stability**, here in the sense of persistence of structural and functional attributes of ecosystems over time. A mature beech forest would be called stable if only minor changes in canopy structure, species composition and soil chemistry are occurring over 20 or 30 yr. If ecosystem change is related to the degree of environmental change or disturbance intensity, the terms **resistance** and **resilience** are used. Ecosystems that show relatively small changes upon disturbance are said to be resistant. For example, *Fagus* forests are more resistant to disturbance by gales and catastrophic insect attack than are *Picea* forests. A severe disturbance is required to change the state of beech ecosystems. Ecosystem characteristics that buffer against disturbance are large storage reservoirs for carbon and nutrients in long-lived stems and roots, and high turnover rates of nutrients with plant uptake and mineralization. On the other hand, resistant ecosystems often take a longer time to return to their initial condition following a severe perturbation than do less-resistant ecosystems.

Resilience expresses the speed at which an ecosystem returns to its initial state. Resilient systems can be altered relatively easily but return to the pre-disturbance structure and function more rapidly; thus, their organisms are adapted to tolerate disturbance. An example of a resilient ecosystem is subalpine alder (*Alnus* spp.) scrub that rapidly resprouts from stumps after mechanical disturbance by ice. Plant productivity and leaf area typically show a high resilience after disturbance, whereas plant biomass and species composition do not. The impact of timber extraction in tropical moist forest thus can be detected centuries later by an altered species composition although forest structure and productivity may have been restored. Ecosystems in cold environments typically have much lower resilience than those in warm climates. In any case, the capacity for self-regulation is an important ecosystem property which may support the restoration of destroyed landscapes.

Self-regulation may also be relevant at higher levels of biological organization. The '*Gaia hypothesis*' (Lovelock 1979) views the whole biosphere of the Earth as one, large ecosystem that evolved over geological time periods and has the capability for self-regulation (Gaia was the Greek Goddess of the Earth). If photosynthesis did not exist, there would be much more carbon dioxide in the atmosphere and the surface temperature of the Earth would be much higher than it presently is. According to this hypothesis, life acts as a stabilizing negative feedback system on the global climate by maintaining the oxygen and carbon dioxide levels within narrow limits. The hypothesis is both attractive and controversial because the nature of the suggested climate control mechanisms of the biota is not sufficiently understood, and there are doubts on the efficiency of biological control of atmospheric chemistry which has experienced large changes of CO_2 concentrations over the past million years (Watson *et al.* 1998).

3.3 Energy flow and trophic structure

3.3.1 Primary productivity

Solar radiation is the direct driving force behind the functioning of nearly all ecosystems on Earth. Only some specialist ecosystems in the deep sea or inside the Earth's mantle which are dominated by micro-organisms are maintained by geothermal energy and, thus, exist independently from the sun's energy (Gold 1992). The quantity of solar energy reaching the surface of the earth is about one-half (*c.* 47%) of the radiation flux at the top of the atmosphere (5.6×10^{24} J per yr). The remainder is either absorbed by the molecules of the atmosphere and heats the air, or is reflected back to space. There are large geographical differences in the annual input of solar radiation: while the tropical regions receive $50-80 \times 10^8$ J per m^2 and yr, the polar regions are only supplied with 25–30. Temperate regions such as Great Britain and the north-eastern United States are somewhat intermediate with $40-50 \times 10^8$ J per m^2 and yr. Only a very small proportion of the solar radiation is captured through the carbon-fixing activity of plants and other autotrophic organisms (photosynthetic bacteria and cyanobacteria) which are called primary producers. Typically, photosynthesis converts not more than 1–2% of the incoming solar radiation to chemical energy in carbohydrate compounds. This is because much of the solar energy is unavailable for use in carbon fixation by plants. In particular, only about 44% of incident solar radiation occurs at wavelengths suitable for photosynthesis (400–700 nm). However, even when this is taken into account, photosynthetic efficiency rarely exceeds 2 or 3%. The bulk of radiation energy either fuels the water cycle (see Chapter 4) or simply warms the surface of plants and soil and, thus, drives the movement of air masses across continents and oceans.

The rate at which solar energy is used by the primary producers to convert inorganic carbon into organic substances is termed the **primary productivity** of the ecosystem. It can be expressed either in units of energy (e.g. $J \cdot m^{-2} \cdot d^{-1}$) or of dry organic matter or carbon (e.g. $g \cdot m^{-2} \cdot d^{-1}$ or $mol \cdot m^{-2} \cdot d^{-1}$). Flows of energy, organic matter and carbon are interrelated because organic matter contains roughly constant amounts of carbon (*c.* 0.48 $g \cdot g^{-1}$), and the energy content of organic matter is reasonably estimated by a value of 18 $kJ \cdot g^{-1}$.

The total fixation of energy by photosynthesis is referred to as **gross primary productivity** (GPP). Plants use a substantial part of this energy to support the synthesis of biological compounds and to maintain themselves, energy which is lost by respiration and is not available for plant growth. The difference between GPP and plant respiration (R) is termed the **net primary productivity** (NPP) of the ecosystem. It gives the production rate of new plant biomass and also indicates how much 'food' is at maximum available for consumption by heterotrophic organisms.

3.3.2 Secondary productivity

Unlike chemical elements, energy does not circulate in the ecosystem but flows unidirectionally through it (Wiegert in Pomeroy & Alberts 1988). The solar energy fixed in carbohydrates by the primary producers is transferred rapidly through several levels of heterotrophs by consumption and predation. Ultimately, all biological energy is converted to heat via respiration and, thus, leaves the system. Heterotrophs (animals, fungi, most bacteria) ingest autotroph or other heterotroph tissues to suit their own respiratory and tissue-building requirements. According to the type of food ingested these organisms are termed primary, secondary and higher-level consumers. The productivity of primary consumers is always much less than that of the plants on which they feed because only a fraction of plant productivity is consumed by plant-eating herbivores, and much energy of the incorporated plant mass is lost to faeces and respiration, and thus does not add to herbivore productivity. The rate of biomass production by all heterotrophs is called the **secondary productivity** of the ecosystem.

3.3.3 Trophic levels and food chains

Ecosystems differ greatly in their trophic structure, i.e. the pattern of energy and matter flow through the different trophic levels of primary, secondary, tertiary and higher-level consumers. Energy and organic substances are transferred from one trophic level to another as living tissue (or bodies), dead tissue, faeces, particles of organic matter (POM) or dissolved organic matter (DOM). According to the preferred type of food, organisms can be grouped as herbivores (consumers of living plant tissue), carnivores (consumers of living animals; consumers of microbial biomass are termed microbivores), detritivores (consumers of dead tissue, POM or faeces, also called decomposers) and DOM feeders (decomposers of dissolved organic substances). The distinction of producers, consumers and decomposers emphasizes the role organisms are playing in the assimilation and release of CO_2 and nutrients: producers (plants and certain micro-organisms) assimilate CO_2 and inorganic nutrients; consumers release CO_2; and decomposers (or mineralizers) release CO_2 and inorganic nutrients (Fig. 3.1). Both classifications are based on organism functions in the matter cycle but do not refer to taxonomic position (Pimm in Pomeroy & Alberts 1988).

The functional groups of organisms assemble into two principal food chains: (i) the live-consumer (or herbivory) chain with the sequence: herbivores – primary carnivores – secondary and higher-level carnivores; and (ii) the detritus chain with the sequence: detritivorous bacteria, fungi and animals – primary carnivores – secondary and higher-level carnivores. The former chain is based on living plant tissue; the latter utilizes dead tissue and bodies, POM and faeces (Fig. 3.2). In a number of ecosystems, a third type of food chain (iii) can be recognized which is based on dissolved organic matter (DOM) that feeds bacteria which themselves are consumed by carnivorous animals ('microbivores'). In the soils of terrestrial ecosystems, DOM originates from carbohydrates that are exudated from living plant roots or that are released during the decomposition process of soil organic matter. All three food chains can have quite a number of cross-links and most often form a complex food web rather than a simple chain.

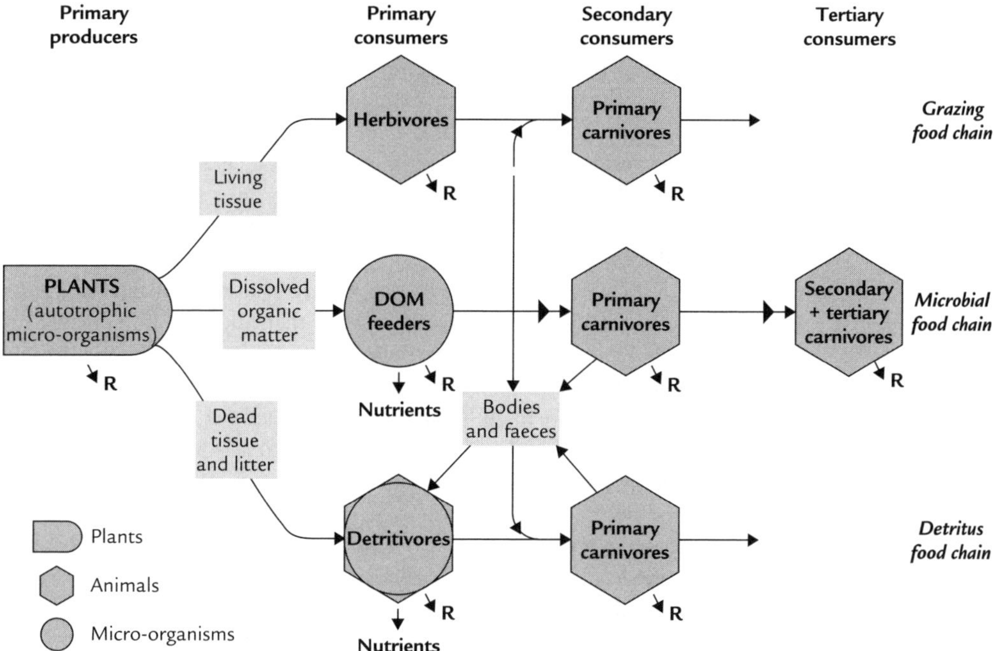

Fig. 3.2 Schematic diagram of the trophic structure of a community in which three principal food chains with different substrates exist. The arrows represent fluxes of organic matter and associated energy. Minor fluxes are omitted. Release of anorganic nutrients and respirative heat (R) is also indicated. DOM = dissolved organic matter.

Terrestrial ecosystems differ greatly with respect to the consumption efficiency (CE) of the herbivore community, i.e. the percentage of plant mass produced that is subsequently ingested by herbivores. CE is very low in many temperate forests (less than 5%; see the beech forest example in Fig. 3.3) where it is only significant in years of moth attacks – occurring every 5 to 10 yr in many temperate *Quercus* forests. Very high consumption efficiencies are characteristic of tropical grasslands where insects (e.g. locusts) and, in Africa, megaherbivores (among them antelopes and elephants) annually ingest 20% to more than 50% of the plant mass produced (Lamotte & Bourlière 1983). Similarly high consumption efficiencies are reached in fertilized pastures of the industrialized countries.

In all terrestrial ecosystems, a large amount of energy-rich plant material is not used by herbivores but dies without being grazed and thus supports a community of detritivorous animals, fungi and bacteria. Indeed, the world is green despite the activity of herbivores. This is a consequence of the fact that, in most terrestrial ecosystems, herbivores are effectively held in check by various mechanisms, among them an effective plant defence by secondary compounds such as lignin, growth-limiting nutrient concentrations in the herbivore's diet, and control of herbivore population size by enemies and intraspecific competition (Hartley & Jones 1997).

3.3.4 Decomposition of organic matter

Perhaps the most important role heterotrophic organisms are playing in the matter cycles of ecosystems is the decomposition of organic substances. This process describes the gradual disintegration of dead organic material by both organisms (detritivorous animals, decomposing fungi and bacteria) and physical agents. Fire can play an important role as a physical disintegrating agent in the dry regions of the world. Decomposition eventually leads to the breakdown of complex energy-rich molecules (such as carbohydrates, proteins and lipids) into carbon dioxide, water, inorganic nutrients and heat. This process which is called mineralization regenerates nutrients for plant uptake. It has the important consequence that nutrients immobilized in organic compounds are eventually released into the soil solution where plant roots and fungal hyphae can assimilate them. Decomposition and mineralization thus complete the carbon and nutrient cycles in ecosystems which started with the fixation of carbon dioxide and the assimilation of nutrients by primary producers.

Over decades and centuries, carbon fixation by primary producers and carbon release through the respiration of heterotrophic organisms will be balanced in most terrestrial ecosystems. Exceptions are many ecosystems that exist on wet soils or in cold climates where decomposition is hampered temporarily or permanently. Here, organic matter gradually accumulates in the soil and eventually may form peat and, in geological time spans, coal and oil. Carbon fixation with primary production and carbon release through heterotrophic respiration may also remain unbalanced in ecosystems that receive input of organic matter from external sources. This occurs in riverine forests and coastal marshes where organic matter and nutrients are deposited during inundation.

Over periods of months to years, carbon gain and carbon loss may differ greatly in most terrestrial ecosystems because photosynthesis and heterotrophic respiration are controlled by different factors and, thus, fluctuate independently. In periods of maximum plant growth, when biomass rapidly increases, carbon fixation largely exceeds carbon release by respiration and the ecosystem will function as a net sink of carbon dioxide. This happens in spring in seasonal climates and, more generally, during the juvenile stages of plant life. In contrast, carbon losses will exceed carbon gain when plant productivity decreases with plant senescence as occurs in autumn and, more gradually, during the senescence phase of plant development (as in ageing forests) when decomposition of accumulated plant biomass dominates over production.

3.3.5 Energy flow in a temperate forest

A good impression of how energy flows through ecosystems is given by the example of the Solling forest, a broad-leaved summer-green forest in temperate central Germany (Fig. 3.3). A team of plant and animal ecologists, soil biologists and micrometeorologists synchronously measured the fixation of solar radiation by trees and herbaceous plants, quantified the growth of plant leaves, branches, stems and roots, and studied food consumption, growth and respiratory activity of all major animal, fungi and bacteria groups in this ecosystem (Ellenberg *et al.* 1986). The primary production of this stand is provided nearly exclusively by European beech

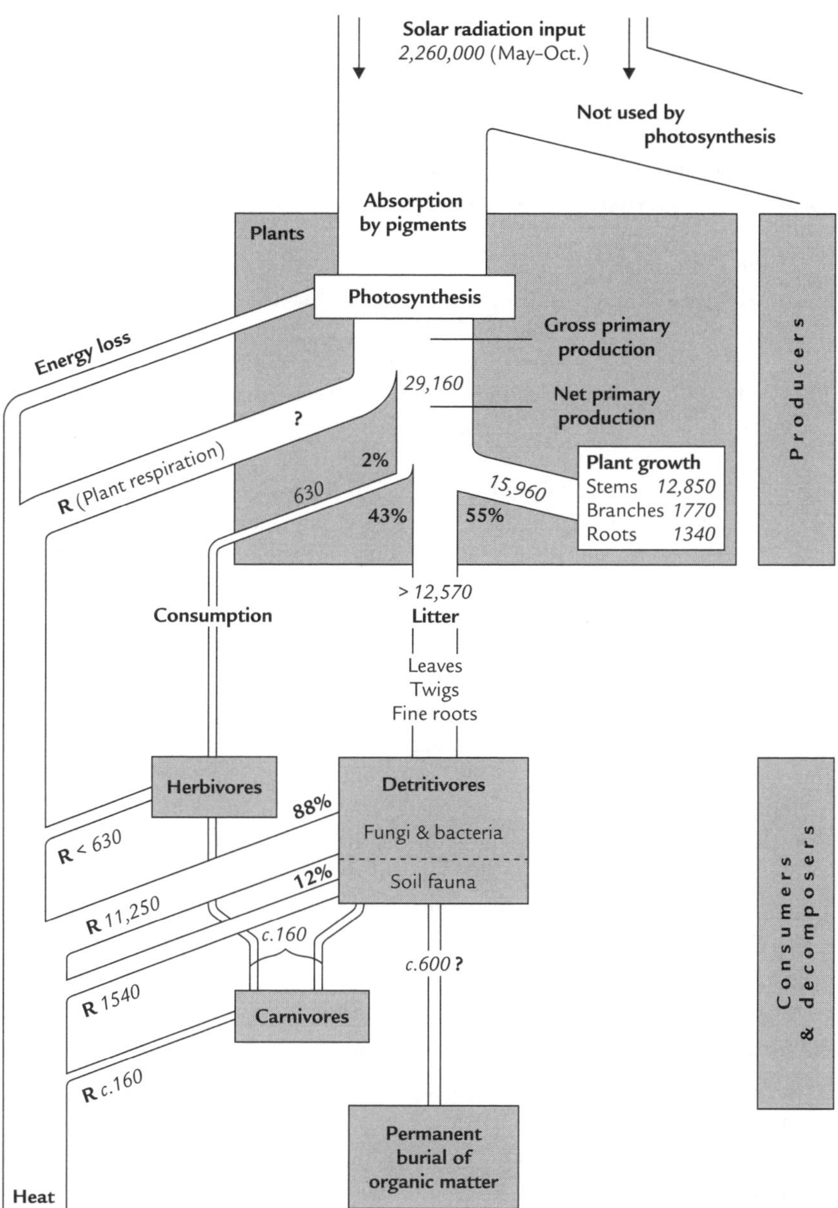

Fig. 3.3 Energy flow through a temperate broad-leaved forest ecosystem with dominating beech (*Fagus sylvatica*; Solling mountains, Germany). All fluxes are in $kJ \cdot m^{-2} \cdot yr^{-1}$; they can be approximately expressed as fluxes of dry matter ($g \cdot m^{-2} \cdot yr^{-1}$) by division with 20. During photosynthesis, an unknown part of the absorbed radiative energy is lost as heat, and through chlorophyll fluoresence and plant respiration. Note that energy flow from dead herbivores and carnivores to detritivores is omitted (after sources in Ellenberg *et al.* 1986).

(*Fagus sylvatica*) that builds pure stands with an only sparse herbaceous layer at the forest floor.

Net primary production (29,160 $kJ\cdot m^{-2}\cdot yr^{-1}$) accounts for *c.* 1.3% of the solar radiation input during the growing season (May–October). Gross primary production (which includes plant respiration) is estimated to be twice as high. The bulk of the incoming radiation energy (more than 80%) is not used in primary production but is consumed by the evaporation of water with transpiration of leaves and rainfall interception of the canopy.

Of the net primary production, 55% are utilized in the growth of long-lived structural organs of beech (mainly stems, but also branches and large roots). These plant organs may be accessible by detritivorous organisms only after 100 or 200 years. Consequently, a substantial net accumulation of biomass still occurs in the 130-yr-old Solling forest which indicates that this ecosystem actually functions as a net sink of carbon dioxide (Dixon *et al.* 1994). Other plant biomass fractions such as beech leaves, acorns, and part of the twig and fine root fractions, are turned over annually and, thus, represent plant litter which is the basis of the detritus food chain (43% of NPP). Although a number of herbivore populations reach high densities, leaf- and root-eating animals consume an only negligible fraction of NPP in this forest: *c.* 2% only of net primary production is channelled through the live-consumer chain which starts with herbivore consumption on leaves and roots.

Shed leaves, twigs and acorns accumulate on the forest floor in autumn; dying roots represent an important additional litter source in the soil. Both components are decomposed by fungi, bacteria and soil animals that are present with high species numbers: more than 1500 soil animal species alone were counted on 1-ha plots in this forest. Decomposition is mainly carried out through the activity of fungi and bacteria which consume *c.* 88% of the energy contained in the detritus material. Soil animals are much less important in terms of energy flow but fulfil important roles in the decomposition process as shredders of dead tissue and by vertically mixing the soil organic layers. A small, chemically inert fraction of the plant litter resists the attack of the detritivorous organisms for years and decades, and appears as dark humic substances in the topsoil.

3.3.6 Global patterns of terrestrial primary productivity

Terrestrial NPP of the globe is estimated at *c.* 60×10^{15} $g\text{-}C\cdot yr^{-1}$ (Houghton & Skole 1990) which is in the same order of magnitude as the known world oil reserves (70×10^{15} $g\text{-}C\cdot yr^{-1}$). NPP varies greatly across the ecosystems of the world (Fig. 3.4). The lowest productivity is not necessarily found in those ecosystems with lowest radiation energy input. For example, tropical deserts have high radiation loads but only very low primary productivities. The understanding of the factors that govern ecosystem productivity, thus, has been a key interest of ecologists during the past 30 yr. This knowledge serves the requirements of agricultural production, and also helps to protect ecosystem functioning.

The most productive terrestrial ecosystems of the globe are found among wetlands (i.e. swamps, marshlands and fens), tropical moist forests, and cultivated lands with typical NPP values in the range of 750–1300 $g\text{-}C^{-2}\cdot yr^{-1}$ of carbon fixed. Low product-

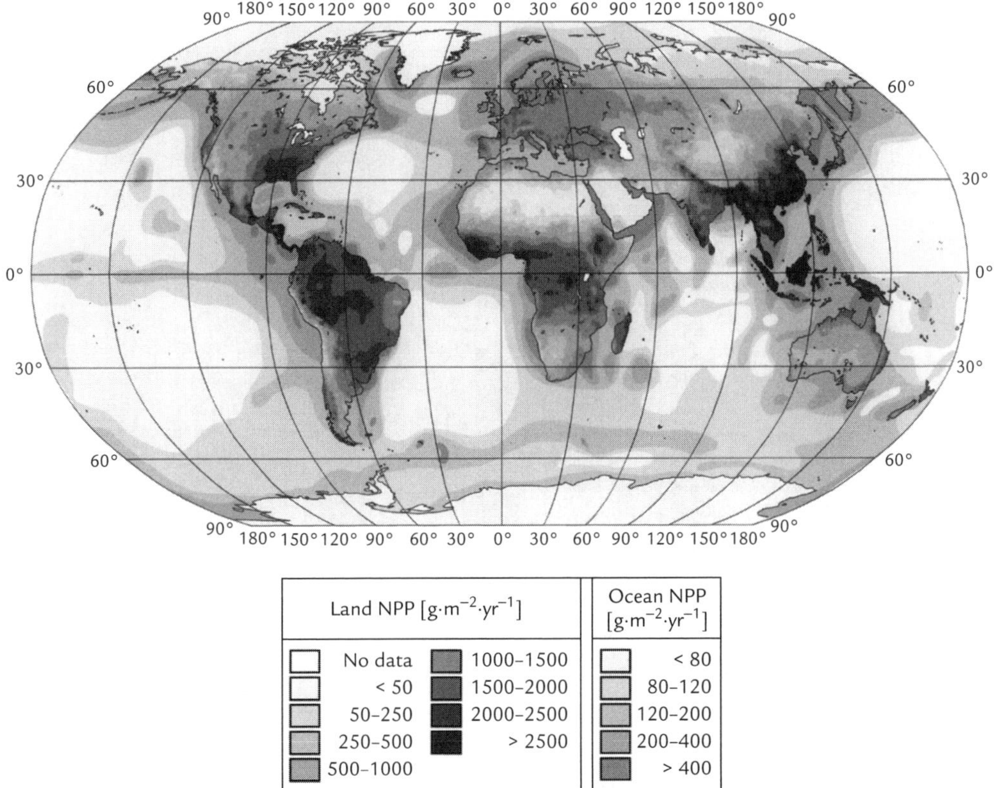

Fig. 3.4 Net primary production of the biosphere (*source*: Berlekamp, Stegemann & Lieth in http://www.usf.uni-osnabrueck.de/~hlieth). Note different scales for terrestrial and marine productivity.

ivities dominate in desert, tundra and arctic ecosystems with 0–150 g-C·m^{-2}·yr^{-1} (circles in Fig. 3.5). A comparison of ecosystem types reveals clear increases in NPP from desert to moist forest, and from tundra to tropical forest. This allows the conclusion that two environmental variables, water availability and temperature regime, are likely to be the principal factors which determine height and global distribution of primary productivity on the continents. A more thorough analysis shows that temperature seems to influence annual NPP mainly through the length of the growing season rather than through a direct dependence of NPP on annual mean temperature. Indeed, tundra, boreal and temperate ecosystems show remarkably high productivities in comparison to tropical moist forest if one considers the different lengths of the growing seasons in these four ecosystem types (1–3, 4–6, 6–8, and 12 months, respectively). Minimum temperature, in particular the occurrence of frost, is another crucial factor for plant productivity.

On a global scale, light and nutrient availability are less important than water and temperature for the height of terrestrial NPP. However, nutrient availability often determines local differences in NPP among sites (Pastor *et al.* 1984).

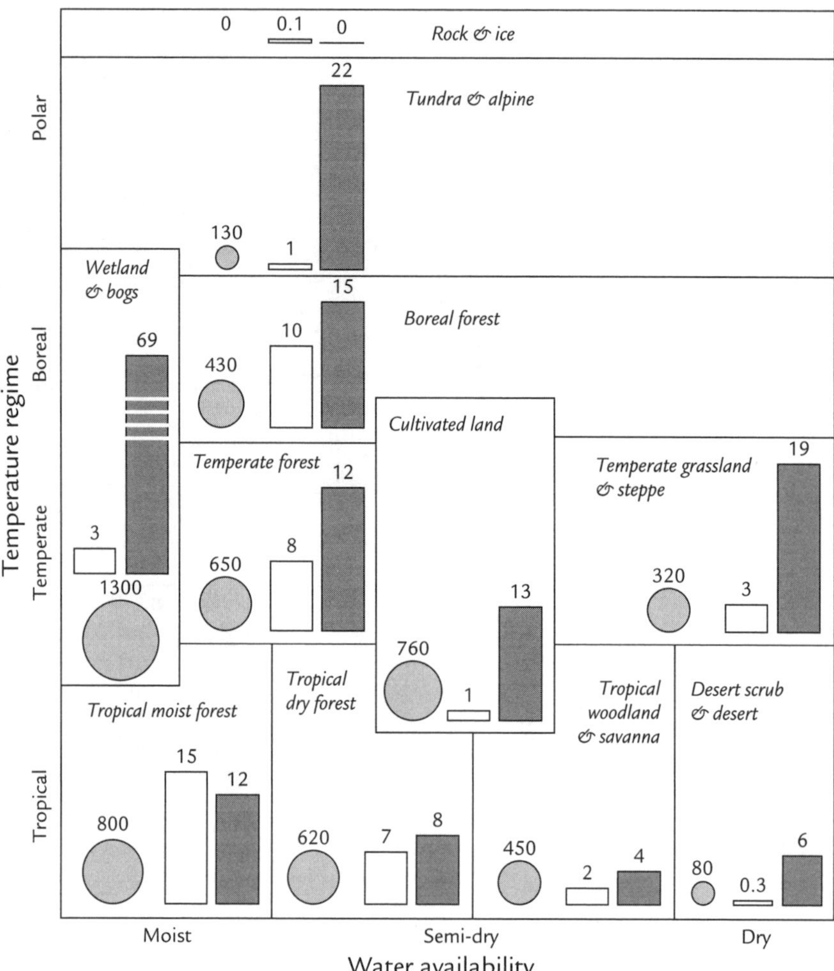

Fig. 3.5 Approximate stores of carbon and net primary production in the major biomes of the world. Soil organic matter (SOM; dark bars), and plant biomass (only above-ground; light bars) are given in kg-C·m^{-2}, NPP (circles) in g-C·m^{-2}·yr^{-1} (after data in Schlesinger 1997). The biomes or ecosystem types are arranged along axes of temperature and water availability, the two key factors that determine terrestrial productivity.

3.3.7 Productivity and energy flow in different ecosystem types

In Fig. 3.6 general patterns of productivity and energy flow in four key ecosystems of the globe are compared, i.e. temperate deciduous forest, tropical moist forest, boreal coniferous forest and temperate grassland. These ecosystems differ greatly with respect to the type of primary producers (summer-green broad-leaved trees, evergreen broad-leaved trees, evergreen needle-leaved trees or grasses) which reflect the contrasting temperature and water regimes. GPP is more variable among the

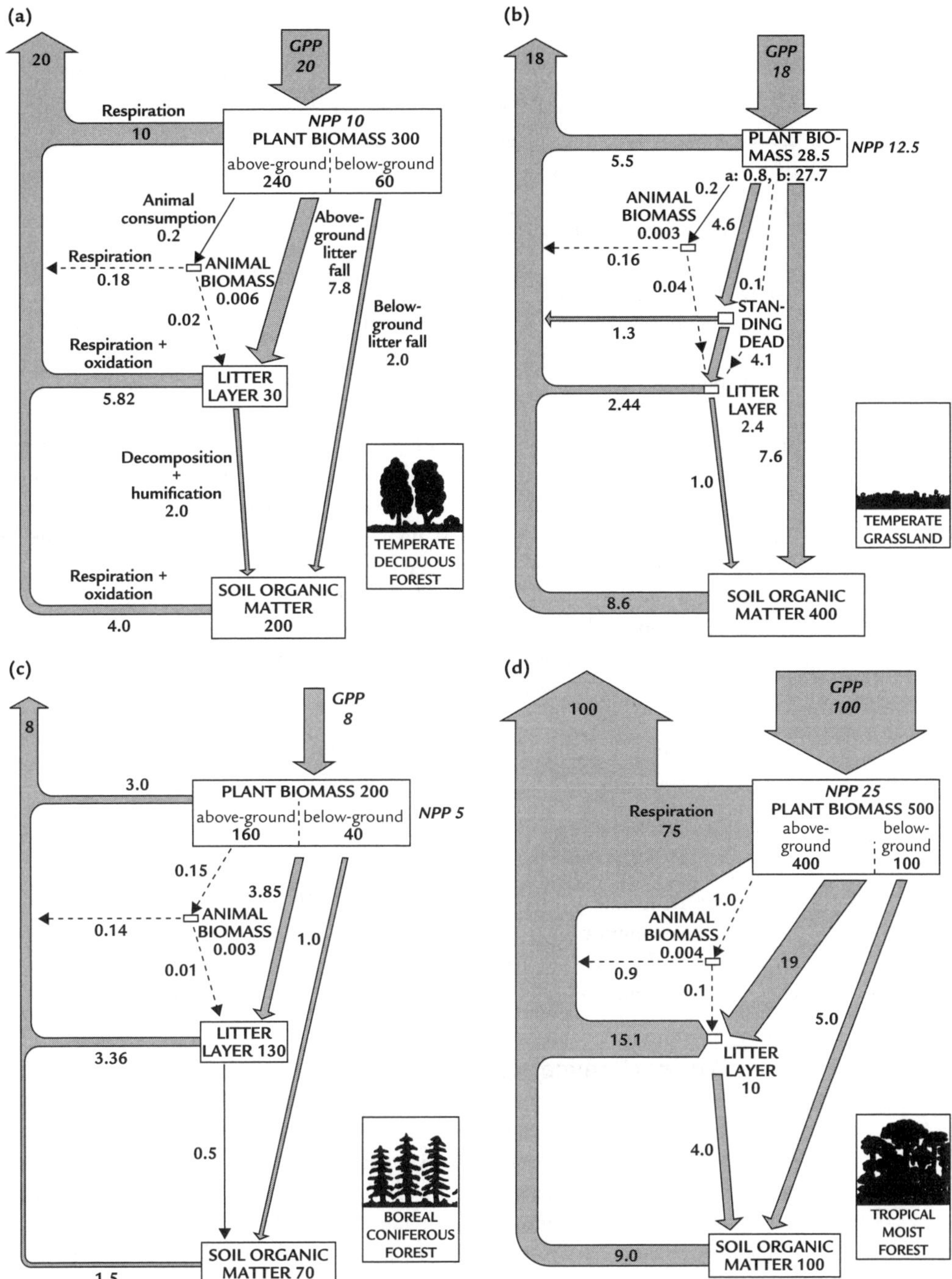

Fig. 3.6 Principal patterns of energy flow through four key ecosystems of the globe. Arrows and related numbers give fluxes of organic matter (in Mg-dry-matter·ha^{-1}·yr^{-1}), squares indicate pools of organic matter (in Mg·ha^{-1}). The relative size of arrows and squares allows a comparison of the four ecosystems. The data are assembled from various ecosystem studies cited in Schultz (2000) or partly based on estimation; thus, only a rough picture is given.

four ecosystems than is NPP, with highest values of both GPP and NPP being found in the tropical moist forest. The high GPP of the tropical forest is partly a consequence of very high plant respiration rates in the hot climate: it is estimated that about 75% of GPP is lost to the atmosphere by plant respiration in this ecosystem. In comparison, plant respiration consumes only about 30–50% of GPP in the cooler temperate and boreal forests.

Plant biomass increases in the sequence temperate grassland – boreal forest – temperate forest – tropical forest (white rectangles at the top of Fig. 3.6). In contrast, soil organic matter (SOM) is much higher in the grassland than in the three forest types (rectangles at the bottom). SOM pools are comparably small under many boreal coniferous and temperate broad-leaved forests because much plant litter accumulates at the forest floor on the soil surface and is not incorporated into the soil profile itself. The contrasting patterns of detritus storage in the four ecosystem types are primarily a consequence of a very rapid litter decomposition in the tropical moist forest where detritus typically is decayed in periods of weeks to several months, and thus is not accumulating on the forest floor. This contrasts with low decay rates in the dry and winter-cold grassland and the cold boreal forest. The size of SOM pools also depends on the amount of litter that is supplied annually above- and below-ground: above-ground litter fall typically decreases in the sequence tropical forest – temperate forest – coniferous forest – temperate grassland. Below-ground litter from decaying roots, however, seems to be much more important in the grassland than in the forest ecosystems.

Animal biomass is negligible in comparison to plant biomass or soil organic matter in all four ecosystems although animals often play important roles in many ecosystem functions such as soil formation, pollination and dispersal. Consumption of plant tissues by herbivorous animals is also quantitatively of minor importance in all three (boreal, temperate and tropical) forest ecosystems. Consumption efficiency is low in the temperate winter-cold grassland as well. This contrasts with intensive herbivore consumption in tropical grassland ecosystems (see Chapter 10). All four ecosystems in Fig. 3.6 are characterized by a detritus food chain which, in terms of energy flow, is much more important than the live-consumer chain.

3.4 Biogeochemical cycles

Nutrients and water tend to circulate along characteristic pathways in ecosystems, in marked contrast to energy. Energy is never recycled but flows through the ecosystem, being finally degraded to heat and lost from the system (Fig. 3.1). Carbon, nitrogen, hydrogen, oxygen and phosphorus are the functionally most important chemical elements in plants. They participate in biogeochemical cycles between living organisms and the environment with movements by wind in the atmosphere and by running waters in soil, streams and ocean currents.

3.4.1 Carbon cycle

The cycle of carbon is predominantly a gaseous one which is driven by two key processes, photosynthesis and respiration (Fig. 3.7a). The only source of carbon available

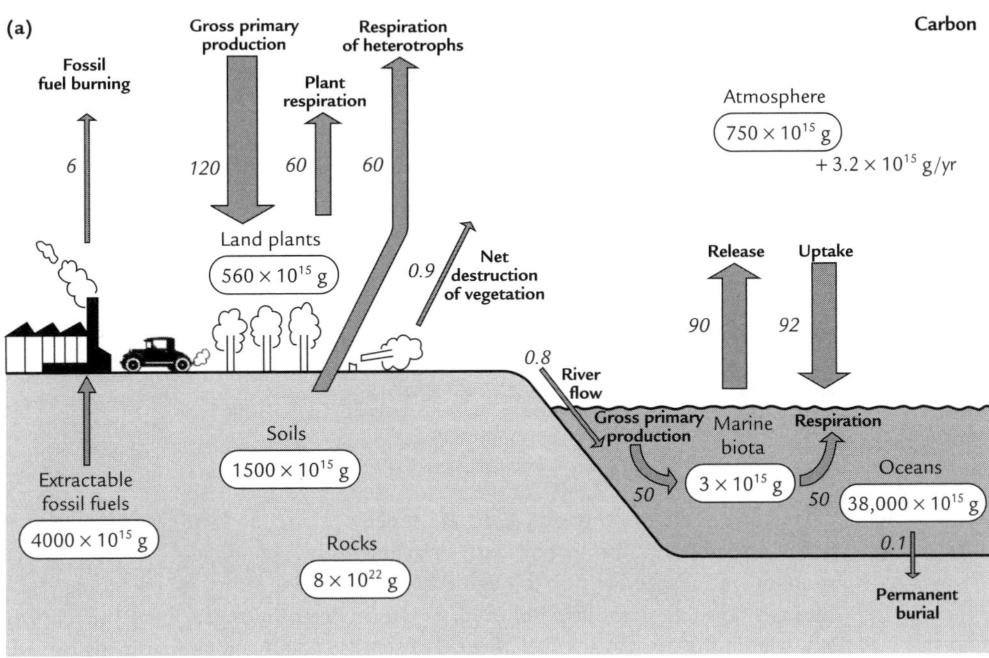

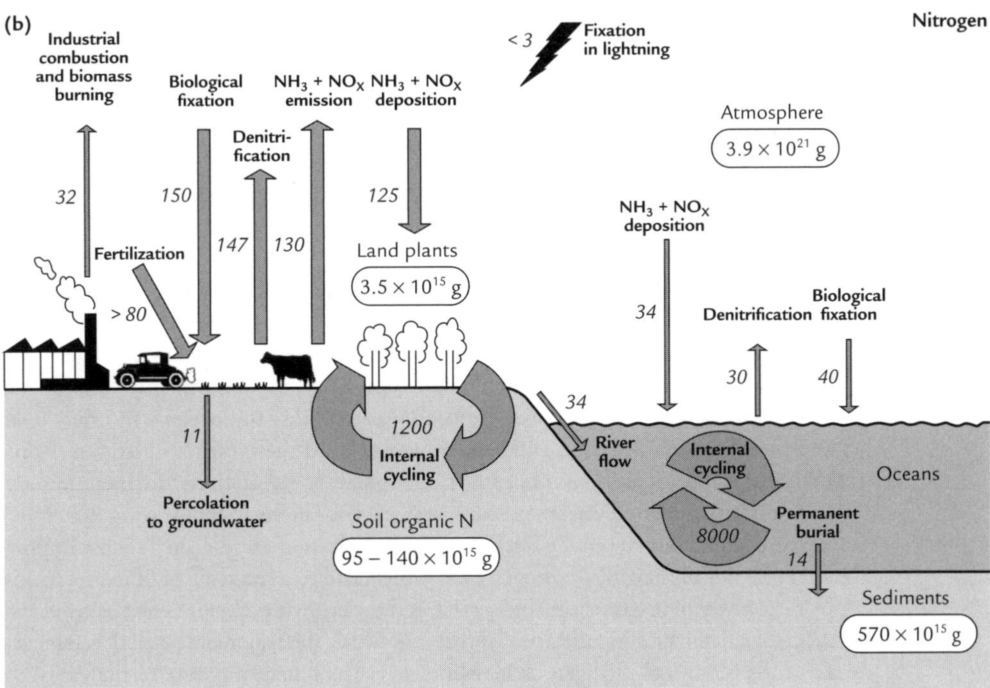

Fig. 3.7 The global cycles of (**a**) carbon and (**b**) nitrogen. The estimated size of pools (white boxes) is given in g of C or N, the transfers between compartments (arrows) in 10^{15} g-C·yr^{-1} and 10^{12} g-N·yr^{-1}, respectively (data from Schlesinger 1997 and Jaffe 1992). Internal cycling refers to plant uptake and release through decomposition.

to land plants is carbon dioxide (CO_2) in the atmosphere, whereas water plants can assimilate CO_2 and/or bicarbonate (HCO_3^-) dissolved in water. The large carbon stocks present in rocks (mainly as carbonates) are not available to plants. Large stores of inorganic and organic carbon also exist in the oceans.

In terrestrial ecosystems, considerable amounts of carbon are sequestered either in plant biomass, or in dead organic matter in the soil (SOM) or on the forest floor (litter layer). The largest stocks of living plant biomass are found in the luxuriant tropical moist forests where, on average, more than twice as much biomass is present than in the temperate summer-green forests of eastern North America or Central Europe. Large biomass stocks are also found in boreal forests. The largest reserves of dead SOM are stored in wetland and mire, grassland and tundra ecosystems where biomass decomposition is inhibited by the lack of oxygen or by low temperatures (Fig. 3.7a).

3.4.2 Nitrogen cycle

The main reservoir of nitrogen is the atmosphere where the inert gas N_2 constitutes *c.* 78% by volume. All nitrogen that is available to biota originally has been derived from this atmospheric pool through nitrogen fixation, either by oxidation of N_2 through lightning in the atmosphere or by the activity of nitrogen-fixing microbes on land and in the seas (Fig. 3.7b). Nitrogen fixers are found among free-living bacteria (e.g. *Azotobacter*, *Azotococcus*), cyanobacteria (e.g. *Nostoc*), and bacteria associated with plant roots (e.g. *Rhizobium*); this process is highly dependent on energy supply in terms of ATP. Plants with symbiotic nitrogen fixation include some legumes (Papilionaceae) and *Alnus* spp. Only relatively small amounts of nitrogen are stored in terrestrial biomass, in soil organic matter and in ocean sediments.

Land plants use ammonium (NH_4^+), nitrate (NO_3^-) or low-weight organic N compounds as primary nitrogen sources for incorporation into their biomass. Organic nitrogen (amino acids and oligopeptides) is an important N source for plants primarily in cold and acidic environments such as arctic and boreal ecosystems where symbiotic fungi (mycorrhizae) shortcut the nitrogen cycle by absorbing organic nitrogen-rich soil compounds and by subsequently transferring amino acids to the plant roots (Smith & Read 1997). In less adverse environments, inorganic nitrogen is released from dead plant biomass by the process of ammoniafication which specifies the mineralization of organic compounds with a production of ammonium mainly through the activity of bacteria. NH_4^+ can be oxidized by autotrophic soil bacteria (*Nitrosomonas*, *Nitrobacter*) to NO_3^- under a sufficiently high pH and the presence of oxygen. Plants with preference of ammonium, nitrate or organic N alternatively dominate ecosystems under different climate and soil chemical conditions (Schulze *et al.* 2002).

Nitrogen assimilation by plants and ammoniafication are the single most important processes of the global nitrogen cycle which lead to a recycling of N in ecosystems by the activity of producers and mineralizers (i.e. detritivores that function as reducers). Microbial denitrification is the only process in which the major end product is removed from the biological nitrogen cycle. Micro-organisms reduce nitrate to the gases N_2, nitrogen monoxide (NO) or dinitrogenoxide (N_2O) in the absence of oxygen, and thus lower the nitrogen load of ecosystems. Denitrification is an important process in wetland and mire ecosystems, but also occurs in other terrestrial ecosystems such as forests after anthropogenic nitrogen input. Undisturbed mature ecosystems typically have more or less equal inputs and outputs of nitrogen. Human impact through elevated atmospheric N-deposition or natural disturbance can result in unbalanced N budgets with either N-accumulation in soil and biomass, or long-term N loss via increased mineralization and leaching of NO_3^- to the groundwater. Increased leaching of N or other nutrients may serve as an indicator of instability in ecosystems (Waring 1989; Likens & Bormann 1995).

3.4.3 Phosphorus cycle

In contrast to nitrogen, the cycle of phosphorus is a sedimentary one with no significant gaseous component. The main phosphorus reservoir exists in rocks which are slowly eroded, releasing P to ecosystems (Fig. 3.8a). The main flux of phosphorus in the global cycle is carried by streams to the oceans where P is eventually deposited in the sediments. Similar to nitrogen, there is an intensive internal cycling of P in ecosystems through plant uptake and assimilation, and subsequent release via microbial mineralization. Unlike nitrogen, however, the major primary source of phosphorus is mineral weathering and not biological fixation from atmospheric sources. Mining of phosphate-rich rocks has greatly increased the phosphorus availability in agricultural ecosystems, and has resulted in P enrichment of adjacent limnic and marine systems.

3.4.4 Water cycle

The global water cycle moves much more substance than any other biogeochemical cycle on earth. Receipt of water through rainfall is one of the key factors controlling primary productivity on land. Water exists in three different states, solid (ice), liquid and gaseous (water vapour). By far the largest reservoir of water is the oceans, comprising *c.* 97% of the total amount of water on earth. Another 2% is bound in polar ice caps and glaciers. Only a very small proportion of the global water resource is available for the growth of terrestrial plants since less than 1% belongs to freshwater reservoirs in rivers, lakes and groundwater (Fig. 3.8b). Water vapour in the atmosphere constitutes only *c.* 0.08% of the total (Chahine 1992; Schlesinger 1997).

These pools are linked by flows of water with evaporation, transpiration, precipitation and overland flow. Water is transferred from land and ocean surfaces to the atmosphere through three processes: (i) evaporation from wet soils and water surfaces; (ii) transpiration from plant leaves; and (iii) evaporation of water from plant surfaces that is intercepted during rain. Among these three evapotranspiration components, only transpiration can be regulated by stomatal aperture according to plant demand. The energy required to move liquid water into the vapour phase is called the heat of vaporization and equals 44 kJ per mol of water. This large amount of heat is consumed when water evaporates from vegetation, water tables or moist soil, and is regained during the process of condensation when clouds form. Therefore, the cycling of water on Earth is a highly effective means to distribute solar energy absorbed by land and water surfaces.

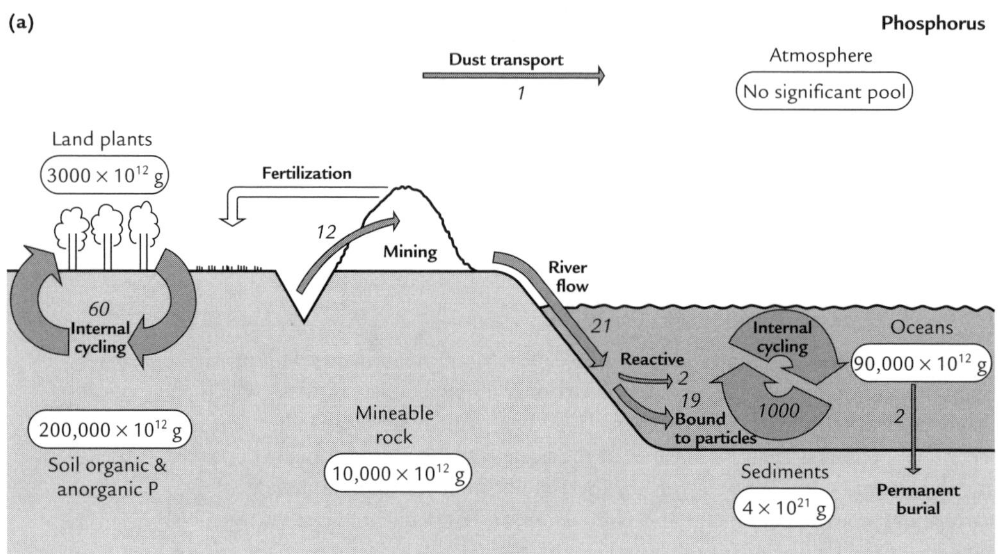

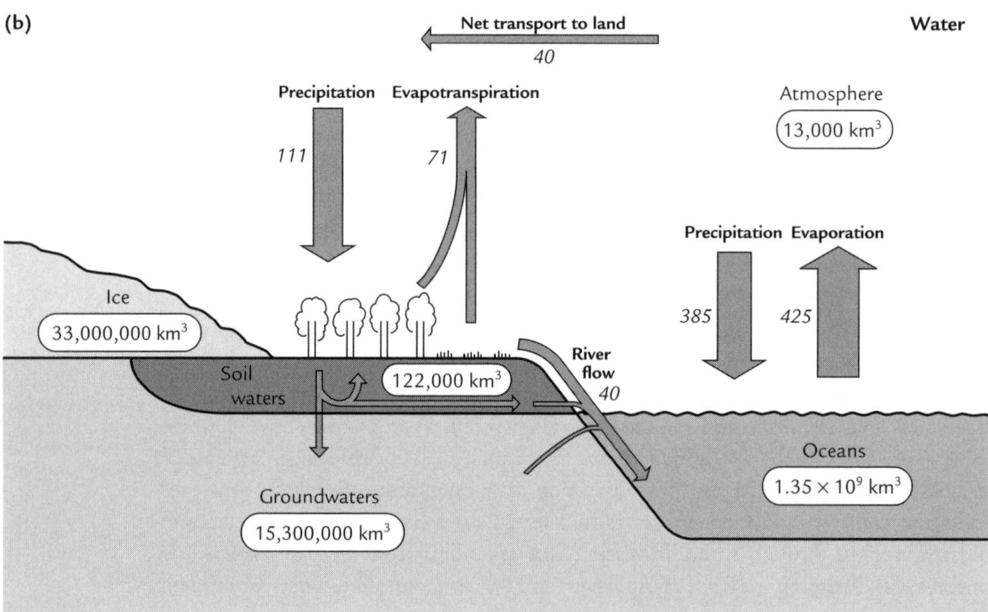

Fig. 3.8 The global cycles of (**a**) phosphorus and (**b**) water. Transfers between compartments are given in 10^{12} g-P·yr^{-1} and 10^3 km^3-water·yr^{-1}, respectively (data from Schlesinger 1997). The net transport of rain water to land is the difference between evaporation and precipitation over the sea, and equals worldwide river flow which results from a precipitation surplus on land.

On the oceans, a surplus of evaporation over rainfall exists whereas the land surfaces, on average, receive more water through precipitation than they loose through evapotranspiration. Consequently, an annual net transport of 40,000 km^3 of water occurs with cloud movement from the oceans to the land. An equal volume flows with rivers from the land to the oceans. River flow also carries the products of mechanical and chemical weathering to the sea and thus is an important agent in nutrient cycles.

Terrestrial vegetation can substantially modify the cycling of water. Vegetation types with large leaf areas such as tropical moist forests may catch much more water through interception of rainfall than, for example, a temperate short-grass steppe. Thus, less water infiltrates into the soil under forests. In addition, deep-rooted forests typically have higher transpiration rates than low vegetation types. This further reduces the water that is available for a recharge of groundwater reservoirs. The principal consequences of large-scale forest destruction, as it occurs in the wet tropics, are a speed-up of erosion and soil nutrient impoverishment, and a lowered transpiration which may result in less cloud formation in the region.

3.4.5 Anthropogenic alterations of biogeochemical cycles

Major perturbations of the biogeochemical cycles have occurred on Earth with 30 to 50% of the land surface transformed by human action. In a few generations, the world population will have exhausted the fossil fuels that were generated over several hundred million years. More than half of the accessible fresh water is already used by mankind.

Rising atmospheric carbon dioxide concentrations are likely to cause global warming and reflect the human perturbation of the carbon cycle (Houghton *et al.* 2001). Compared to pre-industrial values, the atmospheric concentration of carbon dioxide has increased by more than 30%, that of methane by more than 100%. The recent large-scale destruction and burning of both tropical and boreal forests greatly reduces the large biomass and humus carbon pools of these ecosystems and increases the net flux of carbon dioxide to the atmosphere. This source adds significantly to the carbon dioxide emission from the burning of fossil fuel, and leads to an annual increase of the atmospheric carbon dioxide concentration by *c.* 0.5% of its present value. Large stores of inorganic and organic carbon exist in the oceans which most likely act as net sinks of carbon dioxide. They may buffer the anthropogenic carbon dioxide emission to a certain degree (Prentice *et al.* 2001).

Humans also have a dramatic impact on the nitrogen cycle. More nitrogen is now fixed by industrial processes and applied as fertilizers in agriculture (more than 40 million tons of nitrogen) than is fixed by micro-organisms in all terrestrial ecosystems (Vitousek *et al.* 1997). Fertilization results in nitrogen enrichment (eutrophication) in terrestrial, limnic and marine ecosystems, and may lead to dramatic changes in plant species composition. Burning of fossil fuels and biomass, and the emission of ammonia (NH_3) and nitrogen oxides (NO_x) by modern agriculture increases the concentration of these nitrogen compounds in the atmosphere which are subsequently returned to surrounding ecosystems with rainfall deposition. Several ecosystems already have reached 'nitrogen saturation' which results in the increased transport of nitrate with infiltrating water from the soil to groundwater reservoirs which are polluted.

Mankind releases many toxic substances into the environment that often accumulate in organisms. Emission of chlorofluorocarbon gases has led to the ozone hole over the Antarctic and would have destroyed much of the ozone layer if no international measures to end their emission had been taken (Prather *et al.* 2001).

The plundering of resources and the substantial alteration of the global biogeochemical cycles pose a serious threat to the functioning of most ecosystems on earth. To preserve the biosphere with its indispensable life support functions is one of the great future tasks of mankind.

References

Allen, T.F.H. & Starr, T.B. (1982) *Hierarchy: Perspective for Ecological Complexity*. University of Chicago Press, Chicago.

Chahine, M.T. (1992) The hydrological cycle and its influence on climate. *Nature* **359**, 373–380.

Dixon, R.K., Brown, S., Houthon, R.A., Solomon, A.M., Trexler, M.C. & Wisniewski, J. (1994) Carbon pools and flux of global forest ecosystems. *Science* **263**, 185–190.

Ellenberg, H., Mayer, R. & Schauermann, J. (1986) *Ökosystemforschung. Ergebnisse des Sollingprojekts 1966–1986*. Ulmer-Verlag, Stuttgart.

Gold, T. (1992) The deep hot biosphere. *Proceedings of the National Academy of Sciences of the United States of America* **89**, 6045–6049.

Harper, J.L. (1982) After description. In: *The Plant Community as a Working Mechanism* (ed. E.I. Newman), pp. 11–26. Blackwell, Oxford.

Hartley, S.E. & Jones, C.G. (1997) Plant chemistry and herbivory or why the world is green. In: *Plant Ecology*. 2nd ed. (ed. M.J. Crawley), pp. 284–324. Blackwell, Oxford.

Houghton, R.A. & Skole, D.L. (1990) Carbon. In: *The Earth as Transformed by Human Action*. (eds B.L. Turner, W.C. Clark, R.W. Kates, J.F. Richards, J.T. Mathews & W.B. Meyer), pp. 393–408. Cambridge University Press, Cambridge.

Houghton, J.T. *et al.* (eds) (2001) Climate Change 2001. *The Scientific Basis. Contribution of Working Group I to the Third Assessment Report of the Intergovernmental Panel on Climate Change*. Cambridge University Press, New York.

Jaffe, D.A. (1992) The nitrogen cycle. In: *Global Biogeochemical Cycles*. (ed. S.S. Butcher, R.J. Charlson, G.H. Orians & G.V. Wolfe), pp. 263–284. Academic Press, London.

Körner, Ch. (1996) The response of complex multispecies systems to elevated CO_2. In: *Global Change and Terrestrial Ecosystems* (eds B. Walker & W. Steffen), pp. 20–42. Cambridge University Press, Cambridge.

Lamotte, M. & Bourlière, F. (1983) Energy flow and nutrient cycling in tropical savannas. In: *Tropical Savannas* (ed. F. Bourlière), pp. 583–603. Ecosystems of The World 13. Elsevier, Amsterdam.

Lenz, R., Haber, W. & Tenhunen, J.D. (2001) A historical perspective on the development of ecosystem and landscape research in Germany. In: *Ecosystem Approaches to Landscape Management in Central Europe* (eds J.D. Tenhunen, R. Lenz & R. Hantschel), pp. 17–35. Ecological Studies 147. Springer-Verlag, Berlin.

Likens, G.E. & Bormann, F.H. (1995) *Biogeochemistry of a Forested Ecosystem*. 2nd ed. Springer-Verlag, New York.

Lovelock, J.E. (1979) *Gaia: A New Look at Life on Earth*. Oxford University Press, Oxford.

Naaem, S. (2002) Ecosystem consequences of biodiversity loss: the evolution of a paradigm. *Ecology* **83**, 1537–1552.

O'Neill, R.V., deAngelis, D.L., Waide, J.B. & Allen, T.F.H. (1986) *A Hierarchical Concept of Ecosystems.* Princeton University Press, Princeton.

Pastor, J., Aber, J.D., McClaugherty, C.A. & Melillo, J.M. (1984) Aboveground production and N and P cycling along a nitrogen mineralization gradient on Blackhawk Island, Wisconsin. *Ecology* **65**, 256–268.

Pomeroy, L.R. & Alberts, J.J. (1988) *Concepts of Ecosystem Ecology.* Ecological Studies 67. Springer-Verlag, New York.

Prather, M. *et al.* (2001) Atmospheric chemistry and greenhouse gases. In: *Climate Change 2001: The Scientific Basis. Contribution of Working Group I to the Third Assessment Report of the Intergovernmental Panel on Climate Change* (eds J.T. Houghton *et al.*), pp. 239–287. Cambridge University Press, New York.

Prentice, I.C. *et al.* (2001) The carbon cycle and the atmospheric carbon dioxide. In: *Climate Change 2001: The Scientific Basis. Contribution of Working Group I to the Third Assessment Report of the Intergovernmental Panel on Climate Change* (eds J.T. Houghton *et al.*), pp. 183–237. Cambridge University Press, New York.

Schlesinger, W.H. (1997) *Biogeochemistry.* 2nd ed. Academic Press, San Diego.

Schulze, E.-D. *et al.* (2002) Interactions between the carbon and nitrogen cycle and the role of biodiversity: a synopsis of a study along a North-South transect through Europe. In: *Carbon and Nitrogen Cycling in European Forest Ecosystems* (ed. E.-D. Schulze), pp. 468–492. Ecological Studies 142. Springer-Verlag, Berlin.

Schultz, J. (2000) *Handbuch der Ökozonen.* Ulmer-Verlag, Stuttgart.

Smith, S.E. & Read, D.J. (1997) *Mycorrhizal Symbiosis.* Academic Press, London.

Tansley, A.G. (1935) The use and abuse of vegetational concepts and terms. *Ecology* **42**, 237–245.

Vitousek, P.M., Mooney, H.A., Lubchenco, J. & Melillo, J.M. (1997) Human domination of earth's ecosystems. *Science* **277**, 494–499.

Walker, B. & Steffen, W. (1996) *Global Change and Terrestrial Ecosystems.* Cambridge University Press, Cambridge.

Waring, R.H. (1989) Ecosystems: fluxes of matter and energy. In: *Ecological Concepts* (ed. J.M. Cherrett), pp. 17–41. Blackwell, Oxford.

Watson, R.T., Zinyowere, M.C. & Moss, R.H. (eds) (1998) *The Regional Impacts of Climate Change: An Assessment of Vulnerability.* Cambridge University Press, Cambridge.

4

Vegetation types and their broad-scale distribution

Elgene O. Box and Kazue Fujiwara

4.1 Introduction: vegetation and plant community

Vegetation, a term of popular origin, refers to the aggregate of all the plants found in an area. Vegetation involves the species (populations) of the local flora, which in turn involve different genetic, migration, historical or ecological elements. These may further include aliens, escapes, very widespread species and endemic species, that are usually of special interest. Floras and their distribution have been interpreted partly in terms of historical factors, such as shifting continental masses, changing sea levels, and orographic and climatic variations over geologic time, as well as theories of island biogeography and the evolution of the species themselves. All of these have affected the development and perpetuation of regional floras.

The vegetation of a region is also shaped by non-floristic physiological and environmental influences, including climate, substrate, soil microbes and disturbance regimes. To the extent that vegetation constitutes an organized whole, it operates at a higher level of integration than the separate species and may possess emergent properties not necessarily found in the species themselves, such as competition and other biotic interactions. As such, vegetation provides not only the physical structure but also the functional framework of ecosystems (see Chapter 3).

The simplest concepts of vegetation types have been based on physiognomy, i.e. the general physical structure and appearance of the vegetation (Beard 1973). Vegetation types, though, are also often recognizable as and loosely equivalent to plant communities. Plant communities, a concept central to plant ecology and to vegetation science, have been described quite graphically as what one would see as different parts of the pattern as one looks out over a landscape (Kent & Coker 1992). Communities differ visibly by their general appearance, related mainly to the different growth forms of the plant species involved. Plant communities are also usually part of some larger ecosystem that involves the different populations of both plant and animal species that occur together at the particular site. Nevertheless, the concept of a community is very simple, making it one of the most general concepts in all of ecology. If a community recurs frequently with a consistent species composition, it may be recognized as an association (see Chapter 1).

4.2 Plant community types

Plant communities are defined largely by species presence (or absence). Further formal description and classification of communities generally focuses on features such as complete floristic composition, floristic structure and relative species abundances. Although others may first have suggested a formal methodology for such description and classification, the best known and most universally accepted methodology is the phytosociology of Braun-Blanquet (1965). This methodology requires complete inventory of all species composing the vegetation (full floristic inventory) and provides a simple, rapidly applicable method for describing the composition and three-dimensional structure of vegetation in the field (Westhoff & van der Maarel 1973; Fujiwara 1987; Dierschke 1994). It represents the *lingua franca* of vegetation scientists all over the world but has indeed developed in somewhat different ways in different 'schools', as summarized by Whittaker (1962).

Phytosociology can be described in terms of three basic steps. The first is the **field inventory**, in which stand descriptions called relevés (see Chapter 1) are made from representative samples (plots) of the vegetation of interest. The result provides a complete three-dimensional description of the vegetation and provides all the presence/absence data necessary for further classification of types.

The second step involves recognition of regularly occurring species combinations, initially called **communities**, in the collection of all relevés done in a region (or sometimes from similar but other regions as well). The data are arranged in a large matrix, in which each relevé represents a column and each species a row. Species combinations are recognized by means of 'table work', i.e. rearranging rows and columns to obtain clusters. This process can now be aided by interactive matrix software or performed completely automatically by computer clustering algorithms (see Chapter 2).

The final step involves interpretation of the resulting 'communities' as vegetation units at different levels in a system. The units are given Latinate names, with endings reflecting their level within the resulting hierarchy or **syntaxonomy**. This step, and to some extent the second step, involve tacit assumptions about the degree to which the species clusters truly represent integrated units of biological organization and not just recurring range overlappings by species with similar environmental requirements (the continuum-community controversy, see Chapter 2). The main alternative is **ordination**, which involves statistical procedures that separate species in (usually) two-dimensional space based on the similarity of stand samples or of species-environment relationships. Ordination is not, strictly speaking, a classification procedure. It is also much more complex mathematically and is constrained by statistical correctness, which may not permit the use of expert knowledge.

Vegetation patterns and types over larger areas may be studied by phytosociology but were first recognized by simpler distinctions, such as between regions of forest, grassland and desert. Such concepts, and indeed the field of plant geography, developed from the great botanical voyages of the early 1800s, especially by Alexander von Humboldt, sometimes called the father of plant geography. Early efforts identified vegetation types partly by similarity of environmental conditions but also developed new concepts of structure and formations (see section 4.3). Plant geography and plant ecology were originally one field but began to diverge around 1900, with plant ecology focusing more on process and plant and vegetation geography focusing more on distributions and their causes. There were and remain two main perspectives: a historical-floristic perspective concerned with migration, dispersal and the historical development of regional floras; and an environmental perspective concerned with environmental constraints and ecological relations influencing distributions.

Finally, and much more recently, concepts of vegetation and plant community types have been strengthened by the results of broad-scale vegetation surveys. Some of these surveys were specifically for the purpose of inventory and description, while some intended from the beginning to produce formal classifications. An outstanding example is the exhaustive 10-year inventory that resulted in the formal phytosociological system and community identification for Japanese vegetation (Miyawaki 1980–9).

4.3 Formations and biomes

The most recognizable large-area vegetation units correspond to major regional ecosystems and are called **vegetation formations**, such as the Amazon rain forest, Great Plains grassland, Mediterranean maquis or Siberian *Larix* forest. These units occur in somewhat similar form in corresponding locations on different continents and so are called vegetation formation types, such as tropical rain forest or boreal coniferous forest. Vegetation formations are regional units, recognized by a relatively uniform physiognomy but composed of plants from a regional flora, which may differ greatly from region to region. The regional ecosystems roughly corresponding to vegetation formations are called **biomes**, understood as vegetation-faunistic units considering also the regional fauna. Biome types correspond roughly to vegetation formation types, so terms such as tropical rain forest may refer to either one.

The first concepts of vegetation formations and formation types arose from the explorations and subsequent world mapping endeavours by 19th-century Europeans, especially Germans (de Laubenfels 1975; Barkman 1988). While historical biogeography proceeded mainly floristically to identify floristic kingdoms, the task of describing and classifying the huge areas with previously unknown vegetation required a simpler approach, based on what the vegetation looked like, i.e. its physiognomy. Where the vegetation had been seen and adequately described, the units recognized were defined physiognomically, but poorly known regions of the world were often represented on maps by environmental surrogates, such as 'tropical forest'.

These early mappers also recognized that natural vegetation regions are most closely related to climatic conditions. Regions were thus differentiated by obvious features such as wet versus dry climates or permanently warm versus cold-winter conditions. The occurrence or absence of frost was recognized early on as perhaps the most important factor separating the lowland tropical and extra-tropical regions, but the upward decrease of temperatures in mountains complicates the zonation. Landforms, soil types and other environmental factors were sometimes recognized, and early maps generally recognized mountain and lowland areas as distinct. Coastal areas, however, were usually not seen as distinct until much later, perhaps since they are difficult to delimit and often too narrow to be shown easily on global or continental-scale maps.

The best concepts of regional vegetation formations were based partly on similarity of environmental conditions but also on developing physiognomic concepts. Vegetation was described perhaps first in terms of its height, density and dominant plant types. Height criteria varied considerably, since 'tall' in boreal regions or in a desert may be quite different from 'tall' in the humid tropics. Density could be described a bit more clearly, using the terms 'closed' if crowns in the highest vegetation layer were touching, as in a forest, and 'open' otherwise. One could, however, view a dense grassland, for example, as closed (dense sod) or as open (no taller plants such as trees).

Physiognomic concepts like forest, shrubland, grassland and desert were relatively straightforward. New categories, however, were needed for mixes of plant types, such as savanna (a grassland with scattered trees or groves, though usually restricted to the tropics). The idea of scrub was formalized later, as woody vegetation involving a mixture of woody forms, with no one dominant. The term 'woodland' was (and still is) commonly used in Britain for forest but was eventually formalized in the mid-20th century by American ecologists to refer to tree-dominated vegetation that is not both tall and closed, thus not a (tall, closed) forest. Generally accepted concepts of vegetation physiognomic classes, with the necessary plant structural types in each, are summarized in Table 4.1. One of the most detailed, careful classifications of vegetation structural types is by Eiten (1968), for Brazil but applicable to much of the globe.

These general classes of vegetation structure were broken down further by reference to leaf type (broad or needle) and seasonal habit (e.g. evergreen or deciduous). Geographic descriptors were often added to complete the names of the main vegetation formation types, such as tropical rain forest, boreal coniferous forest or temperate grasslands. Where mixes occurred within the dominant plant structural type, the vegetation was often referred to simply as 'mixed', most commonly for mixtures of broad-leaved and coniferous trees in montane or subboreal 'mixed forest'. The most complete classification of vegetation structural types is perhaps that of Mueller-Dombois & Ellenberg (1974, App. A).

A major complication to the recognition and classification of vegetation types is provided by the fact that vegetation is not always stable and unchanging. Concepts of vegetation changes over succession and stability of **climax vegetation** were perhaps first formalized by F.E. Clements (see Chapter 7). Since climate was seen as the overriding control, the ultimate end of vegetation succession would be a stable **climatic climax**. Vegetation that became stable before reaching the climatic climax, due to other environmental constraints, was called **edaphic climax** (under unusual soil conditions), for example, or **fire climax** (under conditions of naturally recurring fire); terrestrial wetlands could be seen as special kinds of **topogenic climax**. Another way of relating dynamics (as recognized in successional sequences) and vegetation types was presented by Daubenmire (1968), who classified and mapped 'habitat types' in the north-western USA based on the ultimate potential dominant tree species. A fundamental concept relating to vegetation dynamics, that of **potential natural vegetation**, was recognized and developed by many but finally formalized by Tüxen (1956) as the vegetation type that would arise on an area if all outside influences were removed now (see Chapter 1). This is not necessarily the 'original' vegetation,

Table 4.1 Physiognomic vegetation structures and their main plant types.

Vegetation structure	Main/dominant plant forms	Stature/closure	Other features
Forest	**Trees**	Tall and closed	
Rain forest		Even taller	Evergreen, with closed evergreen understorey(s)
Woodland	**Trees**	Short or tall open	
Parkland	(+ grass)		With regular openings
Scrub	**Woody forms** (mixed)	Open or closed but not tall	Multiple woody forms, no one dominant
Thicket		Very dense	Usually localized
Dwarf-scrub		Very short	Usually extreme climates
Shrubland	**Shrubs**	Open or closed	(Not sparse)
Shrub-steppe		Quite open	Shrubs regularly spaced, as in semi-desert
Savanna	**Grasses + trees**	Grass closed or nearly so	Trees widely scattered, not in groups
Grove-savanna			Trees in scattered groves
Savanna-woodland		Nearly closed	Trees and grass layer almost equal in cover/importance
Shrub-savanna	Grass + shrubs	Grass more open?	Shrubs scattered, no groups
Grassland	**Grasses**	Tall or short, closed or open	Spreading or bunch grasses, essentially no trees, few if any shrubs
Steppe		Short and open	
Meadow	**Graminoids + forbs**	Tall or short, closed or open	Essentially no woody forms
Tundra	Graminoids/forbs with dwarf-shrubs	Very short, closed or open	As in cold climates with short cool summers
Desert	Any/none	Very short, sparse	Little if any vegetation
Semi-desert	Any	Mostly short, open	Sparse vegetation

since the physical environment may have been altered after the former vegetation was removed (e.g. by soil erosion).

Concepts of world vegetation formation types, regions and regional formations developed with their depiction on early world vegetation maps (e.g. by Schouw and Grisebach) and description in early books on world vegetation (e.g. Schimper, Warming and Rübel; see de Laubenfels 1975). The term biome appeared later, especially in the projects of the International Biological Programme (1964–74) and the series 'Ecosystems of the World' (D.W. Goodall, editor). Comprehensive treatments of world vegetation, generally organized by biome, have been presented, in particular, by Walter (1984, 1985), Eyre (1968, with black-white continental vegetation maps), Schmithüsen (1968) and Archibold (1995).

4.4 Plant functional types in vegetation

Attempts to analyse and interpret functional aspects of plants and vegetation have often been limited by the large numbers of individual species that may be involved. As a result, species that appear to be ecologically similar have often been grouped using concepts such as synusiae, growth forms and most recently plant functional types. A **synusia** is a group of species of roughly the same size, sometimes of similar form, occupying the same layer in a vegetation stand, for example the synusia of ground-layer herbs in a forest. The species in a synusia are similar in that they are subject to similar micro-environmental conditions, such as low light levels on the forest floor.

Another concept of functional similarity is that of vicariance, referring to species that appear similar ecologically and are closely related taxonomically but occur in different, usually distant regions. An example might be the very similar evergreen, sclerophyllous 'live oaks' of the warm-temperate south-eastern US coastal plain (*Quercus virginiana*, *Q. geminata*) and of mediterranean southern California (*Q. agrifolia*, *Q. wislizenii*, *Q. chrysolepis*). These, and many other species of an early Tertiary Geoflora, are thought to have been once a single species but to have separated during the latter Tertiary, when a maritime transgression into the central USA, followed by the rise of the Rocky Mountains, effectively separated the eastern and western parts of their range (Axelrod 1983). Numerous other pairs of vicariant species, thought to have developed in parallel fashion since separation, have been hypothesized, for example, in the floras of eastern Asia and eastern North America (e.g. Graham 1972) and of eastern North America and nemoral Europe (e.g. *Cercis canadensis* versus *C. siliquastrum*, *Ostrya virginiana* versus *O. carpinifolia*).

The most general concepts of ecologically similar plant types are perhaps the ideas of plant growth forms and life forms (see Barkman 1988 and Chapter 3 in Shimwell 1971). Plant growth forms are basic types defined entirely by their structure, such as broad-leaved trees, stem succulents or graminoids. The first classification of such forms is credited by Barkman to Theophrastos of Eresos (371–286 BC), and this convenient concept was already in wide use in the 1800s, as the *Hauptformen* of A. von Humboldt, the 54 types of Grisebach, and the 55 forms of O. Drude, which he eventually called *Wuchsformen* (growth forms) (see Barkman 1988 for references). When structure is interpreted as an ecologically significant adaptation to environmental conditions, the forms involved have been called life forms (e.g. by Warming) and can be interpreted as basic ecological types, since they combine concepts of form, function and their relationship. These types constitute groups of plant taxa with similar form and ecological requirements, resulting from similar morphological responses to similar environmental conditions. Both growth forms and life forms can also be seen as representing the basic building blocks of vegetation and provide a convenient way of describing vegetation structure without having to treat large numbers of species individually.

Concepts of basic plant types also facilitate study of the relationships between form and function. For example, deciduous leaves are generally 'softer' and photosynthesize more efficiently in favourable environments. On the other hand, they may lose more water and require more energy and nutrients for their construction (over the plant's life span) than do 'harder', longer-living evergreen leaves. Also, larger plants with larger total leaf area may be vulnerable to greater water loss but may have more extensive root systems for more effective water uptake. The vegetation of a particular site will be composed of plants with particular combinations of such form characters that permit the plant to function successfully in a particular environment. When similar morphological or physiognomic responses occur in unrelated taxa in similar but widely separated environments, they may be called convergent characteristics (for example, the occurrence of sclerophyll shrubs in the world's five mediterranean-climate regions).

An important early contribution to the idea of ecological plant types was Rübel's (1930) description of world vegetation formation types as the result of the climatic limits of their main structural elements, i.e. growth forms. A more generalized concept of plant life forms was offered by Raunkiær (1934), who defined a system of types based on the location of the renewal buds that must survive the cold winter. The Raunkiær system was expanded by various authors and eventually extended to the tropics. A more purely functional distinction arose in ecology, with notation derived from the sigmoid equation for population growth (with intrinsic growth rate *r*) in a limited environment (of carrying capacity *K*). This concept distinguished '*r*-selected' plants adapted for rapid colonization of disturbed environments (i.e. ruderals such as annual weeds) from '*K*-selected' plants adapted for long-term dominance of vegetation stands (such as oak trees). This idea was expanded and made much more applicable and useful by Grime (1979), who described the '**strategies**' of three basic plant types called ruderals, stress tolerators and competitors (see Chapter 1).

The first truly comprehensive, global classification of plant growth forms (or life forms, depending on interpretation) was that of Mueller-Dombois & Ellenberg (1974, App. B). In order to have a system of growth forms defined purely by structure, for use in global modeling, a system of 90 forms was developed by defining each form in terms of a minimal set of six structural characters: structural type (e.g. tree), leaf form and consistency (e.g. broad-leaved malacophyll), relative plant and leaf size, and seasonal activity pattern (e.g. summergreen) (Box 1981). The occurrence, and proximity to closest climatic limit, of these structurally defined types were then predicted worldwide by constructing climatic 'envelopes' for each growth form (model TVS1). The results suggest not only presence/absence but also relative importance, even dominance, within vegetation stands (cf. Box 1987). The list of plant types was eventually expanded to that shown in Table 4.2, based on field experience in more parts of the world.

The set of particular variables used for the climatic envelopes was also improved, based on better identification of the mechanisms of limitation (see Table 4.3) and on availability of more climatic data to represent these mechanisms. The effect of short-term, extreme minimum temperatures in limiting some plant types was perhaps first expressed quantitatively by Larcher (1973) and was tested geographically by Woodward (1987). One can appreciate this by observing that most of the south-eastern USA has deciduous forest, despite much higher mean winter temperatures – but lower extremes – than corresponding latitudes in East Asia, which has evergreen broad-leaved forest (Box 1988). Inclusion of absolute minimum temperature in climatic envelope models greatly improved their predictive ability in the mid-latitudes as well as tropical and subtropical mountains, both for plant types and for types of potential natural vegetation (Box 1995a,b).

Table 4.2 Main terrestrial plant life forms. This classification represents an expanded version of the original model of 90 plant 'ecophysiognomic' types and their worldwide climatic envelopes (Box 1981). Physiognomic groupings are at the left; numbers of types and short descriptors of the 114 actual types are shown to the right.

Trees (broad-leaved)			
Evergreen	Laurophyll	4	Tropical, tropical-montane, subtropical/warm-temperate, cool-maritime
	Gymnosperm	1	(Sub)tropical broad/linear-leaved
	Microphyll	1	(Sub)tropical coriaceous
	Sclerophyll	4	(Sub)tropical, warm-temperate, mediterranean, tall-temperate
Deciduous	Rain-green	3	Monsoon-mesomorphic, montane, xero-microphyll
	Bottle tree	1	Tropical (rain-green)
	Summergreen	2	Nemoral, short-summer
Needle-leaved	(Evergreen)	6	Tropical-xeric, heliophilic long-needled, submediterranean, temperate, cool-laurophyll, boreal/montane
	(Summergreen)	2	Hydric, boreal
Small trees			
Evergreen	Laurophyll	4	Tropical, cloud-forest, subtropical/warm-temperate, cool-maritime
	Sclerophyll	1	Tropical macro-sclerophyll
	Deciduous	2	Rain-green, summergreen
	Needle-leaved	1	Dwarf-needle/scale-leaved
Tuft-trees and treelets		3	Palm trees, bottle palms, palm treelets
Arborescents			
Evergreen	Laurophyll	2	Tropical, extra-tropical
	Sclerophyll	1	Tropical/subtropical
Deciduous		2	Rain-green, summergreen
Stem-green		2	Mesic-microphyll, xeric/leafless
Tuft-arborescents		4	Tree-fern, tropical-alpine, coriaceous, sclerophyll/succulent
Krummholz (needle-leaved)		2	Evergreen, summergreen
Shrubs			
Evergreen	Laurophyll	2	Tropical, subtropical/warm-temperate
	Sclerophyll	2	Mediterranean, hot-desert
	Succulent	1	
	Needle-leaved	1	
Semi-evergreen xeric		1	Temperate
Summergreen		2	Nemoral-mesomorphic, xeromorphic
Dwarf-shrubs			
Evergreen		4	Mediterranean, temperate, maritime-heath, boreal/tundra
Summergreen		1	Boreal/tundra
Xeromorphic		1	Leptophyll/leafless
Cushion-form		2	Mesic-evergreen, xeromorphic
Rosette-shrubs		2	Trunkless palms, xeric leaf-succulent
Stem-succulents		5	Columnar, branched-arborescent, typical, frutescent, cryptic

Table 4.2 (*cont'd*)

Semi-shrubs		3	Mesomorphic, xeromorphic, xylopodial-xeromorphic
Graminoids			
Broad-leaved (bamboos)		2	Arborescent, dwarf
Tall		3	Cane graminoids, typical-tall, tall-tussock
Short		5	Spreading, bunch, short-tussock, sclerophyll, desert-grass
Forbs			
Evergreen		3	Tropical-arborescent, tropical, temperate
Deciduous	Rain-green	2	Typical, cold-desert
	Summergreen	3	Tall, typical, spring-ephemeral
Xeromorphic		2	Succulent, cushion-form
Ephemeral		2	Desert-ephemeral, cold-desert
Ferns		3	Evergreen, rain-green, summergreen
Vines	Lianas	1	Tropical-evergreen
	Vines	3	Evergreen, rain-green, summergreen
	Epiphytic	1	
Epiphytes	Rosette	2	Large-tropical, stenophyll
	Stem-succulent	1	
	Shrublet	1	Wintergreen-shrublet
	Herbs	3	Tropical-forb, fern, small-fern
Cryptogams		2	Mat-forming, xeromorphic
	Total	114	

The term plant function type (PFT), as most commonly used today (e.g. Smith *et al.* 1997), is largely an outgrowth of the International Geosphere-Biosphere Programme and its emphasis on biosphere response to global climate change. PFTs have been conceived as purely functional plant types, focusing in particular on how their physiologies may respond to such changes as increasing atmospheric CO_2 levels and higher temperatures. It may not be generally possible to separate plant form and function, but there are some aspects of function that do not have structural manifestations (Box 1996). So far, no general, world-applicable set of purely 'functional' (i.e. physiological) plant types, without reference to plant form, has been presented.

4.5 Major ecosystem types of the world

The most general concept of major ecosystem types of the world is the idea of biome types, such as the tropical rain forest, temperate deciduous forest or polar tundra. A list of the main world terrestrial biomes is given in Table 4.4. That these regions occur in a somewhat regular geographic pattern across the world was apparent from even the earliest maps and verbal treatments of world vegetation. Rübel's (1930) *Pflanzengesellschaften der Erde* was probably the first attempt to quantify the climatic ranges of these regions, thus demonstrating the unity of the global system. This unity, and the rough north–south symmetry but also asymmetry in the world system

Table 4.3 Biological significance of upper and lower limits for variables used in climatic envelope models for plant and vegetation types.

Variable	Upper limit	Lower limit
T_y	Maintenance of GPP > R, over whole year	Permafrost tolerance if < 0 °C
BT	Maintenance of GPP > R, during growing season	Cumulative warmth needed for growth functions
T_{max}	Maintenance of GPP > R, especially in summer	Warmth threshold for growth functions
T_{min}	Vernalization, other dormancy requirements	Tolerance to cumulative low temperatures
T_{mmin}	Vernalization, other dormancy requirements	Tolerance to consistently low temperatures
T_{abmin}	Seasonality requirements, re-set 'biological clock'	Physiological tolerance to short-term extreme cold
DT_y	Physiological plasticity of stenothermal metabolisms	Seasonality requirements, functional synchronization
P_y	Soil/leaf aeration needs	Minimal water need for metabolism and growth
MI_y	Soil/leaf aeration; dryness needs by xeromorphics	Minimal water need to cover water loss by particular leaf area/vegetative structure
P_{max}	Rotting tolerance by succulents, etc.	Minimal water need for adequate growing season
P_{min}	Dormancy for rain-greens; tolerance to permanently saturated soil	Drought tolerance, especially by mesomorphic evergreens (e.g. laurophylls)
P_{mtmax}	Dryness needs by xeromorphic plants	Drought tolerance during warm season

Definitions of variables:

T_y = mean annual temperature
BT = biotemperature (Holdridge 1947): sum of mean monthly temperatures above 0 °C, divided by 12
T_{max} = mean temperature of the warmest month
T_{min} = mean temperature of the coldest month
T_{mmin} = mean minimum temperature of the coldest month
T_{abmin} = absolute minimum temperature (coldest ever measured)
DT_y = annual range of mean monthly temperature (= $T_{max} - T_{min}$)
P_y = average annual precipitation
MI_y = annual moisture index: P_y/annual potential evapotranspiration
P_{max} = highest monthly average precipitation amount
P_{min} = lowest monthly average precipitation amount
P_{mtmax} = average precipitation of the warmest month

Other abbreviations:

GPP = gross primary production
R = respiration

Table 4.4 Main world terrestrial biome types. These biome types are recognized by most modern treatments of world vegetation, including classifications, e.g. Eyre (1968), Schmithüsen (1968), Mueller-Dombois & Ellenberg (1974), Walter (1984, 1985) and Archibold (1995), also in conservation efforts, e.g. Udvardy (1975) and Bailey (1989), and in modeling and mapping efforts, e.g. Tateishi & Kajiwara (1991), Prentice *et al.* (1992) and Box (1995b). Some treatments consider alpine and other high-mountain vegetation as part of the larger orobiome (*sensu* Walter & Box 1976).

Biome types
Tropical rain forest (including montane and cloud forests)
Tropical deciduous forest, woodland and thorn-scrub
Tropical savanna
Deserts
Warm desert (subtropical)
Cold-winter desert (continental)
Mediterranean forest, scrub and shrublands
Temperate forests
Deciduous broad-leaved forest
Evergreen broad-leaved forest (incl. laurel forest, warm-temperate mixed forest)
Temperate rain forest
Grasslands (temperate)
Conifer forests
Boreal (including deciduous)
Montane conifer forest (temperate montane and subalpine)
Tundra and alpine vegetation
Polar and temperate-alpine tundra
Tropical alpine vegetation
Terrestrial wetlands (swamp, marsh, bog, fen)

of biomes, was further demonstrated, quite graphically, by the 'Durchschnittskontinent' (average continent) of Troll (1948).

The geographic regularity of the system arises, of course, from the global circulation pattern of the Earth's atmosphere, as suggested in Fig. 4.1. This circulation system consists of a zone of low pressure and frequent precipitation near the equator (the intertropical convergence zone, or ITC), a zone of high pressure and dry conditions in the subtropics of each hemisphere, and prevailing winds in each hemisphere that blow from the high-pressure belts toward the ITC (trade winds) and toward the poles (quickly deflected by coriolis into west–east flows called the westerlies). The fact that these pressure and wind belts migrate north–south with the seasons, the ITC trailing roughly one month behind the latitude of solar zenith (solar declination), insures that some latitudes will experience seasonally alternating effects much more than others. The resulting east–west latitudinal bands of somewhat similar conditions are called zones and give rise to the idea of geographic and bioclimatic zonality. This global geographic zonation represents the fundamental geographic framework for representing the locations of biomes and many other earth features.

The most widely used global system of climate types based on this global zonation is probably that of Walter (1984, 1985). The widely used Köppen system, used in most atlases, separates climates by quantitative indices and thus lends itself readily to mapping. The Walter system, on the other hand, is not quantitative but focuses rather on the atmospheric mechanisms that generate distinct climate types and regions,

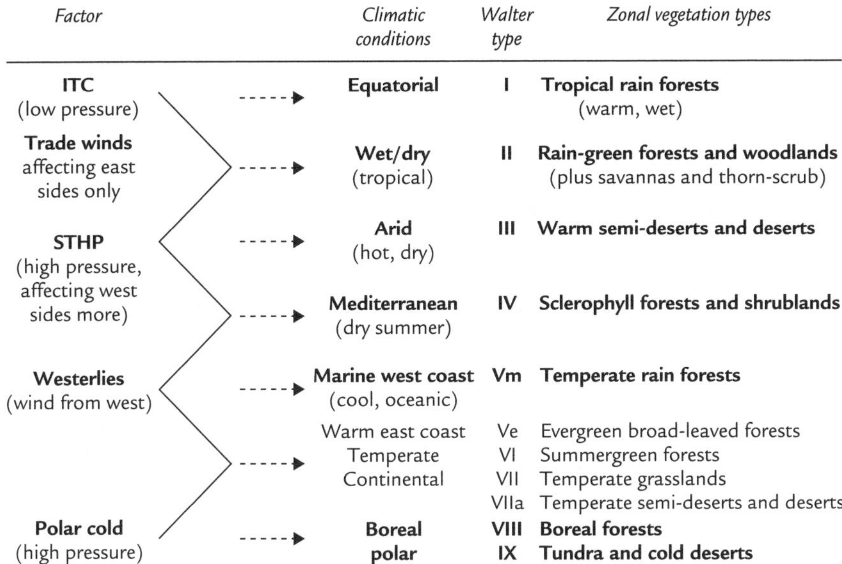

Fig. 4.1 Global climatic zonation resulting from global atmospheric circulation and its seasonal shift. The seasonal north–south shift of the global atmospheric circulation system permits the intertropical convergence zone (ITC), subtropical high-pressure belts (STHP) and other components to affect some latitudes throughout the year and others only seasonally (diagonal lines). Results include, for example, the tropical wet/dry climate, with ITC wet season and STHP dry season. Most major climate types (bold) are direct results of zonal energy inputs and global circulation, but some types in mid-latitudes (non-bold) are determined also largely by their geographic position on the land mass (i.e. windward, interior or leeward).

thus tacitly focusing on climatic core areas rather than boundaries. The relative locations of the Walter climates on an 'ideal continent' are shown in Fig. 4.2, with some simple sub-types and with the type V (warm-temperate) climates separated on the east and west sides of continents, where they occur for very different reasons (the only mechanistic inconsistency in Walter's system).

Due to its good representation of fundamentally different climatic situations and potential mechanisms of limitation to particular plant types, the Walter climatic regions also correspond well to the world's major biomes and their locations. For each Walter climate, one can readily identify up to three major 'zonal' biomes that occur, as natural landscapes, in that climatic region and essentially nowhere else (cf. Fig. 4.1). For example, the zonal (lowland) vegetation of the equatorial climate (climate I) is tropical rain forest, occurring where there is no extended dry period, temperatures never fall below about 10 °C, and plant growth can continue year-round essentially without even any seasonal cues to synchronize plant functions, such as blooming or leaf fall. On the other hand, the tropical summer-rain climate (II), with distinct wet and dry seasons, includes three zonal vegetation types that occupy different portions of this climate's total range of annual precipitation: tropical moist deciduous forest at the wet end (wet season longer than dry season), dry deciduous forest and woodland

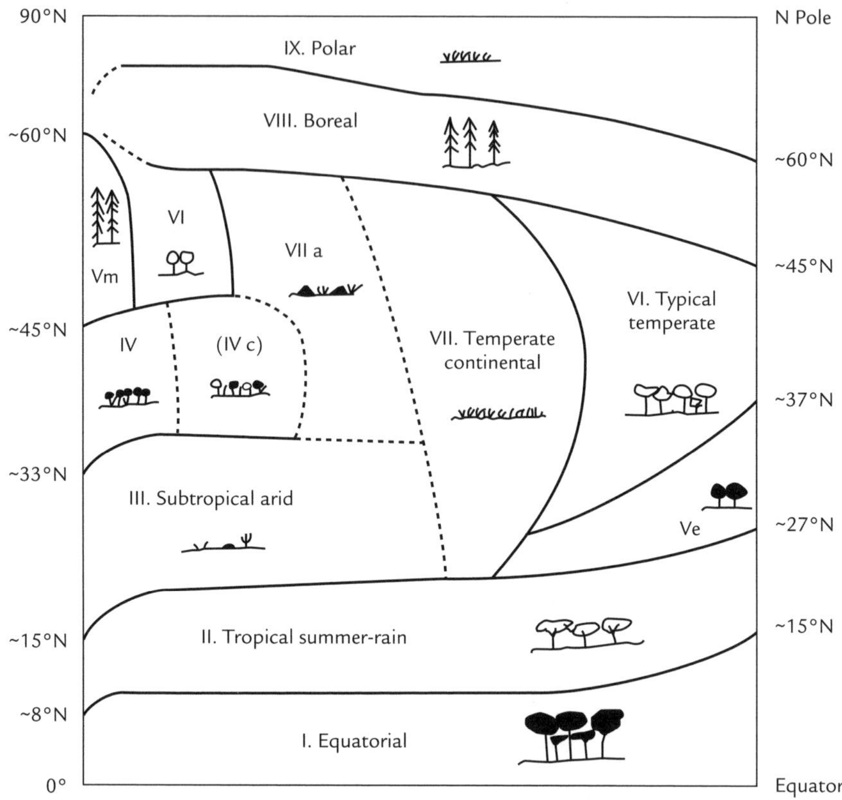

The climates are the genetic climate types of Walter (e.g. 1984, 1985), as modified slightly for mechanistic consistency and expanded with subtypes:

I = Equatorial	II = Tropical summer-rain
III = Subtropical arid	IV = Mediterranean
Vm = Marine West-coast	Ve = Warm-temperate (east side)
VI = Typical temperate (nemoral)	VII = Temperate continental
VIII = Boreal	IX = Polar (including ice caps)

The original climate type V of Walter is split into marine west-coast (Vm) and warm-temperate east-coast (Ve) types, due to the quite different atmospheric mechanisms which produce them.

Possible subtypes include the following:

a = arid:	Ia	= Dry equatorial (e.g. East Africa)
	VIIa	= Temperate arid (interior Eurasia, western North America)
c = continental:	IVc	= Continental mediterranean (e.g. interior Middle East)
	Vc	= Dry warm-temperate (e.g. central Texas, southern Australia)
	VIc	= Dry-winter temperate (northern and north-eastern China)
	VIIIc	= Ultra-continental boreal (eastern Siberia)
m = maritime:	IIm	= Windward tropical wet/dry (e.g. Kerala, Bangladesh)
	IIIm	= Coastal fog deserts (e.g. Namib, Atacama)
	VIm	= Maritime nemoral (e.g. British Isles)
	VIIm	= Maritime dry-temperate (e.g. Patagonian & New Zealand grasslands)
	VIIIm	= Maritime boreal (e.g. Iceland, Kamchatka)
	IXm	= Maritime polar (e.g. sub-Antarctic islands)
x = dry-maritime:	VIIx	= Maritime arid-temperate (Patagonian semi-desert)

In the Southern Hemisphere the boreal climate does not occur at all, and other climates may occur slightly closer to the equator.

Fig. 4.2 Climatic regions on an ideal continent.

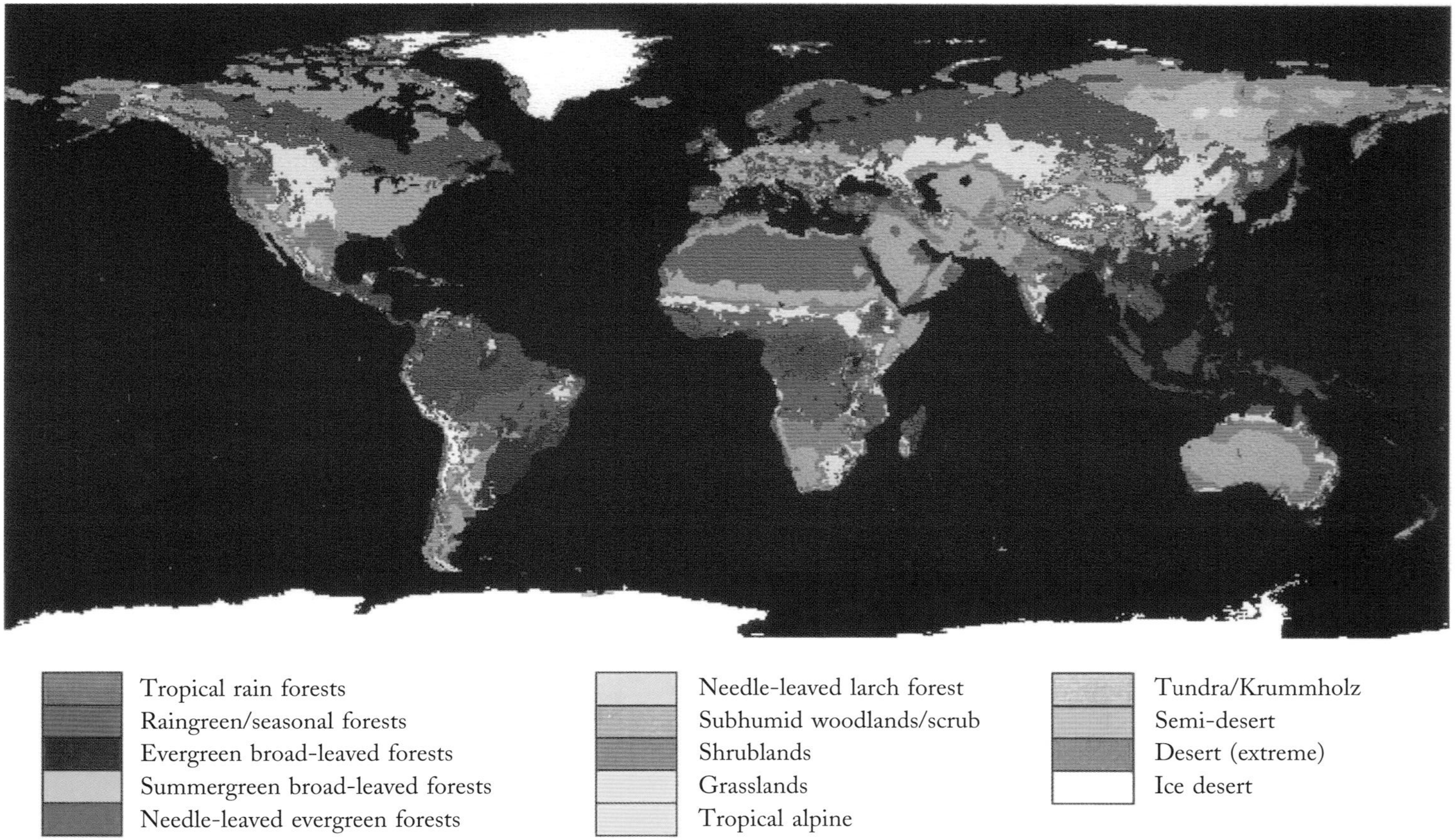

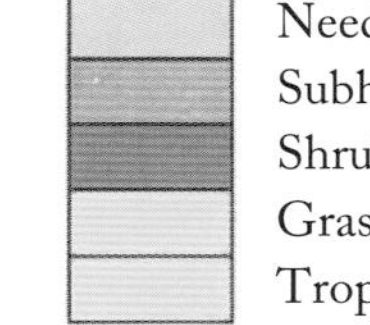
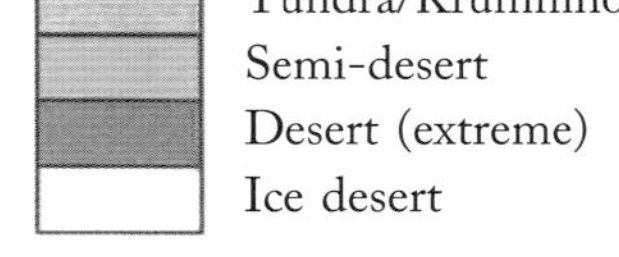

Plate 4.1 World pheno-physiognomic vegetation pattern predicted from climate.

in the middle (e.g. miombo woodland in south-central Africa), and rain-green thorn-scrub or savanna at the dry end (depending largely on topography or other substrate conditions). This logic holds also for the vertical vegetation belts in mountains and is formalized in Table 4.5, which shows the main terrestrial biomes, the corresponding

Table 4.5 Main zonal biomes of the World, their zonal climate type, associated climatic and other vegetation variants, and vertical vegetation zonation in mountains. Climate types are those of Walter, as expanded by Box (see Fig. 4.2). Montane zonation is listed from the alpine belt downward.

Biome	Climate	Climatic variants	Other, azonal vegetation	Montane zonation
Tropical rain forest	I	Semi-evergreen forest, dry evergreen forest/wood, dry scrub (Ia)	Derived woodland, scrub, savanna	Páramo (alpine) Cloud forest Montane rain forest
Tropical deciduous forest and woodland	II	Sclerophyll woodland, thorn-scrub	Derived scrub and savanna, seasonal wetland	Moist puna grassland Evergreen montane forest Semi-evergreen forest
Tropical savanna	II–III		(Thorn-scrub)	
Warm desert	III	Fog desert (IIIm) Xeric shrub-steppe	Oases, wadis, playas	Dry puna grassland Montane scrub
Mediterranean forest, woodland, shrubland	IV	Deciduous forest/ woodland/scrub	Degraded scrub (stable), sclerophyll savanna-woodland	Dry cushion-scrub Montane coniferous forest
Temperate rain forest (evergreen broad-leaved)	Vm	Coniferous rain forest Giant forest (IV–Vm)	Deciduous forest, heath, wetland	Wet alpine 'tundra' Wet montane forest
Evergreen broad-leaved ('laurel') forest	Ve	Evergreen mixed forest Sclerophyll woodland	Pine forest/ woodland, decidous forest, wetlands	Alpine 'tundra' Subalpine/mixed forest
Temperate deciduous (summergreen) forest	VI	Cool-summer deciduous forest	Pine forest, grassland, temperate wetland	Alpine 'tundra' Subalpine/mixed forest
Temperate grassland (prairie, steppe)	VII	Oceanic tussock grassland	Grove belts, riparian forest, degraded steppe, wetland	Dry alpine 'tundra' Dry coniferous forest
Temperate desert	VIIa	Oceanic cushion-steppe	Riparian wood, salt/ rock vegetation	(similar)
Boreal forest (evergreen coniferous)	VIII	Larch forest (VIIIc) Deciduous broad-leaved forest (VIIIm)	Bog forest, open conifer woods, forest-tundra, bog	Alpine tundra
Polar tundra	IX	Maritime 'tundra' Moss-lichen cold-desert	Riparian/other wetland	–

Walter climate types, the main climatic and non-climatic (e.g. edaphic) variants, and the natural vertical zonation in mountains within the zone.

For conservation purposes, a need has been identified for more local, concrete regional ecosystems, generally at sub-biome scale. The first global system of such regions was probably that of Udvardy (1975), developed for the International Union for Conservation of Nature and based on combinations of bioclimatic regions, surface physiography, vegetation associations and local plant and animal ranges. More detail was added in second-generation regionalizations that recognize what have come to be called **ecoregions** or **bioregions** (e.g. Omernik 1987; Bailey 1989). Fairly detailed world maps have been produced of up to 200 such bioregions and are used now as a basis for conservation planning (Olson & Dinerstein 1998).

Comprehensive vegetation classification systems have also been developed, at national and supra-national levels, to serve as the basis for conservation-oriented inventories and mapping programmes. One ambitious such programme is the European Vegetation Survey, which attempts to integrate traditional national approaches into a single classification system for all of Europe (see Chapter 1). Another example is the recent US National Vegetation Classification, which was designed for a large, diverse region and integrates concepts from various earlier approaches to vegetation classification, including physiognomy and phytosociology. In both cases, classification is based at the highest level(s) on pheno-physiognomy but then on concrete plant associations at more local levels in the hierarchy. This permits the larger units in the systems to be identified while more data are being gathered to fill out the smaller, more local units in more floristic detail. Although this appears to work, one wonders whether it would also be applicable in the tropics, where biodiversity is so much greater, phenology (thus some aspects of physiognomy) can be more variable and end-points of landscape development are less clear.

4.6 Vegetation mapping

The first task in vegetation mapping is to decide what to map, the actual vegetation (or other land cover), the 'natural' vegetation, or some other aspect such as successional stage, naturalness, or functional attributes. In the 1800s, what European explorers saw outside of highly disturbed Europe could often be interpreted as 'natural' vegetation, and this is what was depicted on the first global or other large-area vegetation maps. Inside Europe, on the other hand, mapping often focused on actual vegetation (cf. Mueller-Dombois & Ellenberg 1974), including dynamic aspects (e.g. Faliński 1991). Potential natural vegetation (see section 4.3) integrates many environmental factors and can also be a useful mapping objective. It provides a graphic basis for land-use and other environmental planning, and serves as an ever more important 'benchmark' (Box 1995b) as the world's vegetation cover is destroyed more and more.

Mapping of local vegetation might address actual or potential vegetation but generally is based on field data obtained from vegetation surveys, often by the Braun-Blanquet methodology, which is well designed for rapid, extensive, but still full floristic inventory. There is no better example of this than the application of European phytosociology to the problem of vegetation inventory and mapping in Japan by Miyawaki (1980–89). Data on vegetation composition were gathered from remnants of natural or nearly natural vegetation, especially from traditionally protected forest remnants around shrines and temples, and used to infer the general pattern of the potential natural vegetation. These data were also the basis for mapping current actual vegetation, at finer scales, using known relationships between vegetation and topography (as shown on maps). Using far fewer data but some similar concepts, maps of the potential vegetation of large regions have also been produced, for example for the conterminous United States (Küchler 1964), South America (Hueck & Seibert 1972), Africa (White 1983) and tropical Asia (Blasco *et al.* 1996).

For large regions, including the whole globe, resolution of different approaches and development of appropriate vegetation classifications becomes an indispensable first step. The project to produce a European Vegetation Map has been able to use an association basis (within a physiognomic framework) over a large area. This partly floristic approach probably could not be applied over the whole globe, however, with its totally different floristic compositions in the northern versus southern hemispheres, despite some convergence in vegetation adaptations to analogous situations (cf. Troll 1948; Box 2002).

An alternative basis for large-area mapping of potential vegetation is pheno-physiognomy, i.e. definition of vegetation types entirely by their general structure (physiognomy) and seasonal activity pattern (e.g. evergreen versus deciduous; cf. *pheno-morphology*, Orshan 1989). Though not floristic, this approach has the advantages that: (i) the types can be identified from field data; (ii) their geographic occurrence, at least as the potential dominant vegetation (PDV) of the area, can be predicted with considerable accuracy from climate data (Box 1995b; cf. Prentice *et al.* 1992); and (iii) the types can also be recognized remotely, spectrally by satellites and perhaps eventually also by LIDAR (Lefsky *et al.* 2002). A world classification of pheno-physiognomic PDV types, designed to represent world vegetation with as few types as possible, is shown in Table 4.6. This set was derived by a sort of 'geographic regression' that defined types, predicted their occurrence at about 1600 climatic sites worldwide (using a climatic envelope model), and then added or modified types until all sites were adequately described. The resulting 50 PDV types can be grouped into 15 more general pheno-physiognomic groupings, which may be even more readily distinguishable by spectral data. These 15 types represent world vegetation significantly better than do the usually about 10 'types' of the purely spectrally inspired 2-by-2 model (evergreen/deciduous × broadleaf/needleleaf) currently used in most global models (e.g. Myneni *et al.* 1995).

Mapping of any concept of potential vegetation or ecosystems must be based mainly on climate, since climate is the overriding control on the distributions of vegetation and plant types. Problems occur on unusually young or nutrient-poor substrates and in marginal environments where disturbance and stochastic processes may determine which of several possible vegetation types becomes stable. Even in one smaller region with unusual substrates (Florida), however, it was shown that climate could predict the ranges of the main woody species (Box *et al.* 1993).

The first complete global system of ecological types predictable from climate was the system of so-called life zones of Holdridge (1947), which were not strictly defined

Table 4.6 World terrestrial pheno-physiognomic vegetation types. These types were derived by a sort of global geographic regression (see main text; Box 1995b). The individual types on the right can be distinguished by climatic envelopes, and types in both columns can be distinguished spectrally by satellite imagery, due to their pheno-physiognomic definitions.

Grouping	Individual types
1. Tropical rain forests	Tropical rain forest (lowland) Tropical montane rain forest Tropical subalpine (cloud) forest Subtropical rain forest
2. Tropical seasonal woodlands/ savannas	Tropical semi-evergreen forest Rain-green forest Rain-green scrub (incl. montane) Tropical dry evergreen forest
3. Evergreen broad-leaved forests	Evergreen broad-leaved forest Mediterranean evergreen forest Cool-temperate evergreen broad-leaved forest Subpolar evergreen broad-leaved forest
4. Temperate rain forest	(evergreen broad-leaved/mixed/needle-leaved)
5. Summergreen broad-leaved forests and woods	Summergreen broad-leaved forest Summergreen broad-leaved woodland Subpolar summergreen broad-leaved forest
6. Needle-leaved evergreen forests/woods	Dry conifer forest Mediterranean conifer forest Boreal conifer forest (evergreen, incl. dry) Subpolar/subalpine conifer woodland
7. Summergreen needle-leaved (*Larix*) forests/woods	
8. Subhumid woodlands/shrublands	Semi-evergreen dry woodland/scrub Mediterranean woodland/scrub
9. Shrubland/Krummholz (seasonal/evergreen)	Cool-evergreen broad-leaved scrub/Krummholz Subhumid shrubland/low scrub
10. Grasslands	Tropical savanna Temperate grassland Cool-maritime grassland
11. Semi-desert scrub (arid)	
12. Tropical alpine vegetation	Páramo (scrub) Puna (grassland)
13. Tundra and related Krummholz/ cold-desert	Cool-summer-green broad-leaved scrub/Krummholz Subalpine conifer Krummholz Polar/alpine tundra Maritime tundra Cold-desert (moss-lichen semi-desert)
14. Deserts (little/no vegetation: arid desert, polar/subnival cold-desert, ice desert)	

Table 4.7 Bioclimatic factors and resulting vegetation structure. P = annual precipitation; PET = annual potential evapotranspiration; T_{max} = mean temperature of warmest month; T_{abmin} = absolute minimum temperature; T_{mmin} = mean minimum temperature of coldest month. General vegetation structure, seasonality pattern and leaf morphology, at least for zonal vegetation, appear to be related to general climatic indices such as the overall annual climatic water balance (P/PET), summer-mean and winter minimum temperatures, and length of the dry season or growing season. Vegetation types that do not fit these relationships suggest that other factors are controlling, such as edaphic conditions, topography, fire or persistent disturbance.

Water balance + summer temperature	→	**Structural class**
$P/PET \geq \sim 0.95$*	→	forest
0.8–0.95*	→	woodland
0.5–0.8	→	scrub, shrubland, savanna, grassland**
0.2–0.5	→	open shrubland, low savanna, steppe**
$< \sim 0.2$	→	semi-desert
$< \sim 0.05$	→	desert (no vegetation)
$T_{max} < \sim 9$–11 °C	→	tundra
$< \sim 2$ °C	→	cold desert
Growing-season length + winter temperature	→	**Seasonality**
Dry-season > 2–3 months (more deciduous if longer)	→	semi-evergreen/rain-green
$T_{mmin} < \sim 0$ °C	→	summergreen
$T_{abmin} < \sim -15$ °C	→	summergreen
Exception: summers too short for new foliage each year (e.g. boreal)	→	evergreen
Climatic stress (e.g. water, cold)	→	**Leaf morphology**
Water stress, during the growing season (e.g. mediterranean) or winter (e.g. boreal, austral)	→	harder leaves (or needles), often smaller
Cold summers (boreal, polar, etc.)	→	smaller leaves

* P/PET limits lower in tropics (warm dry season)
** Particular structure also strongly influenced by soil or topography

phenologically (evergreen versus deciduous) but were mapped and used many times for land planning in tropical and subtropical countries. By the late 1970s structurally defined vegetation types, at roughly a biome level, could be predictively mapped, using climatic envelopes, and used as a basis for mapping functional aspects, such as photosynthetic efficiency (Box 1979). With the advent of colour printers and better computers, more attractive colour maps of predicted potential vegetation quickly became common (e.g. Prentice *et al.* 1992). A predictive mapping of the potential dominant vegetation types of Table 4.6, using climatic envelopes, is shown in plate 4.1. This and earlier climate-based modeling permitted a general hypothetical quantification of how climate may determine vegetation structure, at least for zonal situations (Table 4.7).

Finally, mapping actual vegetation or other actual land covers, unlike natural or potential, requires satellite data in order to sense what is actually on the ground in increasingly disturbed landscapes over most of the world. The methodology generally involves clustering algorithms to group pixels with similar spectral signatures into 'types', which must then be interpreted as some kind of land cover. In this, it is the managed land covers, such as agricultural fields and field-woodland mosaics, that are usually the most difficult to identify correctly.

The actual land cover of local to sub-national and sometimes national areas have been mapped, mainly using Thematic Mapper (TM) data with pixel resolutions from a few kilometres down to a few tens of metres. An example is the TM-based map of the actual land cover of the USA by Loveland *et al.* (1991). All of North America was eventually mapped, using the newer MODIS data, by Strahler *et al.* (no date: analysis 1999). This latter map illustrates a common problem in high latitudes, namely the low sun angle in winter and consequent difficulty in recognizing the boreal forest. One global map of 'vegetation' (not from a biologist) appeared in Japanese newspapers and some scientific outlets showing no boreal forest at all (!) and only grassland above about 50°N. Such is the danger posed by the enormous amounts of satellite imagery now available to people without the training or geographic experience to interpret it reasonably.

Even when relatively accurate maps of actual land cover can be constructed for large regions, the problem remains to construct an accurate map of the entire globe, because of the limitations of the signature-clustering approach. There is, for example, nothing analogous to the boreal coniferous forest in the southern hemisphere, even if some southern pixels in some imagery are spectrally very similar to some boreal-forest pixels. Perhaps the best global map of actual vegetation/land cover, despite readily identifiable errors, remains that of Tateishi & Kajiwara (1991), precisely because they did not try to do more than the data and clustering methodology would permit. This is where ancillary climatic data could be used (but rarely have been) as a control on interpretation of the satellite imagery. An alternative to the signature-clustering approach is the idea of seasonal 'metrics', by which the shape of the monthly curve of (NDVI) greenness intensity would be analysed quantitatively for indicators of the kinds of factors that are known to control vegetation, such as growing-season length (e.g. DeFries *et al.* 1995).

4.7 Coda

Many global-scale ecological processes appear to follow consistent, even familiar geographic patterns (Box & Meentemeyer 1991). Solar energy inputs and many aspects of temperature, including evaporative power, follow a **thermal** pattern, with highs in the low latitudes and lows near the poles. Annual precipitation inputs and some measures of effective moisture (e.g. versus evaporative power) follow a **moisture** pattern with highs in equatorial regions and on windward temperate-zone coasts (lows in subtropical and mid-latitude continental deserts). Productivity, decomposition, and many other processes of biophysical or ecological 'work' require simultaneous availability of energy and moisture, and thus follow a **throughput** pattern, epitomized by actual evapotranspiration. Some biotic storages follow a less distinct **accumulation** pattern, with highs in cool, humid regions including windward mid-latitude west coasts (total standing biomass in temperate rain forests) and humid boreal climates (total carbon storage in soils, including wetlands). The processes and patterns themselves transcend biome or other vegetation types, but extremes do tend to be located in particular biomes. Most biomes can be characterized by high, medium or low levels for each activity pattern.

Vegetation also performs important functions, at various scales, within ecosystems as well as for human benefit (see Chapter 1). Control, recycling and production are just a few obviously important ecosystem services that must be replaced after extinctions or other ecosystem degradation (Ehrlich & Mooney 1983). Attempts to model changing vegetation dynamics and functional roles under global warming and other global changes are summarized by Solomon & Shugart (1993). Functions of vegetation in built and other modified landscapes include providing green space for wildlife, human recreation and conservation education in urban areas; green screens for localizing industrial effects or buffering schools or residential areas against disturbances; provision of escape routes and refuge during earthquakes; and constitution of barriers to the spread of large urban fires (cf. the 'environmental protection forests' of Miyawaki *et al.* 1987). Such functions of vegetation in both natural and modified landscapes, in Japan, have been classified and mapped on 'functional vegetation maps' by Fujiwara (2001).

The continuously changing nature of landscapes provides much of the basis for their high levels of biodiversity. The prospect of global warming provides a new challenge, however, not only to vegetation and ecosystem function, but also to landscape stability and familiar concepts of vegetation. Some hypothetical patterns of biosphere response to global warming include physiological stresses; changes in metabolic balance, potential biomass and fitness; poleward/upslope shifts of climate spaces and perhaps actual ranges by at least some species; 'weedification' (Ehrlich & Mooney 1983) of some landscapes as invasive secondary (and some alien) species migrate faster than larger, less vagile but potentially stabilizing species; and perhaps long-term instability of landscapes, especially on the equatorward side (cf. Dale *et al.* 2001). Climatic envelopes have been used to project shifting climate spaces and potential range changes of plant species under global warming (e.g. Box *et al.* 1999). An especially disturbing aspect of such range shifts involves the potential for lost integrity and even breakup of familiar plant communities and landscapes if the ranges of their main structural elements diverge (cf. Crumpacker *et al.* 2001). This calls into question the entire concept of plant communities. Perhaps even more threatening is the magnitude of land-use changes and landscape fragmentation resulting from economic globalization and human overpopulation. These human-induced changes are happening much faster than global climate change.

References

Archibold, O.W. (1995) *Ecology of World Vegetation.* Chapman and Hall, London.

Axelrod, D.I. (1983) Biogeography of oaks in the Arcto-Tertiary Province. *Annals of the Missouri Botanical Garden* **70**, 629–657.

Bailey, R.G. (1989) Explanatory supplement to Ecoregions Map of the Continents. *Environmental Conservation* **16**, 307–309, with separate map (1:30 000 000).

Barkman, J.J. (1988) New Systems of Plant Growth Forms and Phenological Plant Types. In: *Plant Form and Vegetation Structure* (eds M.J.A. Werger *et al.*), pp. 9–44. SPB, The Hague.

Beard, J.S. (1973) The Physiognomic Approach. In: *Ordination and Classification of Communities* (ed. R.H. Whittaker), pp. 355–386. Dr. W. Junk, The Hague.

Blasco, F., Bellan, M.F. & Aizpuru, M. (1996) A vegetation map of tropical continental Asia at scale 1:5 million. *Journal of Vegetation Science* **7**, 623–634.

Box, E.O. (1979) Use of Synagraphic Computer Mapping in Geoecology. In: *Harvard Library of Computer Mapping*, Vol. 5, pp. 11–27. Harvard University, Cambridge, MA.

Box, E.O. (1981) *Macroclimate and Plant Forms: An Introduction to Predictive Modeling in Phytogeography*. Dr. W. Junk, The Hague.

Box, E.O. (1987) Plant life forms in mediterranean environments. *Annali di Botanica* **45(2)**, 7–42.

Box, E.O. (1988) Some similarities in the climates and vegetation of central Honshu and central eastern North America. *Veröffentlichungen des Geobotanischen Instituts Rübel* **98**, 141–168.

Box, E.O. (1995a) Factors determining distributions of tree species and plant functional types. *Vegetatio* **121**, 101–116.

Box, E.O. (1995b) Global Potential Natural Vegetation: Dynamic Benchmark in the Era of Disruption. In: *Toward Global Planning of Sustainable Use of the Earth – Development of Global Eco-engineering* (ed. Sh. Murai), pp. 77–95. Elsevier, Amsterdam.

Box, E.O. (1996) Plant functional types and climate at the global scale. *Journal of Vegetation Science* **7**, 309–320.

Box, E.O. (2002) Vegetation analogs and differences in the northern and southern hemispheres: A global comparison. *Plant Ecology* **163**, 139–154. (An appendix to this paper was erroneously not printed; it is available from the author.)

Box, E.O. & Meentemeyer, V. (1991) Geographic Modeling and Modern Ecology. In: *Modern Ecology: Basic and Applied Aspects* (eds G. Esser & D. Overdieck), pp. 773–804. Elsevier, Amsterdam.

Box, E.O., Crumpacker, D.W. & Hardin, E.D. (1993) A climatic model for location of plant species in Florida, USA. *Journal of Biogeography* **20**, 629–644.

Box, E.O., Crumpacker, D.W. & Hardin, E.D. (1999) Predicted effects of climatic change on distribution of ecologically important native tree and shrub species in Florida. *Climatic Change* **41**, 213–248.

Braun-Blanquet, J. (1928) *Pflanzensoziologie: Grundzüge der Vegetationskunde*. Springer-Verlag, Berlin.

Braun-Blanquet, J. (1965) *Plant Sociology*, 3rd ed. Hafner Publishing Co., New York.

Crumpacker, D.W., Box, E.O. & Hardin, E.D. (2001) Potential breakup of Florida plant communities as a result of climatic warming. *Florida Scientist* **64**, 29–43.

Dale, V.H., Joyce, L.A., McNulty, S.R. *et al.* (2001) Climate change and forest disturbances. *BioScience* **51**, 723–734.

Daubenmire, R.F. (1968) *Plant Communities*. Harper & Row, New York.

DeFries, R., Hansen, M. & Townshend, J. (1995) Global discrimination of land cover types from metrics derived from AVHRR Pathfinder data. *Remote Sensing of Environment* **54**, 209–222.

de Laubenfels, D.J. (1975) *Mapping the World's Vegetation*. Syracuse University Press, Syracuse.

Dierschke, H. (1994) *Pflanzensoziologie*. Verlag Eugen Ulmer, Stuttgart.

Ehrlich, P.R. & Mooney, H.A. (1983) Extinction, substitution and ecosystem services. *BioScience* **33**, 248–254.

Eiten, G. (1968) Vegetation forms: a classification of vegetation based on structure, growth form of the components, and vegetative periodicity. *Boletim do Instituto de Botânica* (São Paulo) **4**, 1–88.

Eyre, S.R. (1968) *Vegetation and Soils: A World Picture.* 2nd ed. Arnold, London.

Falińfski, J.B. (ed.) (1991) Vegetation Processes as Subject of Geobotanical Maps. Proceedings 23rd Symposium of the International Association for Vegetation Science. *Phytocoenosis* (Warszawa-Bialowieza) **3-2**: Supplementum Cartographiae Geobotanicae. With map volume.

Fujiwara, K. (1987) Aims and methods of phytosociology or 'vegetation science'. In: *Papers on Plant Ecology and Taxonomy to the Memory of Dr. Satoshi Nakanishi* (ed. Y. Takeda), pp. 607–628. Kobe Geobotanical Society, Kobe.

Fujiwara, K. (2001) *Vegetation-environmental planning based on vegetation function at different scales and in different areas.* Abstracts, 44th IAVS Symposium: Vegetation and Ecosystem Functions, p. 202. Technische Universität München, Abt. Ökologie.

Graham, A. (ed.) (1972) *Floristics and Paleofloristics of Asia and Eastern North America.* Elsevier, Amsterdam.

Grime, J.P. (1979) *Plant Strategies and Vegetation Processes.* Wiley, New York.

Holdridge, L.R. (1947) Determination of world plant formations from simple climatic data. *Science* **105**, 367–368.

Hueck, K. & Seibert, P. (1972) *Vegetationskarte von Südamerika.* 'Vegetation der einzelnen Großräume' Vol. IIa. Gustav-Fischer-Verlag, Stuttgart.

Kent, M. & Coker, P. (1992) *Vegetation Description and Analysis: A Practical Approach.* John Wiley & Sons, Chichester.

Küchler, A.W. (1964) The Potential Natural Vegetation of the Conterminous United States. Americam Geographical Society, New York, Special Research Publ. no. 36 (map + manual).

Larcher, W. (1973) *Ökologie der Pflanzen.* 2nd ed. Uni-Taschenbücher 232. Verlag Eugen Ulmer, Stuttgart.

Lefsky, M.A., Cohen, W.B., Parker, G.G. & Harding, D.J. (2002) Lidar remote sensing for ecosystem studies. *BioScience* **52**, 19–30.

Loveland, T.R., Merchant, J.W., Ohlen, D.O. & Brown, J.F. (1991) Development of a land-cover-characteristics data-base for the conterminous US. *Photogrammetric Engineering & Remote Sensing* **57**, 1453–1463.

Miyawaki, A. (ed.) (1980–9) *Nippon Shokusei-Shi* (Vegetation of Japan). 10 Vols, with separate Vol. on vegetation maps and separate colour maps. Shibundo, Tokyo. (In Japanese, with German or English summary.)

Miyawaki, A., Fujiwara, K. & Okuda, Sh. (1987) The Status of Nature and Re-Creation of Green Environments in Japan. In: *Vegetation Ecology and Creation of New Environments* (eds A. Miyawaki *et al.*), pp. 357–376. Tokai University Press, Tokyo.

Mueller-Dombois, D. & Ellenberg, H. (1974) *Aims and Methods of Vegetation Ecology.* John Wiley and Sons, New York.

Myneni, R.B., Los, S.O. & Asrar, Gh. (1995) Potential gross primary productivity of terrestrial vegetation from 1982–1990. *Geophysical Research Letters* **22(19)**, 2617–2620.

Olson, D.M. & Dinerstein, E. (1998) The Global 200: A representation approach to conserving the earth's most biologically valuable ecoregions. *Conservation Biology* **3**, 502–515.

Omernik, J.M. (1987) Ecoregions of the conterminous United States. *Annals of the Association of American Geographers* **77**, 118–125 (+ map).

Orshan, G. (ed.) (1989) *Plant Pheno-Morphological Studies in Mediterranean-Type Ecosystems.* Geobotany series, vol. 12. Kluwer Academic Publishers, Dordrecht.

Prentice, I.C., Cramer, W., Harrison, S.P., Leemans, R., Monserud, R.A. & Solomon, A.M. (1992) A global biome model based on plant physiology and dominance, soil properties and climate. *Journal of Biogeography*, 19:117–134.

Raunkiær, C. (1934) *The Life Forms of Plants and Statistical Plant Geography.* Clarendon Press, Oxford.

Rübel, E. (1930) *Pflanzengesellschaften der Erde.* Huber, Berlin.

Schmithüsen, J. (1968) *Vegetationsgeographie*. 3rd ed. Walter de Gruyter, Berlin.

Shimwell, D.W. (1971) *The Description and Classification of Vegetation*. Sidgwick & Jackson, London.

Smith, T.M., Shugart, H.H. & Woodward, F.I. (eds) (1997) *Plant Functional Types*. IGBP series, vol. 1. Cambridge University Press, New York.

Solomon, A.M. & Shugart, H.H. (eds) (1993) *Vegetation Dynamics and Global Change*. Chapman & Hall, New York.

Tateishi, R. & Kajiwara, K. (1991) Global Land-Cover Classification by NOAA GVI Data: thirteen land-cover types by cluster analysis. In: *Applications of Remote Sensing in Asia and Oceania* (ed. Sh. Murai), pp. 9–14. Asian Association for Remote Sensing, Tokyo University, Tokyo.

Troll, C. (1948) Der asymmetrische Aufbau der Vegetationszonen und Vegetationsstufen auf der Nord- und Südhalbkugel. *Jahresbericht des Geobotanischen Instituts Rübel* **1947**, 46–83.

Tüxen, R. (1956) Die heutige potentielle natürliche Vegetation als Gegenstand der Vegetationskartierung. *Angewandte Pflanzensoziologie* (Stolzenau) **13**, 5–42.

Udvardy, M.D.F. (1975) A Classification of the Biogeographical Provinces of the World. Occasional Paper 18, Internatl. Union Conserv. Nature Nat. Resources. Morges, Switzerland.

Walter, H. (1984) *Vegetationszonen und Klima*. 5th ed. Verlag Eugen Ulmer, Stuttgart.

Walter, H. (1985) *Vegetation of the Earth and Ecological Systems of the Geobiosphere*. 3rd ed. Springer-Verlag, New York.

Walter, H. & Box, E.O. (1976) Global classification of natural terrestrial ecosystems. *Vegetatio* **32**, 72–81.

Westhoff, V. & van der Maarel, E. (1973) The Braun-Blanquet Approach. In: *Ordination and Classification of Communities* (ed. R.H. Whittaker), pp. 617–726. Dr. W. Junk, The Hague.

White, F. (1983) *The Vegetation of Africa*. Descriptive Memoir to UNESCO/AETFAT/UNSO 'Vegetation Map of Africa'. UNESCO, Paris.

Whittaker, R.H. (1962) Classification of natural communities. *Botanical Review* **28**, 1–239.

Woodward, F.I. (1987) *Climate and Plant Distribution*. Cambridge University Press, Cambridge.

5

Clonal plants in the community

Brita M. Svensson, Håkan Rydin and Bengt Å. Carlsson

5.1 Modularity and clonality

Plants are **modular** – they have a basic structure, the module, which reiterates and repeats itself throughout the plant body. This often results in an immense size variation between individuals of the same species. A sapling and a large tree, for example, differ in size basically because they have different numbers of modules. The module can be seen as the fundamental functional unit of construction in most higher plants, and modularity is the basis for most forms of clonality, or vegetative reproduction.

A vascular land plant needs three structures to function – roots, leaves, and stems connecting the two. When there is only one root, as in a typical tree, we get a mainly vertically oriented plant, where all modules are connected to the same root via the main stem. Such a plant is modular but not clonal. In contrast, many horizontally oriented plants, such as wild strawberry, *Fragaria vesca*, have several main stems which have the ability to root at the nodes. Should the plant break apart, by natural processes or by injury, the fragments can therefore survive by themselves. All fragments are, thus, derived from the same zygote and form a **clone**. The smallest – actually or potentially – independent units of such **clonal fragments** are called **ramets**. This process of fragmentation can be called vegetative reproduction, and the species that exhibit this mode of reproduction are clonal. The longevity of the connection between ramets differs between species, and clonal fragments can, therefore, consist of one or several connected ramets. All ramets from a single zygote collectively make up one clone, also referred to as the genetic individual or **genet** (Fig. 5.1).

Since new ramets belonging to an old genet are in most respects identical to new ramets belonging to a young genet, clonal plants escape the size and age constraints of non-clonal plants. This means that general models and concepts of demographic and evolutionary processes, developed for non-clonal organisms, are sometimes not directly applicable to clonal plants. For example, the concepts of fitness and generation time are ambiguous in clonal plants. In a population, gene frequencies can change over time not only by sexual reproduction, but also by the addition of asexually produced ramets. Among genets, the one that produces most ramets will spread its genes most efficiently by asexual means, and, when the ramets flower, also by sexual reproduction into the next generation. Another twist with clonality is that while ramets are basically genetically identical, somatic mutations may occur in meristems

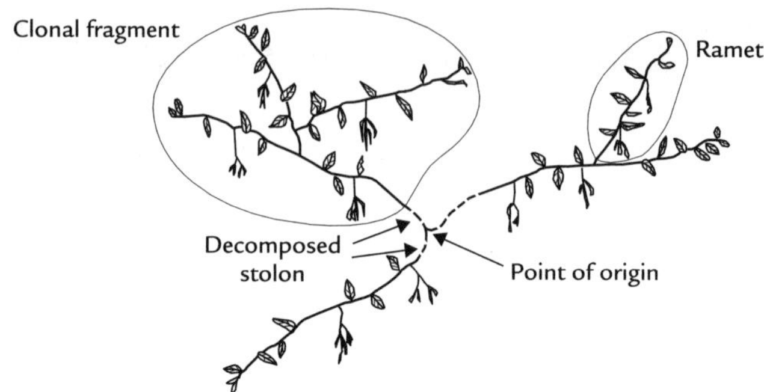

Fig. 5.1 Schematic drawing of a clonal plant to illustrate the terms genet, ramet and clonal fragment. This genet of a stoloniferous plant consists of three clonal fragments and a number of ramets.

and propagate through clonal growth. The somatic mutations are inherited when a mutated meristem forms sexual organs and produces new zygotes.

Does evolution in clonal plants take place at the genet or ramet level? Different answers are obtained if we measure the fitness of a genotype as the number of new genets produced or as the number of ramets it contributes to future generations. One may argue that the potentially unlimited distribution of a genet in space and time, as well as the potential for within-genet genetic variation and selection (Pineda-Krch & Fagerström 1999), makes the ramet the more suitable unit for evolutionary studies in clonal plants. Chessou & Peterson (2002) suggested that fitness should be measured as the relative growth rate of the genet by adding up biomass portions of all ramets of a fragmented genet. This is theoretically appealing but practically difficult. Nevertheless, the hierarchical organization of all modular organisms must be recognized (Tuomi & Vuorisalo 1989) and clonal plants clearly also possess genet characteristics upon which selection can act.

A clone is often more or less fragmented, and we rarely know the genetic identity of the ramets. Therefore, the number (and fate) of genetic individuals in the population is not well known. Instead, the number of ramets and their fates usually form the basis for demographic studies. The fate of an individual ramet is in most cases determined by its size (e.g. number and length of leaves, number of internodes, height), or developmental stage (e.g. seedling, juvenile, adult ramet, flowering versus vegetative ramet). Models exploring the future behaviour of a ramet population are correspondingly size- or stage-based, or based on a combination of the two. Population growth is of course affected by both sexual reproduction (increasing the number of genets) and asexual reproduction (increasing the number of ramets), and the two processes need not go in the same direction. A population characterized by an increasing ramet density may well lose genets in the absence of sexual reproduction; even an expanding population can show genetical impoverishment (Fig. 5.2).

Since clonal plants appear in a huge variety of forms (Table 5.1), and since clonality might serve several functions – growth, vegetative reproduction, dispersal, and a

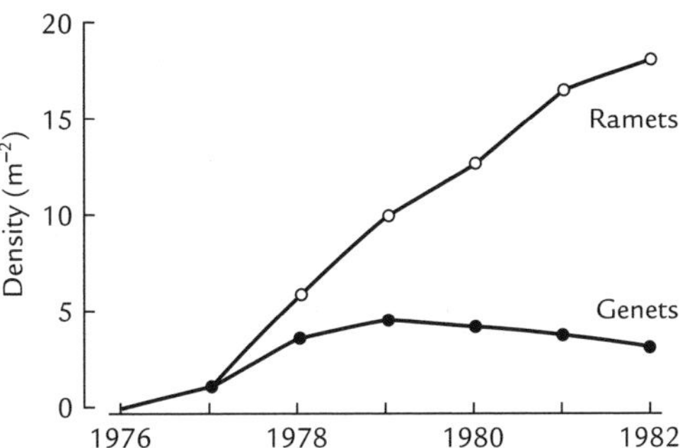

Fig. 5.2 *Solidago canadensis* is a natural invader in abandoned agricultural fields in Illinois, USA. Three years after abandonment the density of ramets (open symbols) and genets (closed symbols) diverge, and the number of ramets continues to increase whereas the number of genets decreases slowly. Redrawn from Hartnett & Bazzaz (1985). Reproduced with permission.

means for increasing genet life span – no clear-cut generalizations can be made as to how clonal plants differ from non-clonal plants in the plant community. We can, however, describe responses and behaviours that only clonal plant species are capable of, and put these in a community context. These behaviours concern the benefits of physiological integration among ramets, the ability to track resources in the habitat, and the ability for short-range dispersal using competitively superior ramets. Many such routes to success have been demonstrated in several species, but for most types of behaviour it is also possible to find species that do not show the expected response, even though they seem to have a suitable clonal architecture.

A very different group of clonal plants are those that produce seeds asexually by various means, but that otherwise are quite similar to 'ordinary' non-clonal plants. Examples include many species of, e.g. *Poa*, *Potentilla* and *Taraxacum*. They will be discussed only briefly in this chapter.

5.2 Occurrence of clonality

Clonality is widespread, both in terms of regional frequency and in local abundance. As an example, the ten most abundant species in Britain are clonal and cover 19% of the ground (Bunce & Barr 1988). In the temperate zone, 65–70% of the vascular plant species are clonal, and in some vegetation types the figure is as high as 80% (Klimeš *et al.* 1997). Exceptions to the widespread dominance of clonality are some forest types, steppes and, especially, man-made habitats (with a strong dominance of non-clonal annuals) (Fig. 5.3). In disturbed sites opportunities for regeneration by seeds are ample, and annuals often dominate because they reproduce quickly after establishment, are predominantly selfing (which makes them independent of pollinators) and often regenerate from the seed bank. Even if clonal plants are not favoured by disturbance, many disturbance-tolerant species exist that recover quickly

Table 5.1 Examples of means of vegetative dispersal in vascular plants, bryophytes and lichens. Tentative estimates of typical annual horizontal dispersal distances are given.

Clonal type	Description	Order of magnitude of annual dispersal (m)
Vascular plants		
Rhizome	Below-ground, horizontally extending stem with adventitious roots, stout, often acts as a storage organ. Similar structure appears above-ground on some epiphytes and woody monocotyledons ('aerial rhizome').[1]	10^{-2}–10^{-1}
Corm	Squat swollen stem, grows vertically in the soil, bears daughter corms (cormels).[1]	10^{-2}
Stem tuber	Swollen shoot with scale leaves each subtending one or more buds. Leaves present. Above- or below-ground.[1]	10^{-2}
Root tuber	As stem tuber, but leaves absent. Below-ground.[1]	10^{-2}
Bulb	Short, usually vertical stem axis bearing fleshy scale leaves.[1]	10^{-3}–10^{-2}
Stolon	Stem growing along the substrate surface or through surface debris. Long, thin internodes, bears foliage and adventitious roots.[1]	10^{-2}–10^{0}
Runner	Thin horizontal stem above-ground, one or more internodes, does not root between mother and daughter plant.[1]	10^{-2}–10^{-1}
Bulbil	A small bulb, e.g. on an aerial stem or developing in the axils of the leaves of a fully sized bulb. Often inaccurately applied to any small organs of vegetative multiplication such as axillary stem tubers.[1]	10^{-2}–10^{-1}
Turion	Detachable bud in water plants for survival during dry or cold periods ('winter bud').[1]	10^{-1}–10^{2}
Dropper	Detachable buds that are transported away from the mother plant at the end of a slender root-like structure.[1]	10^{-2}–10^{-1}
Prolification	Production of vegetative buds instead of flowers.	
	Tiller production in sterile spikelets. 'False vivipary'.[1]	10^{-2}–10^{-1}
Root buds	Buds on roots capable of developing into a new shoot.[1]	10^{-2}–10^{1}
Apomixis	Seed produced without sexual fertilization.	10^{-1}–10^{3}
Fragment	Plant part breaks off and establishes. Particularly long dispersal distances on ice, in water and on sand.	10^{-2}–10^{3}
Layering	Aerial shoot bends down, touches the ground and produces adventitious roots.[1]	10^{-2}–10^{-1}
Bryophytes[2]		
Vegetative growth	Ramets formed after bifurcation or branching of stem or after expansion and separation of thallus parts.	10^{-3}–10^{-2}
Fragment	Detached shoot, leaf, or part of stem or thallus.	10^{-2}–10^{-1}
Bulbil	Axillary, detachable multicellular body, with leaf (or leaf-like) primordia.	
Gemma	Multicellular body on stems or leaves, in splash-cups with dispersal assisted by rain-drops, or carried on specialized stalks.	10^{-2}–10^{-1}
Tuber	Swollen protuberance on rhizoid. Detaches after soil disturbance.	10^{-3}–10^{-2}
Lichens		
Vegetative growth	Ramets formed after expansion and separation of thallus parts.	10^{-4}–10^{-2}
Fragment	Any portion of the thallus.	
Soredia	Bodies of algal cells surrounded by fungal hyphae (25–100 μm). Wind dispersed.	10^{-1}–10^{1}
Isidia	Small thallus outgrowths that break off.	10^{-2}–10^{-1}

[1] Bell (1991)
[2] Shaw & Goffinet (2000)

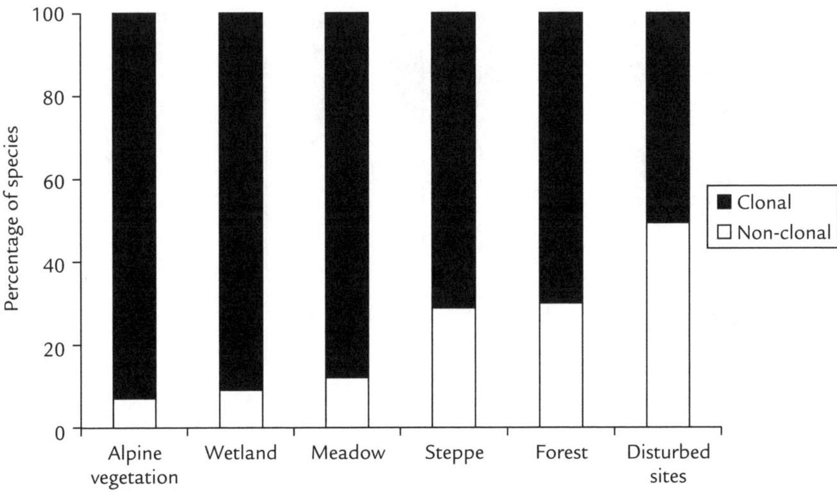

Fig. 5.3 Proportion of species that are clonal or non-clonal in different Central European vegetation types. (Based on data in Klimeš *et al.* 1997.)

after perturbations. One example is the common *Elytrigia repens* which is a serious arable weed in large parts of its distribution area. On shorelines heavily affected by waves or ice-push, disturbances are often too severe for seedlings. Here clonals come to dominate since establishing ramets are initially firmly fixed to subterraneous organs of the mother plant and hereby withstand this kind of disturbance.

Especially in primary succession, dispersal capacities of the pioneers determine the initial species composition. Non-clonal annuals dominate as they have the advantages of a short life-cycle, often combined with selfing. However, the period with annual dominance is generally short, often only a few years (Rydin & Borgegård 1991) and clonals quickly come to dominate, even if the degree of dominance differs among habitat types (Prach & Pyšek 1994).

In secondary succession, many species can emerge from seed banks, and this may favour short-lived, often non-clonal species. Where the vegetation closes quickly, as in old-field succession, it may be easier to expand clonally than to establish from small seeds. If there is a stage with annual or non-clonal dominance this will generally be even shorter than in primary succession. Among grasses, rhizomatous species such as *Elytrigia repens* are successful in old-fields. Among trees, clonal species with root suckers such as *Populus tremula* in north-west Europe or *P. tremuloides* in North America are often the first to arrive. Interestingly, these species are also superior colonizers in primary successions through their small wind-dispersed seeds (Rydin & Borgegård 1991).

It is generally held that seeds are needed for dispersal over long distances. Plants of the genus *Taraxacum* have small, plumed seeds which can travel long distances. Since *Taraxacum* is apomictic, this long-distance seed dispersal in fact represents clonal dispersal. One particular case where clonal long-range dispersal is common occurs

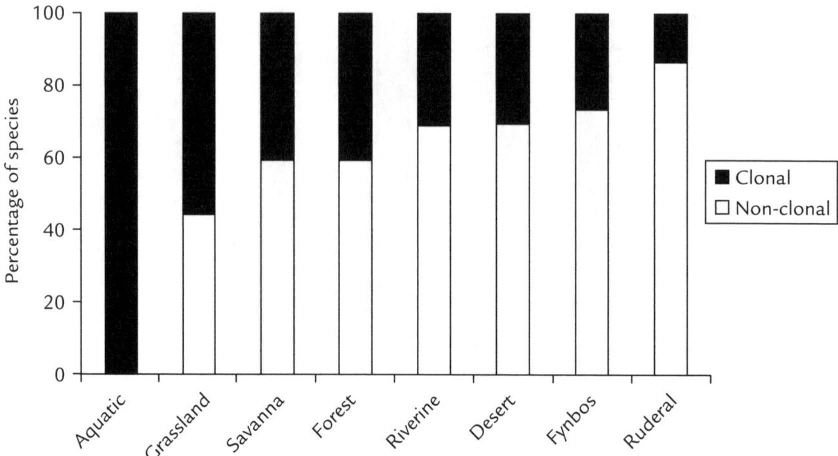

Fig. 5.4 Proportion of species that are clonal or non-clonal among aliens established in different habitat types in South African natural vegetation. Redrawn from Pyšek (1997).

among aquatic and shoreline species which may disperse successfully, also over long distances, with plant fragments. Two dioecious species may serve as examples. The North American aquatic *Elodea canadensis* is a widespread alien. In Europe it appears almost exclusively as female plants, whereas in Australia there are large regions with either only male or female plants (Spicer & Catling 1988). In a similar fashion, *Salix fragilis* occurs almost exclusively as male plants in some regions in Sweden. From medieval times this taxon has escaped (with floating twigs and branches that easily establish downstream) from trees planted at manors where male plants were preferentially used to avoid the heavy seed litter (Malmgren 1982). Another successful invader is the gynodioecious *Fallopia japonica* which in the British Isles probably exists as a single male-sterile clone (Hollingsworth & Bailey 2000). See also Chapter 13.

The successful invasion of different alien species can give a clue to the advantage of clonality in different habitats. In the native vascular plant flora in Central Europe, 69% of the species are clonal, but among the aliens only 36% are clonal (Pyšek 1997). Aliens in natural communities are often clonal, but in man-made habitats they are more often non-clonal, in congruence with the large proportion of non-clonal species in disturbed habitats. We cannot say that successful invaders in general are clonal or non-clonal, but invasion success is probably more likely in man-made habitats and hence clonals are at a disadvantage. As an expected effect of the ease by which clonal plants spread in water, the percentage of clonal invaders is high in wetlands, and they actually dominate among aquatic aliens (Fig. 5.4).

Both dispersal ability and the ability to achieve dominance in the community differ among species, and this affects their regional frequency and local abundance. We can see this among invading plant species: clonal invaders often reach relatively high abundance locally, whereas non-clonal invaders have a higher rate of regional spread.

However, there are latitudinal differences: in the tropics aggressive invaders are often non-clonal, but in temperate zones they are more commonly clonal (Pyšek 1997).

5.3 Habitat exploitation by clonal growth

When diaspores (sexual or asexual) are detached, they are beyond the control of the mother plant. In contrast, clonality potentially allows the plant to exploit the environment via directional growth of, for example, rhizomes or stolons. The plant can move to a more favourable part of the habitat under two conditions: (i) if growth can be directed along gradients; and (ii) if the plant can minimize its elongation growth once it has reached a favourable patch. The latter means that as the plant moves towards favourable patches, it should gradually reduce carbon allocation for elongation.

Habitat exploitation in plants has been compared to foraging in animals, but the differences are obvious. According to the marginal value theorem (Begon *et al.* 1996), an animal will leave a patch at a certain profitability level. In clonal plants the response can be gradual: the lower the resource level, the higher should the tendency be to grow away from the patch. A second difference is that a plant does not leave the patch even if it produces ramets that do so. There is no consensus on the definition of foraging in plants, and Oborny & Cain (1997) suggested that the term should be restricted to cover morphological plasticity that is also selectively advantageous for resource acquisition. Since the focus in this chapter is on the role of clonality in plant communities rather than its role in determining plant fitness we will use the term forage in a wide sense.

A mechanistic problem with habitat exploitation is that in patches where biomass accumulation is lowest, the plant should have its maximum capacity for elongation growth to be able to leave for a better spot. Therefore, search behaviour cannot be achieved by increased growth, only by increased allocation to directional elongation. If the plant's growth is too small, it simply cannot produce stolons or rhizomes that are long enough to reach more productive patches. This constrains the ability of plants to explore their surroundings.

An example of a plant that appears to forage is *Glechoma hederacea*. It branches more sparsely and forms longer internodes under low-light or low-nutrient conditions than in more favourable habitats (Slade & Hutchings 1987a,b). While this leads to habitat exploitation, it is a plastic response that need not be selective. Quite a few clonal species have been tested for their ability to exploit the habitat, and it is clear that far from all species behave like *Glechoma*. Oborny & Cain (1997) found that only two out of 16 species fitted the *Glechoma* model and they offered three reasons:

1 Growth must be financed. The outcome of experiments therefore depends on the range of resource levels tested. At very low resource levels there is very little growth, and long internodes do not occur. As resources increase there may first be a positive relationship – both biomass and internode length increase. At even higher levels biomass will continue to increase, but here internode length may decrease (Fig. 5.5). The *Glechoma* behaviour should therefore only be expected in habitats with a mosaic of patches with high and intermediate resource levels.

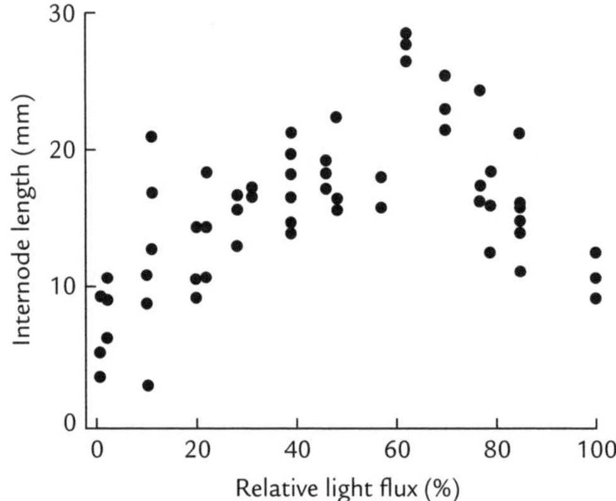

Fig. 5.5 Mean stolon internode length in *Trifolium repens* growing in the field under different levels of photosynthetically active radiation transmitted by natural canopies. Redrawn from Thompson (1993). Reproduced by permission of the publishers.

2 Evolutionary constraints. The morphology may be evolutionarily conservative, which means that not all strategies and responses to environmental heterogeneity can be realized.

3 Physiological integration alters the resource available to the ramet. If a ramet in a poor patch receives nutrients from ramets in richer patches it is not likely to respond to the local resource level.

The mechanisms of habitat exploitation are perhaps easiest to understand when light is the limiting resource. Elongation is promoted by shade and reduced by light. This enables the plant to leave poor patches and also to stay in rich patches. Another plastic response to shade is increased leaf area which means that carbon fixation can be maintained at a higher level than expected from the light flux alone. This certainly helps the plant to allocate more assimilates to clonal escape, and makes this mechanism more probable.

The mechanisms that could enable a plant to leave a patch with low nutrient and water supplies are not so intuitively obvious as the shade response mechanisms. In experimental units with low fertility, *Elytrigia repens* preferentially grew into vegetation-free patches, probably because transport of nutrients through the rhizomes resulted in directional growth into open areas (Kleijn & van Groenendael 1999). The concentration of ramets in favourable patches could happen even without a searching behaviour. The plants may move around in a random fashion, and mechanisms for active growth into good patches are not required, but growth is promoted here (Stoll *et al.* 1998; Piqueras *et al.* 1999).

The next step is to understand how the plants can stay in favourable patches. Some studies have demonstrated that clonal plants concentrate their ramets to fertile patches where they increase productivity without increasing length growth of tubers or rhizomes. The growth morphology of the alpine sedge *Carex bigelowii* may give a clue to how this is possible. In this species, changed growth orientation is more important than internode length growth: higher nutrient levels lead to the produc-

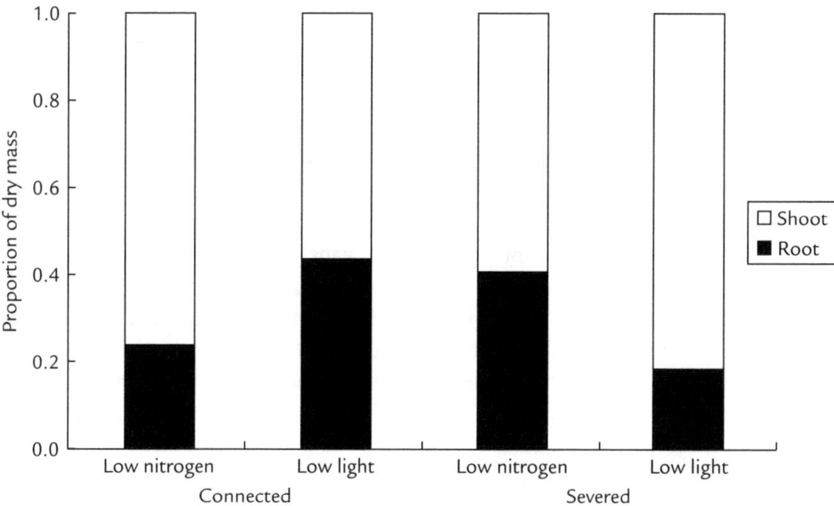

Fig. 5.6 Allocation of biomass within ramets of *Fragaria chiloensis*. When separated, ramets allocated significantly more to root biomass in the low-nitrogen treatments, but when connected, division of labour was induced and the ramets allocated significantly more to root biomass in the low-light treatment. Redrawn from Alpert & Stuefer (1997).

tion of more short-rhizome tillers and less emphasis on horizontal movement (Carlsson & Callaghan 1990). In *Lycopodium annotinum*, longer horizontal segments are produced under favourable light and temperature conditions, and the increased growth makes the plant move away from the favourable patch. However, along the horizontal stem there are vertical segments, which are the main structures for carbon capture. The vertical segments are attached at constant intervals, which means that even if the horizontal apex grows out from the favourable patch the vertical segment stays put (Svensson *et al.* 1994).

A concept often discussed in the clonal literature is 'division of labour' – ramets have different tasks, and resources are exchanged between ramets placed in patches of different quality. This requires a high degree of physiological integration in the clonal fragment and results in a type of habitat exploitation that in many ways is opposite to foraging, and probably often more important than foraging. A non-clonal plant will in general allocate biomass to enable increased uptake of the limiting factor, for instance by increasing the root:shoot allocation ratio in nutrient-poor sites. In contrast, physiologically integrated ramets commonly allocate biomass to increase uptake of resources that are abundant, if these resources are scarce where the other ramets grow. This means that the clonal fragment (i.e. the integrated ramets collectively) follows the same allocation rule as a non-clonal plant, but the individual ramets do not. In the stoloniferous *Fragaria chiloensis*, ramets in a connected system experiencing ample light and low levels of nitrogen specialize in growing leaves whereas ramets in the opposite situation specialize in root growth. When the ramets are no longer connected, each ramet specializes in capturing the scarce resource (Fig. 5.6; Alpert & Stuefer 1997).

A mature plant has a relation between root and shoot mass adjusted to the environment where the plant developed. When shoot parts are removed, the plant allocates resources to restore the relation. The same is true for clonal plants but we have to add a new dimension, the lateral growth of daughter ramets that also must be balanced with the vertical growth of the parent root/shoot system (Pitelka & Ashmun 1985). If the daughter ramet is situated further from the parent, the influence of the parent becomes reduced and the daughter ramet will develop a root/shoot balance for itself.

The degree of physiological integration varies widely among clonal plants, and can partly be explained by differences in vascular architecture (Marshall & Price 1997). Jónsdóttir & Watson (1997) distinguished between integrators (of four grades) and disintegrators. The relationships between longevity of ramet connectivity, longevity of ramets and generation time of ramets determine the degree of integration among species. Jónsdóttir & Watson (1997) suggested that there is a tendency that full integration is more common in stable and low-productive environments but we have too few data for reliable generalizations. In the community the benefits of integration are (i) support to new ramets (which may affect establishment and competition); (ii) recycling of nutrients and assimilates; and (iii) buffering of environmental heterogeneity.

5.4 Competition and co-existence in clonal plants

Since clonal growth is expressed in plant architecture, resource uptake, allocation and size, it is most likely that clonality should affect the competitive ability of the plant. But the relationship between clonality and competitive ability is not easy to generalize: it differs among clonal types and also depends on what aspects of competitive ability we are interested in.

Competition occurs when there is a negative effect on plants as they struggle to capture the same, limiting resource. In addition to resource competition where the struggle is for growth factors such as nutrients, water and light, we can also envisage competition for space (or ground area). In space competition, one plant covers the substrate and thereby prevents germination or rooting of other plants. It is also useful to realize that competitive ability has two components: competitive effect is the ability to take up resources and thereby reduce the amounts available for other plants, whereas competitive response is the ability to perform well even though resource levels are reduced by the competitors (Goldberg 1990).

The competitors are not likely to suffer equally. Space competition is the most asymmetric form of interaction; the winner monopolizes all resources; pre-emptive, or interference competition are terms used to describe the situation. The first species to arrive holds its position and there will be no competitive replacement. If the order of arrival is decisive for species composition even after a long time we say that the community is founder controlled. Competition for light is also asymmetric and in many situations probably the only interaction that leads to competitive exclusion among plant species. In integrated clonal plants young ramets can be supported from other parts of the clone, and will not suffer from being small. There is support for the notion that size-biased asymmetrical competition therefore is relatively unimportant in clonal plants (Suzuki 1994; Pennings & Callaway 2000), but there are also cases where this does not seem to hold.

The competition models of Grime (1979) and Tilman (1985) have bearings on the relationship between clonality and competitive ability. In Grime's CSR model (competitors, stress tolerators, ruderals), the competitors attain dominance in environments with little disturbance and low levels of stress. Stress in this model refers to abiotic conditions that reduce plant growth, for instance low nutrient levels. Since competition is for limiting resources, it is somewhat paradoxical that competition should be important where resources are abundant. The resolution must be that light competition is the dominant process: plants that can grow taller than their neighbours will win. It therefore seems that competitive ability in Grime's model mostly reflects competitive effect when light competition is the structuring force. The CSR classification of species is a synthesis of lateral spread and several other attributes (e.g. plant size, phenology, leaf area; cf. Hodgson *et al.* 1999). For this reason it is clear that there can be no simple relationship between clonality and competitive ability in the CSR space. Among the pure ruderals, species with strong lateral spread are lacking, whereas the pure competitors will, as a rule, be species with rapid clonal expansion (Fig. 5.7). Apart from these rather obvious extreme cases, the cluster of species moves from the ruderal part of the triangle via the stress tolerator part to the competitive part with increasing ability for clonal expansion. Klimeš *et al.* (1997) noted that the proportion of the species that are clonal is higher than the average in habitats with low nutrient levels and low temperature. In our diagram, this is reflected by rather many species with slow clonal expansion among the stress-tolerator species.

In Tilman's competition model (Tilman 1985), the best competitor is the one that can reduce resource levels to a lower level than other species, and maintain population growth at this lower level. Competition is not restricted to fertile patches, but can be important in all sorts of environments. According to this view there is not a single group of globally superior competitors, but, for instance, some species are strong light competitors, others will win when nitrogen is limiting. Root:shoot allocation and uptake efficiency will affect competitive ability, and therefore it is likely that clonality should affect competition. Growth of the plants will lead to resource depletion and ultimately to competitive exclusion. Through physiological integration and foraging, clonal plants may be very efficient competitors.

Various mechanisms may prevent the community from reaching the species composition predicted from equilibrium models based on competitive abilities. Disturbances such as wind, waves and trampling reset the system and open it up for re-colonization, and inferior species can persist if they are good dispersers. A competition-colonization trade-off is often assumed and the competitively weak species doomed to local extinction may disperse at random to occupy patches made available. Such 'escaping dispersal' can of course be through clonal growth as well as by seeds, and we must make a distinction between clonal attributes that confer competitive ability (e.g. large ramets with strong support from their mother plant) and clonal attributes that confer dispersal and escape from competition (e.g. bulbils, apomictic seeds, rapidly disintegrating runners that form seedling-like ramets). The

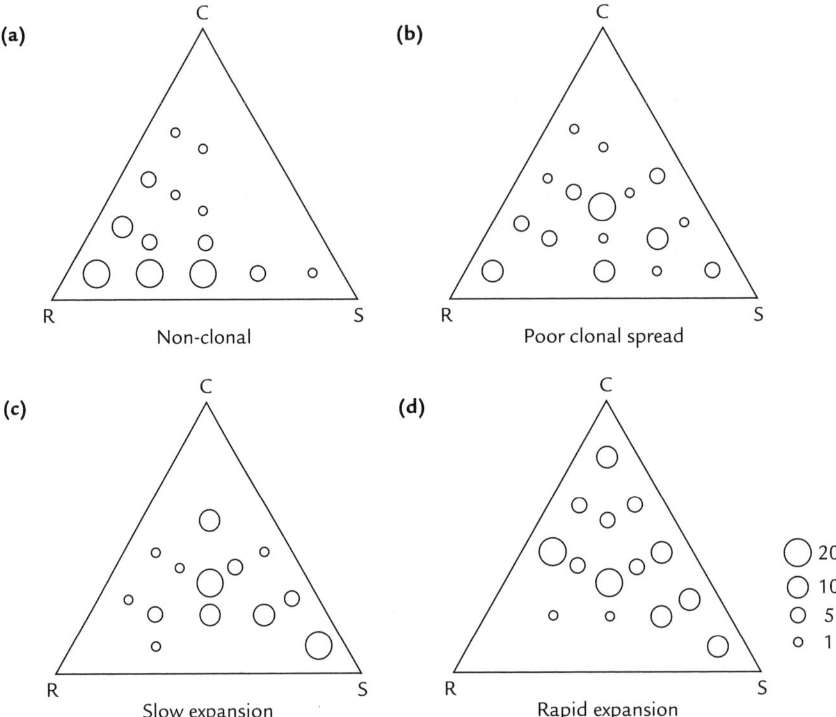

Fig. 5.7 Distribution of 255 herb and graminoid species among the 19 functional types in the CSR triangle. The area of each circle is proportional to the number of species. The position of each species in CSR space is taken from Hodgson *et al.* (1995), but we made the classification of species in clonal types independently, following a scheme modified from Klimeš (1999). **a.** Non-clonal plants. **b.** Plants with poor clonal spread; vegetative reproduction is occasional or does not result in clonal patches. **c.** Plants with a capacity for slow clonal expansion (< 10 $cm{\cdot}yr^{-1}$), or with a limited capacity to form local colonies. **d.** Plants with a capacity for rapid clonal expansion (> 10 $cm{\cdot}yr^{-1}$) or with a capacity to form large clonal patches.

large variation in clonal morphological types has led to the distinction between 'guerilla' and 'phalanx' strategies (Lovett-Doust 1981). Guerilla species have long internodes and typically spread by above-ground runners, whereas phalanx species spread as a front, and the plant will show a clumped distribution. The guerilla-phalanx distinction should be seen as a continuum with many intermediate types.

The presence of guerilla and phalanx species could affect competition in the community in many ways:

- The guerilla strategy could be a way to evade local interspecific competition. Since competition among plants only occurs between immediate neighbours, this will prevent the exclusion from the community of species that are competitively weak but which produce long runners. At the community scale this should slow down or even prevent competitive exclusion.

• The spreading behaviour of the guerilla species leads to increased interspecific contacts and mixing of species. The ramets quickly become independent and suffer from interspecific competition. This should speed up competitive exclusion at the local scale.
• It is generally held that phalanx species are strong in resource competition (they grow bigger to catch light and the new ramets are well fed with resources). Even when they encounter a superior competitor, their aggregation diminishes the degree of interspecific contacts, and it will take a long time for any other species to oust them. Phalanx growth that leads to reduced encounter probabilities among species will eventually result in spatial isolation of local dynamics and render global interactions in the community less important (Oborny & Bartha 1995).
• The longer-dispersing guerilla ramets carry fewer resources from the mother plant, and are most likely not very strong in resource competition. Instead they could capture new space effectively, and may be good at pre-emptive competition. This could potentially lead to founder control in the community. If the phalanx species would gradually take over, the result would be dominance control. Founder control is possible if the guerilla ramet has established itself so well that it can withstand the competition when the phalanx species reaches the patch.
• The shorter-dispersing phalanx ramets may suffer from intraspecific competition (including that from siblings), whereas the guerilla growth form decreases intraspecific competition, but increases interspecific competition.

These mechanisms should be testable for pair-wise species interactions, but their role for community composition is difficult to assess or generalize. Whether a community is founder or dominance controlled is largely dependent on the rate of creation of open patches, the rate at which the species can reach these patches and the rate at which the phalanx species can outcompete the guerilla species. Several factors that counteract dominance control in competition between clonal plants have also been suggested (Herben & Hara 1997). First, there may be architectural constraints that prevent the plant to spread to dominance. Second, in low-nutrient sites there will be more root competition (which generally is symmetrical) and more spatial expansion. Mosaics of species will result. Third, if the competitive abilities of species are ranked as A > B > C, but C > A, there is intransitivity in competitive ability. Experiments have been performed to test several grassland species' ability to invade each others' turfs (Silvertown *et al.* 1994, and other studies quoted therein), and it appears that a competitive hierarchy is more common than intransitivities. However, the rank order could change depending on grazing regime, and there need not be a simple relationship between the ability to invade another species territory and the ability to withstand invasion. In *Sphagnum* mosses, the ability of species to take over occupied area from each other by clonal expansion also varied between years (Rydin 1993).

The effect of aggregation has been modelled by cellular automata (Silvertown *et al.* 1992). A competitive hierarchy rapidly led to loss of species from the community, but only when the starting arrangement was random. Different arrangements with species clumping slowed down the processes and led to different outcomes indicating that the spatial pattern may be more important than the competitive ranking or the abundance of the species.

For many species, clonality enables them to cope with different environments. Species with long-lived rhizomes, root systems or a dormant bud bank in general have broader niches than other species, in the sense that they occur in a wider range of habitats. It also seems that species with several modes of clonal growth have wider niches than those with only one mode (Klimeš & Klimešová 1999). The reasons for this are probably that clonal growth is a way to achieve high phenotypic plasticity and that physiological integration allows individual ramets to survive in suboptimal patches where they would have performed poorly on their own. An example of the plastic response among ramets in a clonal system is *Scirpus maritimus*. In this species, the ramets were shown to be plastically modified to specialize in sexual reproduction, vegetative growth or storage, depending on their position (Charpentier & Stuefer 1999). Such a species would then cover a large portion of the CSR strategy plane, and be able to cope with a large range of circumstances. This, together with a long life span and the presence of a bud bank, enables clonal species to survive as 'remnant populations' which can be sources for expansion when conditions become more favourable (Eriksson 1997).

5.5 Clonality and herbivory

Clonal plants are susceptible to grazing, just as non-clonal plants are, and strategies have evolved to reduce its negative effects. Below we will describe some of these strategies and also some effects that grazing (including insect herbivory) has upon the clonal plant. First of all clonality is a kind of risk-spreading (Eriksson & Jerling 1990); if there are ten ramets in the clonal fragment it is unlikely that all will be eaten.

In clonal as well as non-clonal plant species, the development and function of the different plant parts are restrained by interactions with other parts of the plant. The classical example is apical dominance – the inhibition exerted by the terminal bud on axillary buds. When the terminal bud is removed or damaged, apical dominance is released, and the axillary buds sprout. In *Lycopodium annotinum* (as in many other species) it is the bud closest to the no-longer-existing apex that starts to grow (Svensson & Callaghan 1988). This newly developed apex in turn becomes dominant and exerts apical dominance over buds situated proximally to it. This results in decreased competition between ramets within the clonal fragment (Callaghan *et al.* 1990).

Grazing has a large impact on the architecture of clonal plants by removing dominant apices together with green tissue. When the apex is grazed, trampled or otherwise damaged, some of the buds sprout which results in a proliferation of ramets. In clonal plants with lateral spread herbivory may thus be partly positive and will not kill the genet as in many non-clonal plants. The ability to recover can be amazing: Morrow & Olfelt (2003) suggested that *Solidage missoutieusis* reappeared from rhizomes up to ten years after they disappeared (appearently killed by a specialist herbivorous insect). Another consequence of herbivory may be that grazing speeds up the life-cycle, as in the example with *Carex stans* (Fig. 5.8; Tolvanen *et al.* 2001). Here, grazing not only induced formation of new ramets, but also increased the

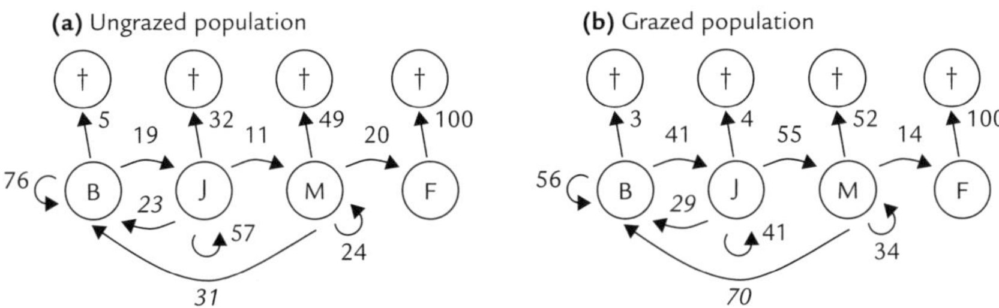

Fig. 5.8 Life-cycle graph of *Carex stans* in Canada. **a.** Population from a sheltered, non-grazed habitat; **b.** Population grazed by musk ox. Values are transition probabilities (%) between the life-stages juvenile tillers (J) and mature tillers (M), and vegetative reproduction, i.e. number of buds (B) produced, in italics. Included are also flowering tillers (F) even though they do not contribute to population growth, and the probabilities to die (†). Based on data in Tolvanen *et al.* (2001).

proportion of buds that developed to juveniles and subsequently to mature ramets. In ungrazed populations most buds remained dormant.

Apart from buds being released from their dormancy, depending on the level of physiological integration, sister ramets within the clonal fragment may or may not compensate for lost tissue by enhanced growth. When such a clone is grazed, resources are transported from the undamaged part to the damaged part and new tissue for photosynthesis is produced. Such compensatory growth increases the chances of survival of the clonal fragment due to re-allocation of resources. Physiological integration also enables grazing-induced secondary metabolites to be transported to ungrazed parts of the clone. A chemical defence can thereby be built up at a lower cost than if the metabolite should always be present (Seldal *et al.* 1994). If heavily grazed, however, the whole clonal fragment may suffer – particularly if there is a high degree of physiological integration between ramets (Pitelka & Ashmun 1985). On the other hand, if the clonal fragment is poorly integrated one ramet may be damaged without any effects on the other ramets, and the genet will not be at risk.

Clonal woody plants have a mixed size and age structure of their above-ground parts, which is helpful in the defence against herbivory (Peterson & Jones 1997). This is because first, clonal woody plants are generally long-lived and have large reserves stored in below-ground tissue such as roots and rhizomes. Second, after disturbance, such as fire or a herbivore attack, ramets sprout from the underground bud reserve. This may happen infrequently and the result is a collection of ramets of different sizes and ages which differ in their attractiveness for herbivores.

Acknowledgements

We thank Leoš Klimeš for providing data for Fig. 5.3, and Camilla Wessberg for compiling data for Fig. 5.7. Eddy van der Maarel and Petr Pyšek gave valuable comments on the manuscript.

References

Alpert, P. & Stuefer, J.F. (1997) Division of labour in clonal plants. In: *The Ecology and Evolution of Clonal Plants* (eds H. de Kroon & J. van Groenendael), pp. 137–154. Backhuys Publishers, Leiden.

Begon, M., Harper, J.L. & Townsend, C.R. (1996) *Ecology. Individuals, populations and communities*. 3rd. ed. Blackwell Science, Oxford.

Bell, A.D. (1991) *Plant form. An illustrated guide to flowering plant morphology*. Oxford University Press, Oxford.

Bunce, R.G.H. & Barr, C.J. (1988) The extent of land under different management regimes in the uplands and the potential for change. In: *Ecological change in the Uplands* (eds M.B. Usher & D.B.A. Thompson), pp. 415–426. Blackwell Science, Oxford.

Callaghan, T.V., Svensson, B.M., Bowman, H., Lindley, D.K. & Carlsson, B.Å. (1990) Models of clonal plant growth based on population dynamics and architecture. *Oikos* **57**, 257–269.

Carlsson, B.Å. & Callaghan, T.V. (1990) Programmed tiller differentiation, intraclonal density regulation, and nutrient dynamics in *Carex bigelowii*. *Oikos* **58**, 219–230.

Charpentier, A. & Stuefer, J.F. (1999) Functional specialization of ramets in *Scirpus maritimus*. *Plant Ecology* **141**, 129–136.

Chesson, P. & Peterson, A.G. (2002) The quantitative assessment of benefits of physiological integration in clonal plants. *Evolutionary Ecology Research* **4**, 1153–1176.

Eriksson, O. (1997) Clonal life histories and the evolution of seed recruitment. In: *The Ecology and Evolution of Clonal Plants* (eds H. de Kroon & J. van Groenendael), pp. 211–226. Backhuys Publishers, Leiden.

Eriksson, O. & Jerling, L. (1990) Hierarchical selection and risk spreading in clonal plants. In: *Clonal Growth in Plants* (eds J. van Groenendael & H. de Kroon), pp. 79–94. SPB Academic Publishing, The Hague.

Goldberg, D.E. (1990) Components of resource competition in plant communities. In: *Perspectives on Plant Competition* (eds J.B. Grace & D. Tilman), pp. 27–49. Academic Press, San Diego, CA.

Grime, J.P. (1979) *Plant Strategies and Vegetation Processes*. Wiley, Chichester.

Hartnett, D.C. & Bazzaz, F.A. (1985) The genet and ramet population dynamics of *Solidago canadensis* in an abandoned field. *Journal of Ecology* **73**, 407–413.

Herben, H. & Hara, T. (1997) Competition and spatial dynamics of clonal plants. In: *The Ecology and Evolution of Clonal Plants* (eds H. de Kroon & J. van Groenendael), pp. 311–357. Backhuys Publishers, Leiden.

Hodgson, J.G., Grime, J.P., Hunt, R. & Thompson, K. (1995) *The Electronic Comparative Plant Ecology*. Chapman & Hall, London.

Hodgson, J.G., Wilson, P.J., Hunt, R., Grime, J.P. & Thompson, K. (1999) Allocating C-S-R plant functional types: a soft approach to a hard problem. *Oikos* **85**, 282–294.

Hollingsworth, M.L. & Bailey, J.P. (2000) Evidence for massive clonal growth in the invasive *Fallopia japonica* (Japanese Knotweed). *Botanical Journal of the Linnaean Society* **133**, 463–472.

Jónsdóttir, I.S. & Watson, M.A. (1997) Extensive physiological integration: an adaptive trait in resource-poor environments? In: *The Ecology and Evolution of Clonal Plants* (eds H. de Kroon & J. van Groenendael), pp. 109–136. Backhuys Publishers, Leiden.

Kleijn, D. & van Groenendael, J.M. (1999) The exploitation of heterogeneity by a clonal plant in habitats with contrasting productivity levels. *Journal of Ecology* **87**, 873–884.

Klimeš, L. (1999) Small-scale plant mobility in a species-rich grassland. *Journal of Vegetation Science* **10**, 209–218.

Klimeš, L. & Klimešová, J. (1999) CLO-PLA2 – a database of clonal plants in central Europe. *Plant Ecology* **141**, 9–19.

Klimeš, L., Klimešová, J., Hendriks, R. & van Groenendael, J. (1997) Clonal plant architecture: a comparative analysis of form and function. In: *The Ecology and Evolution of Clonal Plants* (eds H. de Kroon & J. van Groenendael), pp. 1–29. Backhuys Publishers, Leiden.

Lovett-Doust, L. (1981) Population dynamics and local specialization in a clonal perennial (*Ranunculus repens*). I. The dynamics of ramets in contrasting habitats. *Journal of Ecology* **69**, 743–755.

Malmgren, U. (1982) *Västmanlands flora*. SBT-förlaget, Lund. (In Swedish.)

Marshall, C. & Price, E.A.C. (1997) Sectoriality and its implications for physiological integration. In: *The Ecology and Evolution of Clonal Plants* (eds H. de Kroon & J. van Groenendael), pp. 79–107. Backhuys Publishers, Leiden.

Morrow, P.A. & Olfelt, J.P. (2003) Phoenix clones: recovery after long-term defoliation-induced dormancy. *Ecology Letters* **6**, 119–125.

Oborny, B. & Bartha, S. (1995) Clonality in plant communities – an overview. *Abstracta Botanica* **19**, 115–127.

Oborny, B. & Cain, M.L. (1997) Models of spatial spread and foraging in clonal plants. In: *The Ecology and Evolution of Clonal Plants* (eds H. de Kroon & J. van Groenendael), pp. 155–183. Backhuys Publishers, Leiden.

Pennings, S.C. & Callaway, R.M. (2000) The advantages of clonal integration under different ecological conditions: A community-wide test. *Ecology* **81**, 709–716.

Peterson, C.J. & Jones, R.H. (1997) Clonality in woody plants: a review and comparison with clonal herbs. In: *The Ecology and Evolution of Clonal Plants* (eds H. de Kroon & J. van Groenendael), pp. 263–289. Backhuys Publishers, Leiden.

Pineda-Krch, M. & Fagerström, T. (1999) On the potential for evolutionary change in meristematic cell lineages through intraorganismal selection. *Journal of Evolutionary Biology* **12**, 681–688.

Piqueras, J., Klimeš, L. & Redbo-Torstensson, P. (1999) Modelling the morphological response to nutrient availability in the clonal plant *Trientalis europaea*. *Plant Ecology* **141**, 117–127.

Pitelka, L.F. & Ashmun, J.W. (1985) Physiology and integration of ramets in clonal plants. In: *Population biology and evolution of clonal organisms* (eds J.B.C. Jackson, L.W. Buss & R.E. Cook), pp. 399–435. Yale University Press, New Haven.

Prach, K. & Pyšek, P. (1994) Clonal plants – what is their role in succession? *Folia Geobotanica et Phytotaxonomica* **29**, 307–320.

Pyšek, P. (1997) Clonality and plant invasions: can a trait make a difference? In: *The Ecology and Evolution of Clonal Plants* (eds H. de Kroon & J. van Groenendael), pp. 405–427. Backhuys Publishers, Leiden.

Rydin, H. (1993) Interspecific competition among *Sphagnum* mosses on a raised bog. *Oikos* **66**, 413–423.

Rydin, H. & Borgegård, S.-O. (1991) Plant characteristics over a century of primary succession on islands: Lake Hjälmaren. *Ecology* **72**, 1089–1101.

Seldal, T., Andersen, K.J. & Högstedt, G. (1994) Grazing-induced proteinase inhibitors: A possible cause for lemming population cycles. *Oikos* **70**, 3–11.

Shaw, A.J. & Goffinet, B. (eds) (2000) *Bryophyte Biology*. Cambridge University Press, Cambridge.

Silvertown, J., Holtier, S., Johnson, J. & Dale, P. (1992) Cellular automaton models of interspecific competition for space – the effect of pattern on process. *Journal of Ecology* **80**, 527–534.

Silvertown, J., Lines, C.E.M. & Dale, M.P. (1994) Spatial competition between grasses – rates of mutual invasion between four species and the interaction with grazing. *Journal of Ecology* **82**, 31–38.

Slade, A.J. & Hutchings, M.J. (1987a) The effects of light intensity on foraging in the clonal herb *Glechoma hederacea*. *Journal of Ecology* **75**, 639–650.

Slade, A.J. & Hutchings, M.J. (1987b) The effects of nutrient availability on foraging in the clonal herb *Glechoma hederacea*. *Journal of Ecology* **75**, 95–112.

Spicer, K.W. & Catling, P.M. (1988) The biology of Canadian weeds. *Elodea canadensis*– Michx. *Canadian Journal of Plant Science* **68**, 1035–1051.

Stoll, P., Egli, P. & Schmid, B. (1998) Plant foraging and rhizome growth patterns of *Solidago altissima* in response to mowing and fertilizer application. *Journal of Ecology* **86**, 341–354.

Suzuki, J. (1994) Shoot growth dynamics and the mode of competition of two rhizomatous *Polygonum* species in the alpine meadow of Mt Fuji. *Folia Geobotanica et Phytotaxonomica* **29**, 203–216.

Svensson, B.M. & Callaghan, T.V. (1988) Apical dominance and the simulation of metapopulation dynamics in *Lycopodium annotinum*. *Oikos* **51**, 331–342.

Svensson, B.M., Floderus, B. & Callaghan, T.V. (1994) *Lycopodium annotinum* and light quality: growth responses under canopies of two *Vaccinium* species. *Folia Geobotanica et Phytotaxonomica*, **29**, 159–166.

Thompson, L. (1993) The influence of natural canopy density on the growth of white clover, *Trifolium repens*. *Oikos*, **67**, 321–324.

Tilman, D. (1985) The resource-ratio hypothesis of plant succession. *American Naturalist*, **125**, 827–852.

Tolvanen, A., Schroderus, J. & Henry, G.H.R. (2001) Demography of three dominant sedges under contrasting grazing regimes in the High Arctic. *Journal of Vegetation Science* **12**, 659–670.

Tuomi, J. & Vuorisalo, T. (1989) Hierarchical selection in modular organisms. *Trends in Ecology and Evolution* **4**, 209–213.

6

Plant dispersal potential and its relation to species frequency and co-existence

Peter Poschlod, Oliver Tackenberg and Susanne Bonn

6.1 Aspects of dispersal

6.1.1 What is dispersal?

Dispersal is the movement of dispersal units away from parent plants. Mostly, dispersal units are seeds and fruits (vascular plants) or spores (pteridophytes, bryophytes among others) but they may be also of vegetative origin such as rhizomes, turions, bulbils. Therefore, the terms 'diaspores' (from the Greek *diaspeiro* = I sow) and 'propagules' are often found in the literature (Bonn & Poschlod 1998) but the term 'seed' is often used as an equivalent, particularly in the term 'seed bank'. Dispersal can occur both in space and time. Dispersal in space is the transfer of diaspores over certain distances whereas dispersal in time means that diaspores are able to survive in the soil or the above-ground diaspore bank over long periods of unsuitable conditions before germinating.

6.1.2 Why dispersal?

Dispersal is considered important for many reasons. Dispersal in space is important for escaping from the parent plant, which is necessary (i) to avoid intraspecific competition with the parent plant and with other seedlings from the same individual; (ii) to avoid inbreeding; and (iii) to avoid predation by animals whose activities depend on the density of the individuals which is highest in the vicinity of the parent plant (Hildebrand 1873 in Bonn & Poschlod 1998; Howe & Smallwood 1982; Dirzo & Domingues 1986). Dispersal in space enables species not only to recolonize unoccupied sites but also to colonize new suitable habitats. Dispersal in time, i.e. the maintenance of a diaspore bank, is important because after 'bad' years (environmental changes, catastrophes) a population may become extinct, whereas a persistent diaspore bank may buffer such years (Kalisz & McPeek 1993;Thompson 2000). Therefore, both dispersal potential in space and time are limiting factors within the dynamics of metapopulations (Husband & Barrett 1996; Bonn & Poschlod 1998; Cain *et al.* 2000). Dispersal in space affects the level of gene flow (Young *et al.* 1996) and therefore influences processes such as local adaptation or speciation (Harrison & Hastings 1996). A persistent diaspore bank may at least be a reservoir of genetic variability (Levin 1990; Vavrek *et al.* 1991) to cope with future environmental changes. On the habitat level, dispersal in both time and space are important features especially in ephemeral habitats such as wind-thrown or fire areas within forests or irregularly drained areas in floodplains (Hanski 1987).

On the community level dispersal is related to co-existence – dispersal is a driving factor in the carousel model (van der Maarel & Sykes 1993), and may limit species richness (dispersal in space as the key of the species pool concept; Zobel 1997), diversity and dynamics (Cain *et al.* 2000).

Although there is general agreement that dispersal in space is negatively correlated to dispersal in time (Rees 1993; Ehrlén & van Groenendael 1998), Thompson (2000) concluded that it is 'unreasonable to expect a universal trade-off between dispersal (in space) and persistence (dispersal in time)'. The dispersal potential of a species is too complex to be treated as one discrete variable. Furthermore, dispersal studies often neglect germination and establishment, steps towards successful colonization. Eriksson & Jakobsson (2000) showed that there are also trade-offs between the evolution of dispersal mechanisms and recruitment.

6.2 Brief historical review

Theophrast and Plinius were the first to comment on the dispersal of plants, noting that the germination of seeds from mistletoe is enhanced by birds. It was not until the 18th century that different ways of dispersal were recognized and described in detail by Rumphius and Linné; see Bonn & Poschlod (1998). Linné still thought that all plants had dispersed from a central mountain island which he believed to be a 'paradise' in relation to the actual habitats. His students were probably the first who performed dispersal experiments by feeding propagules of more than 800 (!) plant species to cattle, sheep, goats, horses and reindeer (Bonn & Poschlod 1998). In the first half of the 19th century, dispersal ecology was of little interest since botanists were convinced that species could establish in all habitats where the climatic conditions are suitable – while overestimating the ubiquitous distribution of plants. Only when plant distribution patterns were analysed in detail, discussions on long-distance dispersal started. Darwin did experiments on survival of seeds in sea water and found that out of 87 species, 64 germinated after an immersion of 28 days, and a few survived an immersion of 137 days. He concluded that plants belonging to one country might float across as many as 924 miles of sea to another country and still be able to germinate. Darwin was also one of the first to recognize that dispersal in time was also important, i.e. for the recolonization of sites. From that time on, studies on plant dispersal became frequent (Bonn & Poschlod 1998).

6.3 The dilemma in plant population and vegetation-related dispersal research

Most statements made on 'why dispersal' are still based on theoretical considerations because dispersal is not easy to measure in the field. Detailed data on the diaspore rain of an individual or population within their habitat or even a landscape are extremely rare. As to dispersal in time, burial experiments – the only method to estimate the exact longevity of diaspores in the soil – are rare as well. Additionally, a dispersal event is only successful if germination and establishment occurs, and studies combining this with dispersal are even rarer (Eriksson & Jakobsson 2000). Although dispersal in space and time is acknowledged as a key factor in the survival of subpopulations within metapopulations, it is hardly taken into account in plant metapopulation studies (Husband & Barrett 1996). Recently, the functional analysis of vegetation dynamics and species co-existence has supported the hypothesis that dispersal is a relevant 'filter' next to an abiotic filter (Zobel 1997), but this support is still based on too simple a classification of dispersal types (e.g. Grime *et al.* 1988).

6.4 Dispersal in space

6.4.1 Dispersal vectors, dispersal types, dispersal potential and distances

There is a long history of classifying dispersal types using specific terms, including the monographs of H.N. Ridley, L. van der Pijl, P. Müller-Schneider and H.W. Luftensteiner (see Bonn & Poschlod 1998). These classification systems are based on the morphology of the dispersal unit which is interpreted as an adaptation to a specific dispersal vector (Table 6.1). This may be meaningful for the classification of a single dispersal event; however, the allocation to a certain dispersal type still lacks validation. Furthermore, the simple, binary assignment to a certain dispersal type is limited to correlate species dispersability to population or vegetation ecological issues such as rarity/frequency, species richness of plant communities or species co-existence – every propagule may be dispersed by every vector and a wide range of dispersal distances can be covered. Consequently, Tackenberg *et al.* (2003a) proposed a gliding classification of the dispersal potential by a certain vector. Based on a mechanistic simulation model (Tackenberg 2003) they showed that the dispersal potential of species classified from their morphology of being wind-dispersed range in a scale of 0 (extremely low) to 9 (extremely high; Fig. 6.1). For other vectors, a system based on simple rules was developed which allows the assessment of a dispersal potential ranging from 0 to 3 (Tackenberg 2001).

As to dispersal distance, the terms 'short-distance' and 'long-distance' dispersal are often used, but without definition. Long-distance should be used if populations are connected (Hansson *et al.* 1992) or new habitats are colonized. Dispersal curves are leptocurtic which means that the majority of seeds is deposited within shorter distances. Exceptional long-distance dispersal events cannot be measured. Silvertown & Lovett-Doust (1993) claim that 'the tail of the seed dispersal curve is impossible to

Table 6.1 Classification of dispersal types based on the dispersal vector after Bonn *et al.* (2000); see Bonn & Poschlod (1998) for a literature review. Dispersal types and vectors in bold and italics indicate a high potential for long-distance dispersal; barochory (dispersal by gravitation) is excluded from this system, because the distinction between anemochory and barochory is gradual.

Dispersal type		Dispersal by
Autochory	a. ballochorous	ejection by the parent plant
	b. blastochorous	deposition by parent pant
	c. herpochorous	creeping hygroscopic hairs of the diaspore
Semachory	semachorous	swaying motion of the parent plant caused by external forces (wind, . . .)
Anemochory	*anemochorous*	*wind*
Hydrochory	a. ombrochorous	ejection caused by falling rain drops
	b. *nautochorous*	*water* (moving on water-surface)
	c. *bythisochorous*	flowing *water* (on the ground)
Zoochory	a. myrmekochorous	ants
	b. *ornithochorous* (*epizoo-, endozoo-, dysochorous*)	*birds* (on the body surface, via ingestion, via transport for nutrition)
	c. *mammaliochorous* (*epizoo-, endozoo-, dysochorous*)	*mammals* (on the body surface, via ingestion, via transport for nutrition)
	d. others (epizoo-, endozoo-, dysochorous)	man, other animals (snails, earthworms, . . .)
Hemerochory	a. *agochorous*	human action (work, trade . . .)
	b. *speirochorous*	impure seedcorn
	c. *ethelochorous*	trading as seedcorn

reach' and further 'the occasional seed is carried by chance events quite extraordinary distances, but these seeds are so few that we only know where they end up when they attract attention by starting a new population in an alien site'. This was recently confirmed by field studies (Bullock & Clarke 2000) and simulation models (Higgins & Richardson 1999; Nathan *et al.* 2002).

6.4.2 Measurement of dispersal types and potential

There are many methods to measure dispersal and to assess the dispersal potential of a certain vector which may include direct measurements, indirect measurements (e.g. by genetic analysis) or simulation models. Furthermore, sowing experiments may show if dispersal is a limiting factor (Turnbull *et al.* 2000). Direct measurement in the field includes the catchment of diaspore rain by diaspore traps; these may be funnels in terrestrial habitats or drift nets in aquatic habitats (Bakker *et al.* 1996). A transect sampling design away from the diaspore source allows the assessment of dispersal distances (Bullock & Clarke 2000) as well as the release of diaspore or 'mimicries' at a certain point (Johansson & Nilsson 1993; Bill *et al.* 1999).

Studies on dispersal of diaspore by animals are mostly anecdotal. Diaspore acquisition and transport is passive by attachment to fur or feet or through food consumption.

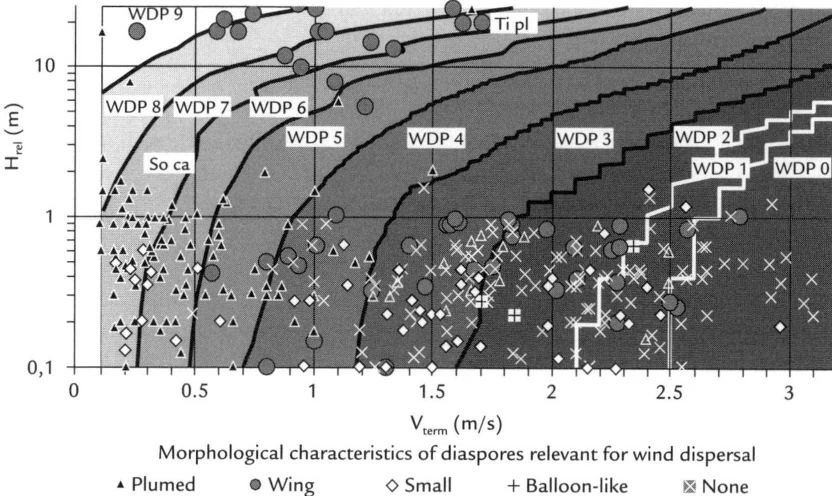

Fig. 6.1 Indicator values of the wind dispersal potential calculated for a reference distance of 100 m (WDP_{100}). 335 plant species are located in the diagram according to the falling velocity of their diaspores (V_{term}) and their releasing height (H_{rel}) and marked by their diaspore morphology. So ca = *Solidago canadensis*; Ti pl = *Tilia platyphyllos*. Wind dispersal potential ranges from 0 = very low to 9 = very high; 0 = < 0.2% of diaspores exceeding the reference distance; 1 = 0.2–0.4%; 2 = 0.4–0.8%; 3 = 0.8–1.6%; 4 = 1.6–3.2%; 5 = 3.2–6.4%; 6 = 6.4–12.8%; 7 = 12.8–25.6%; 8 = 25.6–51.2%; 9 = > 51.2%. After Tackenberg *et al.* (2003a).

Seeds and fruits can also be collected actively for storage or as food (Stiles 2000). Passive transport of diaspores on fur was studied on dead and living animals. However, the collection of diaspores from living animals depends on the 'cooperation' of the animal involved. Alternatively, 'dummies' (Fischer *et al.* 1996) or trained animals such as dogs (Heinken 2000) are used. Dispersal distances can be determined by attachment experiments (Fischer *et al.* 1996). Dispersal distances were derived by looking at the loss of marked diaspores after distinct time periods.

The study of passive transport of diaspores by herbivores has a long tradition. Cattle were fed with seed-containing material and viability of seeds in the dung was tested after excretion by E. Kempski as early as 1906 (Bonn & Poschlod 1998). However, the study of dung deposition is also possible by collecting dung in the field (Welch 1985). Subsequently, dispersal distances can be determined by knowing movements during digestion and time period of digestion (Bakker *et al.* 1996). However, approaches which are suitable for collecting quantitative data in the field can hardly be used for plant communities since they are extremely time consuming and single diaspores dispersed over long distances are only detected by chance. Therefore, experimental and modelling approaches were developed, mainly for wind dispersal. In wind tunnel experiments dispersal distances were measured (Strykstra *et al.* 1998) but these lack reality. Tackenberg (2003) showed that models which do not include the turbulence factor do not give realistic results.

Table 6.2 Potential long-distance dispersal vectors and parameters affecting (+) the dispersal potential.

Dispersal vector	Wind	Water	Animals		Man
			Ectozoo-chorous	Endozoo-chorous	
Parameter affecting the dispersal potential					
Height of infructescence/releasing height	+	–	+	–	–
Seed production	+	+	+	+	+
Time and duration of seed release	+	+	+	+	+
Falling velocity	+	–	–	–	–
Buoyancy	–	+	–	–	–
Attachment capacity	–	–	+	–	–
Digestion tolerance	–	–	–	+	–

Finally, a comparative assessment of the dispersal potential of each vector needs standardized methods and measurements. Therefore, it is necessary to find out which plant parameters affect the dispersal potential such as releasing height and diaspore production (Table 6.2) and which diaspore characteristics may be correlated to the dispersal potential by a certain vector. The availability of such data only allows the testing of different hypothesis if species co-existence and/or richness in plant communities is related to dispersal. However, dispersal may not be successful because of secondary dispersal, predation, death before germination or unsuccessful germination. The only meaningful approach, which may be time consuming but exhibits exact data, is the genetical analysis of populations on different spatial scales (e.g. Willerding & Poschlod 2002). Maternal markers such as chloroplast or mitochondrial DNA allow the identification of mother plants and successfully established offspring (Ouborg *et al.* 1999).

6.4.3 Species traits affecting dispersal potential

There are several plant traits affecting the dispersal potential by the different vectors (Table 6.2). Height of infructescence, seed production and time and duration of seed release are parameters which influence the dispersal potential of every vector. A specific vector such as animals (e.g. migrating birds, grazing livestock), water (e.g. flood events) or man (e.g. seeding, mowing) may be available for the dispersal of diaspores only during distinct periods. Wind dispersal potential is also affected by seasons and the surrounding vegetation since weather, including wind speed, changes during the year and in different environments. A higher diaspore production increases the probability that an exceptional long-distance dispersal event occurs (Tackenberg 2001). The area which can be covered by a certain dispersal vector is larger and the diaspore input per area is higher the more seeds are produced.

6.4.4 Diasporus – a seed dispersal database

Diasporus is a database referring to a vector-based dispersal classification for the Central European flora (Bonn *et al.* 2000), now integrated in BioPop (Poschlod *et al.*

Table 6.3 Seed characteristics correlated to parameters affecting the dispersal potential in space and time.

Seed characteristics	Diaspore mass/ diaspore size	Diaspore shape	Diaspore surface		Diaspore coat	
			Structure	'Hydrophoby'	Thickness	Cells air-filled
Parameter affecting the dispersal potential in space						
Diaspore production	+	–	–	–	–	–
Falling velocity	+	+	+	–	–	–
Buoyancy	+	–	+	+	–	+
Attachment capacity	–	–	+	–	–	–
Digestion tolerance	+	+	+	?	?	–
Parameter affecting the dispersal potential in time						
Persistence	+	+	–	–	–	–

2003). It includes dispersal-relevant traits of the parent plant as well as the diaspore. The application of the database allows a viability analysis and risk assessment for plant species referring to dispersal as a limiting factor as well as the prediction of a local species pool based on the knowledge of the regional species pool and the available dispersal vectors (e.g. Poschlod *et al.* 1998).

6.4.5 Wind

Wind is probably the most common dispersal vector since every propagule may be dispersed by wind (Tackenberg *et al.* 2003a). However, only some have a high wind dispersal potential and are predestined for long-distance dispersal and many species commonly classified as wind-dispersed show low wind dispersal potentials. Propagule traits correlating with a high wind dispersal potential are: (i) low falling velocity; (ii) shape; and (iii) surface structure combined with releasing height (Tables 6.2 and 6.3).

Parameters affecting long-distance dispersal are weather conditions (Tackenberg 2003). The most important parameter is not wind speed but the occurrence of updrafts (Nathan *et al.* 2002; Tackenberg *et al.* 2003b). Plant communities with a high proportion of species with a high wind dispersal potential occur in open landscapes (tundra, alpine belt, grasslands; Fig. 6.2a) or ephemeral habitats (river banks). Many species of these habitats with a high wind dispersal potential have a high seed production but a low dispersal potential in time such as in *Myricaria* spp., *Salix* spp. and *Taraxacum* spp.

6.4.6 Water

Water, especially running water, may transport many propagules (Bill *et al.* 1999). They are transported either on the water surface or along the surface (Table 6.1). Distances which can be covered depend not only on their floating capacity or buoyancy but also on whether they survive transport along the surface which is hampered by sandy or gravel material. Floating capacity depends not only on the specific weight but also on the surface structure and the hydrophoby of the fruit or seed coat. In

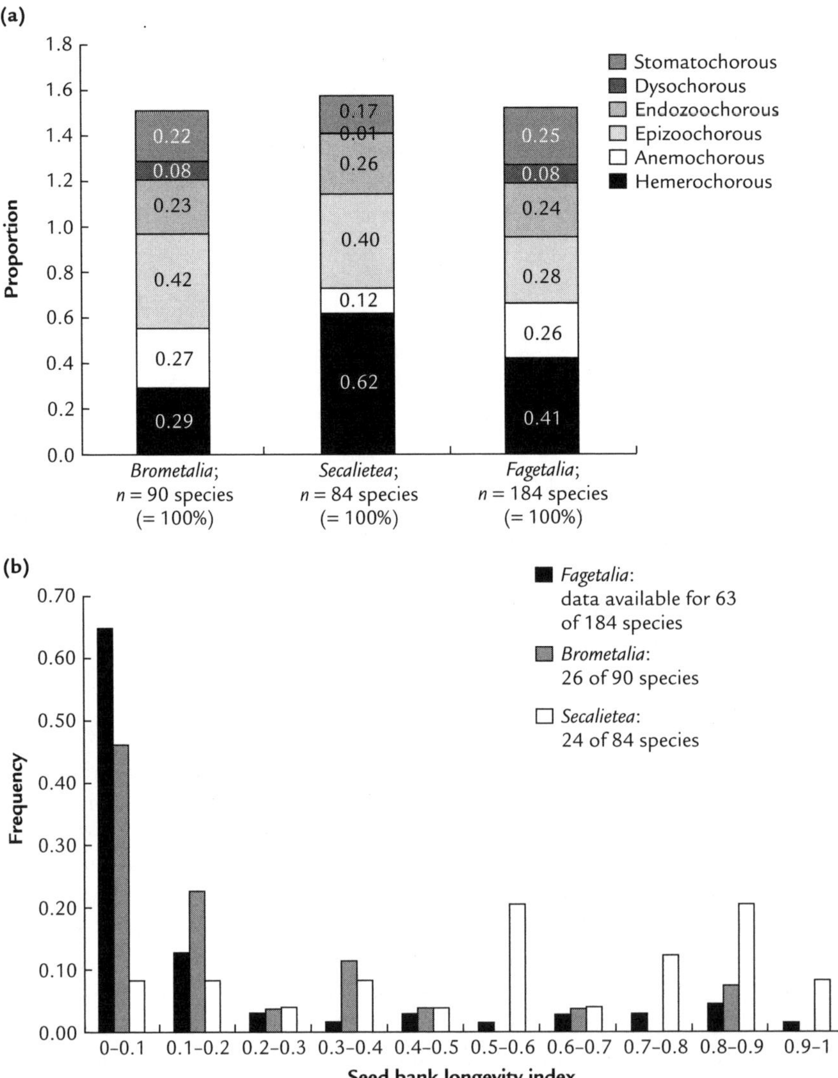

Fig. 6.2 Dispersal spectra **(a)** and diaspore bank longevity spectra **(b)** for agricultural weed communities (*Secalietea*), semi-natural calcareous grasslands (*Brometalia*) and beech forests (*Fagetalia*) of Central Europe. Calculation of the seed bank longevity index (0–1) follows Thompson *et al.* (1998). Phytosociological classification of the species according to Ellenberg *et al.* (1992); dispersal type data according to the Diasporus data base (Bonn *et al.* 2000). Only dispersal types with relevance for long-distance dispersal are considered.

contrast to wind, only specific water-related habitats such as river banks or flooded habitats are connected by this vector. Therefore in this case suitable conditions for germination and establishment are often more limiting for the occurrence of a species, as was shown for *Ranunculus lingua* (Johansson & Nilsson 1993). Plant communities with a high proportion of water-dispersed species are forests, reeds, meadows in floodplains or around lakes, peatlands and aquatic habitats.

6.4.7 Animals

Animals are (beside humans) probably the most important vector for long-distance dispersal. Many animal species, both vertebrates and invertebrates (e.g. ants, earthworms, beetles, snails) transport propagules. However, invertebrates are not able to transport diaspores over large distances (ants: maximum 80 m according to E. Ulbrich, see Bonn & Poschlod 1998). Diaspores of most trees and shrubs are spread by birds. Birds were thought to be responsible for the rapid migration and spread of woody species during the postglacial period; this is known as Reid's paradox (see review by Clark *et al.* 1998). They assume that both birds and mammals have contributed to the rapid migration of fleshy-fruited species, with mammals more likely to have provided (occasionally successful) long-distance dispersal. However, diaspore rain by birds is mostly related to suitable resting places (McClanahan & Wolfe 1987). Therefore, it depends on the vegetation structure of woody species and occurs mostly locally (Kollmann & Pirl 1995) but exceptionally up to several kilometres (Willson & Traveset 2000).

Various birds act as effective dispersal vectors by transporting and burying fruits of various woody species as winter food. Because only some of those buried fruits are recovered during winter, many remain in the soil and may germinate several kilometres away from the parent plant (Johnson & Webb 1989).

Large herbivores are regarded as more effective with respect to the number of species (Janzen 1982; Malo & Suárez 1995; Pakeman 2001), especially non-woody plants. Herrera (1989) supposed that carnivorous species also acted as important dispersal vectors during the postglacial period.

Domestic livestock has probably contributed most to the dispersal of species in man-made landscapes as well as to the increase of species in a local and/or regional flora since man began a settled existence in the Neolithic age (Poschlod & Bonn 1998). In traditionally man-made landscapes herds of domestic livestock migrated over large distances either from summer to winter pastures and back (transhumance; e.g. between 100 and more than 300 km (one way) in SW Germany, and more than 800 km in Spain) or to market places (Poschlod & WallisDeVries 2002). Even dogs are very effective dispersal vectors (Heinken 2000). In two studies on cattle and sheep, it could be shown that more than 50% of the local flora was transported either ecto- or endozoochorously (Fischer *et al.* 1996; Stender *et al.* 1997; see also Fig. 6.2a). From these studies it became also obvious that not only diaspores with a sticky surface are transported ectozoochorously but also those with a coarse or even smooth surface. However, the proportion of sticky diaspores is much higher. Dispersal distances which can be covered do not only depend on attachment capacity but also on the process of attachment and release. Most propagules are lost a short time after their attachment whereas comparatively few are lost after longer time periods (Fischer *et al.* 1996).

Size and shape, as well as a hard seed or fruit coat and the palatability of the species itself, affect the probability of endozoochorous dispersal (Janzen 1971; Pakeman *et al.* 2002). Dispersal distances are a function of migration speed and retention time which in birds is between 1 h and 4–5 days (maximum 12–13 days); for mammals, between 6 h and 10 days (maximum 70 days) (summarized in Bonn & Poschlod 1998). In case of endozoochorous dispersal it was found that the dung heap may act as a safe site for certain species (Malo & Suarez 1995) or increase the probability of the establishment of demanding species in acid and/or nutrient-poor habitats.

6.4.8 Man

With the beginning of the Neolithic age and the settlement of man, the continuous changes of landscape and vegetation were accompanied by a rapid spread of plants leading to an increase of species number all over the world. Di Castri (1989) summarized the most important processes which were the driving forces for the spread of plants since the Neolithic revolution up to the 15th century; from the 15th century until recently; and since the last century with the introduction of, *inter alia*, intensive agricultural practices, trade and traffic. One of his most important conclusions was that, since the beginning of human settlements, the globalization and acceleration of dispersal processes increased. On the other hand, according to Poschlod & Bonn (1998), in the industrial countries of the northern hemisphere, especially in the European man-made landscape, the industrial revolution followed by more and more intensive agricultural and forestry practices led to a tremendous reduction in the diversity of dispersal processes which even explains the extinction of plant populations (Fig. 6.3). Some of the numerous human-related dispersal processes are linked to the development of, and recent changes in, plant communities.

6.4.8.1 Dispersal through agricultural practices. Agricultural practices include processes with a high dispersal potential between the same or different habitats on a regional or local level. They include, *inter alia,* sowing of uncleaned seed, fertilizing, irrigating, harvesting and mowing. From archeological findings (e.g seed from the Thirty Years War period) and the recent extinction of arable weeds (Table 6.4) it became obvious that sowing of uncleaned seed contributed considerably to the development of species-rich arable weed communities and that the recent sowing of cleaned seed has brought about species-poor communities (Poschlod & Bonn 1998). A classical example is the history of *Agrostemma githago*. It was introduced in the Neolithic age and was spread widely by uncleaned seed because its size was similar to that of cereal seed. Today, it is extinct in many places since its dispersal in time is limited as well; it is now re-established by sowing seeds in fields in open-air museums.

Fertilizers in traditionally used landscapes were very diverse (e.g. manure, composts, sods of heathland and/or forest, freshwater-mud) and contained many diaspores whereas today mineral fertilizers are applied. Slurry and sewage waste contain diaspores but of very specific species (e.g. *Chenopodium* spp.) able to survive these specific conditions (Poschlod & Bonn 1998).

Different practices of harvesting cereals since the Neolithic age resulted in a specific arable weed flora due to dispersal adaptations of the respective species (Poschlod & Bonn 1998). Traditional mowing by scythe was not suitable for the dispersal of

Table 6.4 Endangered or extinct arable weeds of Germany, which were often or obligately spread with cereal seed; from Poschlod & Bonn (1998), db = diaspore bank type: ts = transient; sps = short-term persistent; lps = long-term persistent; if more than two sets of data are given the most frequent result is given in bold.

Species	db		db
Spread with cereal seed		**Only in the summer cereal seed**	
Adonis aestivalis	?	*Lolium temulentum*	?
Adonis flammea	?	*Vaccaria hispanica* ssp. *hispanica*	?
*Agrostemma githago**	ts		
Bromus secalinus	ts	**Only in the seed of flax**	
Fagopyrum tataricum	ts?	*Camelina alyssum**	sps?
Galium tricornutum	ts	*Cuscuta epilinum**	?
Lathyrus aphaca	ts?	*Galium spurium* ssp. *spurium**	?
Rhinanthus alectorolophus ssp. *bucalis*	ts	*Lolium remotum**	?
Rhinanthus serotinus ssp. *apterus*	ts	*Silene linicola**	?
		*Silene cretica**	?
		Spergula arvensis ssp. *linicola**	ts-**lps**

* = Extinct according to the German Red Data Book.

seeds. In contrast, modern mowing machineries may transport large quantities of seeds (Strykstra *et al.* 1996). A traditional practice to establish grasslands was the application of hayseed. Species-rich dry grasslands such as calcareous grasslands in abandoned vineyards and even litter meadows were established through this practice (Poschlod & WallisDeVries 2002).

Hemerochory (dispersal by man) plays an important role in many plant communities of the man-made landscape, especially of arable weed communities (Fig. 6.2a).

6.4.8.2 Dispersal by traffic and trade. Nowadays, traffic and trade may be the most effective dispersal vectors over very large distances (Bonn & Poschlod 1998). Through trade (especially of garden plants) many neophytes could spread and establish elsewhere (Di Castri *et al.* 1990). Distribution maps of recently spreading species show their rapid migration along roads (e.g. Ernst 1998). However, diaspores transported in this way mostly germinate along road verges (Hodkinson & Thompson 1997) and many habitats are not connected by roads.

6.5 Dispersal in time

6.5.1 Types of dispersal in time (seed bank longevity/persistence)

Until recently, seed (diaspore) banks were classified with reference to the division proposed by Thompson & Grime (1979) into transient (viable < 1 yr) and persistent (viable > 1 yr). Thompson *et al.* (1997) improved this classification by defining three types:

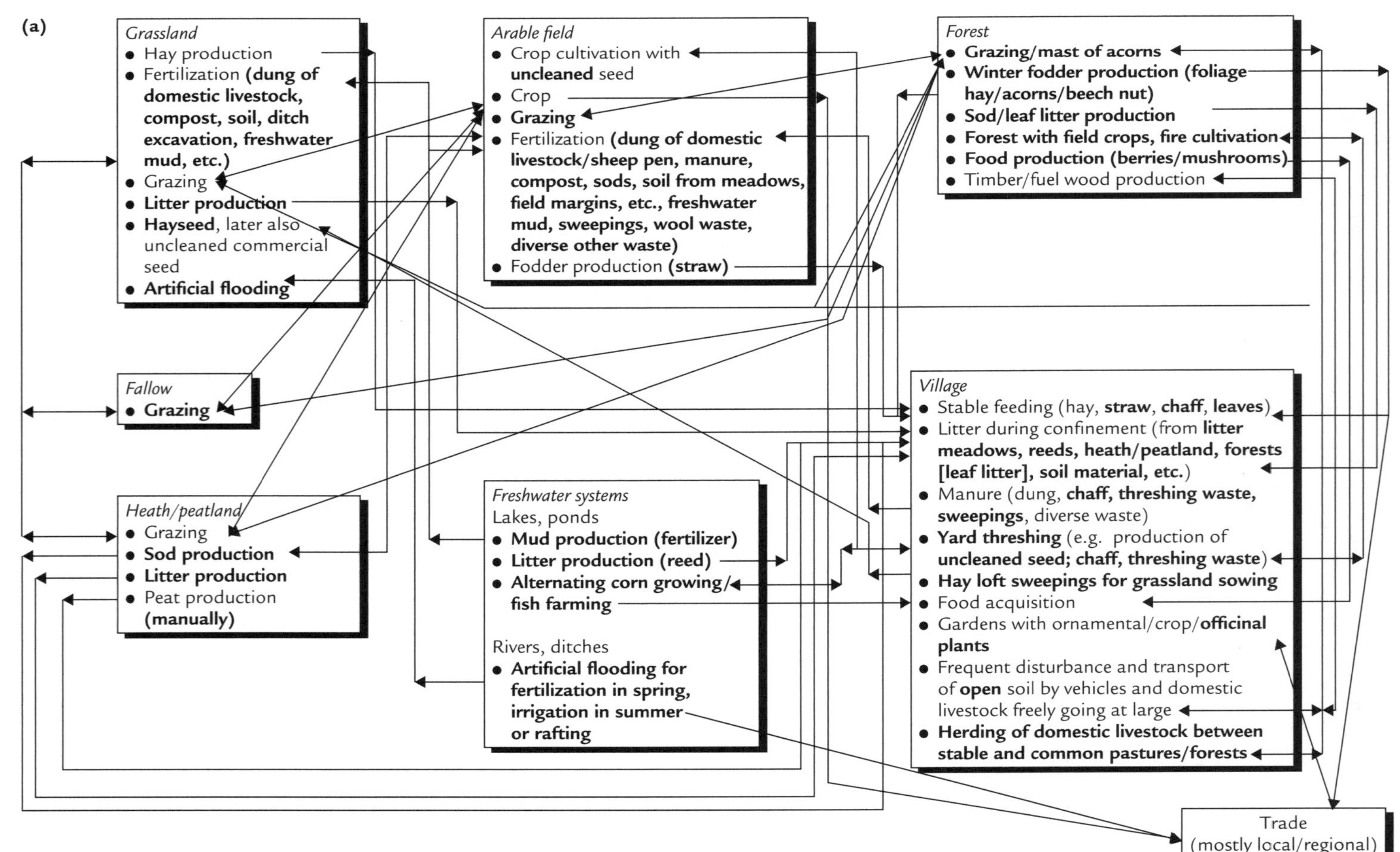
(a)
Grassland
Hay production
Fertilization (dung of domestic livestock, compost, soil, ditch excavation, freshwater mud, etc.)
Grazing
Litter production
Hayseed, later also uncleaned commercial seed
Artificial flooding
Arable field
Crop cultivation with uncleaned seed
Crop
Grazing
Fertilization (dung of domestic livestock/sheep pen, manure, compost, sods, soil from meadows, field margins, etc., freshwater mud, sweepings, wool waste, diverse other waste)
Fodder production (straw)
Forest
Grazing/mast of acorns
Winter fodder production (foliage hay/acorns/beech nut)
Sod/leaf litter production
Forest with field crops, fire cultivation
Food production (berries/mushrooms)
Timber/fuel wood production
Fallow
Grazing
Heath/peatland
Grazing
Sod production
Litter production
Peat production (manually)
Freshwater systems
Lakes, ponds
Mud production (fertilizer)
Litter production (reed)
Alternating corn growing/fish farming
Rivers, ditches
Artificial flooding for fertilization in spring, irrigation in summer or rafting
Village
Stable feeding (hay, straw, chaff, leaves)
Litter during confinement (from litter meadows, reeds, heath/peatland, forests [leaf litter], soil material, etc.)
Manure (dung, chaff, threshing waste, sweepings, diverse waste)
Yard threshing (e.g. production of uncleaned seed; chaff, threshing waste)
Hay loft sweepings for grassland sowing
Food acquisition
Gardens with ornamental/crop/officinal plants
Frequent disturbance and transport of open soil by vehicles and domestic livestock freely going at large
Herding of domestic livestock between stable and common pastures/forests
Trade
(mostly local/regional)

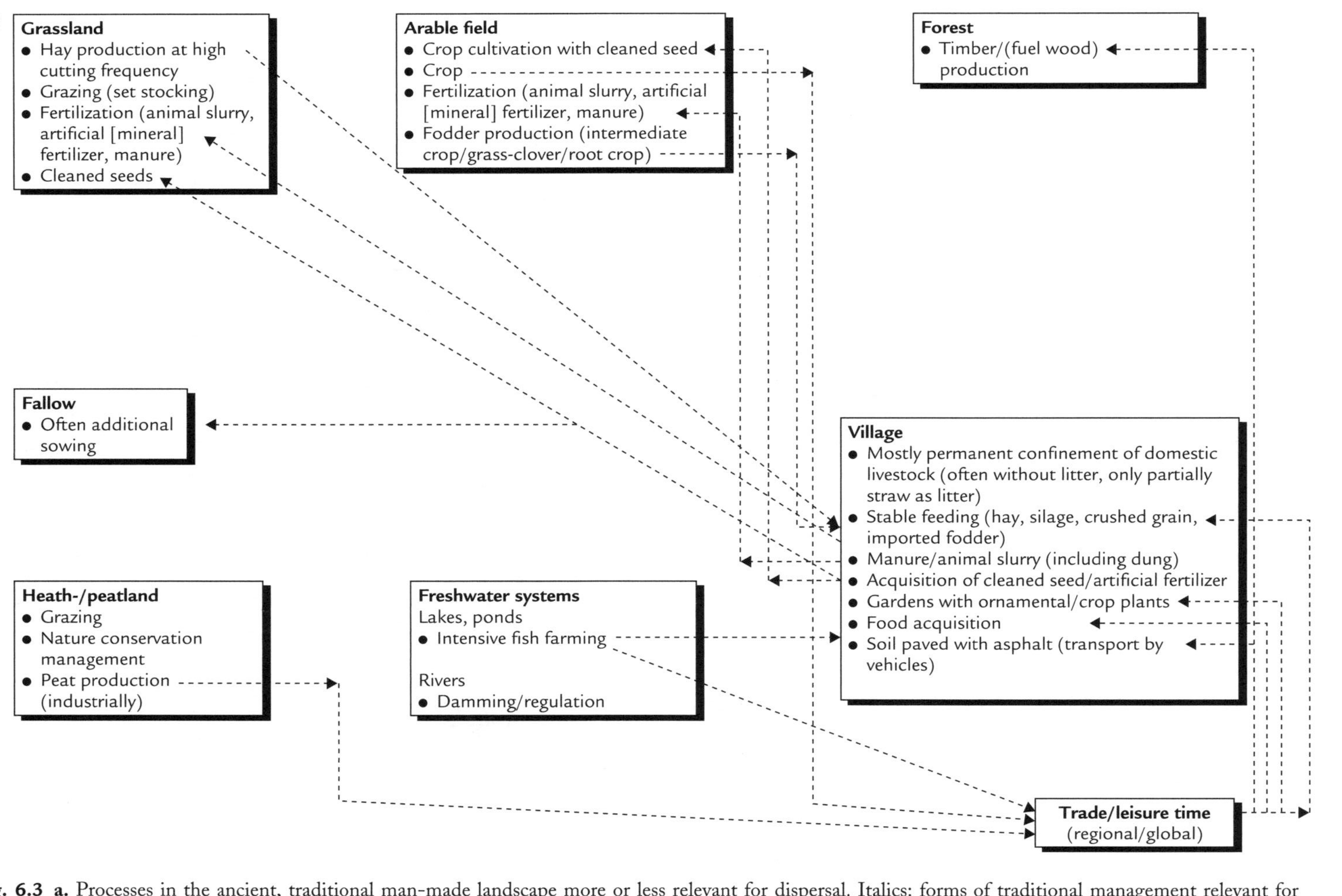

Fig. 6.3 a. Processes in the ancient, traditional man-made landscape more or less relevant for dispersal. Italics: forms of traditional management relevant for dispersal which do not exist any longer; arrows – direction of dispersal (from Poschlod & Bonn 1998). **b.** Processes in the present modern man-made landscape relevant for dispersal. Arrows indicate direction of dispersal; dotted lines indicate reduced dispersal relevance compared to the ancient, traditional man-made landscape. From Poschlod & Bonn (1998).

Transient: species with diaspores that persist in the soil for less than 1 yr, often much less;
Short-term persistent: species with diaspores that persist in the soil for at least 1 yr, but less than 5 yr;
Long-term persistent: species with diaspores that persist in the soil for at least 5 yr.
The cut-off point of 5 yr between the latter two types is arbitrary, and was chosen mainly because it is the end point of a significant number of burial experiments (Bakker *et al.* 1996). The above classification is admittedly very crude, and Poschlod & Jackel (1993) published a further elaboration of the above-mentioned classification which refers to the dynamics of the diaspore bank and seed rain combining seasonal behaviour with depth distribution. However, data are only available for central European dry calcareous grassland communities.

6.5.2 Measurement of dispersal in time/persistence/longevity of diaspore banks

An exact measurement of persistence is only possible by burial experiments like those from W.J. Beal and J.W.T. Duvel which, however, were mostly not performed under natural conditions (Priestley 1986). Another direct, but not equally exact, method is the radiocarbon dating of viable diaspores, ideally of dead parts such as the pericarp or testa (McGraw *et al.* 1991). Since burial experiments are time consuming and radiocarbon dating expensive, several approaches were developed from diaspore bank sampling to estimate persistence which were recently summarized by Bakker *et al.* (1996) and Thompson *et al.* (1997). The most simple method is to compare the diaspore bank with the actual vegetation, but this is not very reliable. Methods to sample the diaspore bank and to estimate the number of viable diaspores are described by Bakker *et al.* (1996) and ter Heerdt *et al.* (1996, 1999).

The degree of persistence can be estimated indirectly by analysing the depth distribution. The higher the proportion of diaspores in deeper layers compared to the number of diaspores in upper layers the more persistent is the diaspore bank. Since diaspore bank sampling on a single date mostly does not catch the transient species, the study of seasonal diaspore bank dynamics gives the best results (Thompson & Grime 1979; Poschlod & Jackel 1993). Thompson *et al.* (1998) proposed the calculation of a longevity index, which is only a provisional tool (see further section 6.5.4).

Another indirect method to estimate the persistence is the study of diaspore banks along successional seres (Poschlod 1993). However, persistence is not entirely specific for a species, population or individual. Environmental factors may affect persistence of soil diaspore banks. Light and, therefore, soil disturbance is probably the most important factor to deteriorate soil diaspore banks. Deterioration of diaspores (of weeds and crops) took place more rapidly in an organic, acidic peat soil than in mineral soil of fairly neutral pH (Lewis 1973). Diaspores may survive for a longer time under wet (Bekker *et al.* 1998a) or dry conditions (Murdoch & Ellis 2000). In sediments of bogs, lakes and ponds the highest amounts of viable diaspores were found (Skoglund & Hytteborn 1990; Poschlod 1995; Poschlod *et al.* 1996). Nutrients (nitrate) reduce persistence as well by affecting dormancy and releasing germination (Bekker *et al.* 1998b).

6.5.3 Diaspore characters correlated with diaspore persistence/longevity

Persistence or longevity of diaspores is correlated with diaspore size and shape (Thompson *et al.* 1993; Bekker *et al.* 1998c). The hypothesis behind these facts is that smaller and round diaspores are buried easier than large and long diaspores or those with attachment such as hooks and bristles.

6.5.4 Diaspore bank database

A first diaspore bank database was published for the Northwest European Flora (Thompson *et al.* 1997). It contains data on more than 1100 species and provides information on the seed bank type, the longevity (if available), density and sampling method. The seed bank type was classified in transient, short-term persistent and long-term persistent. Since data on longevity are scarce, Thompson *et al.* (1998) described the calculation of a longevity index L expressed in a continuous scale from transient to persistent:

L = (short-term + long-term persistent records)/(transient + short-term + long-term persistent records)

The index varies between 0 and 1, where 0 means no persistent records and 1 only persistent records.

6.6 Dispersal spectra of plant communities

Plant communities may have very specific dispersal spectra as was shown for dispersal in space by the classical works of R. Molinier & P. Müller from 1938 and P. Dansereau & K. Lems from 1957, and by more recent studies of Luftensteiner (1982) and Willson *et al.* (1990, who cite the old references) and for dispersal in time by Bekker *et al.* (1998d). Hodgson & Grime (1990) presented dispersal spectra, both in space and time, for different types of habitats. However, studies on dispersal in space neglect the facts that on the one hand the usual classification according to morphology is often wrong and does not reflect reality, and on the other hand that the dispersal of a species often proceeds in different ways (polychory). A comparison of three different vegetation types, using the diasporus database and taking into account these problems, shows that the differences are not as big as was shown by the above-mentioned authors. However, in arable field vegetation, hemerochory is more dominant than in grasslands and in forests whereas in grasslands anemochory and zoochory is more dominant as well as in forests (Fig. 6.2a). On calculating the longevity index clear differences were shown in Dutch plant communities by Bekker *et al.* (1998d). Plant communities of ephemeral, e.g. amphibious or often disturbed habitats such as arable weed communities, exhibit more species with a high longevity index than those of grasslands and forests. Similar results are obtained for central European plant communities (Fig. 6.2b). Arable weed communities contain a higher proportion of species with a long-term persistent diaspore bank than calcareous grasslands and forests where species with a transient diaspore bank are more frequent. However, if there are so many possibilities for species of any plant community to be dispersed, can species richness and composition of plant communities be dispersal limited?

6.7 Dispersal as a limiting factor for species richness in plant communities

6.7.1 Species pools

Theories explaining species co-existence are mostly 'concerned with small-scale (local) processes occurring in ecological time'. However, 'the availability of seeds has been largely disregarded' (Zobel 1997). Therefore, the species pool-concept claims that dispersal is limiting species-richness in plant communities. The local pool which is a set of species occurring in a certain landscape and able to co-exist in a community depends on the long-distance dispersal or large-scale migration ability of the species of the regional pool (set of species occurring in a certain region or, let us say, phytogeographical unit). The actual pool which is the set of species present in the respective community depends on the long-distance dispersal potential on a comparatively small or local scale, the landscape scale (Fig. 6.4). Large-scale migration can only be understood on a historical scale (Clark *et al.* 1998; Cain *et al.* 2000) whereas local-scale migration can occur in very few years or decades (Zobel 1997). As was shown by Pärtel *et al.* (1996), local-scale migration is affected by isolation, successional stage, local management history and other factors (Keddy 1992). A problem in the study by Pärtel *et al.* is that the species pool hypothesis was confirmed only indirectly by correlating the species pool with pools on a different spatial scale (regional, actual, species richness per m^2). The different dispersal potentials of the species were not taken into account.

A different approach was developed by Tackenberg (2001), who tested the occurrence of species in a clearly defined grassland community on dry and isolated outcrops near Halle (Germany), belonging to the Thymo-Festucetum. First the species of the regional species pool – growing only in this community – were selected. Then the number of populations of the species present was correlated with the dispersal potential. This correlation was high, and even higher when only 15 character species were included with similar growth form and a transient diaspore bank (Fig. 6.5).

Other factors, related to germination and establishment or persistence, should not be neglected in concepts of species co-existence (Gigon & Leutert 1996; Eriksson & Eriksson 1997; Eriksson & Eriksson 1998; Fig. 6.4). Local-scale migration processes have changed throughout time in man-made landscapes. Therefore, it is obvious that the re-establishment of plant communities which have developed under historical/traditional land use is not possible if these processes cannot be replaced by other ones; also the dispersal capacity in time should be considered. This hypothesis was

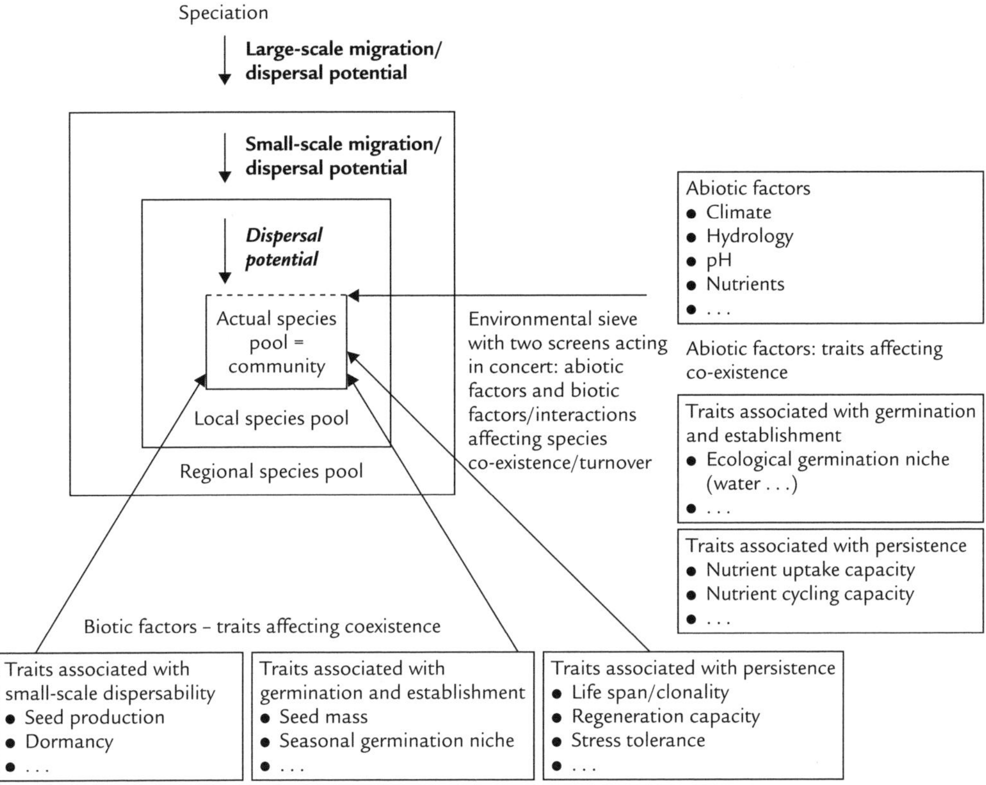

Fig. 6.4 Factors affecting species richness and co-existence in communities. Dispersal-related factors in bold.

validated by several experiments which simulated these processes by the artificial introduction of species by sowing, hay seeding or spreading (Primack & Miao 1992; Poschlod *et al.* 1998; Fig. 6.6). Recently, an exclosure experiment has shown that a community will be more species-rich the more dispersal vectors are operational in the exclosures, with goats and sheep as major agents (B. Bugla & P. Poschlod *in prep.*). Furthermore, vegetation structures may attract seed-dispersing animals. In several savanna-like ecosystems a specific sub-canopy flora occurs, the species of which are characterized by fleshy-fruity species dispersed by birds and small mammals visiting the trees or nesting in them (Guevara *et al.* 1992).

6.7.2 Case study in man-made species-rich wet meadows

Many restoration experiments have shown that success depends on species dispersability in space and time of the target communities (Bakker *et al.* 1996). A community-level experiment in wet meadows in south western Germany based on these experiences is shown in Fig. 6.6. Wet meadows belonging to the Molinion are the most species-rich wet meadows in central Europe. They were drained and fertilized during the

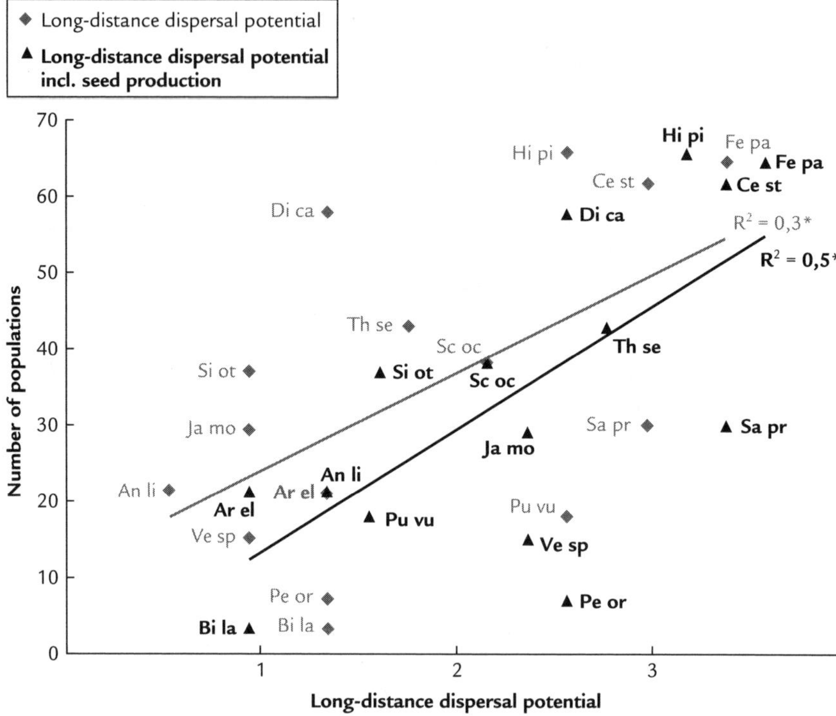

Fig. 6.5 Correlation between number of populations of plant species (a) growing on isolated porphyry hills near Halle (Germany) and (b) not differing in their essential functional traits except the dispersal potential (perennial hemicryptophytes and chamaephytes without long-term persistent seed bank) and species dispersal potential in space (grey line; $R^2 = 0.30$). A stronger correlation (black line; $R^2 = 0.50$) appears when seed production of the species (symbols in bold) is included in the calculation of the dispersal potential (the higher the seed production, the higher the dispersal potential). Classification of the potential for long-distance dispersal (< 1 – very low, > 3 – very high) is built on mechanistic simulation models (wind) or rules (zoochory, hemerochory). After Tackenberg (2001). Species abbreviations are as follows: An li = *Anthericum liliago*, Ar el = *Armeria maritima* ssp. *elongata*, Bi la = *Biscutella laevigata*, Ce st = *Centaurea stoebe*, Di ca = *Dianthus carthusianorum*, Fe pa = *Festuca pallens*, Hi pi = *Hieracium pilosella*, Ja mo = *Jasione montana*, Pe or = *Peucedanum oreoselinum*, Pu vu = *Pulsatilla vulgaris*, Sa pr = *Salvia pratensis*, Sc oc = *Scabiosa ochroleuca*, Si ot = *Silene otites*, Th se = *Thymus serpyllum*, Ve sp = *Veronica spicata*.

1960s and 1970s when the stable management changed from litter to liquid manure management. An attempt was made, therefore, to restore habitat conditions by nutrient impoverishment management and rewetting. However, this was not successful because dispersability of species in time and space was very low and suitable gaps for germination and establishment were lacking.

Both experimental sites were formerly species-rich wet meadow communities in a peatland area, which were drained and fertilized and lost many species. After restor-

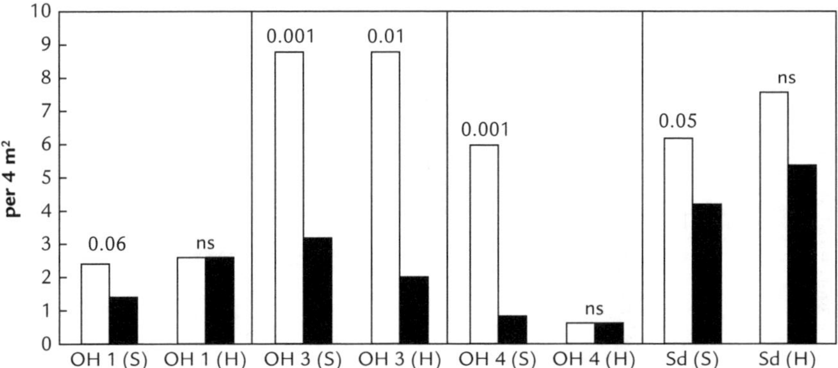

Fig. 6.6 Number of established higher plants re-introduced in 1997 by sowing (S) or hay spreading (H) in restored wet meadow habitats at two locations (OH = Oggelshausen, Sd = Sauden) in the Federsee basin (Baden-Wuerttemberg, Germany). Black columns = without soil disturbance; white columns = with soil disturbance by harrowing; numbers above the columns show significant differences between the two disturbance treatments (n = 5); ns, not significant.

ing the abiotic conditions, there was hardly any re-establishment of the former communities, a *Molinietum* and a *Cirsietum rivulare* meadow, although all target species were still found in the same area. Only the artificial introduction of species by sowing and hay spreading led to the restoration of the former species composition and richness of the communities. The number of characteristic species established after 4 yr was significantly higher when a larger number of germination niches was created by harrowing (Biewer 1997 and unpublished results; Fig. 6.6). This shows once more that dispersal should not be studied independently from establishment (= colonization).

6.8 Dispersal as a limiting factor within plant communities – species co-existence related to short-distance dispersal and seed production

Among the hypotheses to explain species co-existence on a small scale, within communities, the most obvious is the resource-ratio hypothesis which states that species co-exist by niche differentiation using different resources. This was tested in experimental studies by Tilman (e.g. 1988). As an alternative to this hypothesis, van der Maarel & Sykes (1993) indicated on the basis of long-term field studies that existing theoretical models do not fully explain species co-existence in the case of open, dry, species-rich calcareous grassland. They concluded that 'all species of this plant community have the same habitat niche . . . ; the essential variation amongst the species is their individual ability to establish or re-establish by making use of favourable conditions appearing in microsites in an unknown, complex spatio-temporal pattern'

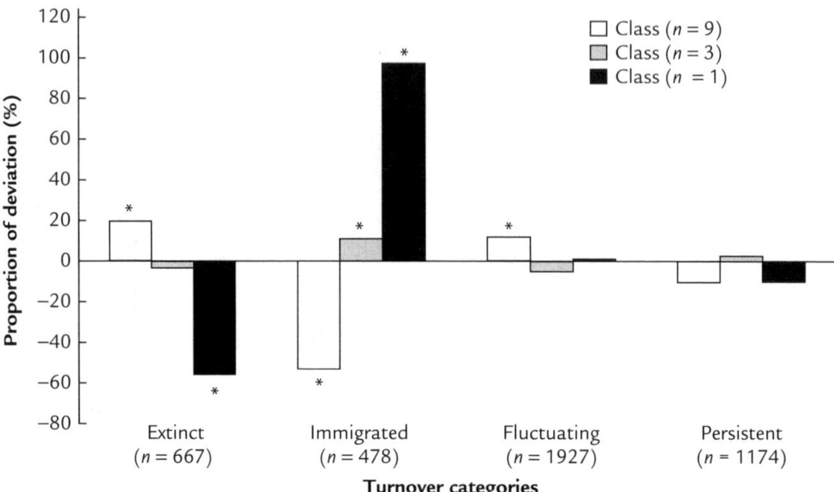

Fig. 6.7 Frequency of species in different turnover categories and with different diaspore production in 65 grassland permanent plots of 5 m × 5 m under 11 different management regimes during 25 yr in the 'fallow experiments Baden-Wuerttemberg' of K.-F. Schreiber, south-western Germany. Frequency is expressed as deviation from the mean frequency. The turnover categories are 'extinct' (disappeared during the experiment), 'immigrated' (arrived during the experiment), 'fluctuating' (showing large year-to-year differences in frequency) and 'persistent' (occurring constantly over the years). Diaspore production classes are 1 = 1–1000 seeds per ramet; 2 = 1000–10,000; 3 = > 10,000. The number of occurrences for each category are indicated. * = Deviation is significant ($P < 0.05$).

and suggested a carousel model to describe the fine-scale mobility of species. This model, which is phenomenological includes a turnover rate, the speed with which a species moves around in the community. Clearly, this rate will depend on the dispersal in space on a fine scale, i.e. the short-distance dispersal capacity, as well as dispersal in time (diaspore bank type). According to Bonn *et al.* (2000) a simple but adequate indicator of short-distance dispersal capacity is diaspore production.

This hypothesis was tested in different grassland communities in experimental plots in a large-scale long-term project on old-field development under different management regimes set up by K.-F. Schreiber in south-western Germany. Indeed, species turnover was strongly related to seed production (dispersal in space), but not to dispersal in time. Amongst the species which became extinct those with a low seed production were overrepresented and species with a high seed production were underrepresented. Amongst the species which immigrated during the experiment, species with a high seed production are strongly overrepresented (Fig. 6.7). On the other hand, Stöcklin & Fischer (1999) showed that plants with longer-lived seeds have lower local extinction rates in grassland remnants over a 35-yr period. In conclusion, as Fig. 6.4 indicates, the co-existence of plant species is determined by many functional traits, but dispersal factors play a major part.

References

Bakker, J.P., Poschlod, P., Strykstra, R.J., Bekker, R.M. & Thompson, K. (1996) Seed banks and seed dispersal: important topics in restoration ecology. *Acta Botanica Neerlandica* **45**, 461–490.

Bekker, R.M., Oomes, M.J.M. & Bakker, J.P. (1998a) The impact of groundwater level on soil seed bank survival. *Seed Science Research* **8**, 399–404.

Bekker, R.M., Knevel, I.C., Tallowin, J.B.R., Troost, E.M.L. & Bakker, J.P. (1998b) Soil nutrient input effects on seed longevity: a burial experiment with fen meadow species. *Functional Ecology* **12**, 673–682.

Bekker, R.M., Bakker, J.P., Grandin, U., Kalamees, R., Milberg, P., Poschlod, P., Thompson, K. & Willems, J.H. (1998c) Seed size, shape and vertical distribution in the soil: indicators of seed longevity. *Functional Ecology* **12**, 834–842.

Bekker, R.M., Schaminee, J.H.J., Bakker, J.P. & Thompson, K. (1998d) Seed bank characteristics of Dutch plant communities. *Acta Botanica Neerlandica* **47**, 15–26.

Biewer, H. (1997) Regeneration artenreicher Feuchtwiesen. *Veröffentlichungen Projekt Angewandte Ökologie* **24**, 3–323.

Bill, H.-C., Poschlod, P., Reich, M. & Plachter, H. (1999) Experiments and observations on seed dispersal by running water in an Alpine floodplain. *Bulletin of the Geobotanical Institute ETH* **65**, 13–28.

Bonn, S. & Poschlod, P. (1998) *Ausbreitungsbiologie der Blütenpflanzen Mitteleuropas*. Quelle & Meyer, Wiesbaden.

Bonn, S., Poschlod, P. & Tackenberg, O. (2000) 'Diasporus' – a database for diaspore dispersal – concept and applications in case studies for risk assessment. *Zeitschrift für Ökologie und Naturschutz* **9**, 85–97.

Bullock, J.M. & Clarke, R.T. (2000) Long distance seed dispersal by wind: measuring and modelling the tail of the curve. *Oecologia* **124**, 506–521.

Cain, M.L., Milligan, B.G. & Strand, A.E. (2000) Long-distance seed dispersal in plant populations. *American Journal of Botany* **87**, 1217–1227.

Clark, J.S., Fastie, C., Hurtt, G. *et al.* (1998) Reid's paradox of rapid plant migration. *BioScience* **48**, 13–24.

Di Castri, F. (1989) History of Biological Invasions with Special Emphasis on the Old World. In: *Biological Invasions: a Global Perspective* (eds J.A. Drake, H.A. Mooney, F. Di Castri, R.H. Groves, F.J. Kruger, M. Rejmánek & M. Williamson), pp. 1–30. Wiley, Chichester.

Di Castri, F., Hansen, A.J. & Debussche, M. (eds) (1990) *Biological invasions in Europe and the Mediterranenan Basin*. Kluwer, Dordrecht.

Dirzo, R. & Dominguez, C.A. (1986) Seed shadows, seed predation and the advantages of dispersal. In: *Frugivores and seed dispersal* (eds A. Estrada & T.H. Fleming), pp. 237–249. Dr W. Junk, Dordrecht.

Ehrlén, J. & van Groenendael, J.M. (1998) The trade-off between dispersability and longevity – an important aspect of plant species diversity. *Applied Vegetation Science* **1**, 29–36.

Ellenberg, H., Weber, H.E. & Düll, R. (1992) Zeigerwerte von Pflanzen in Mitteleuropa. 2nd ed. *Scripta Geobotanica* **18**, 1–258.

Eriksson, A. & Eriksson, O. (1997) Seedling recruitment in semi-natural pastures: the effects of disturbance, seed size, phenology and seed bank. *Nordic Journal of Botany* **17**, 469–482.

Eriksson, O. & Eriksson, A. (1998) Effects of arrival order and seed size on germination of grassland plants: are there assembly rules during recruitment? *Ecological Research* **13**, 229–239.

Eriksson, O. & Jakobsson, A. (2000) Recruitment trade-offs and the evolution of dispersal mechanisms in plants. *Evolutionary Ecology* **13**, 411–423.

Ernst, W.H.O. (1998) Invasion, dispersal and ecology of the South African neophyte 'Senecio inaequidens in The Netherlands: from wool alien to railway and road alien. *Acta Botanica Neerlandica* **47**, 131–151.

Fischer, S.F., Poschlod, P. & Beinlich, B. (1996) Experimental studies on the dispersal of plants and animals on sheep in calcareous grasslands. *Journal of Applied Ecology* **63**, 1206–1221.

Gigon, A. & Leutert, F. (1996) The dynamic keyhole-key model of coexistence to explain diversity in plants in limestone and other grasslands. *Journal of Vegetation Science* **7**, 29–40.

Grime, J.P., Hodgson, J.G. & Hunt, R. (1988) *Comparative plant ecology. A functional approach to British species.* Unwin Hyman, London.

Guevara, S., Meave, J., Moreno-Casasola, P. & Laborde, J. (1992) Floristic composition and structure of vegetation under isolated trees in neotropical pastures. *Journal of Vegetation Science* **3**, 655–664.

Hanski, I. (1987) Colonization of ephemeral habitats. In: *Colonization, Succession and Stability* (eds A.J. Gray, M.J. Crawley & P.J. Edwards), pp. 155–185. Blackwell, Oxford.

Hansson, L., Söderström, L. & Solbreck, C. (1992) The Ecology of Dispersal in Relation to Conservation. In: *Ecological Principles of Nature Conservation* (ed. L. Hansson), pp. 162–200. Elsevier, London.

Harrison, S. & Hastings, A. (1996) Genetic and evolutionary consequences of metapopulation structure. *Trends in Ecology & Evolution* **11**, 180–183.

Heinken, T. (2000) Dispersal of plants by a dog in a deciduous forest. *Botanisches Jahrbuch Systematik* **122**, 449–467.

Herrera, C.M. (1989) Frugivory and seed dispersal by carnivorous mammals, and associated fruit characteristics, in undisturbed Mediterranean habitats. *Oikos* **55**, 250–262.

Higgins, S.I. & Richardson, D.M. (1999) Predicting plant migration rates in a changing world: The role of long-distance dispersal. *The American Naturalist* **153**, 464–475.

Hodgson, J.G. & Grime, J.P. (1990) The role of dispersal mechanisms, regenerative strategies and seed banks in the vegetation dynamics of the British landscape. In: *Species Dispersal in Agricultural Habitats* (eds R.G.H. Bunce & D.C. Howard), pp. 61–81. Belhaven, London.

Hodkinson, D.J. & Thompson, K. (1997) Plant dispersal: the role of man. *Journal of Applied Ecology* **34**, 1484–1496.

Howe, H.F. & Smallwood, J. (1982) Ecology of seed dispersal. *Annual Review Ecology Systematics* **13**, 201–228.

Husband, B.C. & Barrett, S.C.H. (1996) A metapopulation perspective in plant population biology. *Journal of Ecology* **84**, 461–469.

Janzen, D.H. (1971) Seed predation by animals. *Annual Review of Ecology and Systematics* **2**, 465–492.

Janzen, D.H. (1982) Differential seed survival passage rates in cows and horses, surrogate Pleistocene dispersal agents. *Oikos* **38**, 150–156.

Johansson, M.E. & Nilsson, C. (1993) Hydrochory, population dynamics and distribution of the clonal aquatic plant Ranunculus lingua. *Journal of Ecology* **81**, 81–91.

Johnson, W.C. & Webb, T. (1989) The role of blue jays (Cyanocitta cristata L.) in the postglacial dispersal of fagaceous trees in eastern North America. *Journal of Biogeography* **16**, 561–571.

Kalisz, S. & McPeek, M.A. (1993) Extinction dynamics, population growth and seed banks. An example using an age-structured annual. *Oecologia* **95**, 314–320.

Keddy, P.A. (1992) Assembly and response rules: two goals for predictive community ecology. *Journal of Vegetation Science* **3**, 157–167.

Kollmann, J. & Pirl, M. (1995) Spatial pattern of seed rain of fleshy-fruited plants in a scrubland-grassland transition. *Acta Œcologia* **16**, 313–329.

Levin, D.A. (1990) The seed bank as a source of genetic novelty in plants. *American Naturalist* **135**, 563–572.

Lewis, J. (1973) Longevity of crop and weed seeds: Survival after 20 years in soil. *Weed Research* **13**, 179–191.

Luftensteiner, H.W. (1982) Untersuchungen zur Verbreitungsbiologie von Pflanzengemeinschaften an vier Standorten in Niederösterreich. *Bibliotheca Botanica* **135**, 1–68.

Malo, J.E. & Suárez, F. (1995) Herbivorous mammals as seed dispersers in a Mediterranean dehesa. *Oecologia* **104**, 246–255.

McClanahan, T.R. & Wolfe, R.W. (1987) Dispersal of ornithochorous seeds from forest edges in central Florida. *Vegetatio* **71**, 107–112.

McGraw, J.B., Vavrek, M.C. & Bennington, C.C. (1991) Ecological genetic variation in seed banks. I. Establishment of a time-transect. *Journal of Ecology* **79**, 617–626.

Murdoch, A.J. & Ellis, R.H. (2000) Dormancy, Viability and Longevity. In: *Seeds. The ecology of regeneration in plant communities* (ed. M. Fenner), pp. 183–214. CABI Publishing, Oxon, New York.

Nathan, R., Katul, G.G., Horn, H.S., Thomas, S.M., Oren, R., Avissar, R., Pacala, S.W. & Levin, S.A. (2002) Mechanisms of long-distance dispersal of seeds by wind. *Nature* **418**, 409–413.

Ouborg, N.J., Piquot, Y. & van Groenendael, J.M. (1999) Population genetics, molecular markers and the study of dispersal in plants. *Journal of Ecology* **87**, 551–568.

Pakeman, R.J. (2001) Plant migration rates and seed dispersal mechanisms. *Journal of Biogeography* **28**, 795–800.

Pakeman, R.J., Digneffe, G. & Small, J.L. (2002) Ecological correlates of endozoochory by herbivores. *Functional Ecology* **16**, 296–304.

Pärtel, M., Zobel, M., Zobel, K. & van der Maarel, E. (1996) The species pool and its relation to species richness: evidence from Estonian plant communities. *Oikos* **75**, 111–117.

Poschlod, P. (1993) Die Dauerhaftigkeit von generativen Diasporenbanken in Böden von Kalkmagerrasenpflanzen und deren Bedeutung für den botanischen Arten- und Biotopschutz. *Verhandlungen der Gesellschaft für Ökologie* **22**, 229–240.

Poschlod, P. (1995) Diaspore rain and diaspore bank in raised bogs and its implication for the restoration of peat mined sites. In: *Restoration of Temperate Wetlands* (eds B.D. Wheeler, S.C. Shaw, W.J. Fojt & Robertson, R.A.), pp. 471–494. Wiley, Chichester.

Poschlod, P. & Bonn, S. (1998) Changing dispersal processes in the central European landscape since the last ice age – an explanation for the actual decrease of plant species richness in different habitats. *Acta Botanica Neerlandica* **47**, 27–44.

Poschlod, P. & Jackel, A.-K. (1993) Untersuchungen zur Dynamik von generativen Diasporenbanken von Samenpflanzen in Kalkmagerrasen. I. Jahreszeitliche Dynamik des Diasporenregens und der Diasporenbank auf zwei Kalkmagerrasenstandorten der Schwäbischen Alb. *Flora* **188**, 49–71.

Poschlod, P. & WallisDeVries, M. (2002) The historical and socioeconomic perspective of calcareous grasslands – lessons from the distant and recent past. *Biological Conservation* **104**, 361–376.

Poschlod, P., Bonn, S. & Bauer, U. (1996) Ökologie und Management periodisch abgelassener und trockenfallender kleinerer Stehgewässer im schwäbischen und oberschwäbischen Voralpengebiet. *Veröffentlichungen Projekt Angewandte Ökologie* **17**, 287–501.

Poschlod, P., Kiefer, S., Tränkle, U., Fischer, S. & Bonn, S. (1998) Plant species richness in calcareous grasslands as affected by dispersability in space and time. *Applied Vegetation Science* **1**, 75–90.

Poschlod, P., Kleyer, M., Jackel, A.-K., Dannemann, A. & Tackenberg, O. (2003) BIOPOP – a database of plant traits and Internet application for nature conservation. *Folia Geobotanica* **38**, 263–271.

Priestley, D.A. (1986) *Seed Aging*. Cornell University Press, Ithaca.

Primack, R.B. & Miao, S.L. (1992) Dispersal can limit local plant distribution. *Conservation Biology* **6**, 513–519.

Rees, M. (1993) Trade-offs among dispersal strategies in the British flora. *Nature* **366**, 150–152.

Silvertown, J.W. & Lovett-Doust, J. (1993) *Introduction to Plant Population Biology*. Blackwell, Oxford.

Skoglund, J. & Hytteborn, H. (1990) Viable seeds in deposits of the former lakes Kvismaren and Hornborgasjön, Sweden. *Aquatic Botany* **37**, 271–290.

Stender, S., Poschlod, P., Vauk-Henzelt, E. & Dernedde, T. (1997) Die Ausbreitung von Pflanzen durch Galloway-Rinder. *Verhandlungen der Gesellschaft für Ökologie* **27**, 173–180.

Stiles, E.W. (2000) Animals as seed dispersers. In: *Seeds. The ecology of regeneration in plant communities* (ed. M. Fenner), pp. 111–124. CABI Publishing, Oxon, New York.

Stöcklin, J. & Fischer, M. (1999) Plants with longer-lived seeds have lower local extinction rates in grassland remnants 1950–1985. *Oecologia* **120**, 539–543.

Strykstra, R.J., Bekker, R.M. & Verweij, G.L. (1996) Establishment of Rhinanthus angustifolius in a successional hayfield after seed dispersal by mowing machinery. *Acta Botanica Neerlandica* **45**, 557–562.

Strykstra, R., Pegtel, D.M. & Bergsma, A. (1998) Dispersal distance and achene quality of the rare anemochorous species Arnica montana L.: implications for conservation. *Acta Botanica Neerlandica* **47**, 45–56.

Tackenberg, O. (2001) *Methoden zur Bewertung gradueller Unterschiede des Ausbreitungspotentials von Pflanzenarten. – Modellierung des Windausbreitungspotentials und regelbasierte Ableitung des Fernausbreitungspotentials*. Dissertationes Botanicae 347. Schweizerbart'sche Buchhandlung, Stuttgart.

Tackenberg, O. (2003) A model for wind dispersal of plant diaspores under field conditions. *Ecological Monographs* **73**, 173–189.

Tackenberg, O., Poschlod, P. & Bonn, S. (2003a) Assessment of wind dispersal potential in plant species. *Ecological Monographs* **73**, 191–205.

Tackenberg, O., Poschlod, P. & Kahmen, S. (2003b) Dandelion seed dispersal: the horizontal wind speed doesn't matter for long distance dispersal – it is updraft. *Plant Biology* **5**, 451–454.

ter Heerdt, G.N.J., Verweij, G.L., Bekker, R.M. & Bakker, J.P. (1996) An improved method for seed bank analysis: seedling emergence after removing the soil by sieving. *Functional Ecology* **10**, 144–151.

ter Heerdt, G.N.J., Schutter, A. & Bakker, J.P. (1999) The effect of water supply on seed-bank analysis using the seedling-emergence method. *Functional Ecology* **13**, 428–430.

Thompson, K. (2000) The Functional Ecology of Soil Seed Banks. In: *Seeds. The ecology of regeneration in plant communities* (ed. M. Fenner), pp. 215–235. CABI Publishing, Oxon, New York.

Thompson, K. & Grime, J.P. (1979) Seasonal variation in the seed banks of herbaceous species in ten contrasting habitats. *Journal of Ecology* **67**, 893–921.

Thompson, K., Band, S.R. & Hodgson, J.G. (1993) Seed size and shape predict persistence in soil. *Functional Ecology* **7**, 236–241.

Thompson, K., Bakker, J.P. & Bekker, R.M. (1997) *The soil seed banks of North West Europe: methodology, density and longevity*. Cambridge University Press, Cambridge.

Thompson, K., Bakker, J.P., Bekker, R.M. & Hodgson, J.G. (1998) Ecological correlates of seed persistence in soil in the NW European flora. *Journal of Ecology* **86**, 163–169.

Tilman, D. (1988) *Dynamics and structure of plant communities. Monographs in Population Biology* **26**. Princeton University Press, Princeton.

Turnbull, L.A., Crawley, M.J. & Rees, M. (2000) Are plant populations seed-limited? A review of seed sowing experiments. *Oikos* **88**, 225–238.

van der Maarel, E. & Sykes, M.T. (1993) Small-scale plant species turnover in a limestone grassland: the carousel model and some comments on the niche concept. *Journal of Vegetation Science* **4**, 179–188.

Vavrek, M.C., McGraw, J.B. & Bennington, C.C. (1991) Ecological genetic variation in seed banks. III. Phenotypic and genetic differences between plants from young and old seed populations of Carex bigelowii. *Journal of Ecology* **79**, 645–662.

Welch, D. (1985) Studies in the grazing of heather moorland in north-east Scotland. IV. Seed dispersal and plant establishment in dung. *Journal of Applied Ecology* **22**, 46–72.

Willerding, C. & Poschlod, P. (2002) Does seed dispersal by sheep affect the population genetic structure of the calcareous grassland species Bromus erectus? *Biological Conservation* **104**, 329–337.

Willson, M.F. & Traveset, A. (2000) The Ecology of Seed Dispersal. In: *Seeds. The ecology of regeneration in plant communities* (ed. M. Fenner), pp. 85–110. CABI Publishing, Oxon, New York.

Willson, M.F., Rice, B.L. & Westoby, M. (1990) Seed dispersal spectra, a comparison of temperate plant communities. *Journal of Vegetation Science* **1**, 547–562.

Young, A.G., Boyle, T. & Brown, T. (1996) The population genetic consequences of habitat fragmentation for plants. *Trends in Ecology & Evolution* **11**, 413–418.

Zobel, M. (1997) The relative role of species pools in determining plant species richness: an alternative explanation of species coexistence. *Trends in Ecology & Evolution* **12**, 266–269.

7

Vegetation dynamics

Steward T.A. Pickett and
Mary L. Cadenasso

7.1 Introduction

Succession is a fundamental concept in ecology. Simply, it is the change in species composition or in the three-dimensional architecture of the plant cover of a specified place through time. When vegetation dynamics was first codified by ecologists, they focused on three key features: (i) a discrete starting point; (ii) a clear directional trajectory; and (iii) an unambiguous end (Clements 1916). These three assumptions have been associated with the term 'succession'. Succession is a special case of vegetation dynamics.

This chapter takes a broad view of vegetation dynamics that does not always accept the narrower assumptions of succession. The larger focus helps solve many of the arguments and controversies about succession that have emerged from the failure of the narrower concept to portray the variety of patterns and causes of vegetation change in the field (Botkin & Sobel 1975). Using the larger concept of vegetation dynamics, ecologists can appreciate and understand the complexity found in real ecosystems, and can apply it in vegetation management.

7.2 The causes of vegetation dynamics

7.2.1 Vegetation dynamics and natural selection

Vegetation dynamics has many causes (Glenn-Lewin *et al.* 1992). Causes in biology can often be cast in terms of conditional, or 'if-then,' statements (Pickett *et al.* 1994). The general form of a conditional statement is this: if a certain condition holds, then a certain result will follow. Natural selection, for example, is a series of conditionals that leads to a consequence. If: (i) offspring vary; (ii) at least some of that variation is heritable; and (iii) more offspring are produced than can survive; then variation that matches the environmental conditions will tend to accumulate in a population (Mayr 1991). This conditional law requires multiple processes, is probabilistic and sets the bounds of change. The theory of evolution is a contingent, nested and probabilistic theory of a sort that can be adopted for the processes of vegetation change (Fig. 7.1).

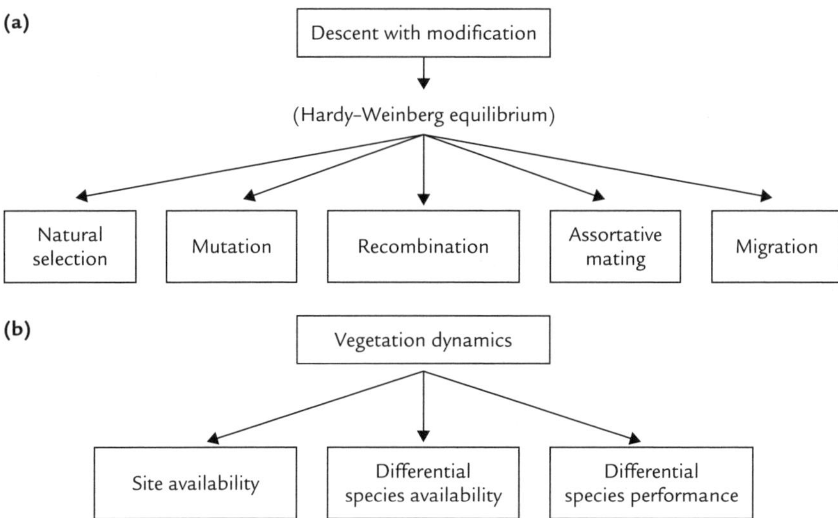

Fig. 7.1 Comparison of the hierarchical structures of the theory of evolution and the theory of succession. Evolution (**a**) is summarized most generally as the phenomenon of descent with modification. The Hardy–Weinberg Law of Equilibrium embodies the processes that can affect evolutionary change between generations. Those factors – natural selection, mutation, recombination, assortative mating and migration, among others – can result in heritable changes between generations. The process of succession (**b**) is represented most generally as vegetation dynamics. Changes in any one or any combination of site availability, differential availability of species or differential performance of species can cause the structure or composition of vegetation to change through time.

As a law, vegetation change is based on the fundamental idea that the different capacities of plants to match the prevailing environment determines the nature of the plant assemblage that will exist in a place. The environment includes both abiotic factors and other organisms. The law of vegetation dynamics also has the form of a conditional statement. It states that if: (i) a site becomes available; (ii) species are differentially available at that site; or (iii) species perform differentially at that site; then the composition and/or structure of vegetation will change through time (Pickett & McDonnell 1989). The causes of vegetation dynamics will be synthesized into a single organizing framework, and related to various vegetation dynamics that ecologists have observed.

7.2.2 Site availability

Sites become available because disturbances disrupt established vegetation, or create new surfaces. **Disturbance** is an event that alters the structure of vegetation or the substrate vegetation is growing on (White & Pickett 1985). Examples include certain kinds or intensities of fire, windstorms, stress-induced mortality of plants or herbivory.

The nature of the open site is governed by how intense the disturbance is, how many layers of the prior vegetation are removed, or how deeply the substrate is stirred or buried. Structures that are disturbed by events capable of starting succession include, for example, forest canopies, grassland root mats and soil profiles. Some disturbance events can be very localized, such as the fall of a single tree in a forest, while others can be quite extensive, such as the opening of the forest canopy by hurricanes or typhoons.

The characteristics of a site following disturbance influence how plants can establish, grow and interact there. Disturbances affect the kinds and amounts of available resources that remain after the event, the degree to which biomass is removed or rearranged at the site, and the water and nutrient holding capacity of exposed substrates. Different disturbances may have contrasting effects on the resources available for colonizing plants. For example, fire may burn much of the organic matter at the soil surface, which will make a poorer resource base for recolonization than a windstorm that blows trees down but leaves the organic matter intact.

7.2.3 Differential species availability

The way vegetation composition and structure changes after disturbance depends on the ability of species to survive the disturbance or their ability to reach the site after the disturbance (Leck *et al.* 1989; Stearns & Likens 2002). Species may become available at the site in two ways. First, species may persist through the disturbance as seedlings, adults, seeds, tubers or the like. Second, they may invade from elsewhere. Therefore, differential species diversity availability depends on the characteristics of species to either survive or disperse to sites.

Differential survival after disturbances is determined by characteristics of both the species and the disturbance. Fires of low to moderate intensities may leave adults above some critical size unscathed due to thick, insulating bark. An example appears in *Sequoia sempervirens* during moderate ground fires. Adult *Sequoia* can be killed by intense fires that spread into the tree crowns, however. Alternatively, in some species, if above-ground parts are killed, lignotubers capable of resprouting may survive in the soil, as in the sprouting shrubs of chaparral or pine barrens (Forman & Boerner 1981). Similarly, a dormant pool of seeds that is triggered to germinate by high temperatures can re-establish vegetation after severe fires. Examples of this mechanism include seeds of annuals in chaparral, grasses in prairie and trees of the pine family having serotinous cones.

Differential species availability can depend on the ability of seedlings to tolerate unfavourable conditions for a time. The seedlings of some tree species are capable of persisting by growing slowly in deep shade, but can take advantage of the altered conditions and resource levels after the canopy and intervening layers of a forest are disturbed by wind. For example, a pool of *Prunus serotina* seedlings on the floor of undisturbed northern hardwood forests is limited by the low light availability beneath the canopy. After a blowdown of the canopy the *Prunus* seedlings are released from suppression because of the increased light near the ground (Peterson & Pickett 1991).

Differential dispersal to open sites is determined by characteristics of species, or the activities of biotic and abiotic vectors that transport seeds. Some seeds disperse readily to open sites due to their small size or their wings or plumes, e.g. *Epilobium angustifolium*. Seeds also move with the help of animals. Dispersal by birds or bats that seek out forest gaps are examples of differential availability that depends on animals (see Chapter 6).

7.2.4 Differential species performance

Species diversity performance refers to the suite of activities that species employ to acquire resources, grow, persist and reproduce (Bazzaz 1996). Life history traits, relative growth rates, age to maturity, competitive ability, stress tolerance and herbivore and predator defence are some of the specific features that will determine differential performance. Ecologists have examined some of the components of differential species performance for more than a century. Examples include the ability to tolerate shade compared to the demand for high levels of photosynthetically active radiation or the possession of thorns or secondary compounds that deter herbivores compared to species that are vulnerable to herbivores.

The first example depends on the high light availability early in many successions compared to the low availability of light in older communities with closed canopies (Bazzaz 1996). Species that dominate early after disturbance in otherwise resource-rich, forest environments often require high levels of light for maximum growth (Fig. 7.2). Such species have high light-saturation levels of photosynthesis. In contrast, high photosynthetic efficiencies characterize the seedlings and juveniles of closed forest dominants. The forest dominants, in contrast to the early field dominants, often cannot tolerate high light levels or the rapid water loss associated with high photosynthetic rates.

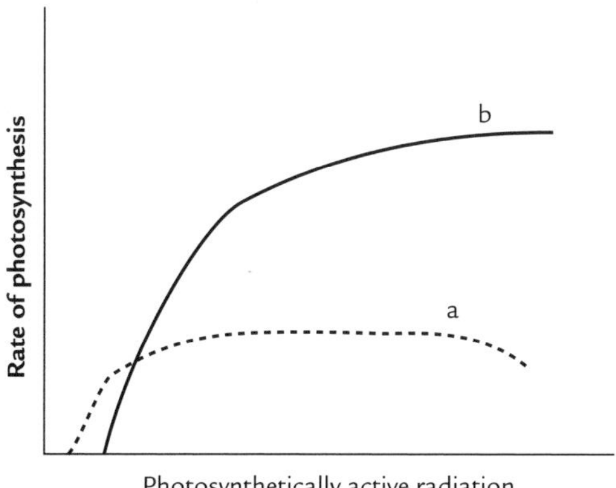

Fig. 7.2 Diagrammatic representation of the contrasts between photosynthetic responses to varying availability of light of early- and late-successional species. **a.** Shade-tolerant species. **b.** Light-demanding species. Following principles in Bazzaz (1996).

Nutrient contrasts also can drive differential species performance (Tilman 1991). On substrates that initially lack a large nutrient pool, successful invaders often have the capacity to fix nitrogen, while species that dominate later exploit the higher levels of available nitrogen that have built up with the accumulation of humus and the increasing soil stratification resulting from initial colonizers (Vitousek *et al.* 1998).

Differential performance can also be illustrated by contrasting capacities to interact with animals. In sites that are exposed to large populations of browsers, species that are chemically or mechanically defensive tend to dominate in a plant community sooner than those woody species that are more palatable. The effects of animals, whether invertebrates or vertebrates, have been relatively neglected over the history of succession studies. Experiments are increasingly showing the importance of herbivores in succession, however (Bowers 1993; Facelli 1994; Meiners *et al.* 2000; Cadenasso *et al.* 2002).

7.2.5 A hierarchical framework of successional causes

Succession results from the interaction of a site, a collection of species that **can** occupy that site and the interactions of the species that actually occupy the site. However, each disturbance event, the resultant characteristics of a site, and the specific characteristics and histories of the mixture of plants on that site can result in a unique trajectory of succession. Because there are so many factors influencing succession at a site it is important to have some way to organize the factors. Ecologists use a hierarchical framework to organize complex areas of study such as succession or evolution (Pickett *et al.* 1989; Luken 1990). Organizing factors into a hierarchical framework means that the general causes of succession must be broken down into more specific events and interactions. In such a causal hierarchy, the more specific processes are nested within the more general causes. The hierarchical framework of succession is similar to the hierarchy of evolutionary mechanisms (Fig. 7.1).

Using the hierarchical approach for succession presents the three general successional processes – site availability, differential species availability and differential species performance – as being composed of more specific causes or mechanisms (Fig. 7.3). The specific mechanisms within each factor are aspects of the abiotic environment, plants, animals and microbes, and their interactions. For succession to occur, at least one of the three general causes must operate. However, not all of the specific mechanisms that can contribute to the general causes will act in every succession. Nor will the detailed factors always act with the same intensity or relative importance. Exactly what factors dominate in a specific succession depends on the history of the site and the specific individuals that reach the site. Yet the fact that we can organize the factors into a hierarchy and generalize them to three broad categories suggests that there are broad expectations that can be drawn from succession (Glenn-Lewin & van der Maarel 1992). The causal hierarchy is a framework for explaining possible specific trajectories, processes, patterns and rates of succession in the field (Fig. 7.3). It also informs experiments that document the role of various factors in specific successions.

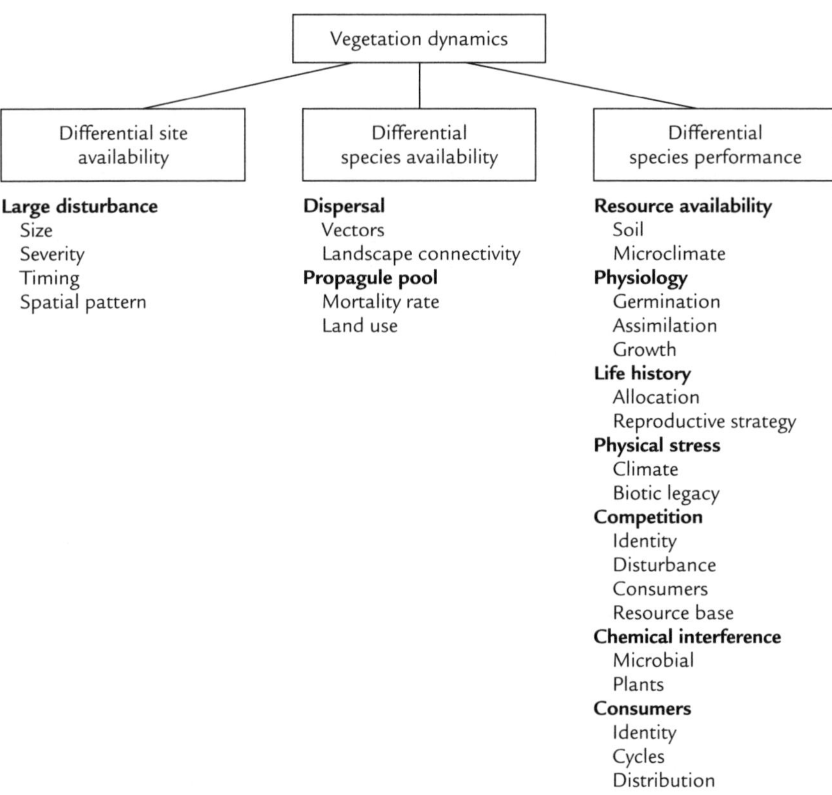

Fig. 7.3 Detailed hierarchy of successional causes, ranging from the most general phenomenon of community change, through the aggregated processes of site availability, and differentials in species availability and performance to the detailed interactions, constraints and resource conditions that govern the outcome of interactions at particular sites. Based loosely on Pickett *et al.* (1989).

7.3 Succession in action: interaction of causes in different places

7.3.1 Complexity of successional patterns

The variety of actual successional patterns is immense. Complexity emerges from the number of different mechanisms that can act in succession and the breadth of conditions that can affect those mechanisms. A second source of complexity in successional patterns is the breadth of conditions that can affect each of the successional causes. The different causes of succession operate across gradients of important factors. For example, gradients may contrast high with low intensities of disturbance, or low to high levels of local species availability. Examples from contrasting environments show the richness of successional causes and trajectories, starting with successions shaped by large natural events continuing with processes that occur in more restricted sites, and ending with sites that experience shifts in management by people.

7.3.2 Vegetation dynamics in large, intensely disturbed sites

Floods and succession. Under the influence of large rivers, vegetation dynamics are driven by the timing, intensity and location of floods. Large floods move large amounts of sediment and organic debris. Such floods tend to occur infrequently. Moderate floods occur more frequently, and at least some flooding will likely occur every year at the beginning of the rainy season.

There are many effects of floods. Some effects are direct, resulting from the presence of water or the energy of moving water and the load of sediment and debris it carries. For example, flooding can kill plants that are intolerant of waterlogging, and the force of water and debris moving downstream can uproot woody plants (Sparks 1996). Other effects of floods can be indirect, such as the alteration of substrates (Moon *et al.* 1997). Substrates in which plants are rooted can be eroded away and sediment can be deposited in other areas. Both the direct and indirect effects of floods provide a heterogeneous template that can start, end, or change the course of vegetation dynamics.

The large rivers that flow through the Kruger National Park, South Africa, provide good examples of the diverse effects of flooding. From south to north in the park, there is a gradient of decreasing rainfall that determines whether the rivers flow continuously or only in the rainy season. In addition, the rivers flow through different substrates, so that in some stretches, the shape of the river channel is determined by bedrock, while in other sections, the flow interacts with deposits of sediment and vegetation (Rogers 1997).

The sequences on bedrock and sediments differ. On bedrock, certain trees can establish in cracks and they can trap sediment. These trees tend to survive moderate floods and form a biological legacy. The clusters of stems and foliage further modify the habitat by trapping additional sediment in subsequent mild floods. With increasing sediment deposition, other trees and associated plants can establish. The earliest dominants have flexible stems and their branches and leaves can adopt a streamlined form if they are submerged in moderate floods (van Coller *et al.* 1997). Such behaviour reduces the likelihood that the pioneering trees will be killed or severely damaged by floods. These early dominants can thus survive modest floods, and continue to influence the site in ways that other species can exploit.

Succession in river channels can also occur on sediment deposited by floods. *Phragmites mauritianus* is the common colonist on newly deposited sediment. In many cases, plants are established from surviving rhizomes and buried stems. Once established, they trap additional sediment and further modify conditions. Some early woody dominants can resprout from stems of large trees that remain rooted although they are toppled by floodwaters and buried by sediment. The build-up of sediment has two effects. First, higher surfaces will be affected only by larger floods. Second, higher surfaces have a deeper water table. Tree species respond differently to both these effects of sediment accumulation and new species typically dominate as the sediment collects. Eventually spiny shrubs characteristic of upland vegetation and trees that cannot tolerate waterlogging emerge. In all these cases, conditions are modified by the early colonists and other species are better matched to the new environments. This net effect is called **facilitation**.

Floods of different intensities have different effects on succession. The typical sequences outlined above are those that are associated with floods of modest to high intensity but which occur relatively frequently. However, in 2000, extreme floods removed vegetation from both bedrock- and sediment-controlled sections of some of Kruger National Park's rivers. Following these more severe floods the 'typical' sequence of vegetation dynamics that had been elucidated was shown to be associated with only a particular part of the flooding regime. The 2000 floods were large, infrequent disturbances that made new sorts of site available by increasing the amount of bedrock available, and set up new templates of woody debris that had not been observed earlier.

The intense floods of 2000 also set up unusual patterns of species availability because they removed some established 'upland' trees from the upper terraces near the rivers. Species availability associated with recent floods may also support novel successions because of the increase in exotic species as a result of activities upstream of the park. All kinds of flood, representing the entire range of the temporal and spatial extent and volume, affect vegetation dynamics. What sequence of vegetation actually appears, what patterns of species availability are expected and how differential species performance plays out, depends entirely on what part of the long-term flood regime has been studied. In all cases the disturbances set up a spatially heterogeneous distribution of vacant substrates, surviving plants and living or dead biological legacies.

One of the key insights from the Kruger floods is that the pattern of vegetation dynamics observed depends on when one starts looking, and how long the observations last. The heterogeneity of substrate types available for the vegetation dynamics varies by flood intensity. Notably, the study of succession in these rivers continues to discover new patterns and interactions as observations encompass rarer events. However, the insight that what succession looks like depends on when the observations start, and where observations are framed in a complex disturbance regime, can also guide our exploration of other cases of vegetation dynamics. There is no unambiguous point zero.

Tornado blowdown. Large areas of forest canopy can be blown down by hurricanes, large tornadoes and by downdrafts associated with large thunderstorms (Dale *et al.* 1999). Hurricanes or typhoons tend to be associated with coastal regions. Tornadoes may be spawned by hurricanes in some cases, but tornadoes are more commonly associated with convective storms located in interior regions of continents. Tornadoes tend to be temperate phenomenona, while downdrafts also affect tropical habitats. Extremely severe windstorms of all types can occur on the order of once in a century or several centuries in mesic, closed canopy forest sites.

An example of a large forest blowdown is the result of the class 4 tornado in the Tionesta Scenic Area and adjacent Tionesta Research Natural Area in western Pennsylvania, USA, in spring of 1989 (Peterson & Pickett 1991). This forest had been free of large blowdowns for several hundred years, as indicated by the ages and architectures of many of the canopy trees. The canopy at the time of the storm was dominated by *Acer saccharum*, *Fagus grandifolia*, *Betula allegheniensis* and *Tsuga canadensis*.

The blowdown created a heterogeneous template for vegetation dynamics. Bare soil was exposed in the pits and on tip-up mounds created by the uprooting of canopy trees. The stacked boles of uprooted and snapped trees created a jumble of debris. Crowns of fallen trees created a cover of fine- and medium-sized woody debris and litter composed of broad leaves and needles. Dense patches of the fern *Dennstaedtia punctilobula* survived the storm in many places. The various habitats differentially favoured different species (Peterson *et al.* 1990). *T. canadensis* seedlings died from desiccation on exposed soil but were protected from the depredations of deer beneath crown debris. In contrast, *B. allegheniensis* seedlings sprang up on bare soil patches and survived where soil was not waterlogged or unstable. Unfavourable sites for *B. allegheniensis* seedlings included pits, which accumulated standing water, and large soil plates eroding from tipped up root mats. Conspicuously absent were the pioneer species expected in northern hardwood forest in the eastern United States after large forest-clearing events, such as *Prunus pensylvanica*. This species produces hard seeds that can survive in soil for a long time so that when a disturbance opens the canopy, the seeds are ready to germinate. However, the age of the pre-disturbance forest at Tionesta exceeded the life span of the dormant seed pool of *P. pensylvanica* and there were no viable seeds left in the soil to germinate after the disturbance. In younger forest areas near Tionesta affected by the same storm system *P. pensylvanica* was important in the regenerating vegetation (Peterson & Carson 1996).

The Tionesta example highlights specific cases of site availability, differential species availability and differential species performance. In particular, the role of spatial heterogeneity created by the interaction of the tornado with the species composition and size of the pre-disturbance canopy was important. Differential species availability was expressed in the appearance of some species in a seedling pool on the forest floor following the disturbance, such as *T. canadensis* and *A. saccharum*. Seed banks were not important in Tionesta, although they were important for pioneers in a nearby, younger stand of similar forest. Seed rain was important for *B. allegheniensis* and some few individuals of *P. serotina* that colonized tip-up mounds. Differential species performance was expressed in drought tolerance, growth rate, interactions with herbivores and interactions between plant species. For example, sites that were protected from deer browsing by branch debris tended to support more *T. canadensis* and *F. grandifolia* seedlings than other sites, and dense patches of hay-scented fern inhibited growth of *P. serotina* and *B. allegheniensis* seedlings.

Volcanic eruption. Another example of a large, infrequent disturbance is the eruption of Mount St Helens, USA, in 1980 (Anderson & MacMahon 1985). This cone-shaped volcano had been dormant for centuries. The 1980 eruption blew off a large volume of one side of the mountain, created mud and ash flows, displaced volumes of water from Spirit Lake in tidal-wave proportions and deposited ash and coarser airborne debris widely. This single event thus produced a great variety of substrates on which subsequent vegetation dynamics would play out. Although there were many sites in which all adult plants and seeds were killed, there were patches in some sites, such as the pumice plains, in which fast-growing nitrogen fixers emerged from a surviving seed source. Other, wind-borne invaders such as *Epilobium angustifolium*, colonized other sites. In a few places, animals, such as gophers (*Thomruya talpoides*) survived in their burrows and were available to interact with plants early in the succession.

The succession at Mount St. Helens shows the great heterogeneity of initial conditions created by the eruption (del Moral 1993). Differential species availability likewise played a role, although ecologists were not expecting there to be a pool of surviving seeds. Differential species performance appeared in the role of nitrogen fixers and the different life histories available immediately after the disturbance. Some patterns in the dynamics were expressions of different degrees of clonal growth and tolerance of relatively low versus high nutrient availabilities. In all cases, interaction of the plants with the heterogeneous template created by the disturbance was key. Different vegetation trajectories appeared on different patches, as was the case in the South African rivers and the Tionesta blowdown.

7.3.3 Fine-scale vegetation dynamics

Vegetation dynamics can also respond to finer-scale and less intense events. Such events are often referred to as producing gaps in vegetation or substrate. The two examples emphasize that disturbances can affect both the above-ground architecture of vegetation as well as the below-ground organization of a system. The phenomenon of fine-scale vegetation dynamics brings up the concept of 'pattern and process', first introduced by Watt (1947). He focused on vegetation dynamics resulting from the loss of individual plants from a closed canopy and the subsequent invasion of new individuals or release of seedlings that had been stagnant beneath the closed canopy. This idea has been enlarged to include openings of any scale (e.g. Prentice & Leemans 1990), as indicated by the concept of patch dynamics (Pickett & Thompson 1978). Another enlargement of the spatially dynamic pattern and process approach includes the movement of plants through a community in the concept of the 'carousel model' (van der Maarel & Sykes 1993). This approach has been a major stimulus for improving the understanding of vegetation dynamics at any scale as a spatial phenomenon, not just as an interaction of neighbouring plants in small areas (van der Maarel 1996). Below, we present several examples of the fine-scale vegetation dynamics that are most closely aligned with Watt's (1947) original conception of pattern and process.

Forest canopy gaps. One or a few trees can be removed from a forest canopy by several kinds of event. Wind may uproot or snap trees, lightning may kill trees, old trees may die or parasites may kill one species in a mixed species stand, leaving a gap in the canopy. In such gaps the resource availability may be altered and environmental signals may change. For instance in treefall gaps, substrate may be turned over by uprooting. Furthermore, water may be either more or less available – depending on whether the rainfall can better reach the forest floor compared to the rate of soil moisture removal by roots of neighbouring canopy trees or understorey plants that remain in the gap. Nutrients, such as nitrogen, can become more available in the gap due to altered conditions for soil microbes or reduced root demand. Soil temperature extremes may increase, altering soil moisture availability or acting as a signal for germination of dormant seeds.

In an experiment conducted by cutting trees to create canopy gaps in the Kane Experimental Forest in western Pennsylvania, USA, differential species performance was the general cause of successional dynamics following the experimental cut (Collins & Pickett 1987, 1988). The experimental treatment mimicked a windstorm that snapped off the trees rather than uprooted them. Interactions between the causal factors of vegetation dynamics appeared in the experiment. Site availability was governed by the size and type of the experimental disturbance. Gaps were created without disturbing the forest floor and no new substrate was exposed, although the resources and regulators of the sites were altered. Greater alteration occurred in the larger (10 m diameter) experimental gaps than in the smaller (5 m diameter) gaps.

Differential species availability was based primarily on the existence of a pool of suppressed woody seedlings in the forest understorey (Collins & Pickett 1982). The altered conditions in the gaps changed the performance of understorey species and altered the rate of growth of tree seedlings that had been present before gap creation. The experimental treatment did not increase species richness of the understorey layer, indicating that species availability was not influenced by the experimental manipulation. Growth of some understorey broad-leaved species increased but, in general, the spread of the ferns and increase in height and cover of *Prunus serotina* seedlings far outstripped the enhanced performance of broad-leaved angiosperm herbs. The *P. serotina* seedlings had been 'idling' in the forest floor layer before the gaps were created. The change in resources, primarily light, as a result of opening the canopy allowed these seedlings to grow more rapidly. Therefore, the degree of differential species performance observed was modified by more specific mechanisms in the causal hierarchy. These more specific causes were competition with ferns, browsing by deer, the head start enjoyed by certain woody seedlings and the greater range of resources that were released in the larger gaps.

Desert soil disturbance. An example of fine-scale gap dynamics in which the substrate is disturbed comes from the Negev Desert of Israel (Boeken *et al.* 1995; Boeken & Shachak 1998; Shachak *et al.* 1999). In areas where soil lies downslope of rocky outcrops that supply runoff water, perennial geophytes – bulb-bearing plants like tulips – can establish. Porcupines (*Hystrix indica*) exhume the bulbs of geophytes for food. In the process, they create a pit measuring 10–15 cm in diameter and 15–20 cm deep. Such pits concentrate runoff water that flows from the rocks and intact soil upslope. Runoff water generally does not penetrate the surface of intact soil because the surface is cemented into a microphytic crust by the secretions of cyanobacteria, mosses and lichens. Pits collect the water and are, therefore, hot spots for water availability in this arid system. In addition, seeds and organic matter accumulate in the pits. As a result, the diversity and productivity of desert annual plants is greatly enhanced in the pits. The structure of the community in the pit changes through time as the pit fills with sediment carried by runoff water and wind. After it fills in, the pit again supports a microphytic crust and becomes a less effective trap for water and organic matter. Therefore the small site undergoes succession until it is indistinguishable from adjacent intact soil.

Sometimes the porcupines do not consume all of a given bulb, and the geophyte can resprout once the pit fills in somewhat with soil. At a later time the same site may be dug up by another porcupine. Geophytes also establish in new spots in the desert. The interaction of the porcupines, microphytic crust and filling of the pits creates a shifting mosaic of pits with their associated altered resource levels and enhanced availability of seeds of annual plants. This kind of dynamic is a pattern-and-process cycle like Watt (1947) envisioned.

7.3.4 Vegetation dynamics under changing management regimes

Vegetation dynamics are increasingly affected by human activities. As human societies modify or construct more and more systems, it becomes important to understand how human activities affect succession. Human domination of systems spans a range of management or control. One extreme is in national parks where managers control the nature and frequency of fire or the population densities and movements of herbivores. An example of a more intensively managed system is agriculture in which fields left fallow permanently or for various lengths of time exhibit change in vegetation. The planted, managed and volunteer vegetation in urban areas shows perhaps the strongest influence of humans on succession. In all these cases, some aspects of vegetation dynamics may be purposefully managed while other aspects are only indirectly influenced by human actions or the built environment. However, in all cases, managing vegetation for any purpose, whether aesthetic, productive or for ecosystem services, is essentially managing succession (Luken 1990).

Succession and management in riparian vegetation. The relationship of management and succession in riparian zones involves changes in the amount and timing of flow in rivers and the role of introduced species. The sources of the major rivers in Kruger National Park lie well outside the park. They arise in the uplands of Mpumalunga and Limpopo Provinces. The flow of water in the rivers is influenced not only by the seasonal patterns and amounts of rainfall, but also by removal of water for various purposes upstream of the park. On the escarpment, water is removed by transpiration from forest plantations. Once the rivers reach the lowlands at the base of the escarpment, they are subject to use by orchard and row crop agriculture and by an increasing number and density of settlements.

In arid environments such as Kruger National Park, the band of structurally or compositionally distinct forest vegetation adjacent to the stream – the riparian – is an important component (Rogers 1995; Pickett & Rogers 1997). A successional trend that is viewed as problematical in the riparian zones of the park is an encroachment of upland savanna plants into riparian habitats due to accumulation of sediment and alteration of depth to the water table, the process of **terrestrialization**. This process is driven by a reduction of flow in the rivers, due to the upstream removal of water and the attempts to control high-flow events, and causes altered sediment dynamics. A retreat of the water table and an absence of flood-related mortality of upland-adapted species can alter the successional trajectories in riparian zones. This trend, evidenced by the invasion of small-leaved, spinescent trees and shrubs in the upper ranges of riparian zones, is the result of terrestrialization. The practical management concern is that fires may spread into the now drier riparian zone resulting in a decline in primary productivity that supports certain key herbivore species dependent on these riparian zones. Such a shift in vegetation would also shift the pattern of movement and diversity of animal species in the park.

In addition to influencing water flow, human influence has also increased the availability of exotic species in South Africa. Such exotic species have broad tolerances and may alter the successions and contribute to terrestrialization regardless of the flood regime (see also Chapter 13). In particular, changes in the availability or performance of species that typically do best on different riparian and in-channel geomorphologic features may be altered by exotic species, such that successions in the future, even those starting after severe floods, may have different trajectories than those of the past. Hence, human influence in this system has multiple layers and effects.

Post-agricultural succession. Post-agricultural successions have served as a model system for understanding succession, especially in the United States (Bazzaz 1986). This is because abandoned fields have been common, are easily manipulated and change relatively rapidly. In the eastern United States, land abandonment was especially common in the late 19th and early 20th centuries as farms were exhausted, or as more fertile or more easily tillable land became available farther west. Such post-agricultural successions may become more common now in Europe than in the past with the alterations of agricultural policy in the European Union. Old-field succession has been most often studied by observing vegetation on fields having similar soils and that were abandoned after the same kind of crop and cultivation, but which differ in age. This strategy is used because it is difficult, costly and slow to observe vegetation change in one place over a long period of time. By substituting differences in age across fields separated in space for changes through time in one field, ecologists can study succession more quickly and conveniently. This research strategy is called space-for-time substitution or, more technically, **chronosequence** (Pickett 1989). The patterns derived from such space-for-time substitutions have been a staple of community ecology. In addition, studying replicate fields of the same age served to reduce the variation in the patterns. However, such studies also excluded the understanding of spatial heterogeneities in succession because the variations from place to place were assumed to be noise and only the mean was considered. Of course, space-for-time substitution assumes that fields abandoned at different times experience the same conditions through time. In spite of the limitations of space-for-time substitution, expectations of increasing species richness, a decline of exotic, often weedy, species and orderly transitions from herb to shrub and forest tree cover were often concluded from such studies.

The direct studies of successional change through time in specific fields are now yielding detailed information not available using space-for-time substitution (Pickett *et al.* 2001; Meiners *et al.* 2001). Information is now emerging from long-term studies of post-agricultural succession that use the same, permanently marked plots studied through time. One of these studies, the Buell–Small Succession Study (BSS) named after its founders, Drs Helen Buell, Murray Buell and John Small of Rutgers University, was begun in 1958. The same 48 plots in each of 10 fields have been studied continuously since then. Because the study is continuous in time and extensive over

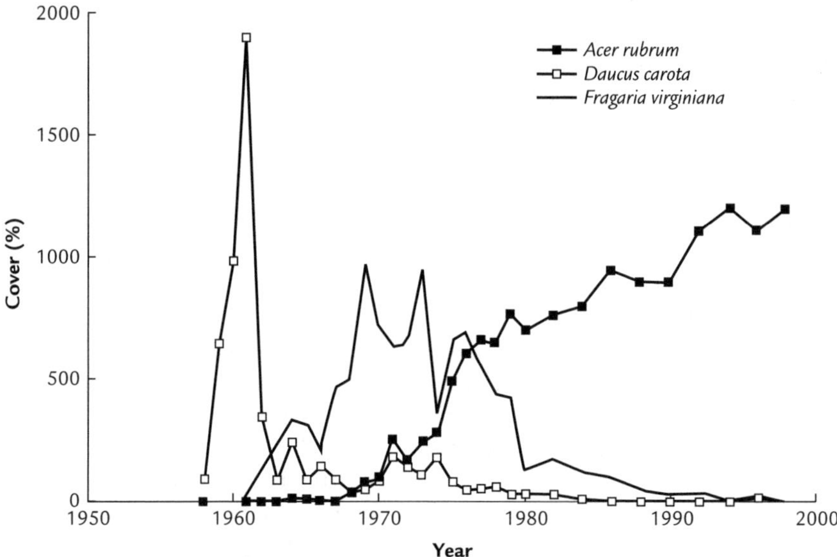

Fig. 7.4 Distribution through time of *Acer rubrum*, *Daucus carota* and *Fragaria virginiana* in field C3 of the Buell–Small Succession Study (BSS), illustrating the early arrival and long persistence of species common in the succession. Details of the succession are described in Myster & Pickett (1990).

space, different spatial scales of the process can be assessed and processes of plant species turnover be observed directly rather than inferred (Bartha *et al.* 2000).

Because of direct observation over 40 yr, the BSS can discriminate among hypotheses that have been persistently controversial. An example is the controversy between the initial floristic composition hypothesis compared to the relay floristics hypothesis (Pickett *et al.* 2001). These two hypotheses deal with differential species availability in succession. The initial floristics hypothesis predicts that species that will later dominate the community will be present from the start of the succession, while the relay floristics hypothesis posits that pioneer species dominate early but disappear, to be replaced by a flora of mid-successional species, which are in turn replaced by late successional species. Indeed, this ability to discriminate between these two competing hypotheses was one of the principal motivations for starting the study (H. Buell pers. comm.). In fact, aspects of both hypotheses have been supported over the 40-yr study. In short, some species are present either long before or long after their period of dominance and some expected turnovers do not occur in specific plots.

Some species characteristic of later successional communities (in the context of the *c.* 40 yr of record) do invade early. Herbaceous species that dominate in mid or late portions of the record are often present early (Fig. 7.4). In some cases, the rise to dominance is an expression of life history traits. For example, short-lived perennials that dominate in years 5–10 are in fact present in low abundance earlier in the succession. Woody species, for example the wind-dispersed *Acer rubrum*, are often early invaders. However, not all individuals that invade early persist through succession. There is considerable turnover in individuals, inferred from periods of presence versus absence in specific plots through time. Though individuals are replaced, the species as a whole is present from the first or second years. In a somewhat more mesic field than those included in the permanent plot study, the ages of all woody stems present during year 14 of the succession were determined. The vast majority of *A. rubrum* individuals were themselves 14 yr old. In other words, most surviving *A. rubrum* individuals had invaded early in the succession (Rankin & Pickett 1989). Other species, such as the wind-dispersed *Fraxinus americana*, showed increasing densities with age of the field, such that most individuals present in year 14 of the succession were younger than 14 yr old.

Another instance of differential species availability is shown by the legacies of different abandonment treatments. Fields in the BSS varied by the last crop before abandonment and by treatment at time of release. Fields abandoned as hayfields maintain grass dominance for a long time, compared to fields that supported row crops at the time of abandonment. The legacy of the last crop can be detected in plant assemblages for *c.* 10 yr after abandonment of the hayfields. In the abandoned hayfields, the grass species remain available to contribute to the succession, while in the ploughed fields species availability depends more heavily on dispersal to the site from external sources (Fig. 7.5).

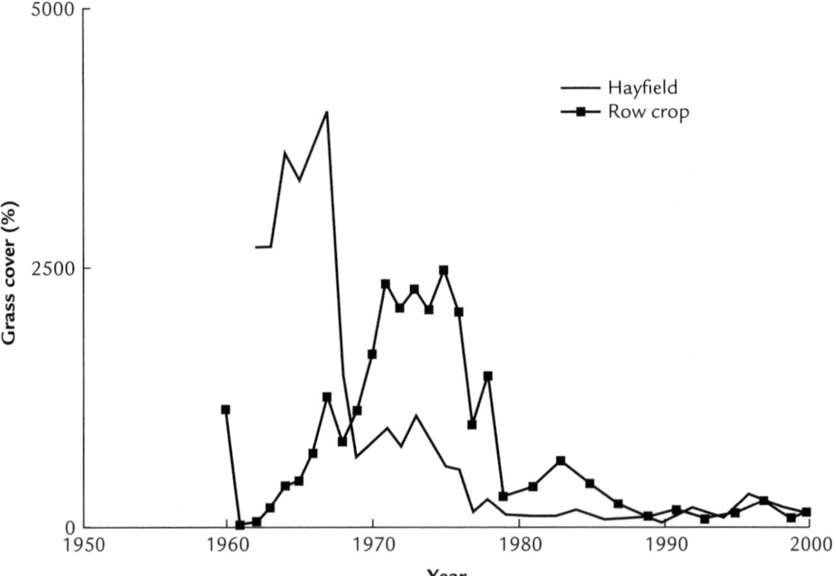

Fig. 7.5 The role of legacies as illustrated by grass dominance in an abandoned hayfield (BSS Field E1) compared to a plowed field abandoned from a row crop (Field D3). An analysis of all hayfields and fields abandoned from row crops indicated a significant legacy effect of the hayfield grasses persisting for 10 yr after abandonment (see Myster & Pickett 1990 for details).

In the permanent plot study, other woody invaders tend to be delayed. For example the bird-dispersed *Rosa multiflora* and *Rhus glabra* first appear in intermediate years. The dominance of tree species in plots is often in an order that reflects dispersal mode, with bird- and wind-dispersed species establishing first followed by species that may be scattered or hoarded by mammals, for example. The order of dominance may also reflect differential sensitivity to browsing by mammals. Among *Juniperus virginiana*, *Acer rubrum* and *Cornus florida* the order of dominance is inversely related to the sensitivity of the species to browsing by mammals, so that browsing-resistant species dominate earlier in the fields (Cadenasso *et al.* 2002). In spite of such orderly patterns in differential sensitivity to browsing, forest canopy species can be present relatively early in the succession. These observations address processes that are features of both differential species availability and differential species performance.

One surprising feature in the long-term data is how commonly species remain present in the fields long after they decline in dominance. *Ambrosia artemisiifolia*, the most dominant herbaceous species in the early record in fields ploughed at abandonment, recurs in low abundance throughout the record. This is true also of some perennial species. Understorey species such as *Poa canadensis* or *Hieracium caespitosum* can be encountered late in the sequence. Some herbaceous species experience a second period of dominance when shrub canopies decline without being overtopped by trees. *Solidago* spp. usually dominate from 5–10 yr after abandonment; however, in cases where *Rhus glabra* shrubs decline precipitously *c.* 20 yr after they became dominant, *Solidago* assumed dominance again (Fig. 7.6). *Rosa multiflora* also declined either with or without an overtopping tree canopy. Because of its architectural complexity and the persistence of its dead stems, *R. multiflora* appears to have a more substantial legacy than *Rhus glabra*, and herbaceous species are slow to regain dominance in plots vacated by *R. multiflora*. Often the introduced invasive vine, *Lonicera japonica*, replaces the declining *R. multiflora*. A new invader, a disease of the *R. multiflora*, may play an increasing role in the shift of *R. multiflora* from dominance. The complex patterns of entry, persistence and demise of species in old-field succession combine aspects of differential availability and differential performance. In the realm of differential availability, mode of dispersal – whether wind, bird or mammal – possession of seed dormancy and landscape position all feature prominently. In differential performance, life cycle, competition and interaction with consumers stand out in the examples above.

The differential performance of exotic and native species further characterizes post-agricultural succession. The proportion and cover of exotic species is expected to decline with succession. This is because many exotics that appear in early succession are crop weeds, adapted to the disturbance regimes of row crop agriculture. These species exploit relatively open sites and, as the layering and cover of successional communities increase with time, these species are restricted in abundance and frequency (Fig. 7.7). During the middle portion of the 40-yr record, native perennials become more dominant than the agricultural weeds and ruderal plants (Meiners *et al.* 2002a). However, as the fields begin to support a closed canopy of trees, a new exotic herb is appearing in the understorey. *Alliaria petiolata*, characteristic of forest edges in Europe and increasingly common in deciduous forest of the eastern United

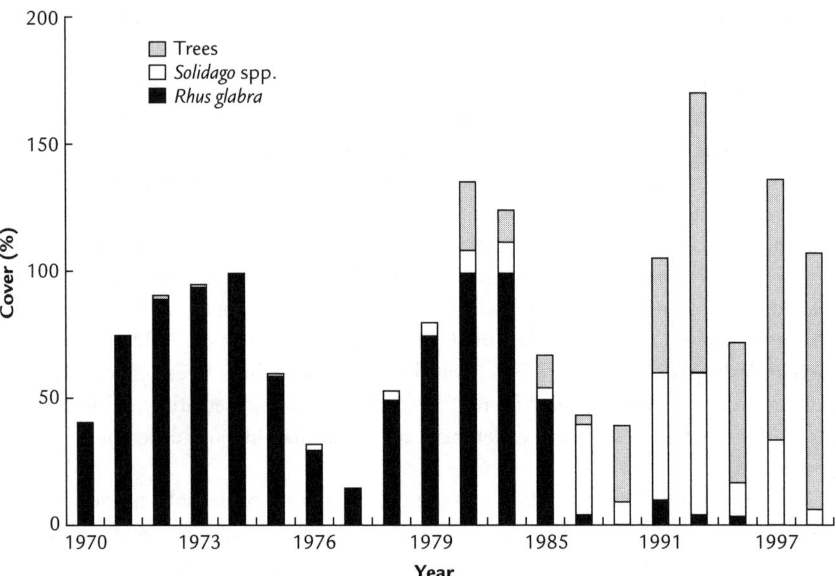

Fig. 7.6 Distribution of *Rhus glabra* and species of *Solidago* in plot 10 of field D3 of the Buell–Small Succession Study, from 1970 on. The field was abandoned as plowed bare ground after a row crop in 1960. The expected replacement of a dominant shrub, *R. glabra*, by overtopping trees does not appear. Instead, *R. glabra* declines without being overtopped, and the plot is subsequently dominated by patches of *Solidago* species and trees.

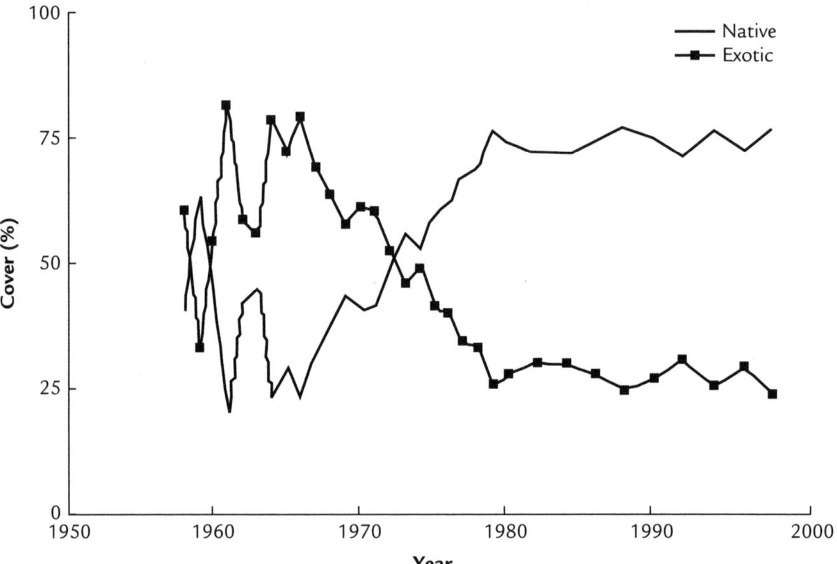

Fig. 7.7 Shift of dominance between exotic and native species over 40 yr of the Buell–Small Succession Study in field C3. The proportion of each species group, as defined by Gleason & Cronquist (1991), is shown for each year.

States, is beginning to increase in those plots having a tree canopy. It may be that forest-dwelling exotics, such as *Acer platanoides* or *Berberis thunbergii*, as well as *Alliaria*, will increase in the community in the future (Meiners *et al.* 2002a).

The spatial pattern of vegetation in fields is an important aspect of succession (Gross *et al.* 1995), just as it was in the other cases of succession examined, such as the rivers or the tornado blowdowns at Tionesta. Spatial heterogeneity can exist within fields and also between fields based on the landscape context in which they exist. We will exemplify within-field heterogeneity first and then indicate a role for the larger landscape context.

In plots at the edge of fields nearest the remnant forest, woody species tend to invade earlier than in plots farther away from the forest (Myster & Pickett 1992). The forest may have both a direct and an indirect effect on species availability and species performance. Direct effects of forest edges likely result from altering species availabilities of both wind- and bird-dispersed species as edges provide a seed source for both. Other influences of the forest are indirect. For example, leaf litter from the forest reduces light availability at the surface in old fields, affecting establishment of light-sensitive species (Myster & Pickett 1993). In addition, the presence of tree leaf litter in the fields affects competition between herbaceous dominants, and also the sensitivity of different species to predation. Meiners & Pickett (1999) examined both field and forest 'sides' of an old-field edge. They discovered that the boundary affected all major characteristics of the ground, shrub and seedling layers of both communities. Species richness and diversity increased from the forest to the edge and decreased slightly with distance into the field. Exotic species were most abundant in the forest within 20 m of the edge. Between-plot heterogeneity was greatest at the field edge (Meiners & Pickett 1999). Establishment probabilities of *Acer saccharum* and *Quercus palustris* increased with distance into the old field (Meiners *et al.* 2002b). Spatial heterogeneity in the case of old-field-forest edges affects both differential species availability and differential performance.

The effect of herbivores and predators on differential performance of plants is proving to be important in succession. However, few studies have addressed the role of herbivores and predators in succession. In fields near the permanent plots, experimental fences that exclude large to medium herbivores have altered several features of successional communities. Soon after abandonment, exclusion of mammals by fine-meshed fences, with metal skirts sunk into the ground, affected plant species richness and evenness and substantially increased the success and survival of tree seedlings (Cadenasso *et al.* 2002). In addition, the architecture of the community was affected, with the maximum height of woody and herbaceous elements increased by the exclusion of mammals. The structure of the ground layer was also reduced in exclosures (Cadenasso *et al.* 2002).

Spatial heterogeneity also affects predation and herbivory. Predation upon *Quercus rubra* seedlings was concentrated at the forest-field edge (Meiners & Martinkovic 2002). Insect herbivory (Meiners *et al.* 2000), as well as mammalian herbivory (McCormick & Meiners 2000), was important for various species.

Differential species performance is an especially complex kind of successional cause since so many different processes can interact to affect it. In addition to herbivores and predators, the kind and interaction of resource types may be important. Resources such as space, light, water and nutrients influence the performance of species differentially. Experiments on the role of resources in population performance show that a mixture of resources governs the organization of old-field communities (Carson & Pickett 1990). The cover of different species in the understorey is significantly affected by different resources. For instance, *Fragaria virginiana* is affected by light, *Rumex acetosella* by nutrients and *Hieracium caespitosum* by water. Water availability has a major impact on community richness and composition late in the season in years with normal rainfall, while in drought years, water is a key controller of old-field community structure in general.

The richness of the record from the spatially extensive permanent plots of the BSS reveals a great deal about successional pattern. Experiments with different ages of successional communities in the same environment have exposed important interactions and mechanisms of succession. In addition, the insights from permanent plot and associated experimental studies relate to the general causes of succession. Site availability is clearly controlled by the agricultural history of the site and the season and action of disturbance at the time of abandonment. Although abandonment is often taken as a zero point in successional studies, clear legacies of the prior management and composition of the communities on the fields persist into succession (Myster & Pickett 1990). The presence of crop residue and survival of perennial species or propagules are important aspects of legacy. Differential species availability reflects the abandonment treatment, the distance to forest and adjacent field edges, and the season of abandonment. Differential species performance is based on life-history attributes and longevity, different sensitivities to disturbance, light, water and nutrients, sensitivity to browsing and herbivory and competitive ability among others. In other words, the same kinds of causes of succession seen after large, infrequent disturbances over which people have little control and which are also found at the fine-scale natural disturbance events, also act in post-agricultural old fields. Exotic species, management decisions on and off site and landscape features are important elements in all the successions discussed.

7.3.5 Succession in forests with large herbivores

The classical trajectories of succession in mesic, temperate environments all recognize the tendency of a closed canopy forest to establish. This tendency was enshrined in the term 'climax' by ecology's early theorists (Clements 1916), suggesting that only a single possible state exists for late successional vegetation on fertile, well-watered sites. Although the idea has been deposed by the realization that natural disturbances, historical contingencies and variety in site conditions may result in considerable and persistent heterogeneity in late successional vegetation, most ecologists still accept that closed canopy forest was the dominant vegetation cover in the eastern United States and western lowland Europe before colonially and industrially subsidized agriculture consumed most land in those regions. Vera (2000) pointed out that this trajectory to closed forest was developed by observation of sites which lacked large non-specialist grazers. If large generalist herbivores were common, tree regeneration may have been restricted to areas dominated by thorny shrubs, rather than beneath closed canopies (Olff *et al.* 1999).

Vera (2000) hypothesized that many of Europe's large, old forest reserves were originally mosaics of grass-dominated, woody and scrub patches. He suggested that heterogeneity was obliterated with the extinction of the large generalist herbivores and the exclusion of domestic grazing from the reserves. His alternative hypothesis for forest succession is still too recent to have received wide evaluation although criticisms have already appeared (Svenning 2002).

7.4 Common characteristics across successions

The preceding sections have exemplified a wide variety of successional pathways and causes of succession. Within this variety, there are themes and insights that are common to all the examples.

1 Exactly what the succession pathway looks like depends on when in time one starts the observations. Starting a successional series after a large, infrequent disturbance yields a pathway that is affected by highly altered substrate availability and perhaps resources and propagules that remain in the site. Beginning with a smaller or less intense disturbance often leaves greater biological legacies. Succession studies frequently invoke an arbitrary zero point.

2 Few successions begin with a completely clean slate. There are two ways in which sites affect the subsequent successions on them. First are legacies that persist through the initiating disturbance. There are legacies that remain from the prior state of the community. Far from being the empty site suggested by the term that the earliest generation of theoreticians used – nudation – newly opened successional sites reflect structures, resources and reproductive potential left by some previously dominant community. Second are heterogeneities created by characteristics of the site or the disturbance, or the interaction of the two.

3 Sites can range from those that are relatively depauperate to those that are relatively rich in legacies. Classical terms identify the endpoints of this continuum (Fig. 7.8). The term primary succession is assigned to sites having new or newly exposed substrates, while secondary succession is assigned to sites that had previously supported a community. While it is true that a completely new site can exist, such as a volcanic island that emerges from the open ocean, it may be useful to think of most sites as having characteristics that are some mix of primary and secondary.

4 How a successional trajectory is described depends on how long observations last. Classically, ecologists have called a succession complete when certain features, such as diversity or productivity, were maximized. However, important community dynamics continue in almost all communities beyond some idealized maximum state. For instance, continuing observations of forests well beyond the period of canopy closure typically exposes successions that take place in smaller patches, such as gaps opened by disturbance within the community.

7.4.1 Refining the concept

By combining the themes that have emerged from the examples in this chapter, a refined view of succession becomes clear. When ecologists first began to cement their

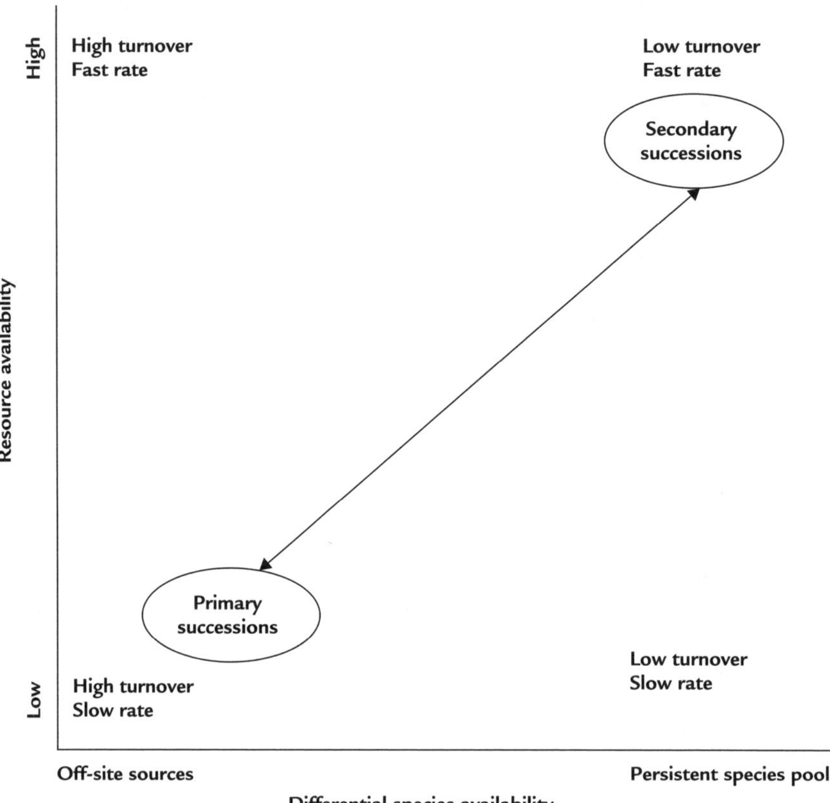

Fig. 7.8 Primary and secondary successions as extremes in a multidimensional space representing continua of differential site resources and differential species availabilities. Breaking down the process into controls by propagule sources and control by resources exposes complexities not apparent from the simple one-dimensional contrast between primary and secondary succession. Low resources are assumed to result in low rates of competitive interaction and low impact by herbivores and browsers. Local sources of propagules are assumed to make differential interactions more apparent while off-site propagule sources emphasize any time lags in the arrival of different species. This simple classification of successional patterns may be confounded by other specific mechanisms in the successional hierarchy (Fig. 7.3).

growing knowledge of succession, they emphasized the directional and irreversible nature of the process. In part they did this because it matched the linear and seemingly goal-oriented patterns that were then being articulated in evolution and in geomorphology. The first theories of succession emphasized the development of a community to be analogous to the growth and maturity of an individual organism.

In contrast, the contemporary literature emphasizes a different conception of succession. First, the pathways of succession do not necessarily follow a prescribed order. There is a high probability that short-lived, fecund species with extensive dispersal capacities will dominate sites soon after a disturbance. Likewise, species that grow slowly and allocate much of their assimilated resources to growth and structure as opposed to species that allocate resources to producing many, widely dispersable seeds, are more likely than not to dominate in communities not recently disturbed. In between disturbances, biomass tends to accumulate, spatial structure of the community tends to become more heterogeneous and richness of species tends to increase as early successional and late successional species overlap in time. However, not all specific locations experience these probabilistically described trends. In some cases, pioneer species are not present in the seed bank to capture recently disturbed sites. In other cases, expected linear increases in species richness do not appear. Herbivorous animals can alter the patterns of succession that would appear if the only interactions were those among plants.

In many cases, the soil resources are more limiting at the beginning of succession, while later, as plant biomass and structure accumulate, light becomes the limiting resource. This shift in limiting resources may be mirrored by a shift in species from those that can deal with low nitrogen levels to those that can deal with low light levels during their establishment phases (Tilman 1991). Events that reorganize the environment – its resources and the signals that govern species growth and reproduction – can appear with differing intensities through time. Thus observing a community through time shows the relationship of successional processes to both episodic events that originate from outside the community, as well as interactions within the community.

7.4.2 Net effects in succession

The multiplicity of successional trajectories have been summarized in 'models' of succession. Connell & Slatyer (1977) proposed that succession pathways could express three distinct kinds of turnover: facilitation, tolerance and inhibition. These alternative models helped expose the richness of successional processes, rather than a monolithic series of events. Early invaders have sometimes inhibited the invasion or performance of species that generally dominate later in succession. The third so-called model of succession, tolerance, is not only a simple neutral case. It can either reflect the meshing of life histories or the playing out of different environmental tolerances of the species without substantial interaction. Tolerance of the non-interactive kind is seen in those plots in which shrubs die with no apparent effect on the dominance of trees. Another complexity in models of succession is the mediation of effects of plant–plant interaction by herbivores or seed predators. In all cases, the models of succession are in fact complex, net effects of many interactions.

7.4.3 The differential processes of vegetation dynamics

The picture of succession that seems clear now is much more subtle than the classical view of succession (Odum 1969). The first subtlety applies to site availability. In order to understand why succession differs across the variety of kinds of conditions ecologists want to understand, it is clear that site characteristics differ considerably. Therefore, it seems wise to add 'differential' to the process of site availability as one of the fundamental successional processes. The recognition of differentials in site availability reminds ecologists that even if the availability and kinds of interactions among species are held constant, successions can differ in rate and composition simply because of differences in resource availability, landscape context and biological legacies present in different sites.

Succession has often been narrowly defined as the change in species composition of a community through time. More broadly, it is the change in both composition and structure. Therefore, the more inclusive definition emphasizes that communities can differ vastly because the architecture of the plants making them up differs.

Another generalization that emerges from the examples we have used is that the communities involved in succession are spatially heterogeneous. This is not merely an inconvenience to investigators and managers but is part of the fundamental nature of communities. The ability of different species or different species groups to contribute to different trajectories in succession adds to the richness of those communities. It allows species to specialize on different resources or reflects their dependence on different kinds of sites and interactions. Spatial heterogeneity is also the result of the rich variety of kinds and intensities of disturbance that affect successions. Finally, spatial heterogeneities result from the landscape context in which successional sites are located. All these kinds of heterogeneity – internal and externally generated – are part and parcel of succession. Succession is as much a spatial phenomenon of extensive landscapes as it is a temporal process in local communities.

7.5 Summary

In summary, vegetation dynamics is governed by three general processes – differential site availability, differential species availability and differential species performance. These three processes interact in a spatially heterogeneous array that reflects the nature of the disturbance that punctuates community dynamics and the spatial neighbourhood of the landscape in which succession occurs. The general processes themselves are composed of more specific mechanisms that describe the detailed characteristics of sites, species dispersal and interactions. The pathways of succession that ecologists actually observe result from the specifics of each of these kinds of processes and how those processes interact. Wild and managed sites all support successions that combine these different processes in specific ways. This chapter has provided the conceptual tools that can be used to understand successional pathways observed in any specific situation. The study of succession is an example of how ecologists have to link generality of process with the specific constraints and opportunities that different sites provide. After more than 100 yr of study, new combinations of factors and events are still being discovered that change our view of how succession occurs.

Acknowledgements

We are grateful to Sarah Picard for careful and efficient assistance with the long-term data set and analyses from the Buell–Small Succession Study. We thank Kirsten

Schwarz for constructing most of the data figures used in this chapter. The Buell–Small Succession Study has been supported in part by student help provided by the Hutcheson Memorial Center of Rutgers University and by a grant for Long-Term Research in Environmental Biology from the National Science Foundation, DEB 9726992. Our understanding of insights from the Kruger National Park were made possible by support of the Andrew W. Mellon Foundation of the River/Savanna Boundaries Programme.

With great affection and respect, we dedicate this chapter to Prof. Fakhri A. Bazzaz on the occasion of his retirement.

References

Anderson, D.C. & MacMahon, J.A. (1985) Plant succession following the Mount St. Helens Volcanic Eruption: Facilitation by a burrowing rodent, *Thomruya talpoides. American Midland Naturalist* **114**, 62–69.

Bartha, S., Pickett, S.T.A. & Cadenasso, M.L. (2000) Limitations to species coexistence in secondary succession. In: Vegetation Science in Retrospect and Perspective. *Proceedings 41th IAVS Symposium* (eds P.J. White, L. Mucina & J. Lepŝ), pp. 55–58. Opulus Press Uppsala.

Bazzaz, F.A. (1986) Life history of colonizing plants: Some demographic, genetic, and physiological features. In: *Ecology of Biological invasions of North America and Hawaii* (eds M.A. Mooney & J.A. Drake), pp. 96–110. Springer-Verlag, New York.

Bazzaz, F.A. (1996) *Plants in changing environments: Linking physiological, population, and community ecology.* Cambridge University Press, New York.

Boeken, B. & Shachak, M. (1998) Colonization by annual plants of an experimentally altered desert landscape: source-sink relationships. *Journal of Ecology* **86**, 804–814.

Boeken, B., Shachak, M., Gutterman, Y. & Brand, S. (1995) Patchiness and disturbance: plant community responses to porcupine diggings in the Central Negev. *Ecography* **18**, 410–422.

Botkin, D.B. & Sobel, M.J. (1975) Stability in time-varying ecosystems. *American Naturalist* **109**, 625–646.

Bowers, M.A. (1993) Influence of herbivorous mammals on an old-field plant community: years 1–4 after disturbance. *Oikos* **67**, 129–141.

Cadenasso, M.L., Pickett, S.T.A. & Morin, P.J. (2002) Experimental testing of the role of mammalian herbivores on oldfield succession: Community structure and seedling survival. *Journal of the Torrey Botanical Society* **129**, 228–237.

Carson, W.P. & Pickett, S.T.A. (1990) Role of resources and disturbance in the organization of an old-field plant community. *Ecology* **71**, 226–238.

Clements, F.E. (1916) *Plant succession: an analysis of the development of vegetation.* Carnegie Institution of Washington, Washington.

Collins, B.S. & Pickett, S.T.A. (1982) Vegetation composition and relation to environment in an Allegheny hardwoods forest. *American Midland Naturalist* **108**, 117–123.

Collins, B.S. & Pickett, S.T.A. (1987) Influence of canopy opening on the environment and herb layer in a northern hardwoods forest. *Vegetatio* **70**, 3–10.

Collins, B.S. & Pickett, S.T.A. (1988) Demographic responses of herb layer species to experimental canopy gaps in a northern hardwoods forest. *Journal of Ecology* **76**, 437–450.

Connell, J.H. & Slatyer, R.O. (1977) Mechanisms of succession in natural communities and their role in community stability and organization. *American Naturalist* **111**, 1119–1144.

Dale, V.H., Lugo, A.E., MacMahon, J.A. & Pickett, S.T.A. (1999) Ecosystem management in the context of large, infrequent disturbances. *Ecosystems* **1**, 546–557.

del Moral, R. (1993) Mechanisms of primary succession on volcanoes: a view from Mount St. Helens. In: *Primary succession on land* (eds J. Miles & D.W.H. Walton), pp. 79–100. Blackwell Scientific Publications, Boston.

Facelli, J.M. (1994) Multiple indirect effects of plant litter affect the establishment of woody seedlings in old fields. *Ecology* **75**, 1727–1735.

Forman, R.T.T. & Boerner, R.E.J. (1981) Fire frequency and the pine barrens of New Jersey. *Bulletin of the Torrey Botanical Club* **108**, 34–50.

Gleason, H.A. & Cronquist, A. (1991) *Manual of Vascular Plants of Northeastern United States and Adjacent Canada.* The New York Botanical Garden, Bronx, New York.

Glenn-Lewin, D.C. & van der Maarel, E. (1992) Patterns and processes of vegetation dynamics. In: *Plant succession: theory and prediction* (eds D.C. Glenn-Lewin, R.K. Peet & T.T. Veblen), pp. 11–59. Chapman and Hall, New York.

Glenn-Lewin, D.C., Peet, R.K. & Veblen, T.T. (eds) (1992) *Plant succession: theory and prediction.* Chapman and Hall, New York.

Gross, K.L., Pregitzer, K.S. & Burton, A.J. (1995) Spatial variation in nitrogen availability in three successional plant communities. *Journal of Ecology* **83**, 357–367.

Leck, M.A., Parker, V.T. & Simpson, R.L. (eds) (1989) *Ecology of soil seed banks.* Academic Press, San Diego, CA.

Luken, J.O. (1990) *Directing ecological succession.* Chapman and Hall, New York.

Mayr, E. (1991) *One long argument: Charles Darwin and the genesis of modern evolutionary thought.* Harvard University Press, Cambridge.

McCormick, J.T. & Meiners, S.J. (2000) Season and distance from forest – old field edge affect seed predation by white-footed mice. *Northeastern Naturalist* **7**, 7–16.

Meiners, S.J. & Martinkovic, M.J. (2002) Survival of and herbivore damage to a cohort of *Quercus rubra* planted across a forest – old field edge. *American Midland Naturalist* **147**, 247–256.

Meiners, S.J. & Pickett, S.T.A. (1999) Changes in community and population responses across a forest-field gradient. *Ecography* **22**, 261–267.

Meiners, S.J., Handel, S.N. & Pickett, S.T.A. (2000) Tree seedling establishment under insect herbivory. *Plant Ecology* **51**, 161–170.

Meiners, S.J., Pickett, S.T.A. & Cadenasso, M.L. (2001) Effects of plant invasions on the species richness of abandoned agricultural land. *Ecography* **24**, 633–644.

Meiners, S.J., Pickett, S.T.A. & Cadenasso, M.L. (2002a) Exotic plant invasions over 40 years of old field successions: Community patterns and associations. *Ecography* **25**, 215–223.

Meiners, S.J., Pickett, S.T.A. & Handel, S.N. (2002b) Probability of tree seedling establishment changes across a forest-old field edge gradient. *American Journal of Botany* **89**, 466–471.

Moon, B.P., van Niekerk, A.W., Heritage, G.L., Rogers, R.H. & James, C.S. (1997) A geomorphological approach to the ecological management of the rivers in the Kruger National Park: the case of the Sabie River. *Transactions of the Institute of British Geographers* **22**, 31–48.

Myster, R.W. & Pickett, S.T.A. (1990) Initial conditions, history and successional pathways in ten contrasting old fields. *American Midland Naturalist* **124**, 231–238.

Myster, R.W. & Pickett, S.T.A. (1992) Effects of palatability and dispersal mode on spatial patterns of trees in oldfields. *Bulletin of the Torrey Botanical Club* **119**, 145–151.

Myster, R. & Pickett, S.T.A. (1993) Effects of litter, distance, density. *Oikos* **66**, 381–388.

Odum, E.P. (1969) The strategy of ecosystem development. *Science* **164**, 262–270.

Olff, H., Vera, F.W.M., Bokdam, J., Bakker, E.S., Gleichman, J.M., de Maeyer, K. & Smit, R. (1999) Shifting mosaics in grazed woodlands driven by the alternation of plant facilitation and competition. *Plant Biology* **1**, 127–137.

Peterson, C.J. & Carson, W.P. (1996) Generalizing forest regeneration models: the dependence of propagule availability on disturbance history and stand size. *Canadian Journal of Forest Research* **26**, 45–52.

Peterson, C.J. & Pickett, S.T.A. (1991) Treefall and resprouting following catastrophic windthrow in an old-growth hemlock-hardwoods forest. *Forest Ecology and Management* **42**, 205–217.

Peterson, C.J., Carson, W.P., McCarthy, B.C. & Pickett, S.T.A. (1990) Microsite variation and soil dynamics within newly created treefall pits and mounds. *Oikos* **58**, 39–46.

Pickett, S.T.A. (1989) Space-for-time substitution as an alternative to long-term studies. In: *Long-term studies in ecology: Approaches and alternatives* (ed G.E. Likens), pp 110–135. Springer-Verlag, New York.

Pickett, S.T.A. & McDonnell, M.J. (1989) Changing perspectives in community dynamics: A theory of successional forces. *Trends in Ecology and Evolution* **4**, 241–245.

Pickett, S.T.A. & Rogers, K.H. (1997) Patch dynamics: the transformation of landscape structure and function. In: *Wildlife and landscape ecology* (ed J.A. Bissonette), pp. 101–127. Springer-Verlag, New York.

Pickett, S.T.A. & Thompson, J.N. (1978) Patch dynamics and the design of nature reserves. *Biological Conservation* **13**, 27–37.

Pickett, S.T.A., Kolasa, J., Armesto, J.J. & Collins, S.L. (1989) The ecological concept of disturbance and its expression at various hierarchical levels. *Oikos* **54**, 129–136.

Pickett, S.T.A., Kolasa, J. & Jones, C.G. (1994) *Ecological understanding: the nature of theory and the theory of nature.* Academic Press, San Diego.

Pickett, S.T.A., Cadenasso, M.L. & Bartha, S. (2001) Implications from the Buell-Small Succession Study for vegetation restoration. *Applied Vegetation Science* **4**, 41–52.

Prentice, I.C., & Leemans, R. (1990) Pattern and process and the dynamics of forest structure a simulation approach. *Journal of Ecology* **78**, 340–355.

Rankin, W.T. & Pickett, S.T.A. (1989) Time of establishment of red maple (*Acer rubrum*) in early oldfield succession. *Bulletin of the Torrey Botanical Club* **116**, 182–186.

Rogers, K.H. (1995) Riparian wetlands. In: *Wetlands of South Africa their conservation and ecology* (ed G.I. Cowan), pp. 41–52. Department of Environmental Affairs, Pretoria.

Rogers, K.H. (1997) Operationalizing ecology under a new paradigm: an African perspective. In: *The ecological basis of conservation: heterogeneity, ecosystems, and biodiversity* (eds S.T.A. Pickett, R.S. Ostfeld, M. Shachak & G.E. Likens), pp. 60–77. Chapman and Hall, New York.

Shachak, M., Pickett, S.T.A., Boeken, B. & Zaady, E. (1999) Managing patchiness, ecological flows, productivity, and diversity in drylands. In: *Arid lands management: toward ecological sustainability* (eds T.W. Hoekstra & M. Shachak), pp. 254–263. University of Illinois Press, Urbana.

Sparks, R.E. (1996) Ecosystem effects: positive and negative outcomes. In: *The great flood of 1993: causes, impacts, and responses* (ed. S.A. Changnon), pp. 132–162. Westview Press, Boulder.

Stearns, F. & Likens, G.E. (2002) One hundred years of recovery of a pine forest in northern Wisconsin. *American Midland Naturalist* **148**, 2–19.

Svenning, J.C. (2002) A review of natural vegetation openness in north-western Europe. *Biological Conservation* **104**, 133–148.

Tilman, D. (1991) Constraints and tradeoffs: toward a predictive theory of competition and succession. *Oikos* **58**, 3–15.

van Coller, A.L., Rogers, K.H. & Heritage, G.L. (1997) Linking riparian vegetation types and fluvial geomorphology along the Sabie River within the Kruger National Park, South Africa. *African Journal of Ecology* **35**, 194–212.

van der Maarel, E. (1996) Pattern and process in the plant community: fifty years after A.S. Watt. *Journal of Vegetation Science* **7**, 19–28.

van der Maarel, E. & Sykes, M.T. (1993) Small-scale plant species turnover in a limestone grassland: the carousel model and some comments on the niche concept. *Journal of Vegetation Science* **4**, 179–188.

Vera, F.W.M. (2000) *Grazing ecology and forest history.* CABI Publishing, New York.

Vitousek, P.M., Hedin, L.O., Matson, P.A., Fownes, J.A. & Neff, J. (1998) Within-system element cycles, input-output budgets, and nutrient limitation. In: *Successes, limitations, and frontiers in ecosystem science* (eds M.L. Pace & P.M. Groffman), pp. 432–451. Springer, New York.

Watt, A.S. (1947) Pattern and process in the plant community. *Journal of Ecology* **35**, 1–22.

White, P.S. & Pickett, S.T.A. (1985) Natural Disturbance and Patch Dynamics: An Introduction. In: *The Ecology of Natural Disturbance and Patch Dynamics* (eds S.T.A. Pickett & P.S. White), pp. 3–13. Academic Press, Orlando.

8

Diversity and ecosystem function

Jan Lepš

8.1 Introduction

Many ecologists are convinced that species diversity is important for the stability and proper functioning of ecosystems (Schläpfer *et al.* 1999). Indeed, the well-known Shannon formula (H') for species diversity was introduced to ecology as a stability index (MacArthur 1955). The relation between diversity and stability has been partially based on well-known ecological patterns. For instance, population outbreaks are more common in species-poor boreal regions than in the species-rich tropics, and they are also more common in species-poor man-made communities (agro-ecosystems and planted monocultures of woody species) than in the species-rich natural communities. However, in all those comparisons, variation in species richness is linked to many other characteristics. Also, the causality of observed patterns could well be reversed: the tropics are so rich in species because they have experienced long-term environmental stability, which enabled the survival of many species.

Since the 1990s, the global loss of biological diversity has become a major concern, not only of ecologists but also of the general public. Could indeed the decline of biodiversity impair the functioning of ecological systems? And do we have sound evidence of ecological consequences of declining biodiversity? These are matters of much concern and controversy (Naeem *et al.* 1999; Wardle *et al.* 2000; Kaiser 2000), but also of a growing consensus (e.g. Loreau *et al.* 2001). This is not a simple matter. Pimm (1984) and others showed that there are many aspects of stability as well as many aspects of diversity, and any response to the stability-diversity issue will depend on which aspects we choose. In natural ecosystems, community diversity is a 'dependent variable', i.e. it is a result of evolutionary and ecological processes, which also affect ecosystem functioning. For any statistical relationship several causal explanations can be found. All this renders answering questions about diversity effects very difficult.

If we want to study diversity and its effects, we must be able to quantify diversity and we need to understand the factors that affect diversity in nature. Then, we will have to quantify the ecosystem functions that are expected to be affected. Next we must find interrelationships and find ways to test the causality behind the statistical relationships.

8.2 Measurement of species diversity

Ecologists often use the term diversity, species diversity, richness and, in recent years, the terms biodiversity and complexity have become very popular. However, the concepts underlying those terms differ among ecologists. There are several choices that influence our evaluation of ecological diversity.

8.2.1 Which organisms to include

In most studies, the community is defined taxonomically. In many cases only the vascular plants are counted. In some studies, the bryophytes are also included (this decision usually depends on whether the researcher is able to determine bryophytes rather than on functional viewpoints). Such restrictions are unavoidable. It would have little sense to sum up the number of species of vascular plants and of soil algae. However, one should be aware that the diversity of vascular plants needs not necessarily be a good indication of diversity of all the plant species, and the diversity of all plant species needs not be a good indicator of richness of the whole ecosystem. Also, for pragmatic reasons, usually only the plants that are recognized above-ground are counted as present in the community and species present in the seed bank are ignored. This might cause some problems because a seed is a substantial part of the species' life cycle, particularly in arid systems. For the study of some processes (e.g. reaction to certain perturbations), the seed bank might be very important. There is no general rule for what should and what should not be included in the community. Decisions are made by the researcher, often on pragmatic grounds. One should be aware, however, that the relationships found might be crucially dependent on the decisions made.

8.2.2 Number of species and diversity

Let us assume that we have a plant community growing on a well-delimited area (even if this is usually not the case). The number of species itself is not a sufficient characteristic of community diversity. Even if two communities are formed by the same number of species, one of them can be dominated by a single species, whereas in the other one the species are more equally represented. This leads to the idea of two components of species diversity: species richness and equitability, also called evenness. Species richness is the number of species on the area delimited as the community area, and equitability expresses how equal the species are represented in the community. (However, many ecologists use the term diversity even when they are dealing with the number of species.)

Several diversity indices has been devised, the two most popular being the Shannon index and the reciprocal of the Simpson dominance index. Let us define p_i as the proportion of i-th species, i.e. $p_i = N_i/N$, where N_i is the quantity of the i-th species (e.g. its abundance, biomass), and $N = \sum N_i$, i.e. total of values for all the species (e.g. total abundance, total biomass). Then, the Shannon index is defined as:

$$H' = -\sum_{i=1}^{S} p_i \log p_i \tag{8.1}$$

where S is the number of species. The Shannon index is based on information theory; this is why, originally, $\log_2$ was used. In ecology, with N being the number of individuals, H' was interpreted as bits per individual; this interpretation is not used any longer. As a matter of fact, all three variants, $\log_2$, natural log and $\log_{10}$ are used, and all functions are called Shannon (or Shannon–Wiener, or Shannon–Weaver) index. To avoid confusion, it is always necessary to state which logarithm was used. Some problems might be avoided by using the antilogarithm of H' (i.e. $2^{H'}$, $e^{H'}$ or $10^{H'}$ for $\log_2$, natural log and $\log_{10}$ respectively). This value can be interpreted as the number of species needed to reach diversity H', when the species are equally represented. The value of H' equals 0 for a monospecific community, and $\log S$ for a community of S equally represented species.

The second most frequently used diversity index is the reciprocal of the Simpson dominance index, $1/D$. The Simpson index itself is defined as:

$$D = \sum_{i=1}^{S} p_i^2 \tag{8.2}$$

Similarly as for antilog H', the minimum value of $1/D$ equals 1 for a monospecific community, and its maximum is S for a community of S equally represented species.

Hill (1973) has shown that the common indices of diversity are related to each other (and to Rényi's definition of generalized entropy) and suggested a unifying notation. His general diversity index can be written as:

$$N_a = \left(\sum_{i=1}^{S} p_i^a \right)^{1/(1-a)} \tag{8.3}$$

N_a is a general numerical diversity of 'order' a – do not confuse this with N_i, which is used for the quantity of i-th species in a community! By increasing a an increasing weight is given to the most abundant species. The following series arises:

$N_{-\infty}$ Reciprocal of the proportion of the rarest species;
N_0 Number of species;
N_1 Antilog of H', the Shannon index (asymptotically);
N_2 Reciprocal of the Simpson index, $1/D$;
N_∞ Reciprocal of the proportion of the most abundant species, also known as the Berger–Parker Index.

Equitability is usually expressed as the ratio of the actual diversity and the maximum possible diversity for a given number of species. More complicated evenness indices were also suggested. However, the interpretation of evenness indices is problematic (see, e.g. Routledge 1983; Magurran 1988) and the on-going discussion on the appropriateness of the various evenness indices has so far been inconclusive.

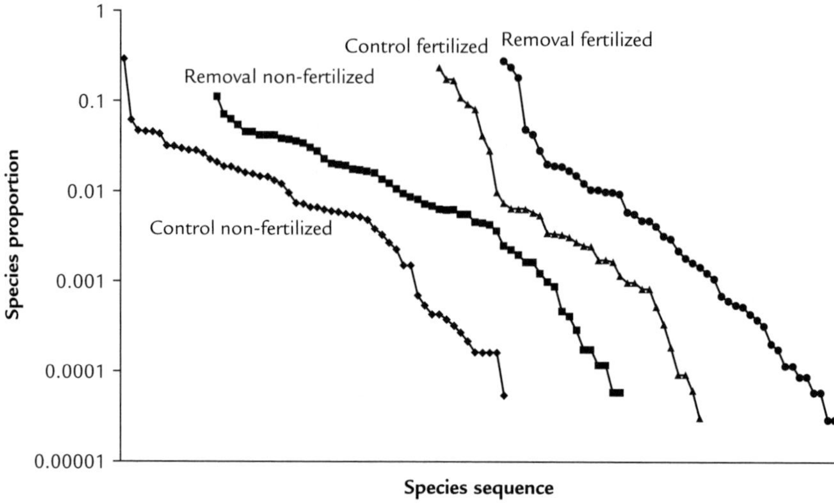

Fig. 8.1 Diversity-dominance curves for the biomass of four experimental treatments in a wet oligotrophic meadow in Central Europe (Lepš 1999). From experimental plots, the dominant species *Molinia caerulea* was removed, control plots are without any removal, and plots were either fertilized or not. The figure shows the situation in 2000, i.e. 6 yr after the start of the experiment. Each curve displays a sequence of species (x-axis); proportion of species is the ratio of species biomass to biomass of the whole community. Each curve is based on pooled biomass from three quadrats, 0.5 m × 0.5 m each. The respective values of reciprocal of the Simpson index are (from left to right) 9.2, 22.6, 7.0 and 5.9, the values of antilogarithm of H' are 19.7, 30.0, 9.6 and 9.7 and the numbers of species 54, 57, 37 and 47 respectively.

It is clear that the whole complexity of distribution of importance values among species in a community cannot be described by a single parameter. In a graphical form, the species structure of a community is probably best described by the so-called dominance-diversity curves (also called rank/abundance curves or importance-value curves, see Whittaker 1975; for the other possibilities of graphical representation see Magurran 1988 and Hubbell 2001). Here, we first rank the species from the most important to the least important one (as importance value, biomass or number of individuals are used). Then, the relative importance (e.g. proportion of community biomass) is plotted on a logarithmic scale against the species rank number. In this way, we obtain decreasing curves, differing in their shape and length, which characterize the community species structure (Fig. 8.1). Sometimes, various importance value distribution models are fitted to the data, notably the geometric series, the log series, the lognormal distribution and the broken stick model (see, e.g. Whittaker 1975; Magurran 1988; Hubbell 2001). Their parameters are also used as diversity indices, in particular Fisher's α, characterizing the log series, and z, characterizing the lognormal distribution. Some of them were derived from assumptions about modes of partitioning the niche space among the species. However, later on it became clear that the same distribution can be derived from various underlying assumptions, and consequently the distribution could hardly be used to support mechanistic assumptions of the models (Magurran 1988).

Nevertheless, the shape of the dominance-diversity curves may change in a predictable way along gradients, or among community types. In the example from an experiment in a mown meadow (Fig. 8.1), one can compare the structure of four experimental combinations of the removal of the dominant grass (*Molinia caerulea*) and fertilization. The basic difference is between fertilized and non-fertilized plots: the slope in fertilized plots is much steeper – which corresponds to a higher degree of dominance – and the curves are shorter – fewer species. The non-fertilized, non-removal plots are strongly dominated by *Molinia*, but the remaining species occur in rather equal proportions. After 6 yr none of the remaining species had developed a strong dominance in the non-fertilized removal plots (incidentally, this fact cannot be learned from any of the diversity indices). Because *Molinia* was suppressed by fertilization, even in the non-removal plots, the dominance relations of the major species are very similar for non-removal plots and removal fertilization plots. Still, more of the minor species are found in the removal plots. Comparison of the curves with values of $1/D$ and H' shows that the reciprocal of the Simpson index is much more affected by the presence of the single dominant than the antilogarithm of H'.

In most plant communities, regardless of their species richness, the community consists of relatively few dominant species and many subordinate species, most of which have a low abundance – and consequently have a small effect on community productivity or nutrient cycling. In the example in Fig. 8.1, 90% of the biomass was made up by 27 out of 57 species (47%) in the non-fertilized plots with the dominant removed, by 12 out of 47 (25%) in removal/fertilization plots, by 23 out of 54 species (42%) in non-fertilized control plots, and by 8 out of 37 (21%) in fertilized control plots. In the last case, the 14 least abundant species made up less than 1% of the community biomass. One can reasonably expect that those species have a very limited effect on ecosystem processes like total productivity or nutrient retention, which will be determined mostly by ecological features of the dominant species (the mass ratio hypothesis of Grime 1998). On the other hand, even low-abundance plants can support populations of specialized herbivores. For example, in a wet alder forest, even species with biomass as low as $0.27\ g{\cdot}m^{-2}$ (dry weight) supported large populations of monophagous insects (Lepš *et al.* 1998). So, low-abundance plant species, although not affecting ecosystem processes, may be crucial for the maintenance of diversity on higher trophic levels.

8.2.3 Spatial characteristics of diversity

In the preceding section a community occupying a fixed, well-delimited area was considered. However, we usually sample only part of a much larger community. With increasing spatial extent of a community, the number of species found also increases. The rate of increase differs among communities. The dependence of the number of species S found on the size of the investigated area A is usually described by the species-area curve (see Rosenzweig 1995 for discussion). Two functional forms are employed: the power curve, usually written as $S = c \cdot A^z$, and the logarithmic curve $S = a + b \cdot \log(A)$; c, z, a and b are parameters estimated by the methods of regression analysis. The power curve implies a linear dependence of the variables after log transformation of both; the logarithmic curve is linearized by log transformation of the area only. The power curve starts in the origin – there are no species present at plot size zero; c is the species number in a plot of unit size; z measures the rate of increase: when doubling the plot size, the number of species increases 2^z times (z usually ranges from 0.15 to 0.3). According to the logarithmic curve a sample plot of unit size contains a species, and when doubling the area, $b \cdot \log(2)$ new species are added. The number of species at zero area is not defined; actually, for very small plot sizes S becomes negative. Note that both c and a depend on the units in which area is measured, whereas z and b do not. Theoretical arguments supporting either of the relationships were suggested, and also, when fitted to real data, neither of the two is systematically better than the other one; consequently, the decision which of them should be used usually depends on their fit to actual data sets. Functions with three parameters were also suggested – however, they are seldom used.

Species-area curves are used on widely varying spatial scales, from within-community areas of cm^2 to tens of m^2, to whole continents. However, each curve should be interpreted solely for the scale at which it was derived, and not used for extrapolations. Indeed, it was shown (Rosenzweig 1995; Crawley & Harral 2001) that the slope of the relationship changes when based on different ranges of spatial scales. Similarly, it was shown (Lepš & Štursa 1989) that the estimate of the species number in the Krkonoše (Giant) Mountains in Central Europe extrapolated from the within-habitat species area curve for mountain plains would be 30.3, and for avalanche paths 8225 species; the real value is *c.* 1220. The relationships are governed by various mechanisms at various scales. At within-community scales, the increase of the number of sampled individuals is decisive, together with the ability of species to coexist. The number of sampled individuals is negatively related to the mean size of an individual – a 1 m^2 plot may host thousands of individuals of tiny spring therophytes, but not a single big tree. With increasing area, the effect of environmental heterogeneity increases. This can be biotically generated heterogeneity – e.g. the variability between the matrix of dominant species and the gaps between them occupied by competitively inferior species – or small-scale heterogeneity in soil conditions at the within-habitat scale, or heterogeneity of habitats at the landscape scale. At continental scales the evolutionary differentiation between subareas starts to play a role.

Species-area curves are often used to describe the increase in species number with increasing area within a habitat, or predict the number of species in a county, region or state; the estimates are often based on nested data – and one has to accommodate the statistical analysis to this layout. However, the same functional forms are used when studying the species-area dependence in an archipelago (i.e. in a set of non-overlapping islands). Within the island biogeography framework, this approach was used to study both the real islands, and also the habitat islands (like nature reserves in a cultural landscape, mountain tops with open vegetation in a forested area). Usually, in small isolated (habitat) islands, the species richness is lower than in a plot of corresponding area within a continuous habitat, and the species–area relationship is consequently steeper.

Species-area curves are also used to estimate the species extinction rate on a global scale (Jenkins 1992), particularly in the tropics. The knowledge of the taxonomy of tropical organisms is far from complete. Invertebrates are notoriously underrepresented, but even the knowledge of vascular plants is fragmental. Botanists from temperate zones are not always aware that only the inventory of trees in a hectare of tropical forest may take weeks in the field and months of subsequent work in herbaria. Hence it is extremely difficult to document extinction of tropical rain forest tree species, because their distribution is mostly unknown, particularly that of rare species, which are prone to extinction; and, of course, extinction of species not yet described would proceed unnoticed.

When we are able to estimate the species-area curve for rain forest trees, the reduction in rain forest area could be used as a predictor of the reduction in the number of species, and thus of the extinction rate. As an example, we could estimate the loss of species between 1990 and 2000 from the expression: $c \cdot A_{1990}^{z} - c \cdot A_{2000}^{z}$, where A_{1990} and A_{2000} are the areas of rain forest in the respective years. It is clear that this approach has many weak points: First, we have to assume that the species are not able to grow outside the rain forest. Second, the species-area relationship is usually estimated on the basis of known numbers of species in continuous areas, whereas the area of remaining rain forest may be distributed over patches. Consequently, the extinction rates based on this method might be overestimated.

When comparing various communities, the species area curves sometimes cross each other. In these cases it will depend on the area chosen for comparison which of the communities compared is richer in species. This problem can be solved pragmatically by adopting a spatial scale which is most appropriate for our investigations, or compare the species area relationships as such for the communities.

As the number of individuals is linearly dependent on the area sampled, we can expect a similar functional dependence of species number on the number of individuals sampled. Methods based on counts of individuals are more obvious in animal ecology; in plant ecology, they are mainly used in communities where the definition of an individual does not cause problems, e.g. communities of annuals in deserts or of trees in forests. When the number of individuals of each species in the community is known, a method known as rarefaction can be used. This analytical procedure enables the estimation of the number of species expected in a random draft of individuals from the original sample according to the formula:

$$E(S) = \sum_{i=1}^{SO} \left[1 - \frac{\binom{N - N_i}{n}}{\binom{N}{n}} \right] \tag{8.4}$$

where $E(S)$ is the expected number of species in a subsample containing n individuals, SO is the number of species in the original sample, N is the total number of individuals, and N_i is the number of individuals of the i-th species in the original sample. In this way, we can estimate the species richness of subsamples of fixed numbers of individuals from various communities, even when the original samples differ in size. Unfortunately, the individuals are not distributed randomly in space, and individuals of the same species are very often clumped. Consequently, the subsample from a continuous subarea of the original sample is not a random draft of individuals from the original sample and will usually contain fewer species than suggested by rarefaction. Another solution of comparing samples containing various numbers of individuals is, as above, to adopt the log series species abundance model, and to calculate Fisher's α as a measure of diversity. This is determined by the ratio of number of species and number of individuals, but has to be calculated using an iterative procedure – see Magurran (1988).

To characterize the spatial aspects of diversity, the terms α- or within-habitat diversity and β- or between-habitat diversity are sometimes used. Whereas α-diversity can be measured by the number of species or any of the diversity indices within a limited area, β-diversity is characterized by differences between species composition in different habitat types, or by species turnover along environmental gradients. A simple straightforward way for measuring β-diversity was suggested by Whittaker (e.g. 1972; see Magurran 1988) as $\beta_w = S/\alpha - 1$, where S is total number of species in the system studied and α is the α-diversity, expressed as the mean number of species per fixed sample size (e.g. a quadrat). This method is intuitively appealing. It would provide a good diversity estimate if we have a good estimate of S, the total number of species in the system studied. Usually, the number of all species in all quadrats is used as an estimate of S. The mean number of species per quadrat is independent of the number of quadrats investigated, but the total number of species increases with the number of quadrats in the study, and thus β_w will increase with the number of quadrats. A better approach to β-diversity is based on a (dis)similarity measure. The distribution of (dis)similarity values between all pairs of samples is a good indication of β-diversity (Magurran 1988). We can base (dis)similarity measurements on both presence-absence and quantitative data. If the samples are taken along known ecological gradients, the species turnover along those gradients can be studied. Multivariate analysis, notably cluster analysis or ordination can also be used to get insight into the β-diversity of our system.

8.2.4 Taxonomic diversity versus functional diversity

A community composed of four annuals will be less diverse from a functional point of view than a community composed of two annuals and two perennial herbs and this again is less diverse than a community composed of one annual, one perennial herb, one shrub and one tree. Functional diversity is usually measured through species diversity but that is only a surrogate (Loreau 2000).

The measurement of functional diversity is not an easy task. The most common approach is based on the recognition of functional groups of species (see Chapter 1). Community diversity can be described in a hierarchical way – as diversity of functional groups, and as species diversity within functional groups. The definition of functional group is crucial here, and there is wide range of possible approaches. One can use the well-known plant life forms of Raunkiær, taxonomically based units such as grasses versus forbs, C_3 versus C_4 plants, clonal versus non-clonal plants, etc.; very often, legumes are considered a separate functional group because of their nitrogen-fixing ability. Recently, the attempts are made to formalize the definition of plant functional types, which should also reflect species responses to environmental constraints (e.g. Smith *et al.* 1997; McIntyre *et al.* 1999). The plants can be functionally classified in a hierarchical manner (e.g. the clonal plants can be further divided into several functional types, see Klimeš & Klimešová 1999). Similarly, taxonomic diversity can be seen in a hierarchical manner: the number of higher taxonomic units (genera, families) is an important characteristic, particularly when studying the diversity on large spatial scales or from a historical perspective.

8.2.5 Intraspecific diversity

Populations are not homogeneous groups of individuals. Each population is composed of different genotypes. The genotype composition (the genetic structure of the population) depends on the mating system of the population, on the clonality of plants, and also on the population size. In community studies, the genetic diversity is mostly ignored; however, recent studies suggest that the fitness of a population and its ability to cope with environmental variability can be dependent on its genetic structure. Also, the population decline is usually correlated with a loss of genotype diversity.

8.3 Determinants of species diversity in the plant community

8.3.1 Two sets of determinants

The species diversity of a plant community is determined by two sets of factors. The first one is the pool of species, i.e. the set of species propagules of which are able to arrive at the site. The second factor comprises the local ecological interactions, selecting from the species pool those species that are able to co-exist in a community (Zobel 1992; Eriksson 1993; Pärtel *et al.* 1996). In this 'community filter', two main forces can be distinguished: the abiotic conditions and the biotic interactions. In other words, the plants have to be able to survive in a given physical environment (concerning both the climate and site conditions such as soil, moisture, but also the disturbance regime, e.g. avalanches, fire), and also to withstand the competition of other plants, the grazing pressure of herbivores and effects of pathogens. The situation is even more complicated, because in some cases other plants modify the environment in a way favourable to a plant species. Also, many plants depend on animals for the dispersal of their pollen or seeds and are dependent on symbiotic microorganisms (bacteria, fungi), particularly in the soil. In some cases, the absence of a species in a certain area can be caused by the absence of its specialized dispersal agent, or absence of mycorrhizal fungi in a site. Note that the definition of species pool used in this chapter is the broadest one; some authors use a much narrower definition, e.g. species able to reach the site and survive in given conditions (e.g. Zobel *et al.* 1998).

8.3.2 The species pool

The species pool is affected mostly by historical factors: the place where the species evolved, and whether they were able to migrate to a certain site. With increasing distance from the Equator, more and more of the plants that at present are found in the species pool migrated into the area after the retreat of the glaciers (Tallis 1991). The species pool is affected by the proximity of glacial refugia, and by the migration barriers between the refugium and the site. The barriers are either physical (e.g. mountains), or biological. For example, the most important barrier for the dispersal of heliophilous mountain plants are the forested areas between the mountains, where the plants are not able to establish themselves, because they loose from the trees in the competition for light (most mountain plants grow well in lowlands when released of the competition). The species pool is thus affected not only by abiotic factors, but also by past and present competition (including competition having happened on migration pathways). However, even during postglacial time, the micro-evolutionary processes modified the species so that they are better adapted to the life in existing habitats, and new species developed during this time. It is probable that there are more species adapted to the life in habitats that were abundant during the postglacial period (Taylor *et al.* 1990; Zobel 1992).

In the ecological literature, the species pool is generally described as a set to which a species does belong or not. This is a simplification. Plant establishment from seed is a highly improbable event: often, one of many thousands or millions of seeds gets established. Consequently, the amount of seeds (or other propagules) has to exceed some species-specific threshold for a species to have a reasonable chance to invade a community and establish a viable population there. Not all populations found in nature are viable by themselves. Metapopulation theory (Hanski 1999) distinguishes source and sink populations, where source populations are donors of propagules to other populations, and sink populations are recipients of propagules. Sink populations are found in suboptimal habitats and are not viable when restricted to themselves, but are often able to keep a stable population size, thanks to a constant influx of propagules from the source populations (Cantero *et al.* 1999). Grime (1998) called those species transitional; they are probably not rare and may substantially increase the species richness of some communities. Clearly, there is a mass effect in the species pool: the probability that a species will pass through the community filter increases with its abundance in the influx of propagules, in fact, with its abundance in surrounding communities. (This effect is also called vicinism; see Chapter 1.)

Sometimes, local and regional species pools are distinguished. Clearly, at any spatial scale, the species present in a wider surrounding of the area under study are considered to be the corresponding species pool. When studying a plant community in 50 m × 50 m plots in tropical forest, all the species in the forest patch of the size of several km^2 can be considered its species pool. However, we can study the species composition of the whole patch; in this case, all the species found in the region are considered to be the corresponding species pool. The first one would be called local, the second regional; however, their distinction is quite arbitrary (see Chapters 1 and 6).

8.3.3 Species co-existence

Classical theory predicts that the number of co-existing species should not exceed the number of limiting resources. The competitive exclusion principle of Gause (see

Chapter 9) states that two species cannot co-exist indefinitely in a homogeneous environment, if they are limited by the same resource. Plant species compete for light, water and a few mineral nutrients. In spite of this, there are plant communities where scores of species co-exist on a square metre, and hundreds on a hectare. This co-existence can be seen as a violation of assumptions of the competitive exclusion principle (Palmer 1994). There are many possible explanations for species co-existence – and these are not mutually exclusive. They can be conceptually divided into two broad categories: the equilibrium and non-equilibrium ones. The equilibrium explanations explain species co-existence by the violation of the spatial homogeneity assumption together with the ability of species to use different parts of the resource gradient, 'niche differentiation'. They include the differentiation of rooting depth of plants, differential ability to use light, and phenological differentiation.

The non-equilibrium explanations challenge the assumption of 'indefinite' time. If the environment changes, the competitive equilibria will change as well. If the rate of competitive displacement is low, environmental variability may be sufficient to prevent the competitive displacement. For example, competitive hierarchies in grassland communities have been shown to change from year to year, depending on the weather (Herben *et al.* 1995). Recruitment of seedlings is affected much more by both spatial heterogeneity, temporal variability and their interaction than the occurrence of established plants. The theory of the regeneration niche (Grubb 1977) assumes that co-existence is promoted through the differentiation of species requirements for successful germination and establishment. Many species are dependent in their recruitment on gaps in otherwise closed canopy, being it in forests, or in grasslands. The gaps can be seen as highly variable resources: they differ not only in size, but also in the time of their creation – and plants differ in their seedling phenology (Kotorová & Lepš 1999). Plant competition is highly pre-emptive: the plant which established first in a gap is usually the winner in any competition with newly arriving seedlings. Which will be the first individual occupying the gap depends also on many random circumstances. All this might lead to postponement of competitive exclusion and species co-existence. Indeed, the small-scale species composition in a community patch changes in time, whereas the species composition on a larger scale is fairly constant (a key element in the carousel model of van der Maarel & Sykes 1993; see Chapter 1).

Part of the explanation is based on effects of organisms from higher trophic levels. In particular, both pathogens and specialized herbivores are expected to have greater effects on dominant species. This is expected to be of particular importance in the tropics.

Hubbell (2001), using mathematical models, demonstrated that species co-existence could be maintained under the species 'neutrality' hypothesis – i.e. when all the species have the same competitive abilities, in case there is some constant influx of new species (by immigration, or by speciation). However, because species in any community differ in their competitive abilities, it is difficult to see the relationship between such neutral models and ecological reality. Nevertheless, the role of 'lottery recruitment', implying that the identity of a species filling a gap is determined solely by chance (which is one of the bases of Hubbell's model) is increasingly accepted by ecologists. The chances to be the winner, however, differ among species, and a 'weighted lottery' (Busing & Brokaw 2002) is probably a more realistic model.

8.3.4 Distinguishing the effect of the species pool from local ecological interactions

In the literature, the relative importance of historical factors (reflected in the species pool) and local ecological factors is often discussed. It is very difficult to separate those two effects, particularly because the actual species pool is also affected by local species interactions. The positive correlation between the actual species richness of communities and the set of species able to grow in them (i.e. the species pool in a narrow sense) in a region was repeatedly demonstrated. However, the set of species able to grow in a habitat is clearly determined by the species which actually grow in the communities, and consequently by local ecological factors (Herben 2000). For example, one can argue that calcareous grassland communities in Central Europe are rich in species because there are many species adapted to the conditions in this community in this region. But one can also expect that the reason for the occurrence of many species in calcareous grasslands here is that local ecological processes in calcareous grasslands enable the co-existence of many species.

Probably the best way to separate the effects of local ecological interactions and general historical effects is to compare patterns of species richness between geographical regions. Schluter & Ricklefs (1993) suggested a procedure for the decomposition of the variance in species richness into parts attributable to habitat, geographical region and their interaction. The method is analogous to the decomposition of the sum of squares in two-way ANOVA. Repeated patterns in geographical regions differing in their history suggest effects of local conditions, while differences point out the effects of history.

Indeed, some patterns in species richness are repeated in various geographical regions; they are very probably based on local mechanisms. For instance, tropical rain forests are always much richer in plant species than the adjacent mangroves. This is probably the consequence of the fact that only a limited number of species is able to withstand the physiologically extreme conditions in mangroves. But there are very pronounced anomalies, where ecologically similar communities markedly differ in their species diversity, depending on geographical region. For instance, mangroves in West Africa are poor in comparison with the relatively species-rich mangroves in Malaysia. The latter difference is thought to have historical reasons.

8.4 Patterns of species richness along gradients

8.4.1 Introduction

Ecologists have long known that species richness of plant communities changes along environmental gradients in a predictable way (reviews in Huston 1994; Rozenzweig 1995; Pausas & Austin 2001). However, few general patterns have emerged from these studies. This might be caused by several factors. First, species diversity is a compound characteristic consisting of the richness of functional groups and the number of species within these groups. The mechanisms of competition within a functional group may be different from mechanisms of competition of species from different functional groups. Mechanisms which enable species co-existence within a group by preventing competitive exclusion can be the same as those that limit the presence of certain functional groups (Huston 1994). For example, disturbance (like periodical avalanches) may promote the co-existence of herbaceous species, but at the same time exclude all functional groups of woody species. The net effect of disturbance in terms of the number of species in a community is then based on the relative richness of the species pools of woody and herbaceous species.

Here, we will discuss the three well-known patterns: the latitudinal gradient in species richness, and the diversity response to productivity and to disturbance. The latter two gradients are considered to be the most important axes determining the habitat templet (Grime 2001; Southwood 1988).

8.4.2 Species diversity along latitudinal gradients

The decrease of species diversity from the Equator to the poles is one of the most universal patterns in nature (Fig. 8.2). This decrease does not only hold for species, but also for higher taxonomic levels (genera, families). The fossil records show that this pattern can be traced back at least to the Cretaceous (Crane & Lidgard 1989; note that original latitude of fossil records could be determined by geophysical methods). At present, tropical rain forests are the richest plant communities on Earth at larger spatial scales – on smaller areas the size of single trees naturally causes the diversity to be low. On a square metre the record species diversity will rather be in semi-deserts with thousands of tiny therophytes, or in meadows with equally numerous plant units. Also, tropical forests are unsurpassable in their diversity of plant life forms (i.e. their functional diversity), and the diversity of genera and families. The analysis of 1-ha plots is becoming the standard way of plant biodiversity assessment in tropical forests, where the records are usually limited to trees with DBH (diameter at breast height) > 10 cm; some detailed analyses record smaller trees. Typical numbers of tree species are between 100 and 300. For example, in the Lakekamu basin alluvial plot in Papua New Guinea, 182 species belonging to 104 genera and 52 families were identified (Reich 1998). Typically, many species have a low abundance; 86 species, i.e. almost half the species in this inventory were represented by only one single individual. When smaller trees would be included, the number of species would be even higher and the species-area curve suggests that many additional species would be found with a further increase of the area. To explain this extraordinary species diversity, we have to explain both the extremely large species pool and the ability of species to co-exist. There is little doubt that the high number of tropical species has historical reasons – the relatively stable environment minimizes extinction rates. Although the glacial periods also affected the tropics, the island of rain forest pertained through all the 'full-glacial' periods in the tropics of Africa, South America, Southeast Asia and Oceania (Tallis 1991). Unlike in temperate regions, the changes were probably not of an extent causing mass extinction. It has also been shown that low temperature and water stress are limiting factors for many life forms (Woodward 1987). Thus ecological conditions are favourable for an exceptional diversity of life forms.

We can hypothesize that a long period of environmental stability and of interspecific interactions triggered the adaption of species to tolerance of competition of other

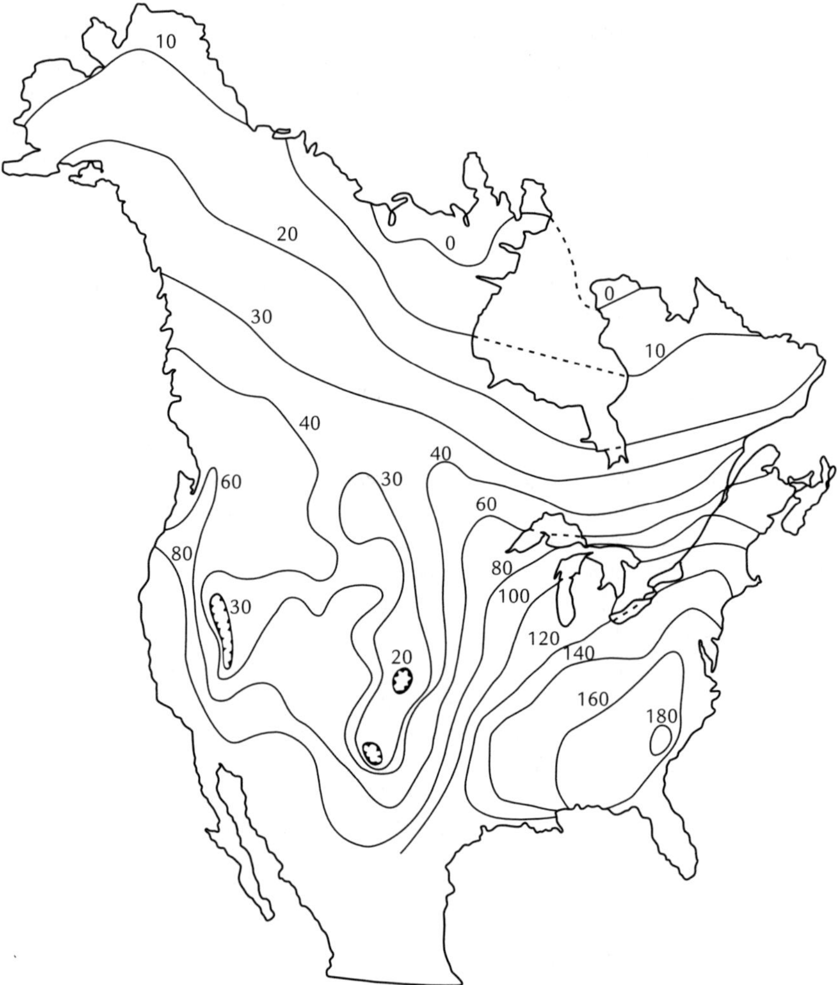

Fig. 8.2 Tree species richness in Canada and the USA. Contours connect points with the same approximate number of species per quadrat. Quadrat size is 2.5° × 2.5° south of 50°N, and 2.5° × 5° north of 50°N (Currie & Paquin 1987). Reproduced by permission from *Nature*, copyrighted 1987, Macmillan Publishers Ltd.

species. However, all this does not explain, *how* the species are able to co-exist. In particular, how are these hundreds of tree species able to co-exist? The high photon flux enables the diversification of the tree canopy (emergent trees, several canopy layers), so there is some niche differentiation. Many explanations have been suggested for the high species diversity in the tropics (reviewed by Hill & Hill 2001), which are often not mutually exclusive. Denslow (1987) stressed the importance of the internal rain-forest dynamics, with a variety of tree gaps and differential colonization of gaps. However, when such a high diversity should be maintained, we need mechanisms that selectively suppress the establishment or survival of individuals in the vicinity of their conspecific adults. It seems that biotic factors are more probable agents able to differentiate between species. Janzen (1970) suggested a mechanism, based on the effect of herbivores (the mechanism is referred to as the Janzen, or Janzen–Connell hypothesis): in the close vicinity of an adult tree, seed predation by specialized herbivores is high, which leads to a comparative advantage of other species. A similar reasoning can be applied to pathogens. Unfortunately, this hypothesis has long lacked strong empirical support. Only recently, patterns consistent with the idea of 'biotic density dependence' were found in tropical forest in the Barro Colorado Island, Panama, by Wills *et al.* (1997). Their analyses are based on the repeated detailed inventory of each tree with DBH > 1 cm in a 0.5 km^2 plot of tropical forest (*c.* 240,000 trees and large shrubs were recorded, together with their location), and demonstrated that for many species, the establishment is negatively correlated with the basal area of their conspecifics, and overall recruitment and rate of increase is positively correlated with the evenness of surrounding vegetation. The authors strengthened the original Janzen hypothesis – the 'pests' are postulated not to be strictly species-specific, but oligophagic, which leads to a more complicated model – and obtained the data consistent with biotic density dependence; nevertheless, the effect of herbivores or pathogens are still to be directly tested. Trees in the tropical forest form a complex spatial structure, with sufficient vertical and horizontal heterogeneity, enabling (together with the benign environment) the existence of many species of other life forms (e.g. climbers and epiphytes).

The terrestrial plant species diversity decreases from the tropical rain forests toward higher latitudes. The decreasing species pool reflects both the historical reasons (reflected, e.g. by the decreasing richness of genera and families), but also increasing harshness of the environment (reflected by the decreasing richness of life forms). This decrease is, however, not always monotonous; there are extremely species-rich communities in various parts of the subtropical and temperate zones. For example, at a scale of 1 m^2, the temperate grasslands in various parts of the world, or even semi-deserts are among the most species-rich communities, hosting close to 100 vascular plant species per m^2 (e.g. Cantero *et al.* 1999). Also, the species richness of some mediterranean heathlands (e.g. fynbos in South Africa) is extraordinary. However, neither of those is comparable to tropical rain forest as to functional diversity and diversity of higher taxonomic units.

8.4.3 How is species richness affected by productivity?

At the global scale, the productivity of terrestrial vegetation decreases from the Equator to the poles, and therefore, at this scale, species richness is positively correlated with productivity. At the local scale, however, unimodal (humped) relationships have often been found (Fig. 8.3) between species richness and productivity. Similar relationships were also found with disturbance intensity. In both cases of productivity and disturbance, the 'adverse' end of the gradient is self-evident: the environment can be so unproductive and/or the disturbance so frequent, that no organism survives. The increase of species diversity from zero to 'improving' conditions (i.e. increasing productivity or decreasing disturbance) is then rather trivial. What puzzles ecologists, is what happens at the other side of the hump: why does species richness decrease at high productivity levels and at low levels of disturbance?

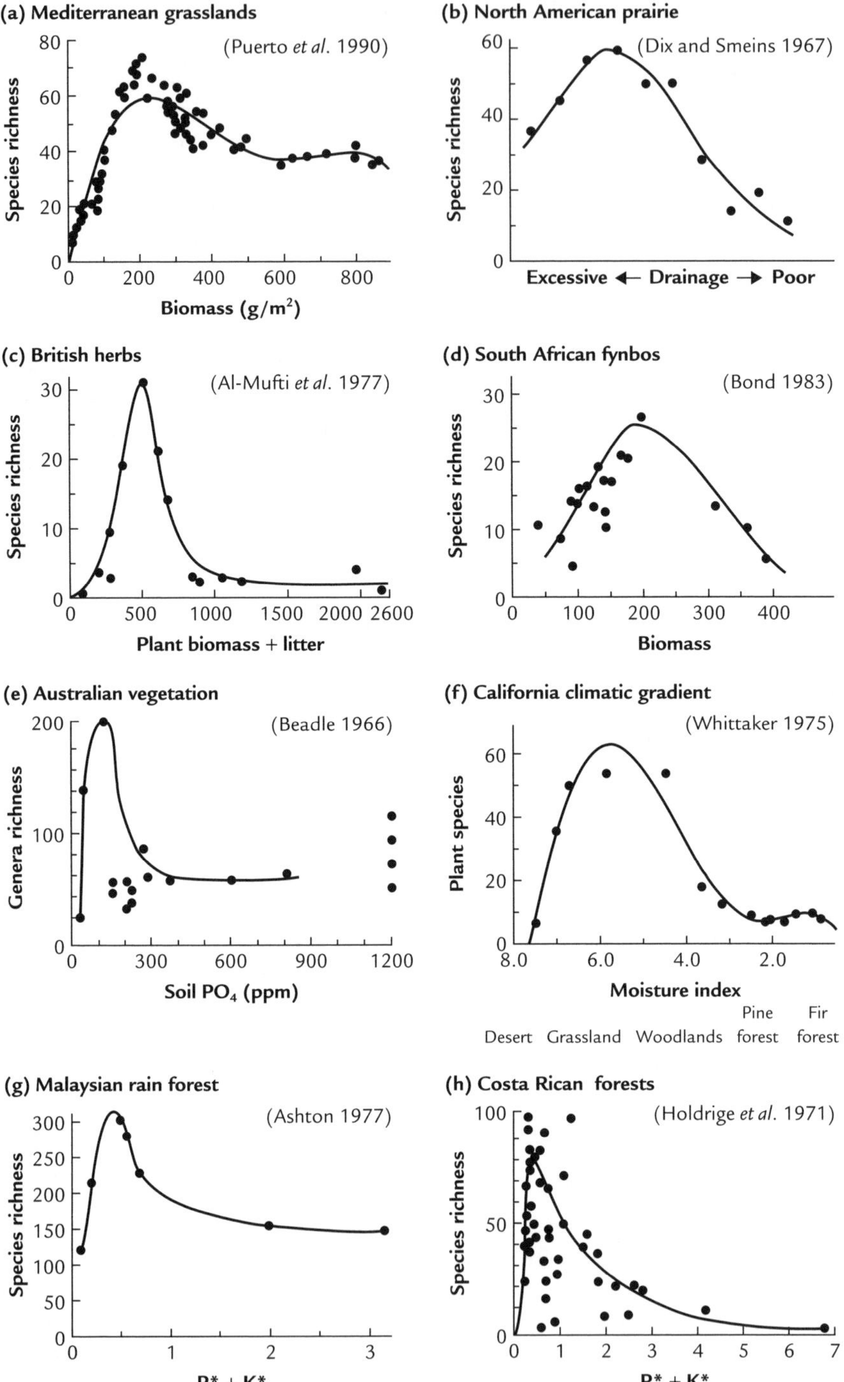

Fig. 8.3 Examples of unimodal relationships between species richness and measures of habitat productivity in plant communities. P* and K* are normalized concentrations of soil phosphorus and potassium, which were summed to give an index of soil fertility. From Tilman & Pacala (1993), where also references to the original sources can be found. Reproduced by permission of the University of Chicago Press.

Most studies of the relationships between productivity and species diversity have been carried out in temperate grasslands (both natural and semi-natural), i.e. communities dominated by a single (herbaceous) life form; however, the decrease in species richness in highly productive environments was found also in gradients including the woody vegetation (Fig. 8.3). In grasslands, the decline was demonstrated many times in fertilization experiments (e.g. Fig. 8.1); indeed, the increased nutrient load is considered one of the most important factors in the recent loss of diversity in European semi-natural and natural grasslands.

The explanation of the decline of species diversity at high nutrient levels is not trivial. It has been shown that most species of oligotrophic grasslands are able to grow under (and often take advantage of) high nutrient levels, when released from competition. This means that the species are outcompeted under these conditions. Why is this the case when they are able to co-exist under low-nutrient conditions? There are several explanations for this phenomenon, again not mutually exclusive. Increased productivity means an increased population growth rate and consequently also an increased rate of competitive displacement (Huston 1994). Very important is also the switch from competition for nutrients to competition for light with increasing productivity. With increasing nutrient level, plants are released from root competition, increase their growth rate, and (at least some) are able to grow faster and to become taller. This increases competition for light. For driving a species to competitive exclusion from a community, not just the strength of the competition, but rather its asymmetry is important. Competition for light is more asymmetric than the belowground competition. Soil heterogeneity, together with varying supply rates and varying rooting depth of plants probably enable better niche differentiation and also less asymmetric competition, than is competition for light. To monopolize light, it is enough that a species arranges its leaves into a single layer above all the other species. On the contrary, to monopolize soil resources, it would be necessary to fill completely the three-dimensional below-ground space with roots (with a density high enough to prevent other species from entering). This can be illustrated by the fact that the best predictor of species response to fertilization in a grassland experiment was the potential height of species. The taller species responded more positively to nutrient amendments (Lepš 1999), supporting the idea that with increased nutrient levels (or productivity), the importance of competition for light increases. Another explanation (Tilman & Pacala 1993) considers that the effective heterogeneity decreases when plant size increases with the increased nutrient supply. However, many other explanations of this phenomenon exist and various explanations may apply in various cases. The decrease of diversity at high productivity was also demonstrated in phytoplankton (where it is known as the paradox of enrichment) and other aquatic systems.

The pattern of change along the productivity gradient also depends on whether the differences in productivity are of an extent, where the more productive part of the gradient supports the woody vegetation, whereas the less productive does not. The presence of woody species may create further niches, enabling the survival of additional species. The woody plants usually create environmental heterogeneity, which may promote the species co-existence. Generally, if the increase in productivity enables the existence of a more demanding functional group, which is able to 'create further niches', the diversity-productivity dependence would become very complex: co-existence of functional groups is often governed by other mechanisms than co-existence of species within functional groups (Huston 1994).

8.4.4 How is the species richness affected by disturbance?

As with the productivity gradient, species diversity often exhibits a unimodal response to disturbance intensity, with the maximum diversity found in the middle of the disturbance gradient. For our discussion, we will accept Grime's (2001) definition of disturbance: partial or complete destruction of plant biomass. According to this definition, the avalanches, fire, windstorms, but also grazing and mowing are all types of disturbance. There are at least three features that characterize the disturbance regime: severity (what proportion of biomass is destroyed), frequency (how often the disturbance appears) and spatial extent (or spatial pattern). As with the productivity, the low diversity at extremely high levels of disturbance is not surprising – there is some disturbance which no species is able to withstand. Why some (medium) disturbance intensity promotes diversity has been a question of debate for several decades.

The 'Medium disturbance hypothesis' was suggested by Huston (1979), although the pattern was known earlier. Using simple models Huston demonstrated that in systems where a competitively strong species prevails in the absence of disturbance, a medium frequency of disturbance leads to prolonged co-existence, but under a high frequency of disturbance, only the fast growing species do survive (the trade-off between growth rate and competitive ability is envisaged). It can be demonstrated (e.g. Huston 1994) that what is 'medium' depends on the system's productivity (i.e. on the growth rates in the equations); in the more productive environment, the maximum diversity is achieved at higher disturbance intensities (Fig. 8.4). What is a 'medium disturbance', promoting community diversity in temperate meadows, will lead to a total destruction of an arctic tundra ecosystem.

The suggested mechanisms are many, and again not mutually exclusive. In his original paper, Huston (1979) showed that the destruction of a constant proportion of each species could postpone competitive exclusion. However, most of the disturbances are more destructive to the dominant species: even by simple mowing, the proportion of tall species removed is higher than the proportion of low or creeping species. As the higher species are better competitors for light, mowing prevents exclusion of the short species by competition for light. Similarly, an avalanche destroys the trees completely, affects the shrubs negatively, and the herbs often remain unaffected. The proportionally higher removal of dominants prolongs co-existence. Further, the disturbance is usually not spatially homogeneous and the spatio-temporal heterogeneity of disturbance is an important mechanism promoting species co-existence. It has been shown that the spatial extent of a disturbance and the turnover time (average time between two subsequent disturbances) are inversely related in nature (single tree falls appear often, large windbreaks happen maybe once in a century). Medium disturbance leads to a mosaic community structure, with patches of various successional stages – and the resulting complex community is species-rich. In communities with many species dependent on regular seedling recruitment, disturbance provides the 'safe sites' for seedling recruitment.

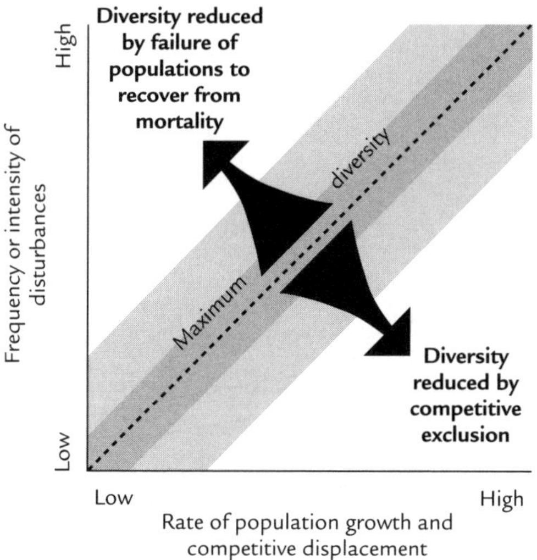

Fig. 8.4 Conceptual model of domains of the two primary processes that reduce species diversity. Diversity is reduced by competitive exclusion under conditions of high rates of population growth and competitive displacement and low frequencies and intensities of disturbance. Diversity is also reduced by failure of small and slowly growing populations to recover from mortality under conditions of low population growth rates and high frequencies and intensities of disturbance. Note that the frequency or intensity of disturbance supporting maximum diversity increases with population growth rate (i.e. with system productivity). From Huston (1994). Reproduced by permission of Cambridge University Press.

The pattern of diversity change along a disturbance gradient is echoed in changes in species richness during succession. Although the total species diversity often increases through the succession at large, the richness of vascular plants often reaches its maximum in intermediate successional stages, at least in many temperate ecosystems (Fig. 8.5). In the tropics, however, species richness increases steadily towards the undisturbed mature forest. Denslow (1980) suggested an explanation based on the different disturbance regimes in temperate and tropical climates: in temperate regions, succession starts usually from larger disturbed patches (large fires, large portions of fallen forest after windbreaks) than in the tropics, where the typical disturbance is a gap created by the fall of a single large tree. Consequently, there are numerous species in temperate regions adapted to large-scale disturbance and the species richness of intermediate successional stages is high. On the contrary, in the wet tropics, the small gaps are short-lasting, and there are not enough adapted species to form a species-rich community in the succession after a large-scale disturbance.

Each community type has its typical disturbance regime. Fire regimes are typical of some forest types (and fire frequency and intensity differs among the types); gap dynamics is typical for many community types, but most typical examples are gaps after tree falls. The extraordinary richness of some temperate semi-natural meadows is dependent on regular mowing or grazing (Bakker 1989; see also Chapter 12). Changes in the disturbance regime usually lead to a shift in competitive relationships

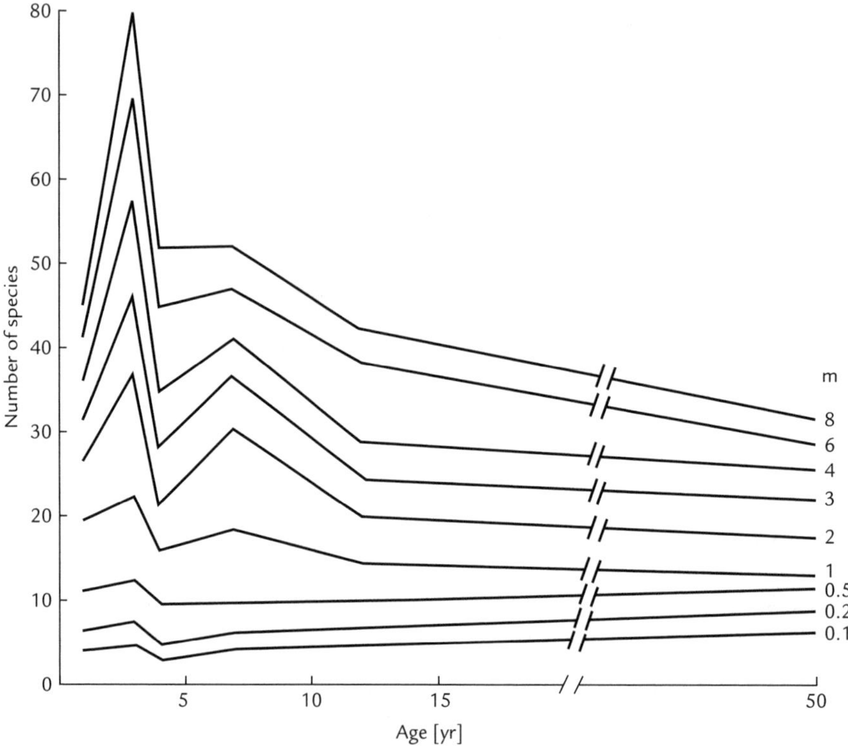

Fig. 8.5 Changes in species richness during an old-field succession, measured on various spatial scales. The numbers on the right side are sizes of sampling plots expressed as the lengths of the quadrat side. From Osbornová *et al.* (1990). Reproduced by permission of Kluwer Academic Press.

accompanied by extinction of species dependent on the regime. The results may lead to a considerable decrease in species richness: examples are the results of the fire suppression policy in North American forests (e.g. Hiers *et al.* 2000), and the loss of species in European managed meadows after cessation of grazing and/or mowing (Lepš 1999).

8.5 Stability

8.5.1 Ecological stability

Ecologists have long believed that diversity begets stability (e.g. MacArthur 1955). On the other hand, conservationists know how difficult it is to restore species-rich communities once they have been destroyed. Species-rich tropical forests are considered to be highly vulnerable – and thus less stable. This (apparent) contradiction is partially caused by the fact that the term 'stability' has been used in a very vague manner in ecology, often referring to very different concepts.

In the early 1970s, May (1973) demonstrated that mathematical models predict a negative relationship between stability and diversity. However, the results were based on highly unrealistic models: contrary to the model assumptions, ecological communities are far from random assemblages of species, and by analysing a linearized model close to system equilibrium one does not learn too much about the many-sided behaviour of ecological systems. What the model in fact has demonstrated is that the probability of an equilibrium of a randomly generated community matrix (i.e. a matrix of interaction coefficients between all species in the community), which is stable in terms of Liapunov stability, decreases with the size of the matrix, i.e. with the number of species in the model community. Also, Liapunov stability as used in mathematics – and in models of theoretical ecology – is not an ideal reflection of what ecologists consider to be ecological stability. One of the most positive effects of May's book was that ecologists realized that it is necessary to define clearly what ecological stability is, and how we should measure it in real ecological systems. In mathematical models, we have various analytical tools that enable the analysis of system equilibrium (equilibria), and its (their) stability. The only way to assess ecological stability in nature is to follow a real system trajectory in a 'state space' defined by selected measured variables. In other words, we have to select the variables of interest (such as total biomass, population sizes and rates of ecosystem processes), measure them in the course of time, and evaluate their temporal changes. From this follows that our evaluation of stability is dependent on which variables we selected for measurement, and also on the length of the period and the frequency of the measurements. In plant communities, we might be interested in species composition, in total biomass, or we can measure the nutrient retention by the community. Each of these characteristics can give different results; for instance, total community biomass can be fairly constant over the years, even when the species composition exhibits pronounced fluctuation. On the contrary, the relative proportions of species can be stable, even when total biomass changes over time.

8.5.2 Characteristics of ecological stability

Various characteristics of ecological stability are distinguished (e.g. Harrison 1979; Pimm 1984, 1991; Fig. 8.6). The first two concepts are based on system behaviour under 'normal conditions':

1 *Presence of directional changes in the state of the system.* The lack of directional changes is usually interpreted as 'stability' (the system is considered to be in a state of 'equilibrium'), systems undergoing directional changes are called transient or unstable. This concept roughly corresponds with the existence of a stable equilibrium in mathematics. It is linked to that of succession – the successional communities are (by definition) unstable, the climax communities (or 'blocked successional stages') are stable. There is a third possibility: the system may be subjected to cyclic changes (cyclic succession of Watt 1947; see Chapter 1). In real ecosystems, we have no other possibility to judge this aspect of stability than to observe a trend in a series of measurements. In praxis, the temporal trend is impossible to distinguish from a cycle with a long period (any trend with a finite time interval could be part of a cycle with a long period). The slow and small directional changes might be masked by random

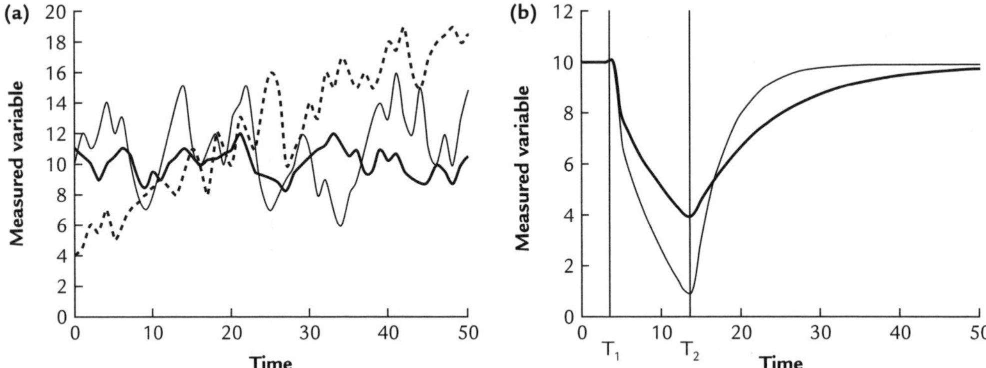

Fig. 8.6 Concepts of ecological stability. The measured variable is the choice of the researcher, e.g. community biomass, photosynthesis rate or population size. **a.** Community fluctuation in a constant environment. Broken line: unstable transient community. Full lines: communities in some steady state. Heavy line: the more constant (less variable) community. **b.** Community stability when facing a perturbation (sometimes called a stress period), which starts at T_1 and ends at T_2. Heavy line: the community which is more resistant, but less resilient than the community indicated by the light line. Adapted from various sources. The time scale depends on the rate of ecosystem dynamics; for terrestrial plants, it is usually measured in years.

variability. This is why we will always use quotation marks when speaking about 'equilibrium' in real communities. The concept is also scale-dependent. Depending on spatial and temporal scales, even climax communities undergo local successional changes. This characteristic is a qualitative one – the community is either 'stable', or is undergoing successional changes (directional or cyclic). The next three characteristics are quantitative ones:

2 *Temporal variation* (also indicated as variability) or, its opposite, *constancy*, determines how much the system fluctuates under 'normal conditions'. Standard measures of variability are used (e.g. standard deviation, SD, in the temporal sequence), usually standardized by the mean. For example, for total biomass, the coefficient of variation (CV = SD/mean) or the SD of log-transformed data would be appropriate measures of temporal variability. When the data are counts rather than a continuous variable, use of Lloyd's index of patchiness: $L = 1 + (SD/mean - 1)/mean$, will lead to a reasonable standardization by mean. However, there are many possible methods how to analyse a temporal series: we can be interested in temporal autocorrelation, focus on differences between subsequent stages, search for cyclic changes, etc. When we are interested in species composition, the measured variable is multivariate; for the evaluation of such data, we should apply methods of multivariate analysis. For example, we can follow the community trajectory in ordination space, or measure the average (dis)similarity between subsequent measurements, or use multivariate analogues of variance, standardized by corresponding means.

Ecological stability is often defined as the ability to remain in a state ('equilibrium') when facing some perturbation, and to return to the original state after the perturbation ceases. The next two characteristics are coined to measure the reaction of a system to external perturbation:

3 *Resistance*, the ability to resist a perturbation;

4 *Resilience*, ability to return to a pre-perturbation state.

Both *resistence* and *resilience* suppose that there is some period of 'normal conditions' (during which the system is in some sort of equilibrium), after which a temporarily limited perturbation period comes, which will displace the system from its 'equilibrium'; after the perturbation period ceases, the system returns to its original state. (In mathematical models, it is no problem to study the system behaviour when displaced from its equilibrium; in nature, we need some environmental change to change the state of a community.) Each community exists in a variable environment. Which variability is still considered 'normal', and which not, is undoubtedly an arbitrary decision. In experimental studies (particularly in a controlled environment), the 'perturbation' is introduced experimentally, whereas in observational studies, climatic extremes will usually be considered as perturbations (as in Lepš *et al.* 1982). Nevertheless, it is clear that resistance and resilience should always be related to the perturbation under study (it is not likely, for instance, that a community resistant to extreme drought will also be resistant to flooding).

Resistance is measured by the proximity of a system displaced by perturbation to its original state, i.e. by the similarity between the pre-perturbation and post-perturbation states. For example, it was demonstrated that during the extreme drought in 1976, young fallow decreased its standing crop by 64% in comparison with the 'normal' year 1975, but the old fallow only by 37% (Lepš *et al.* 1982; Fig. 8.7). Consequently, we concluded that the old fallow was more resistant. When the species composition of a community is of interest, then we can use various (dis)similarity measures between the original and perturbation states.

Resilience means the ability of a system to return to its original state after perturbation. It can be measured by the time, when the displacement immediately after the pertubation has decreased to 50%. In many cases a return is neither smooth nor monotonous. Then, *ad hoc* measures of resilience have to be used, e.g. reduction of the displacement after a fixed period of time. The concepts of resistance and resilience can be applied to communities which are stable according to the first definition, i.e. being in 'equilibrium', a state towards which the system returns after the perturbation. However, the concepts can be used also when the original community is not stable according to the first definition, i.e. not in equilibrium. For instance it can be applied to a successional community, provided that the rate of successional development is considerably slower than the response to perturbation.

The above-mentioned four characteristics are frequently used. Additional, less frequently used characteristics describing the ecological stability include persistence, defined as 'the ability of a system to maintain its population levels within acceptable ranges in spite of uncertainty of the environment' (Harrison 1979). Again, the definition is dependent on another definition, in this case of the term 'acceptable'. Often the community is considered persistent when no species are lost during the observed time period.

Use of resilience as a measure of stability supposes that the system will return to the original state after cessation of perturbation. However, after some large (or

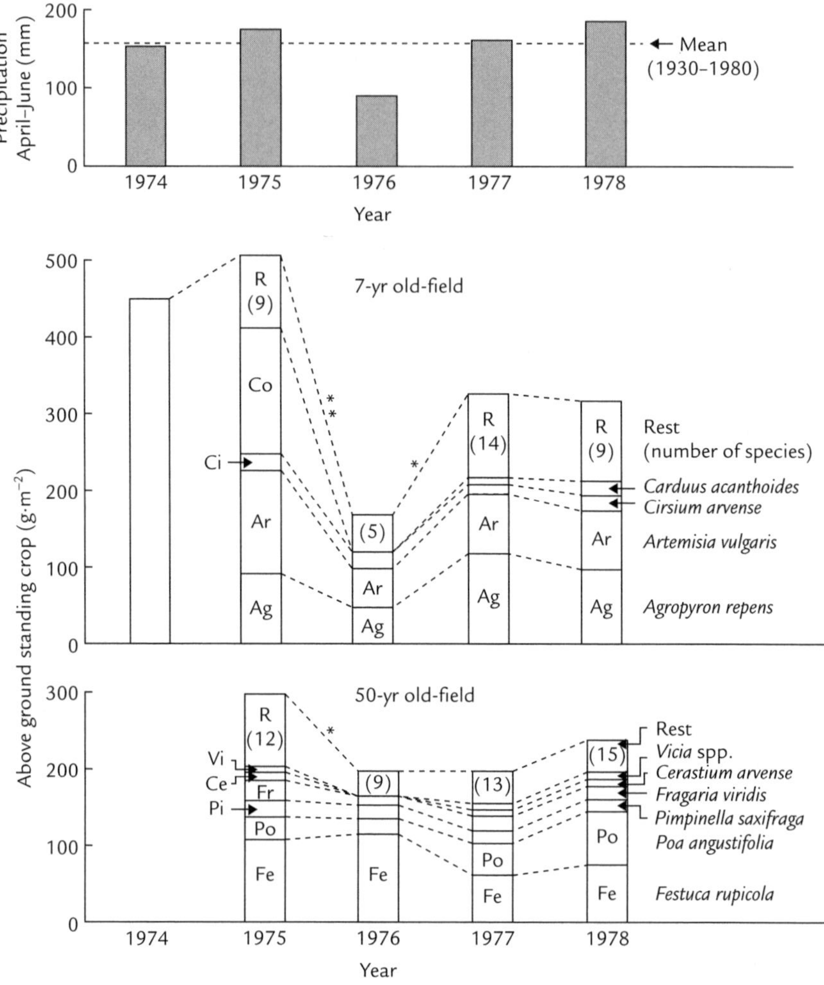

Fig. 8.7 Comparison of resistance and resilience of a 7-yr old-field and a 50-yr old-field. The top panel shows the course of the spring precipitation in 1974–1978 suggesting that 1976 was an extreme year and can be considered as a 'perturbation period'. The decrease in the total productivity from 1975 to 1976 was considerably higher (and also more significant) in the younger field (so the younger field has a lower resistance). However, the younger field started earlier to return to the 'normal' state – so it has a higher resilience. Note that the characteristics were used for the successional stages; we expected that the successional development would be much slower than the response to drought. However, in the younger field, there is some decrease of productivity, that should be taken into account – the standing crop never returned to the 1974 value in this plot. Differences between subsequent years were tested using the student's t test (* = $P < 0.05$, ** = $P < 0.01$). The number of species constituting the rest (in parentheses) has indicative meaning only. From Lepš *et al.* (1982). Reproduced by permission of Kluwer Academic Press.

long-lasting) perturbations, ecological systems have no tendency to recover. The floristic composition of a system cannot recover when species have become locally extinct without the possibility of re-immigration. Or, the soil profile under a forest on mediterranean mountains can be destroyed through large-scale clear-cutting, after which the forest cannot return. These are examples of irreversible changes. Thus, an important characteristic of stability is the range or/and the intensity of perturbations from which the system is able to return to its original state.

For all the variation, resistance and resilience, the temporal scale is very important. In comparative studies, we may use a time scale which is absolute, but it would be more realistic to relate recovery time to the generation time of the constituent species. The number of trees in a forest varies less on a year-by-year basis than the number of individuals of the annual *Erophila verna* on a sand dune. However, with respect to the difference in the generation times of the species, this observation is rather trivial. A burnt grassland would return to the pre-fire state much faster than a boreal forest on an absolute scale, but on the scale of the generation time of constituent species (trees versus grasses) the difference may be small.

Temporal variation, resistance and resilience reflect the community response to environmental fluctuation. An important part of the response by individual species is found in the physiological tolerance of their populations. In order to construct a realistic model, one would need to quantify the response of populations to environmental fluctuations (Yachi & Loreau 1999). Also, there is a physiological trade-off between growth rate and resistance to extreme events (MacGillivray *et al.* 1995); consequently the species that are highly resistant are usually not highly resilient. Similarly, communities composed of highly resistant species will not be highly resilient.

8.6 On the causal relationship between diversity and ecosystem functioning

8.6.1 On correlations and causes

As discussed above, diversity changes in a (more or less) predictable way along several ecological gradients. We may expect that the functional characteristics, such as primary productivity, nutrient retention and stability, will change along such gradients as well. Consequently, diversity and function may often be found to be correlated. Such a correlation, however, is not sufficient to support the belief of ecologists that diversity **causes** a better functioning of ecosystems. Both diversity and function are dependent on the same set of environmental constraints. And, of course, diversity might also be the consequence rather than the cause of stability.

8.6.2 Biodiversity experiments

To demonstrate possible effects of diversity, ecologists have carried out experiments where diversity is experimentally manipulated – and consequently considered an independent variable – and the ecosystem response is measured and considered to be a dependent variable in statistical models. This approach seems to be straightforward:

by manipulating one variable (diversity), a significant statistical relationship with the other (functional) variable is strong evidence for a causal relationship. Still, the approach has a weak point: the diversity can only be changed by changing the species composition. However, as demonstrated many times (e.g. Lepš *et al.* 1982; MacGillivray *et al.* 1995; Rusch & Oesterheld 1997; Grime *et al.* 2000), the identity (and plant functional types) of the constituent species is the basic determinant of ecosystem functioning. Is it possible to separate in such experiments, the effect of species composition from an effect of diversity? This is a difficult question; it seems that carefully designed experiments can provide some insight here, but there is no agreement among ecologists on this matter – compare Huston (1997) and Tilman (1999a,b). We will try to illustrate the problem with a simple example (Fig. 8.8).

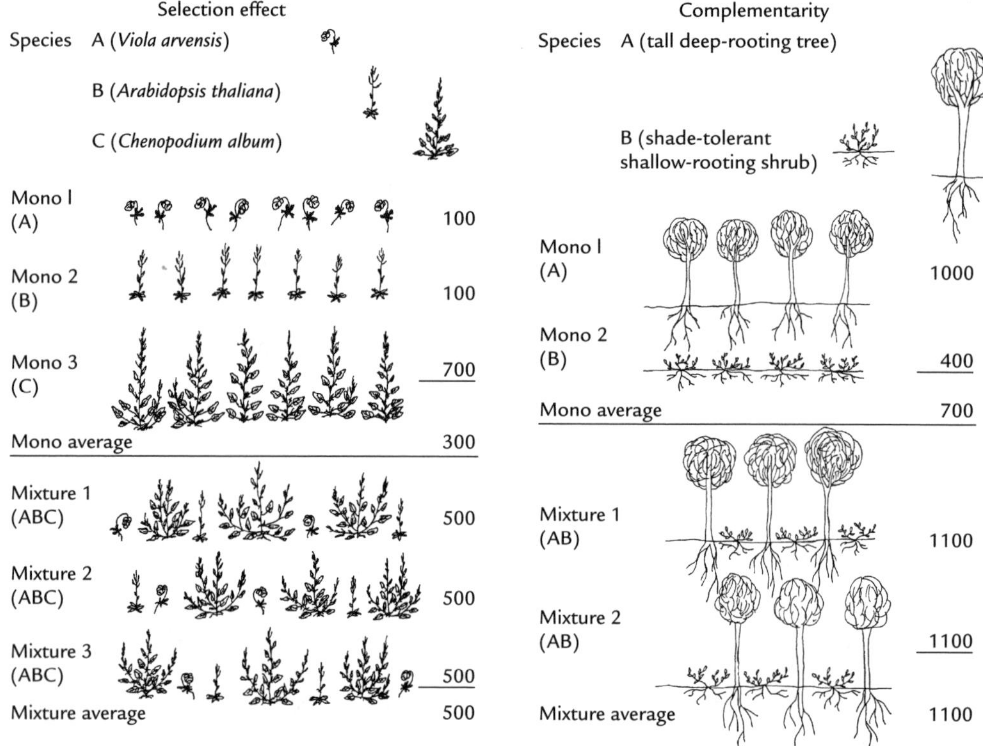

Fig. 8.8 Selection (sampling, chance) effect and complementarity affecting the final yield in biodiversity experiments. Comparison of three monocultures (Mono 1, Mono 2,..) with mixture of the three species is shown for selection effect, of two monocultures and their mixture, for complementarity. Mixture 1, Mixture 2, are replications of the mixture. Final biomass (in arbitrary units) is taken as the response variable. In the selection effect, the most productive species (such as *Chenopodium album*) prevails in the mixtures, and suppresses the less productive species. When sown in sufficient density, the mixture biomass approaches the biomass of the most productive species, but does not surpass it. In the complementarity effect, the species use resources in a different way, and consequently the biomass of the mixture might (but need not) exceed that of the most productive monoculture. Figure drawn by Eva Chaloupecká.

The example concerns a biodiversity experiment with three species, A, B and C, focusing on the effect of species richness (S) of a mixture on the final yield (biomass) of a mixture (Y). The productivity, usually measured as standing crop biomass, is often considered a parameter of 'ecosystem function'. This is in part because it is one of the most easily measurable characteristics. However, when plants are grown from seeds, biomass is a reasonable measure of productivity, and many of the functional characteristics, being it nutrient retention or CO_2 assimilation, are usually correlated with biomass and/or productivity. Most of the reasoning presented below for biomass can be equally applied for any other ecosystem function. By choosing productivity we can, of course, also take advantage of the enormous amount of earlier experimental studies, both ecological and agronomical (e.g. Trenbath 1974; Austin & Austin 1980; Vandermeer 1989).

We can imagine two designs: In the first design, all replications at $S = 1$ are composed of species A, at $S = 2$ of mixtures of A and B, and at $S = 3$ of mixtures of A, B and C. This type of design, where the species composition is constant in all replications of given species richness, which is a subset of the composition at higher levels of species richness, was used in the pioneering Ecotron experiment (Naeem *et al.* 1994). In the second design, at $S = 1$, the replications are monocultures of all three species A, B and C, which are equally represented; at $S = 2$, all three possible pairs (i.e. AB, AC and BC) are equally represented, so that we have three species mixtures. In the first design, the effect of increasing species richness from 1 to 2 cannot be distinguished from the specific effect of species B, and of increasing richness from 2 to 3, from the effect of species C.

Imagine that A and B are small annual weeds, e.g. *Viola arvensis* and *Arabidopsis thaliana*, and C is a highly productive species, e.g. *Chenopodium album*. In design 1 this means that we get a tremendous increase in productivity when increasing the species richness to 3, which is a species-specific effect of *C. album*. Such a design should of course be avoided. However, there are problems even with design 2 (Fig. 8.8). We can reasonably expect that *C. album* will dominate all mixtures where it is present, and those communities will have much higher biomass than the other ones. We can also expect that – if the sowing density is not very low – *C. album* will achieve a biomass in the mixtures which is close to its biomass in a monoculture. As *C. album* is present in one-third of the monocultures, in two-thirds of the two-species mixtures and in all the three-species mixtures, the average biomass will increase with species richness. Thus, we can expect a positive, highly significant regression of Y on S. This effect is called sampling or chance effect (Aarssen 1997; Huston 1997); it caused much controversy in the recent biodiversity debate. Loreau (2000) suggested the name (positive) selection effect, to stress the fact that the most productive species has to prevail in the mixture to produce this effect. The average biomass increases as a consequence of the positive selection effect, but the biomass of the mixture never exceeds the biomass of the most productive monoculture. When the mixture exceeds the biomass of its most productive constituent species monoculture, we speak of overyielding. This is how the term was defined in the classic paper on the productivity of mixtures (Trenbath 1974) and is used in many other papers; we will follow this usage. However, the term is sometimes used also in a much wider sense: e.g. Tilman (1999b) used the term for the situation where the productivity of a species in a mixture is higher than its yield in a monoculture divided by the number of species in a community. Consequently, overyielding strongly suggests that it is not only a selection effect that plays a role. Possible effects are complementarity and facilitation. Complementarity means that various species are limited by different resources, or differ in the mode of their use of a resource. A typical example are different rooting depths of different species or temporal separation of species (spring versus summer species). A typical example of facilitation is increase of soil nitrogen as a consequence of the presence of legumes, leading to increased productivity of other species. Complementarity cannot be distinguished from facilitation on the basis of the final outcome of a response – we also need a knowledge of the biology of the constituent species and supplementary experiments focused directly on the mechanisms of interactions. It should be noted that both complementarity and facilitation could, but need not, lead to overyielding. Consequently, the absence of overyielding is no proof of the absence of complementarity or facilitation. Indeed, the complementarity is equivalent to the niche differentiation – which is considered a necessary condition for species co-existence by the equilibrium theory, and is probably very common in nature. However, it probably seldom leads to the overyielding (Trenbath 1974).

Most biodiversity experiments demonstrated that, on average, the productivity increases with the species richness (e.g. Tilman *et al.* 1996; Hector *et al.* 1999; van der Putten *et al.* 2000, Špačková & Lepš 2001). Similarly, species-rich communities are on average more efficient, for example in nutrient uptake (Tilman *et al.* 1996), or on overall catabolic activity of soil bacteria (Stephan *et al.* 2000). There is little doubt that this is a general pattern (but see Kenkel *et al.* 2000, who found no effect of species richness on either standing crop or nutrient uptake in mixtures of perennial grasses). It also seems that this effect is saturating, with most pronounced changes at low richness (Fig. 8.9). However, evidence that species-rich communities are better than the best of monocultures or species-poor communities is missing. This leaves room for contradictory interpretations – compare Garnier *et al.* (1997) with Loreau & Hector (2001). Fig. 8.9 presents an example of how different graphical presentations based on the same data might be.

The problem here is that species richness is basically a 'dependent variable', the result of ecological forces. Many species combinations are highly unlikely to occur in certain environments: for example, a *Thero-Airion* ephemeral community of acid sand would never be found in a productive environment. In experiments, such 'unrealistic' communities could be maintained only by careful weeding. Still, in the experiments with manipulated diversity that include a species such as *Aira praecox*, low productivity of its monoculture will decrease the mean productivity of the monocultures. Because many ecosystem characteristics are positively correlated with primary productivity (e.g. abundance of insect herbivores or activity of soil biota), similar effects can be expected also for other studied characteristics. We can expect that the monoculture of *A. praecox* will also support a lower number of insect herbivores than more productive species, and will have a low below-ground biomass, resulting in low activity of soil biota.

Generally, plant communities can be species-poor for three basic reasons: (i) lack of species in the species pool; (ii) an extremely harsh environment (low productivity or high disturbance); and (iii) a highly productive environment, where competitive

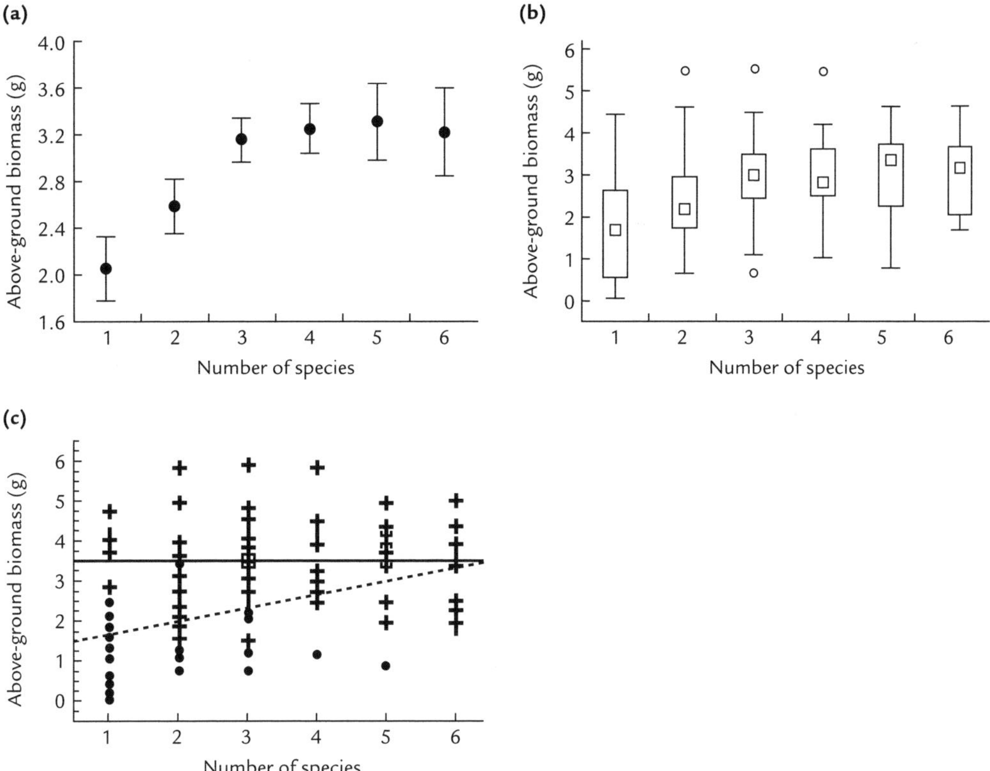

Fig. 8.9 The perception of the results of a biodiversity experiment can be affected by the manner of statistical analysis and graphing (data from the low sowing density in the pot experiment of Špačková & Lepš 2001). **a.** Mean and standard error of mean showing that mean biomass increases with species richness. **b.** Median values shown by squares, interquartile ranges by boxes, non-outlier extremes by whiskers, outliers by circles; outliers are more than 1.5 × the interquartile range from the quartiles. Median values increase, minimum values increase as well, but the maximum is more or less independent of species richness. **c.** Biomass value for each pot is shown separately. Data set divided into pots containing the most productive species, *Holcus lanatus* (+; regression shown by full line), and those without this species (•; broken line). When the most productive species is absent, the average biomass is lower and increases with the number of species, as the probability that the second most productive species will be present increases.

exclusion is fast. The low-diversity treatments in biodiversity experiments reflect only the first case.

The biodiversity experiments usually work with highly simplified artificial systems of limited spatial extent (for obvious practical reasons), which further restricts their interpretability. The mechanisms in natural communities, having developed their composition over long periods of time, might be very different from the *ad hoc* synthesized assemblages. Other approaches have to be developed. One is the removal of species from communities (Zobel *et al.* 1994).

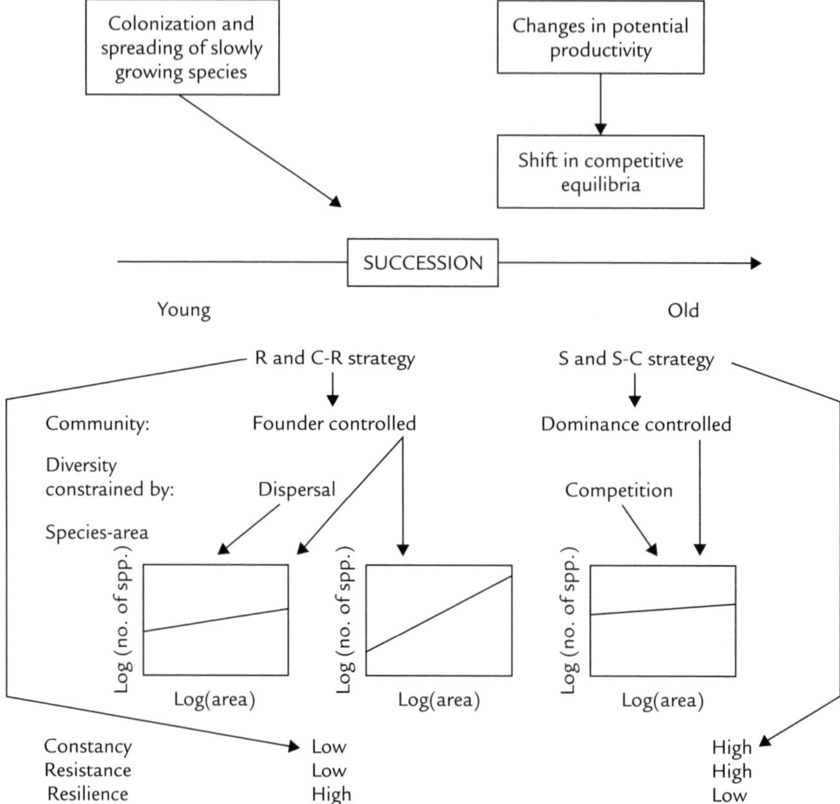

Fig. 8.10 Mechanisms behind changes in the species-area curve and stability characteristics during secondary succession. Based on an old-field succession in Central Europe (Lepš *et al.* 1982; Lepš & Štursa 1989; Osbornová *et al.* 1990). For strategy types see S.1.16.2.

8.6.3 Does diversity beget stability?

Just like ecosystem functioning, ecological stability is also mostly determined by the prevailing species life history (Fig. 8.10). A community of cacti will be highly drought resistant, regardless of its species richness. However, when damaged (for whatever reason), the recovery (i.e. the resilience) will be slow, regardless of the species richness. As the species (and life history) composition is determined by the habitat characteristics, we can expect that the habitat characteristics are the main determinants of both species richness and stability. Ecosystem functioning (energy flow and matter cycling) is crucially dependent on a limited number of dominant species. The minor (subordinate) species need not be very important for the actual functioning of the community, but might play an important role when the conditions change (Grime 1998). As the communities exist in a variable environment, species richness might help to cope with changes in the environment; the communities richer in species should be more stable. What are the possible mechanisms?

Explanations follow different formal approaches but all of them are in principle based on the explanation suggested by MacArthur (1955): each species has a role in a community, and failure of a single species to fulfil its role has no detrimental effect on community functioning, when the species function can be replaced by another species. Naturally, the probability that such a species is present increases with community species richness. At present, this effect is known as the portfolio or insurance effect. The name portfolio effect is based on the analogy with economics, where it is accepted that more diversified portfolios are less volatile.

Various mathematical models were used to demonstrate this effect. Doak *et al.* (1998) have suggested that aggregate community characteristics such as total biomass should be less variable as an effect of statistical averaging. The coefficient of variation (CV; see section 8.5.2) is usually used as a measure of variability. It follows from basic probability laws that the CV of the sum of independent random variables should generally decrease with the number of variables included. Doak *et al.* (1998) called this the 'statistical inevitability of stability diversity relationships'. However, the decrease in the coefficient of variation depends on two relationships: (i) on the way the variance is scaled with the mean; and (ii) on the independence of the variables (Tilman *et al.* 1996; Tilman 1999b). The stabilizing effect weakens when the more abundant species are more stable than the low-abundant ones. This is often the case, however. Nevertheless, the relationship is never strong enough to fully compensate for the averaging effect. The other assumption is even more important: the independence of the variables. There is a nice analogy with economics here: probably, no portfolio was resistant to the detrimental effects of Black Friday on the New York stock market – the variation of the share prices was not independent. Similarly, biomass variation will not be damped by diversity when all the species will respond to environmental variation in a concordant way. However, only a perfect positive correlation could counteract the averaging effect and there are two reasons why this rarely happens. First, species are different and we can expect different sensitivities to the variation in the environment. Moreover, due to interspecific competition in a community, we might expect some negative correlations between species, i.e. with the decline of one species, the others will make advantage of the empty space and increase.

Yachi & Loreau (1999) used different mathematical expressions to describe the insurance effect. Their model is based on explicitly stated species responses to environmental variables, and shows that different species response curves are a necessary condition for the insurance effect. However, the more different species responses are, the lower will be the correlation of species abundances in a variable environment. This means that their explanation is more mechanistic, but in principle it corresponds to that of Doak *et al.* (1998). Community resistance will be higher in species-rich communities. We can also expect a higher resilience because with an increasing number of species there is a greater chance that a fast-regenerating species will be present. Consequently, with a natural environmental variation, total community biomass will be less variable in species-rich communities. However, this stability is reached by a changing species composition; consequently, species richness does not support compositional stability (Tilman 1999b). However, with the empirical data supporting this effect alternative interpretations are also possible; see Loreau *et al.* (2001) for a thorough discussion.

The above predictions hold when there is no systematic shift in species properties between rich and poor communities. But this need not be the case. Species in species-poor communities might have broader niches. A single species with a wide ecological amplitude, i.e. with a high tolerance to environmental fluctuation, can function similarly well as the species-rich community.

In between-habitat comparisons, there are systematic patterns in richness along environmental gradients. Then the effect of species life history on the stability characteristics usually overrules the effects of species richness. This can be illustrated by community changes in the course of an old-field succession (Lepš *et al.* 1982; Figs. 8.7, 8.10). In the course of this successional series, the community productivity decreased, which together with successional development resulted in the shift from R-strategy (*sensu* Grime 2001) to C-R strategy and finally to S-strategy (see section 1.16.2). The species-area curve changes in a predictable way during succession. More small S-strategists are able to co-exist at a very small scale; at larger spatial scales, however, the community dominated by R-C strategists forms a mosaic of diverse species-poor patches, which is more species-rich, (Lepš & Štursa 1989; Osbornová *et al.* 1990). The R-C-dominated community is more productive, less resistant, and more resilient than the S-strategist community. If there is some direct effect of species richness on ecosystem functioning, it is probably negligible in comparison with the effects of prevailing life-history strategies.

Resistance to invasion of alien species is a special case of stability. Species-rich communities were traditionally considered more resistant to invasions of exotic species (Elton 1958). Whereas this comparison was partially based on a comparison of species-rich tropical forests with much poorer extra-tropical communities (and consequently prone to the effect of many confounding factors), there are also good theoretical reasons for this pattern. The more species, the less 'empty niches' there are available for the newcomers. However, the empirical support from observational data is not unequivocal (Rejmánek 1996). Undisturbed tropical forests are both extremely rich in species and highly resistant to invasions; on the other hand, extra-tropical centres of diversity such as the South African Cape Floral region are very vulnerable to plant invasions. Clearly, some of the factors promoting species co-existence also promote invasions. When species richness is manipulated, species-poor communities were shown to be more susceptible to invasions (Naeem *et al.* 2000). It seems that there is a difference between communities that are species-poor because of the habitat effect, and unsaturated communities, where the low diversity is a consequence of the lack of propagules, or the artificial removal of some species. The latter two community types are more vulnerable to invasions. This corresponds to the larger invasibility of island ecosystems (Rejmánek 1996; see Chapter 13).

8.6.4 The selection effect in biodiversity experiments and the functioning of real communities

Whereas the complementarity is generally accepted as a legitimate 'diversity effect', the relevance of the selection or sampling effect for real communities is often questioned. Is the selection effect a mechanism playing a role in nature (Tilman 1999b), or is it merely an artefact of the design of biodiversity experiments, as suggested by

Huston (1997)? It is true that, in its simplest form, it usually brings trivial results. With an experiment in a stable productive environment, and a set of species, few of which are highly productive in monoculture, we can reasonably expect that the productive species will prevail in the mixtures, and that the mixtures containing the productive species will be productive as well. It is then a trivial probabilistic issue that the probability of the mixture containing a productive species increases with the number of species. However, it seems that similar effects might be important in nature, particularly in variable or unpredictable environments. If the issue is to maximize the yield, it is often easily predictable which species will be the most productive one, and the monoculture or low-diversity mixture might seem an appropriate procedure. However, by sowing a species-rich mixture, the probability is increased that at least some species establish even when the weather would be unfavourable (selected by a mechanism equivalent to a 'selection effect'), or that at least some species will be resistant to attacks by pathogens or herbivores. Indeed, many of the 'pests' causing ecological catastrophes were highly species-specific. The Dutch elm (*Ulmus hollandica*) disease in Europe or the chestnut (*Castanea dentata*) blight in North America (see Chapter 11) usually eliminated their hosts from the plant communities they occurred in. As these species were parts of species-rich communities containing species of a similar ecological function, the functioning of these communities (total biomass, photosynthesis, habitat for other organisms) was not changed drastically. On the contrary, elimination of *Picea abies* in Central European mountain forests, where the species was a single dominant tree with no functional analogues, led to tremendous changes in the whole ecosystem, regardless of whether the spruce was planted or indigenous, and regardless of whether the spruce dieback was caused by emissions (acid rain) or by a barkbeetle outbreak. Species richness functions as an insurance against complete failure in unpredictable conditions (the insurance hypothesis of Yachi & Loreau 1999, or Grime 1998), and the mechanism is comparable to the selection effect.

8.6.5 Consequences of species losses

The biodiversity debate is fuelled with practical problems concerning the global loss of diversity. Will the loss of species impair the functioning of ecosystems? And do biodiversity experiments provide an answer to this question?

In most biodiversity experiments, the species present in low-diversity communities are random subsets of species present in high-diversity communities (which corresponds to random loss of species from high- to low-diversity communities). On the contrary, in natural conditions, species are not lost from the community at random. The loss is nearly always connected with changes in competitive relationships in a community. This change might be caused by a plethora of factors, however, some of them being highly species-specific, and some not. Highly species-specific factors are usually direct and negative in their effects; a typical example is the introduction of a new pathogen or a specialized herbivore, or a population outbreak (e.g. the two diseases mentioned above or a barkbeetle outbreak). Also, human exploitation is often highly species-specific, e.g. selective logging and the collection of plants for pharmaceutical use. Although some of those exploitation measures are sufficient to exterminate the species, the final exclusion is usually pursued by the competition of other species. (There are different ultimate causes, but the competition is often the proximate cause.) The species affected by species-specific negative effects are often 'selected at random', i.e. independently of their function in a community, independently of their competitive strength. Dutch elm disease eliminated elms from part of the European forest based on species taxonomy – the species of the genus *Ulmus* and not other functionally similar species were affected; the decline of *Gentiana pannonica* in the Bohemian forest in the first half of the 20th century was caused by selective digging of its roots for a local liqueur – their functionally analogous species were not affected at all.

When environmental conditions (e.g. productivity or disturbance regime) are changed, many (all) species are affected simultaneously. Typical examples include consequences of land use changes (e.g. land abandonment), large-scale pollution (e.g. nitrogen deposition) or climate change. Under such circumstances, some species are eliminated, some are affected negatively, but for some the change will be beneficial, and some might even newly invade the community. Regardless of the final net change in species richness, the most pronounced effect is the change in life history spectra, and its effect will probably overrule any possible diversity effect. Functionally similar species will be affected in similar ways. Although changes in environmental conditions can lead to both increase and decline of species richness, most of the recent changes result in a net decline of species richness. Typical examples are the recent loss of species due to eutrophication, where few productive species prevail in a community, excluding competitively less productive species, and loss of species due to the abandonment of previously extensively managed grasslands (Bakker 1989) – cessation of regular mowing or grazing leads to extinction of many species, usually less productive, weak competitors.

The effects of species loss for community functioning are fundamentally different when the species are lost due to species-specific agents or when due to changes in habitat. In the first case, some lessons from the biodiversity experiment may be relevant: if the lost species has functionally analogous species in the community (redundancy within the functional group), then the effect of the species loss might be negligible (the effect described by MacArthur 1955, when proposing the Shannon H' to be an index of stability). However, with every species lost, the redundancy decreases, and consequently the capacity to buffer is lost as well. Also, the probability of occurrence of complementarity or even facilitation decreases, which might lead to decrease of productivity. On the contrary, when the species are lost due to habitat changes, the systematic shifts in the species strategy will overweigh possible effect of species richness. The tremendous loss of biodiversity in European meadows is partially caused by the increasing nutrient load. As the productive species are selected in this way (Lepš 1999), the loss of species diversity in these meadows is connected with increased productivity. Conservationists in several European countries have tried for decades to persuade farmers to keep productivity of species-rich grasslands low in order to keep diversity high. Under those circumstances, the use of the argument that keeping diversity high might be economical because of increased productivity (as suggested by Tilman 1999a) might be contra-productive.

This all does not mean that the loss of species does not impair community functioning at all. However, it is probably more important which species are lost, than the number of species as such (Aarssen 2001). Even when a monospecific or low-diversity community could be as efficient as a species-rich community under given environmental conditions, with each species lost it would lose some functional properties that can be useful, when the environment would change. Stampfli & Zeitler (1999) documented that the decline in species richness in neglected meadows cannot be reversed by the re-introduction of mowing, because the species lost would not return. This means that under the renewed mowing regime, the species composition will not return to the earlier 'optimal state', because the propagules are no longer available. As the total number of species lost is large when whole regions are considered, this can impair the possibilities for selection of the 'best' species combination in a community, and in this way its functioning. The ability of a community to respond to environmental change is probably a function of the richness of the species pool rather than the number of species already growing in the community. However, with the exception of species with a permanent seed bank, and species with long-distance dispersal mechanisms, the species pool is determined by species growing in nearby communities in the landscape (Cantero *et al.* 1999). From this point of view, the simultaneous loss of species in many vegetation types in the landscape can have detrimental consequences, not envisaged by small-scale biodiversity experiments.

The relationship between biodiversity and ecosystem functioning – or with an even more anthropocentric term, ecosystem services – is not only of academic interest, but has important consequences for environmental policy. It is probably why the topic is now not only so popular, but also so contentious. Great research efforts resulted in newly published books, attempting to provide some synthesis (e.g. Kinzig *et al.* 2001), but also to reconcile the contrasting interpretations (Loreau *et al.* 2002). Plant ecologists should, however, never forget that the vascular plants (and their diversity) are just one of the ecosystem components and that particularly the linkage to below-ground components and their diversity is of basic importance for ecosystem functioning (Wardle 2002).

Acknowledgements

This chapter is to a large extent based on experiences gained in the European projects TERI-CLUE (ENV4-CT95-0002) and TLinks (EVK2-CT–2001-00123). I am grateful to Marcel Rejmánek and Eddy van der Maarel for invaluable comments on earlier drafts of the chapter.

References

Aarssen, L.W. (1997) High productivity in grassland ecosystems: effected by species diversity or productive species? *Oikos* **80**, 183–184.

Aarssen, L.W. (2001) On correlation and causation between productivity and species richness in vegetation: predictions from habitat attributes. *Basic and Applied Ecology* **2**, 105–114.

Austin, M.P. & Austin B.O. (1980) Behaviour of experimental plant-communities along a nutrient gradient. *Journal of Ecology* **68**, 891–918.

Bakker, J.P. (1989) *Nature Management by Grazing and Cutting*. Kluwer, Dordrecht.

Busing, R.T. & Brokaw, N. (2002) Tree species diversity in temperate and tropical forest gaps: the role of lottery recruitment. *Folia Geobotanica* **37**, 33–43.

Cantero, J.J., Pärtel, M. & Zobel, M. (1999) Is species richness dependent on the neighbouring stands? An analysis of the community patterns in mountain grasslands of central Argentina. *Oikos* **87**, 346–354.

Crane, P.R. & Lidgard, S. (1989) Angiosperm diversification and paleolatitudinal gradients in cretaceous floristic diversity. *Science* **246**, 675–678.

Crawley, M.J. & Harral, J.E. (2001) Scale dependence in plant biodiversity. *Science* **291**, 864–868.

Currie, D.J. & Paquin, V. (1987) Large scale biogeographical patterns of species richness of trees. *Nature* **329**, 326–327.

Denslow, J.S. (1980) Patterns of plant-species diversity during succession under different disturbance regimes. *Oecologia* **46,** 18–21.

Denslow, J.S. (1987) Tropical rain-forest gaps and tree species-diversity. *Annual Review of Ecology and Systematics* **18**, 431–451.

Doak, D.F., Bigger, D., Harding, E.K., Marvier, M.A., O'Malley, R.E. & Thomson, D. (1998) The statistical inevitability of stability-diversity relationships in community ecology. *American Naturalist* **151**, 264–276.

Elton, C.S. (1958) *The Ecology of Invasions by Animals and Plants*. Methuen, London.

Eriksson, O. (1993) The species-pool hypothesis and plant community diversity. *Oikos* **68**, 371–374.

Garnier, E., Navas, M.L., Austin, M.P., Lilley, J.M. & Gifford, R.M. (1997) A problem for biodiversity-productivity studies: how to compare the productivity of multispecific plant mixtures to that of monocultures? *Acta Oecologica* **18**, 657–670.

Grime, J.P. (1998) Benefits of plant diversity to ecosystems: immediate, filter and founder effects. *Journal of Ecology* **86**, 902–910.

Grime, J.P. (2001) *Plant Strategies, Vegetation Processes, and Ecosystem Properties*. Wiley, Chichester.

Grime, J.P., Brown, V.K., Thompson, K. *et al.* (2000) The response of two contrasting limestone grasslands to simulated climate change. *Science* **289**, 762–765.

Grubb, P.J. (1977) The maintenance of species richness in plant communities: the importance of the regeneration niche. *Biological Reviews of the Cambridge Philosophical Society* **52**, 107–145.

Hanski, I. (1999) *Metapopulation Ecology*. Oxford University Press, Oxford.

Harrison, G.W. (1979) Stability under environmental stress: resistance, resilience, persistence, and variability. *American Naturalist* **113**, 659–669.

Hector, A., Schmid, B. & Beierkuhnlein, C. *et al.* (1999) Plant diversity and productivity in European grasslands. *Science* **286**, 1123–1127.

Herben, T. (2000) Correlation between richness per unit area and the species pool cannot be used to demonstrate the species pool effect. *Journal of Vegetation Science* **11**, 123–126.

Herben, T., Krahulec, F., Hadincová, V. & Pecháčková, S. (1995) Climatic variability and grassland community composition over 10 years – separating effects on module biomass and number of modules. *Functional Ecology* **9**, 767–773.

Hiers, J.K., Wyatt, R. & Mitchell, R.J. (2000) The effects of fire regime on legume reproduction in longleaf pine savannas: is a season selective? *Oecologia* **125**, 521–530.

Hill, J.L. & Hill, R.A. (2001) Why are tropical rain forests so species rich? Classifying, reviewing and evaluating theories. *Progress in Physical Geography* **25**, 326–354.

Hill, M.O. (1973) Diversity and evenness: a unifying notation and its consequences. *Ecology* **54**, 427–432.

Hubbell, S.P. (2001) *The Unified Neutral Theory of Biodiversity and Biogeography.* Princeton University Press, Princeton.

Huston, M.A. (1979) A general hypothesis of species diversity. *American Naturalist* **113**, 81–101.

Huston, M.A. (1994) *Biological Diversity. The Coexistence of Species on Changing Landscapes.* Cambridge University Press, Cambridge.

Huston, M.A. (1997) Hidden treatments in ecological experiments: re-evaluating the ecosystem function of biodiversity. *Oecologia* **110**, 449–460.

Janzen, D.H. (1970) Herbivores and the number of tree species in tropical forests. *American Naturalist* **110**, 501–528.

Jenkins, M. (1992) Species extinction. In: *Global Biodiversity. Status of the Earth Living Resources* (ed. B. Groombridge), pp. 192–205. Chapman & Hall, London.

Kaiser, J. (2000) Rift over biodiversity divides ecologists. *Science* **289**, 1282–1283.

Kenkel, N.C., Peltzer, D.A., Baluta, D. & Pirie, D. (2000) Increasing plant diversity does not influence productivity: empirical evidence and potential mechanisms. *Community Ecology* **1**, 165–170.

Kinzig, A.P., Pacala, S.W. & Tilman, D. (eds) (2001) *The Functional Consequences of Biodiversity. Empirical Processes and Theoretical Extensions.* Princeton University Press, Princeton.

Klimeš, L. & Klimešová, J. (1999) CLO-PLA2 – a database of clonal plants in central Europe. *Plant Ecology* **141**, 9–19.

Kotorová, I. & Lepš, J. (1999) Comparative ecology of seedling recruitment in an oligotrophic wet meadow. *Journal of Vegetation Science* **10**, 175–186.

Lepš, J. (1999) Nutrient status, disturbance and competition: an experimental test of relationships in a wet meadow. *Journal of Vegetation Science* **10**, 219–230.

Lepš, J. & Štursa, J. (1989) Species-area relationship, life history strategies and succession – a field test of relationships. *Vegetatio* **83**, 249–257.

Lepš, J., Osbornová, J. & Rejmánek, M. (1982) Community stability, complexity and species life-history strategies. *Vegetatio* **50**, 53–63.

Lepš, J., Spitzer, K. & Jaroš, J. (1998) Food plants, species composition and variability of the moth community in undisturbed forest. *Oikos* **81**, 538–548.

Loreau, M. (2000) Biodiversity and ecosystem functioning: recent theoretical advances. *Oikos* **91**, 3–17.

Loreau, M. & Hector, A. (2001) Partitioning selection and complementarity in biodiversity experiments. *Nature* **412**, 72–76.

Loreau, M., Naeem, S., Inchausti, P. *et al.* (2001) Biodiversity and ecosystem functioning: Current knowledge and future challenges. *Science* **294**, 804–808.

Loreau, M., Naeem, S. & Inchausti, P. (eds) (2002) *Biodiversity and Ecosystem Functioning. Synthesis and Perspectives.* Oxford University Press, Oxford.

MacArthur, R.H. (1955) Fluctuations of animal populations and a measure of community stability. *Ecology* **36**, 533–536.

MacGillivray, C.W., Grime, J.P., Band, S.R. *et al.* (1995) Testing predictions of the resistance and resilience of vegetation subjected to extreme events. *Functional Ecology* **9**, 640–649.

Magurran, A.E. (1988) *Ecological Diversity and Its Measurement.* Princeton University Press, Princeton.

May, R.M. (1973) *Stability and Complexity in Model Ecosystems.* Princeton University Press, Princeton.

McIntyre, S., Diaz, S., Lavorel, S. & Cramer, W. (1999) Plant functional types and disturbance dynamics – Introduction. *Journal of Vegetation Science* **10**, 604–608.

Naeem, S., Thompson, L.J., Lawler, S.P., Lawton, J.H. & Woodfin, R.M. (1994) Declining biodiversity can alter the performance of ecosystems. *Nature* **368,** 734–737.

Naeem, S., Chapin III, F.S., Costanza, R. *et al.* (1999) Biodiversity and ecosystem functioning: maintaining natural life support processes. *Issues in Ecology* **4**, 1–14.

Naeem, S., Knops, J.M.H., Tilman, D., Howe, K.M., Kennedy, T. & Gale, S. (2000) Plant diversity increases resistance to invasion in the absence of covarying extrinsic factors. *Oikos* **91**, 97–108.

Osbornová, J., Kovářová, M., Lepš, J. & Prach, K. (eds) (1990) *Succession in Abandoned Fields. Studies in Central Bohemia, Czechoslovakia.* Geobotany 15, Kluwer, Dordrecht.

Palmer, M.W. (1994) Variation in species richness – towards a unification of hypotheses. *Folia Geobotanica & Phytotaxonomica* **29**, 511–530.

Pärtel, M., Zobel, M., Zobel, K. & van der Maarel, E. (1996) The species pool and its relation to species richness: Evidence from Estonian plant communities. *Oikos* **75**, 111–117.

Pausas, J.G. & Austin, M.P. (2001) Patterns of plant species richness in relation to different environments: An appraisal. *Journal of Vegetation Science* **12**, 153–166.

Pimm, S.L. (1984) The complexity and stability of ecosystems. *Nature* **307**, 321–326.

Pimm, S.L. (1991) *The Balance of Nature?* University of Chicago Press, Chicago.

Reich, J.A. (1998) Vegetation Part 1: A comparison of two one-hectare tree plots in the Lakekamu basin. In: *A Biological Assessment of the Lakekamu Basin, Papua New Guinea.* RAP Working Papers 9. (ed. A.L. Mack), pp. 25–35. Conservation International, Washington, DC.

Rejmánek, M. (1996) Species richness and resistance to invasions. In: *Biodiversity and Ecosystem Processes in Tropical Forests.* Ecological Studies 122. (eds G.H. Orians, R. Dirzo & J.H. Cushman), pp. 153–172. Springer, Berlin.

Rosenzweig, M.L. (1995) *Species Diversity in Space and Time.* Cambridge University Press, Cambridge.

Routledge, R.D. (1983) Evenness indexes – are any admissible? *Oikos* **40**, 149–151.

Rusch, G.M. & Oesterheld, M. (1997) Relationship between productivity, and species and functional group diversity in grazed and non-grazed Pampas grassland. *Oikos* **78**, 519–526.

Schläpfer, F., Schmid, B. & Seidl, I. (1999) Expert estimates about effects of biodiversity on ecosystem processes and services. *Oikos* **84**, 346–352.

Schluter, D. & Ricklefs, R.E. (1993) Convergence and the regional component of species diversity. In: *Species Diversity in Ecological Communities. Historical and Geographical Perspectives* (eds R.E. Ricklefs & D. Schluter), pp. 230–240. The University of Chicago Press, Chicago.

Smith, T.M., Shugart, H.H. & Woodward, F.I. (eds.) (1997) *Plant Functional Types. Their Relevance to Ecosystem Properties and Global Change.* Cambridge University Press, Cambridge.

Southwood, T.R.E. (1988) Tactics, stategies and templets. *Oikos* **52**, 3–18.

Špačková, I. & Lepš, J. (2001) Procedure for separating the selection effect from other effects in diversity-productivity relationship. *Ecology Letters* 4: 585–594.

Stampfli, A. & Zeiter, M. (1999) Plant species decline due to abandonment of meadows cannot easily be reversed by mowing. A case study from the southern Alps. *Journal of Vegetation Science* **10**, 151–164.

Stephan, A., Meyer, A.H. & Schmid, B. (2000) Plant diversity affects culturable soil bacteria in experimental grassland communities. *Journal of Ecology* **88**, 988–998.

Tallis, J.H. (1991) *Plant Community History. Long-term Changes in Plant Distribution and Diversity.* Chapman and Hall, London.

Taylor, D.R., Aarssen, L.W. & Loehle, C. (1990) On the relationship between r/K selection and environmental carrying-capacity – a new habitat templet for plant life-history strategies. *Oikos* **58**, 239–250.

Tilman, D. (1999a) Diversity and production in European grasslands. *Science* **286**, 1099–1100.

Tilman, D. (1999b) The ecological consequences of changes in biodiversity: a search for general principles. *Ecology* **80**, 1455–1474.

Tilman, D. & Pacala, S. (1993) The maintenance of species richness in plant communities. In: *Species Diversity in Ecological Communities. Historical and Geographical Perspectives* (eds R.E. Ricklefs & D. Schluter), pp. 13–25. The University of Chicago Press, Chicago.

Tilman, D., Wedin, D. & Knops, J. (1996) Productivity and sustainability influenced by biodiversity in grassland ecosystems. *Nature* **379**, 718–720.

Trenbath, B.R. (1974) Biomass productivity of mixtures. *Advances in Agronomy* **26**, 177–210.

van der Maarel, E. & Sykes, M.T. (1993) Small-scale plant-species turnover in a limestone grassland – the carousel model and some comments on the niche concept. *Journal of Vegetation Science* **4**, 179–188.

Vandermeer, J. (1989) *Ecology of Intercropping.* Cambridge University Press, Cambridge.

Van der Putten, W.H., Mortimer, S.R., Hedlund, K. *et al.* (2000) Plant species diversity as a driver of early succession in abandoned fields: a multi-site approach. *Oecologia* **124**, 91–99.

Wardle, D.A. (2002) *Communities and Ecosystems. Linking the Aboveground and Belowground Components.* Princeton University Press, Princeton.

Wardle, D.A., Huston, M.A., Grime, J.P., Berendse, F., Garnier, E. & Setälä, H. (2000) Biodiversity and ecosystem functioning: an Issue in Ecology. *Bulletin of the Ecological Society of America* **81**, 235–239.

Watt, A.S. (1947) Pattern and process in the plant community. *Journal of Ecology* **35**, 1–22.

Whittaker, R.H. (1972) Evolution and measurement of species diversity. *Taxon* **21**: 213–251.

Whittaker, R.H. (1975) *Communities and Ecosystems.* 2nd ed. Macmillan, New York.

Wills, C., Condit, R., Foster, R.B. & Hubbell, S.P. (1997) Strong density- and diversity-related effects help to maintain tree species diversity in a neotropical forest. *Proceedings of the National Academy of Sciences of the United States of America* **94**, 1252–1257.

Woodward, F.I. (1987) *Climate and Plant Distribution.* Cambridge University Press, Cambridge.

Yachi, S. & Loreau, M. (1999) Biodiversity and ecosystem productivity in a fluctuating environment: The insurance hypothesis. *Proceedings of the National Academy of Sciences of the United States of America* **96**, 1463–1468.

Zobel, K., Zobel, M. & Rosén, E. (1994) An experimental test of diversity maintenance mechanisms by a species removal experiment in a species-rich wooded meadow. *Folia Geobotanica & Phytotaxonomica* **29**, 449–457.

Zobel, M. (1992) Plant-species co-existence – the role of historical, evolutionary and ecological factors. *Oikos* **65**, 314–320.

Zobel, M., van der Maarel, E. & Dupré, C. (1998) Species pool: the concept. Its determination and significance for community restoration. *Applied Vegetation Science* **1**, 55–66.

9

Species interactions structuring plant communities

Jelte van Andel

9.1 Introduction

A plant community in a habitat is composed of individuals of different species that have arrived and established at the site and persist there until they become locally extinct. This is called the dispersal-assembly perspective on the community (Hubbell 2001). Each of the established species can be considered to occupy a realized niche. The latter results from the species' fundamental or potential niche and interactions among the individuals of the component species (the niche-assembly perspective) (Hubbell 2001). The presence of species in a plant community depends, apart from the availability of propagules and safe sites, on the abiotic resources (nutrients, water, light) and conditions (climate, soil pH, human impact), whereas the species' abundances in the community can be modified by a variety of interspecific interactions structuring the community, both in space and in time. Interactions between species are generally considered as mechanisms that affect community structure and provide the community with emergent properties as compared to the sum of the individual plants (cf. Looijen & van Andel 1999).

First, different types of interaction will be presented and discussed and their importance evaluated in structuring plant communities, in space and time. Interactions between organisms can be measured or estimated in terms of gains or losses in fitness at the level of individual organisms. Several fitness components have been used to measure species interactions, such as survival, biomass and reproductive capacity. These gains or losses may or may not have an effect on community structure.

9.2 Types of interaction

When individuals of two species meet, the interaction between the two results in a positive (advantageous), negative (disadvantageous) or indifferent effect on either or both species. The effect can be measured in terms of an increase or decrease in fitness (components), as compared to a control situation with no interaction (see Table 9.1).

Competitive interactions are either mutually disadvantageous, or one species is inhibited and the other is not. The situation of the latter species is sometimes called amensalism. This term should be avoided because amensalism can be the result of

Table 9.1 Simplified presentation of different interactions between two species (A and B), when they meet or do not meet: disadvantage (–), advantage (+) or indifference (0).

	Meeting		Not meeting	
	Species A	Species B	Species A	Species B
Competition	–	–	0	0
Allelopathy	0	–	0	0
Parasitism	+	–	–	0
Facilitation	0	+	0	0
Mutualism	+	+	–	–

either asymmetric competition or allelopathy which are quite different interactions: competition is mostly two-sided and concerns space or resources, while allelopathy is one-sided and results from the release of toxic organic compounds from one of the species into the environment.

Parasitism, like predation and herbivory, is a relationship where one of the species (the consumer) benefits, whereas the other (the resource species) suffers. Effects of herbivores are the subject of the next chapter. Parasitism will be dealt with mainly as far as plant–plant interactions are concerned.

Facilitation and mutualism are interactions where (respectively) one or two species benefit. The term commensalism for a one-sided advantage should be avoided, because it is difficult to know if the facilitator is unaffected or suffers in any way, for example by becoming outcompeted or parasitized.

9.3 Competition

9.3.1 Direct and indirect competition

Competition between organisms can be direct (for space or territory) or indirect (for resources). Direct or interference competition is advantageous for one and disadvantageous for the other, or the two parties remain in balance, but there is a cost of energy anyway which implies a relative loss as compared to having no interference. Indirect or exploitation competition is usually disadvantageous for both parties. If a large plant competes for nitrogen with a small plant, each of them has a negative impact on the other. In view of the absolute amount of nitrogen taken up, exploitation competition can be asymmetric – **contest competition** – but in terms of the relative loss of fitness of each of the participating plants it can be symmetric: **scramble competition**. Mostly, loss of fitness between two parties is measured in a relative way, as compared to the potential fitness of an organism, and not in terms of the total amount of resource captured.

Begon *et al.* (1996) presented a useful working definition of exploitation competition: 'An interaction between individuals, brought about by a shared requirement for a resource in limited supply, and leading to a reduction in the survivorship, growth and/or reproduction of the competing individuals concerned'. The latter part could be summarized by stating that the process leads to a reduction in one or more fitness components.

9.3.2 Early studies: phenomenology

A large amount of literature is available on interspecific resource competition as a phenomenon, starting in the early 1930s with Gause's famous experiments with three *Paramecium* species, two by two feeding on either the same food resource (bacteria) or on different food resources (bacteria and yeasts). From these experiments Gause's principle of competitive exclusion was derived, implying that the number of species that can co-exist cannot exceed the number of limiting resources. Competition theory was at that time mathematically related to population growth, in terms of the Lotka–Volterra logistic growth curves, which suggest *r* and *K* being important parameters of success. MacArthur & Wilson (1967) elaborated on this approach by formulating the concept of *r*- and *K*-selection, resulting in a colonizing strategy and a competitive or maintenance strategy, respectively. In this concept, competitive ability is assumed to have been evolved at the expense of the colonizing ability of species; there would be a trade-off between *r*- and *K*-characteristics. Fitness of individuals is considered a compromise between these contrasting choices, expressed in the relative importance of different fitness components (e.g. generative versus vegetative reproduction of plants).

Another line of research originates from agronomy. Yoda's self-thinning law, developed in the early 1960s, describes the relationship between plant density and plant biomass in monospecific stands of annual crops. This resulted in the notion of optimal sowing or planting densities in agriculture and forestry. As a follow-up, the question was asked to what extent mixed cropping could be advantageous as compared to monocultures. De Wit (1960) developed an experimental technique, the so-called replacement series, and a mathematical analysis to investigate niche overlap and niche differences, by comparing yields under intra- and interspecific competition in annual crops. This approach stimulated worldwide research in this field, both in agriculture and in ecology. It implicitly recognized Gause's ecological concept of niche separation or differentiation between co-existing species or varieties; if they occupy different niches, they can together exploit the resources to a larger extent than can be done by each of the species or varieties alone. In this case, biomass of harvestable parts of the crop could easily be used as measure of interaction effects. The replacement series have been varied in several ways (Fig. 9.1), e.g. by applying non-homogeneous mixtures of plant species in short-term greenhouse experiments (van Andel & Dueck 1982) and in long-term field experiments (van Andel & Nelissen 1981), which have shown that mosaic patches almost behave as monocultures (Fig. 9.2; Table 9.2).

9.3.3 Recent debates on mechanisms

Research and discussions on mechanisms of plant competition date back only to the last two decades. Two frequently cited books mark this period and the ongoing debate thereafter: Grime (1979) and Tilman (1982). Gause had noticed already that the order of competitive abilities between two species of *Paramecium*, both feeding

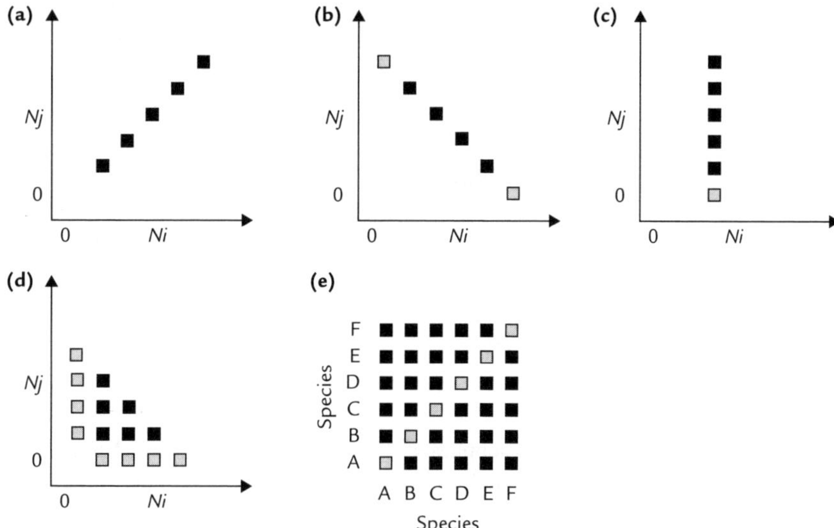

Fig. 9.1 Five designs of competition experiments. **a.** Simple pair-wise design at multiple densities; **b.** Replacement series at a single total density; **c.** Target-neighbour design with a constant density; **d.** Additive series; **e.** Diallel design of competition between six species A–F. In designs a–d *Ni* and *Nj* are densities of species *i* and *j*. Black squares: mixtures of two species; grey squares: monocultures. After Gibson *et al.* (1999). Reproduced by permission of Blackwell Publishing Ltd.

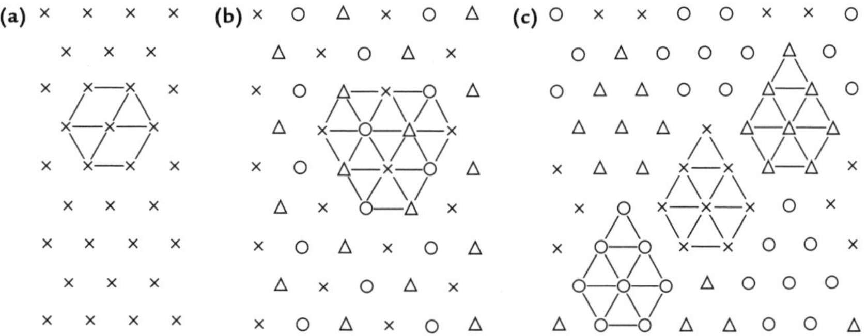

Fig. 9.2 Different patterns of replacement series, applied to three annual *Senecio* species ×, O, Δ: (**a**) Homogeneous mixture (× only); (**b**) Heterogeneous mixture; (**c**) Mixture of clusters. After van Andel & Dueck (1982). Reproduced by permission of the publisher.

on one and the same resource, did not depend on the densities at the start of the experiment, which implied that the population growth rate could not easily explain which species would win the competition. While Grime's explanations of competitive ability are largely based on differences in relative growth rates and plant morphology, Tilman was able to explain competitive hierarchies on the basis of resource depletion, in addition to population growth. In the case of two limiting resources, it

Table 9.2 Yield of achenes (mean mg of dry weight) from plant individuals of three annual *Senecio* species arranged in similar density but different patterns of a replacement series (see Fig. 9.2). The relative growth rate is highest in *S. vulgaris*, intermediate in *S. sylvaticus* and lowest in *S. viscosus*. After van Andel & Dueck (1982). Reproduced by permission of the publisher.

Treatment	Yield of achenes		
	S. vulgaris	*S. sylvaticus*	*S. viscosus*
Monoculture	61.6 ± 16.6	6.4 ± 6.0*	158.1 ± 50.2
Homogeneous mixture	149.3 ± 16.6	11.1 ± 2.7	75.3 ± 4.9
Mixture of clusters	59.5 ± 11.5	36.4 ± 3.9	162.1 ± 41.9

* Monoculture attacked by mildew.

was actually the ratio between the supplies of the two resources that determined which species would be favoured after equilibrium conditions had set, and under which conditions the two species could co-exist (Tilman 1985). In theory, the benefiting species has the lowest minimum resource requirement, expressed as the resource concentration at which growth and mortality are equal. This point of view seems to contrast with Grime's view that the species with the highest resource capture will be favoured. However, the two are not mutually exclusive; species A can gain dominance in the early phase of an experiment, at non-equilibrium transient conditions, because it has the highest growth rate by capturing a greater amount of the resource, whereas species B can ultimately be favoured at equilibrium conditions, because it utilizes the resource more efficiently, i.e. it has a lower minimum requirement. Eventually, species B may competitively exclude species A; that is, if it survived as a subordinate species in the transient period, and if an equilibrium can be achieved anyway.

9.3.4 Inhibition in a transient state

Most of the theoretical predictions have been tested and proven by using unicellular organisms of different species in continuous flow cultures, which have the opportunity to arrive at one equilibrium state. A few experiments with terrestrial plants have shown that the theory may be valid for root competition (for nutrients and water), but should be modified if light is the resource to compete for (Huisman *et al.* 1999). While competitive abilities among unicellular organisms can change in such a way that species which benefit first are at a disadvantage in later stages of succession, communities of higher plants can become fixed halfway the competition process, e.g. due to positive-feedback switches (Wilson & Agnew 1992; Aerts 1999). This may imply that dominant plant species promote their own performance. In plant communities, it frequently happens that succession is inhibited before any potentially final equilibrium stage can become established as it generally happens to occur in communities of unicellular organisms. This has been shown for slow-growing pioneer species in nutrient-poor wet dune slacks, where the pioneer plants inhibit succession towards more competitive species, due to radial oxygen loss from the roots which prevents the accumulation of organic matter (Grootjans *et al.* 1998).

However, succession may also become inhibited in stages when strong competitors for light have taken over dominance and do not leave sufficient room for co-occurring, slow-growing species to survive until they eventually could benefit through their more efficient resource utilization system. Examples are *Molinia caerulea* in wet heathlands (Berendse & Elberse 1990), *Brachypodium pinnatum* in calcareous grasslands (Bobbink *et al.* 1989) and *Elymus athericus* in coastal salt marsh (Bakker 1989). Apart from the problem of whether equilibrium conditions may occur in terrestrial plant communities (may be not even at the climax stage of succession), it should be kept in mind that terrestrial plants are living partly in the soil environment, and partly under atmospheric conditions. This is not to say that roots and shoots act independently, but that there is also room for trade-offs between root and shoot competition. Tilman (1985) proposed a trade-off between below-ground and above-ground competitive abilities as related to a resource ratio gradient. Berendse & Elberse (1990), extending the classical competition theory of de Wit (1960), suggested a trade-off between nutrient acquisition efficiency and nutrient use efficiency; for nitrogen these efficiencies are defined as the efficiency with which the acquired nitrogen is used for carbon assimilation of dry matter production, and the efficiency with which the assimilated carbon is used for the acquisition of nitrogen.

9.3.5 Effects and responses

Which plant properties determine the competitive ability of the plant under various environmental conditions and how are these plant features affected by an increasing nutrient supply and increasing biomass of the plant's neighbours? Goldberg (1990) proposed a distinction between effect and response: in case of exploitation competition, one or both plants have an effect on the abundance of the resource and a response to changes in abundance of the resource. Individual plants can be good competitors by rapidly depleting a resource, or by being able to continue growth at depleted resource levels. Both the effect and response components of competition must be significant and of appropriate sign for competition to occur. Individuals of different species can be ranked in competitive ability either by how strongly they suppress other individuals (net competitive effect) or by how little they respond to the presence of competitors (net competitive response). When we search for mechanisms of competition between organisms of different species, both competing species should be taken into account. If we are interested in the effects only, the response of the 'target species' to competition from its 'associates' is sufficient as a measure. According to Goldberg *et al.* (1999), this could be an individual-level measure (e.g. a behavioural descriptor or a component of individual fitness such as growth rate, survival, reproduction) or a population-level measure (e.g. population size or growth rate, relative abundance). Gaudet & Keddy (1988) used 44 wetland plant species to test whether competitive ability could be predicted from plant traits. Plant biomass explained 63% of the variation in competitive ability, and plant height, canopy diameter, canopy area and leaf shape explained most of the residual variation. From these results we conclude that measures of plant architecture are as important as measures of plant biomass. This does not only hold for shoot architecture as related to the light regime in a canopy, but also for root architecture. Indeed, resources are seldom homogeneously distributed. Keddy (1990) suggested that species responses to gradients in soil fertility might be considered part of their fundamental niches, whereas species responses to light gradients may be part of their realized niches, because nutrient gradients can occur in the absence of neighbours, whereas light gradients are generally produced by the presence of neighbours.

Studies on the issue of plant interactions focus on only a few dominant species in plant communities under more or less homogeneous environmental conditions. The increasing interest in biodiversity issues motivated researchers to wonder how (subordinate) species can remain co-existing in competitive plant communities. This resulted in a renewed interest in environmental heterogeneity. A potentially important, but – as compared to abiotic influences – less considered determinant of soil nutrient heterogeneity is the horizontal distribution of the bottom moss and lichen layer in plant communities (Zamfir *et al.* 1998). Cryptogams are able to quickly intercept nutrients from litter fall and throughfall; they compose slowly, thus limiting nutrient availability for the roots of vascular plants. In this way, mosses and lichens have an important effect on vascular plant community structure, i.e. on species richness and diversity. While cryptogam-induced environmental heterogeneity may reveal a higher potential co-existence among vascular plant species, the cryptogams themselves may also act as competitors that reduce the potential for species richness of vascular plant communities. These counteracting effects have not been investigated so far. Below, we will consider a few of such complex interactions that structure plant communities.

9.4 Allelopathy

9.4.1 Definition

Allelopathy can be considered a form of interference competition among plants. It is an unidirectional process, where plants of one species negatively affect plants of other species, generally by releasing toxic (mostly organic) compounds. Examples are tannins from *Pteridium aquilinum* and volatile oils from *Eucalyptus* species, substances that are supposed to have an anti-herbivore function as long is the plant organ is alive, and which in plant communities simultaneously may prevent changes in species composition and thus inhibit vegetation succession. Phenolic compounds in plant litter may have further effects on ecosystem functioning, by retarding litter decomposition rates, thus contributing to the accumulation of organic matter. The definition given by Calow (1998) may be adopted, because it is in concert with the aforementioned definition of competition: 'Allelopathy is a form of interference competition, brought about by chemical signals, i.e. compounds produced and released by one species of plants which reduce the germination, establishment, growth, survival or fecundity of other species'. This is in contrast with what has been named positive-feedback switches, which means that plants of a species modify their abiotic environment in such a way as to promote their own persistence in the area; these plants benefit from their own fitness promotion, thus also potentially reducing the competitive abilities of other component species.

9.4.2 Early studies

The classic work of Rice (1974) gives an overview of the state of affairs as resulting from early research on plant–plant, plant–micro-organism and micro-organism–micro-organism interactions, both in terrestrial and aquatic systems. Most chemical inhibitors are compounds that have been termed secondary substances. There are many thousands of such compounds, but only a limited number of them have been identified as toxins involved in allelopathy. Among them, Rice mentioned several organic acids and alcohols, fatty acids, quinones, terpenoids and steroids, phenols, cinnamic acids and derivates, coumarins, flavonoids and tannins, amino acids and polypeptides, alkaloids and cyanohydrins. Not all allelochemicals with potentially detrimental effects cause allelopathy in reality.

9.4.3 Phenolics

In more recent papers, emphasis has been put on effects of phenolic substances. Kuiters (1990) reviewed their role in forest soils. They include simple phenols, phenolic acids and polymeric phenols (condensed tannins, flavonoids). Once released in the soil environment, they influence plant growth directly by interfering with plant metabolic processes and by effects on root symbionts, and indirectly by affecting site quality through interference with decomposition, mineralization and humification. Effects of phenolics on plants include almost all metabolic processes, such as mitochondrial respiration, rate of photosynthesis, chlorophyll synthesis, water relations, protein synthesis and mineral nutrition. Interestingly, phenolic substances affect plant performance especially under acidic, nutrient-poor soil conditions. In calcareous soils, most phenolic compounds are rapidly metabolized by microbial activity and adsorption is high.

Nilsson *et al.* (1998 and earlier work) investigated allelopathic effects of the dwarf-shrub species *Empetrum hermaphroditum*, a widespread species in arctic and boreal Europe. In the Swedish boreal forest, the ground-layer vegetation in late post-fire successions is frequently dominated by dense clones of this species, which produces large quantities of phenolics, in particular batatasin-III, which is held responsible for the strong negative effects exerted by *E. hermaphroditum* on tree seedling establishment and growth, microbial activity and plant litter decomposition rates, thus contributing to humus accumulation and reduced nitrogen availability. Further, this plant species is conspicuously avoided by herbivores. In earlier work, Nilsson tried to determine the relative impacts of chemical inhibition (allelopathy) and resource competition by *E. hermaphroditum* on seedling growth of *Pinus sylvestris*, by adding fine powdered pro-analysis activated carbon as an adsorbent to the soil surface to remove the allelopathic effect, while exclusion tubes were used to subject seedlings to allelopathy in absence of below-ground competition by *E. hermaphroditum*. Both allelopathy and root competition had a strong, negative influence on seedling growth of *P. sylvestris*.

9.4.4 Allelopathy and fire

The interaction between allelopathy and fire is an interesting issue. According to Zackrisson *et al.* (1996), with prolonged absence of fire the boreal forest can become dominated by *Picea abies* and *E. hermaphroditum*, while fire at intervals of 50 to 100 yr is conducive to dominance of *P. sylvestris* and the ground layer species *Vaccinium vitis-idaea* and *V. myrtillus* in mesic and nutrient-poor sites. Indeed, charcoal was shown to absorb phytotoxic active phenolic metabolites from a *E. hermaphroditum* solution, and wildfires were shown to play an important role in the boreal forest. Zackrisson *et al.* proposed that charcoal particles can act as foci for both microbial activity (biodegradation) and chemical deactivation of phenolic compounds through adsorption.

9.5 Parasitism

9.5.1 Definition

A parasite depends on a host for its fitness, whereas the host can live without the relationship; it only suffers from the parasite if it is present. It is not in the interest of the parasite to kill its host, but it may occur, for example in the case of *Cuscuta* species. Among plants, holoparasites (such as *Orobanche* and several orchid species) exploit both root and shoot products from the host, while hemiparasites (such as *Rhinanthus* and *Striga* species) exploit the root products only and are capable of photosynthesis themselves. Hemiparasites contain chlorophyll, but among them there are many differences with regard to their host-dependence. Holoparasites do not contain chlorophyll and are heterotrophic; they do not only depend on the host for water and minerals, but also require its organic compounds. Calow (1998) defined parasitism as an intimate and usually obligate relationship between two organisms in which, essentially, one organism (the parasite) is exploiting resources from the other organism (the host) to the latter's disadvantage. It is, therefore, an interaction between individuals of different species, brought about by a consumer–resource relationship, leading to a gain in fitness of the consumer species and a decrease in fitness of the resource species.

9.5.2 Inter-plant parasitism

Parasitism between plants is a widespread phenomenon (Kuijt 1969), with over 3000 species of parasitic plants occurring worldwide. Parasitism in the plant kingdom does occur among trees, shrubs, long-lived perennials and annuals, and all parasites are dicotyledons in only a few lineages. The parasitic plant species which are known as weeds belong to the families Cuscutaceae (with *Cuscuta*), Loranthaceae (with *Loranthus*, *Viscum*, mistletoes), Viscaceae (dwarf mistletoes), Lauraceae (with *Cassytha*), Orobanchaceae (with *Orobanche*, broomrapes) and Scrophulariaceae (with *Striga*). *Santalum album* (Santalaceae) is a well-know parasitic tree (producing sandalwood) from Indonesia and Malaysia.

All holo- and hemiparasites are connected with roots or shoots of host plants by means of a haustorium. Water, minerals, and a wide variety of organic substances are transported through this organ. It is always a one-way flow, but the degree of dependence varies; some species can be grown to flower and set seed without a host, whereas others do not even germinate without a host stimulus (after germination, *Striga* can only survive for 4–5 days without host). The effect on the host is variable, too; it can be dramatic or hardly measurable and difficult to detect in other cases. Strict host specificity does not seem to exist.

The role of inter-plant parasitism in determining plant community structure is poorly understood. Pennings & Callaway (1996) investigated the impact of *Cuscuta salina*, a common and widespread obligate parasitic annual in saline locations on the west coast of North America. Their results suggested that the parasite is an important agent affecting the dynamics and diversity of vegetation. Because it prefers to parasitize the marsh-dominant *Salicornia virginica*, *C. salina* indirectly facilitates the rare species *Limonium californicum* and *Frankenia salina*, thus increasing plant diversity, and possibly initiating plant vegetation cycles. For other hemiparasites such as species of *Rhinanthus*, *Odontites*, *Euphrasia* and *Melampyrum*, it is clear that the parasites depend on a host vegetation to some extent, but in which way do they affect the vegetation? Is the vegetation open because of the presence of the parasite, or is the parasite present because of the vegetation is rather low and open? The data on the role of vegetation structure show that the latter certainly has some effect (ter Borg 1985), which may be negative, neutral or positive (Pennings & Callaway 2002).

Damaging effects on crop plants are known for the annual hemiparasitic species of the genus *Striga* and the holoparasitic species of the genus *Orobanche* (see, e.g. Pieterse *et al.* 1994). *Orobanche ramosa* reduced the biomass of its host (*Nicotiana tabacum*) by 30% with main effects on the leaves.

9.5.3 Fungal parasites on plants

Zadoks (1987) mentioned a few studies which showed that pathogenic fungi, selectively parasitizing plant species in a plant community, may accelerate vegetation succession. For example, the willow rust *Melampsora bigelowii* killed many seedlings of *Salix pulchra* and *S. alexensis*, pioneer species which formed nearly pure stands on gravel banks in the river Yukon in Alaska, once the ice had receded. This might have accelerated a succession to *Betula* and *Picea*. Another example he referred to is the massive wane of submarine *Zostera* beds in Dutch estuaries in the early 1930s, partly due to the pathogen *Labyrinthula macrocystis*. Several other examples were given by Dobson & Crawley (1994), showing that removal of a dominant tree species by a plant pathogen may lead to the development of forests dominated by less competitive species from earlier successional stages.

9.5.4 Nematodes feeding on plants

Nematodes of the genus *Longidorus* are capable of damaging stands of *Hippophae rhamnoides* in coastal dune areas. They damage the root system, including N_2-fixing nodules, and the related mycorrhizal system, thus reducing the uptake of phosphate and other nutrients. This damage may result in acceleration of the succession to *Sambucus nigra*, *Ligustrum vulgare* and *Rosa rubiginosa* on calcareous soils, or to *Empetrum nigrum* on acidified soils. Endoparasitic nematodes appeared also responsible for reduced vitality of *Ammophila arenaria* in coastal dunes, thus favouring *Festuca rubra* ssp. *arenaria* (van der Putten *et al.* 1993).

9.6 Facilitation

9.6.1 Definition

Vegetation may have a huge impact on abiotic conditions. Such changes may be to the own benefit of the species or other species may profit from it. Facilitation implies that plants of a species modify the abiotic environment in such a way that it becomes more suitable for the establishment, growth and/or survival of other species. This may result in a change in plant species composition or even in vegetation succession in case the beneficiary species outcompetes its facilitator (a mechanism of vegetation succession), but facilitation may also result in co-existence of different species (e.g. nursery among plants, or one herbivore species facilitating grazing by another herbivore species in savanna grassland). It does not seem useful to distinguish between direct and indirect facilitation, as in the case of competition, because the effects are always indirect, via an impact on the abiotic environment, i.e. by transformation of the physical or chemical soil conditions (aerating a wet anoxic soil, increasing the moisture content of a dry soil, or soil formation due to N_2-fixing plants) or by acting as a shelter against harsh above-ground conditions. Facilitation may be defined as an interaction between individuals of different species, where one species changes the environment in such a way that it is beneficial to the other, either in space or in time.

While recognizing the importance of facilitation in primary successions through plants with N_2-fixing micro-organisms in their root systems, be it *Rhizobium* or *Frankia* species (see Chapter 10), appearing in almost every textbook (e.g. Krebs 2001), here we focus on lesser known examples.

9.6.2 Nursery phenomena: nutrients, light and temperature

Maillette (1988) showed that the biomass of each of three *Vaccinium* species, which each in turn dominates a section of a gradient of light and temperature, with some overlap, was significantly reduced outside its own area of dominance, except in direct contact with the locally dominant congener. On the other hand, the biomass of the locally dominant *Vaccinium* species was either unchanged or somewhat greater in direct contact with a congener, compared to non-contact zones. The author interpreted this phenomenon as apparent commensalism – which is in fact facilitation – or even mild mutualism among the three congeneric species.

Positive spatial associations between seedlings of one species and sheltering adults of another species are common, and have been widely referred to as the 'nurse plant syndrome' (see review by Callaway & Walker 1997). According to the latter authors, the importance of facilitation of seedlings by adults of other species has been supported by studies in deserts, savannas and woodlands, tropical forests, mediterranean-climate shrubland, salt marshes and grasslands. In many of these cases, seedlings of beneficiary species are found spatially associated with nurse plants, whereas adults are not, which suggests that the balance of competition and facilitation shifts among the various life stages of the beneficiary and the benefactor.

Nursery is a phenomenological expression of facilitation. The mechanisms that may act have in some cases been discovered through field manipulations, showing that there are effects on nutrients, light, temperature, humidity and other abiotic factors. Raffaele & Veblen (1998) showed experimentally that two shrub species facilitate the vegetative resprouting of herbaceous and woody plants in a post-fire matorral shrubland in northern Patagonia (Argentina). *Schinus patagonicus* proved to be the most favourable nurse, due to it producing more shade and humidity. The magnitude of facilitation may be reduced as a result of cattle browsing.

In a fragmented agricultural landscape in the humid tropics of Mexico, Guevara *et al.* (1992) demonstrated that large isolated trees enabled the germination and establishment of woody species which otherwise would not succeed in open pasture conditions. After flying animal seed dispersers have contributed to the availability of propagules under the trees, the tree-induced conditions of shade, soil moisture and soil fertility further facilitate a high species richness under the canopy (Fig. 9.3). In

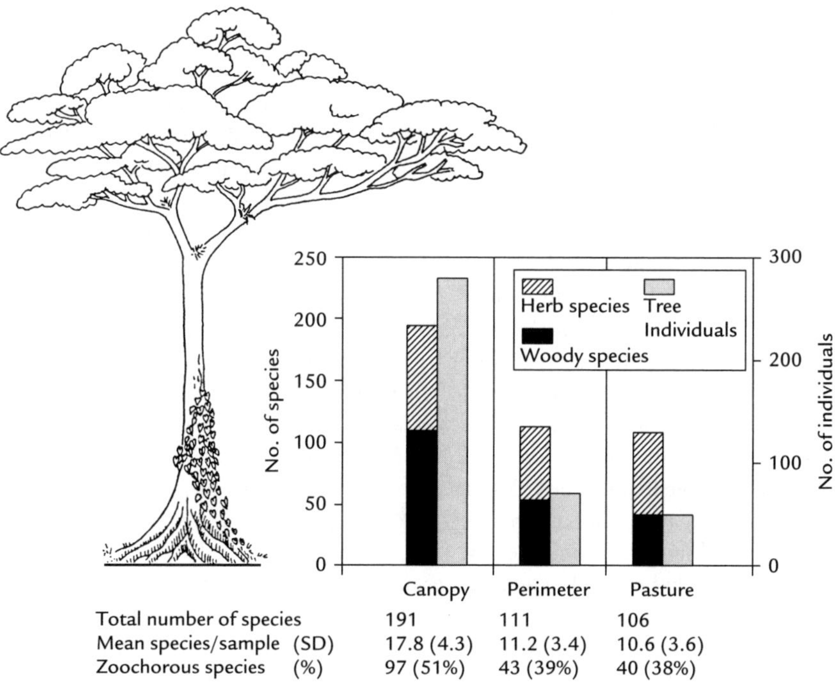

Fig. 9.3 Large isolated trees facilitate species richness in a former rain forest site in Mexico, now used as pasture. After Guevara *et al.* (1992). Reproduced by permission of the publisher.

an experimental field study in a sand dune succession in Ontario, Canada, Kellman & Kading (1992) showed that establishment of *Pinus strobus* and *P. recinosa* was facilitated by trees of *Quercus rubra* of at least 35 yr of age. This effect could be attributed to shading effects, which might imply an improved moisture and temperature regime for seed germination and early seedling survival.

A major role of facilitation between higher plant species, particularly in stressful environments (salt marshes, dune slacks, arctic regions and desert environments) has been reported during the last decade; see references in Pugnaire *et al.* (1996). The latter authors have shown, in south-eastern Spain, that the leguminous shrub *Retama sphaerocarpa* strongly improved its own environment, facilitated the growth of *Marrubium vulgare* and other understorey species, and at the same time obtained benefits from sheltering herbs underneath. The interaction between these two species was indirect, associated with differences in soil properties and with improved nutrient availability under shrubs compared with plants grown on their own.

9.6.3 Water: hydraulic lift

Horton & Hart (1998) reviewed the phenomenon of hydraulic lift. Deep-rooted plants take water from deeper, moister soil layers at daytime and transport it through their roots to upper, drier soil layers, where the roots release it at night, after the stomata have closed. Hydraulically lifted water can benefit the plant that lifts it, but might also benefit neighbouring more shallow-rooted plants, which was proven by using deuterated water (Caldwell & Richards 1989; Dawson 1993). The phenomenon has been demonstrated in, for example the desert shrub *Prosopis tamarugo*, the semi-arid sagebrush *Artemisia tridentata* and others as well, and *Acer saccharum* in mesic forest (references in Horton & Hart 1998). The nocturnal increase in soil water potential could be several orders of magnitude greater than that expected from simple capillary water movement from deep to shallow soil. In a field study in a mesic forest, water hydraulically lifted by sugar maple supplied up to 60% of the water used by neighbouring shallow-rooted species. Plants that used hydraulically lifted water were able to maintain higher transpiration rates and experienced less water stress than plants that did not and *A. saccharum* seedlings that performed hydraulic lift were able to achieve higher daily integrated carbon gain than plants in which hydraulic lift was experimentally suppressed (Dawson 1993). Recently, hydraulic lift was documented for the first time for a CAM (crassulacean acid metabolism) species, *Yucca schidigera*, a native plant in the Mojave Desert (Yoder & Nowak 1999). The pattern of diel flux in soil water potential for the CAM species was temporally opposite to that of the C_3 species investigated. The authors suggested that, because CAM plants transport water to shallow soils during the day when surrounding C_3 and C_4 plants transpire, CAM species that hydraulically lift water may influence water relations of surrounding species to a greater extent than hydraulically lifting C_3 or C_4 species.

Several authors recently pointed at the phenomenon of 'reverse hydraulic lift'. Burgess *et al.* (1998) measured sap flow in the roots of *Grevillea robusta* and *Eucalyptus camaldulensis* that could be interpreted as hydraulic lift. After this, however, hydraulic redistribution of water occurred, facilitating root growth in dry soils and modifying resource availability. Water can move down the taproot of trees when the surface soil layers are wetter than the deeper soil layers. Similarly, Smith *et al.* (1999) utilized measurements of reverse flow in tree roots to demonstrate the opposite process to hydraulic lift: the siphoning of water downwards by root systems of trees spanning the gradient in water potential between a wet surface and dry subsoil. They suggested a competitive advantage for trees over their neighbours in dry environments where plants are reliant on seasonal rainfall for water. Reverse hydraulic lift has been suggested to facilitate root growth through the dry soil layers underlying the upper profile where precipitation penetrates, thus allowing roots to reach deep sources of moisture in water-limited ecosystems.

9.7 Mutualism

9.7.1 Definition

In the case of mutualism, the facilitation is bidirectional between two species. Mutualism among higher plants is a rare phenomenon, but mutualisms among higher plants and micro-organisms have a huge impact on plant community performance. Mutualistic relationships can be facultative (leguminous plants can live with or without *Rhizobium*), or obligate, i.e. a condition for survival, as in many lichens which are a symbiosis between a fungal and an algal component. For vascular plants in general, mutualistic relationships with mycorrhizal fungi are of utmost importance. Many experimental investigations have shown that both plant and fungal symbionts benefit from the reciprocal exchange of mineral and organic resources. Mutualism can be defined as an interaction between individuals of different species that lead to an increase of fitness of both parties, based on mutual assistance in resource supply.

9.7.2 Mycorrhizae

The majority, *c.* 80%, of species of temperate, subtropical and tropical plant communities are colonized by vesicular-arbuscular mycorrhizal (VAM) fungi; the number of fungal species is only *c.* 200. Since a part of these fungi never form vesicles the term arbuscular mycorrhizal is preferred nowadays (see Chapter 11). In this chapter the well-known abbreviation VAM will be maintained. There is ample evidence that in such communities a vigorous semi-permanent population of fungal symbionts with low 'host' specificity is involved in an infection process which effectively integrates compatible species into extensive mycelial networks (Francis & Read 1994); see Chapter 11.

Ectomycorrhizal (ECM) fungi occur mainly on woody plants and only occasionally on herbaceous and graminoid plants and involve more than 5400 fungal species. Ericoid mycorrhizae (EM) occur mainly in Ericales and are physiologically comparable with ECM fungi. Non-mycorrhizal plants occur mainly in very wet or saline ecosystems and in ecosystems with a high nutrient availability and/or with recently disturbed soil; see references in Ozinga *et al.* (1997). Orchid–mycorrhiza relations are a special case (Dijk *et al.* 1997; see below).

VAM fungi are presumably especially efficient in the uptake of inorganic P (and other relatively immobile ions such as Cu^{2+}, Zn^{2+} and ammonium) and are capable of increasing the P uptake more in nutrient-rich patches than in soils with a uniform P-distribution (Cui & Caldwell 1996), while ECM and EM fungi are more efficient in N-limited ecosystems. Enzymatic degradation by ECM and EM fungi has been shown for proteins, cellulose, chitin and lignin. Changes in the proportion of nutrients in inorganic or organic form may create changes in the competitive abilities provided by different mycorrhiza types; see Chapter 11.

During vegetation succession on nutrient-poor soils, the proportion of N and P bound in organic form can increase, as with allelopathic substances and pathogen concentrations. Hence, during succession on nutrient-poor soils, the benefit for plants of C-allocation to the relatively expensive ECM and EM fungi probably increases (Francis & Read 1994).

Dijk *et al.* (1997) reviewed nutritional relationships of orchid species. In the first heterotrophic and subterranean phase of orchid development, the growth of the protocorm is entirely dependent on mycorrhizal fungi. The nutrient metabolism of developing orchid individuals is adapted to this symbiosis: reductions in orchid nitrogen metabolism are permitted which can be considered adaptations to the parasitic habit during at least this phase. Mycorrhizal infection is restricted to subterranean tissues only, i.e. to the subepidermal zone of the protocorm and root parenchyma. After the initial infection the development of mycorrhizae can easily derail, and in symbiotic cultures a range of interactions can be met from a loss of mycorrhiza via normal mycorrhizal infections to pathogenic effects. The primary function of mycorrhizal infection in the juvenile phase lies in the transport of C compounds to the developing seedlings. Translocation of sugars towards protocorms has been demonstrated by radioactive labelling in classic studies. Apart from interfering with the carbon metabolism, mycorrhizal infection has a pronounced influence on the uptake of mineral macronutrients (P and N); see Chapter 11.

Johnson *et al.* (1997) evaluated cost:benefit analyses of plant–fungus mutualisms. Resource limitation is a key component of cost:benefit analyses of mycorrhizal effects on plant fitness. Carbon allocated to a fungus is only a cost if it could otherwise have been allocated to increase plant fitness, and resources gained through the activities of a fungal symbiont are only beneficial if those resources are in limiting supply. The authors considered mutualism as the normal state of mycorrhizal functioning, but tried also to understand causes of 'parasitic' exceptions to this norm. The state of interaction, in the spectrum between mutualism and fungal parasitism, can be hypothesized to result from: (i) developmental factors; (ii) environmental factors; or (iii) genotypic factors.

9.7.3 Plant–pollinator interactions

Plant–pollinator interactions can be considered a non-symbiotic mutualism. In many cases, there is a mutual exploitation interest, sometimes a consumer–resource relationship between the two parties. In their review on 'endangered mutualisms', Kearns *et al.* (1998) pointed out the phenomenon that over 90% of 250,000 modern angiosperm species are pollinated by animals. Among the nearly 300,000 flower-visiting animals are insects, birds, bats and small marsupials. Habitat fragmentation and other effects of land use (agriculture, grazing, herbicide and pesticide use), and the introduction of non-native species, have a crisis-like impact on plant–pollinator systems. Specialist relationships are, of course, much more vulnerable than generalist relationships, but Kwak *et al.* (1998) illustrated that pollen and gene flow in fragmented habitats not only depend on the investigated plant populations as such, but also on the neighbouring species of the plant communities and the flowering phenologies of the component species. Plant–pollinator interactions are only seldom specific to the species level; relatively few plant–pollinator interactions are absolutely obligate in a strict sense. Mostly, mutualism is an interest of the plant community and the pollinator community, rather than that of two specific plant–pollinator populations.

9.8 Complex species interactions affecting community structure

9.8.1 Interactions change during succession

It has been pointed out by several authors that it is not useful to distinguish between succession types, but rather between mechanisms of succession, which may occur simultaneously or one after the other in any particular successional sere (Connell & Slatyer 1977; Miles 1987; Glenn-Lewin & van der Maarel 1992; van Andel *et al.* 1993; McCook 1994; see Chapter 7). In the context of this chapter, it should be emphasized that the intensity of a particular interaction (be it competition or facilitation or other relationships) may change during succession, and that several interactive mechanisms may be at work simultaneously, structuring plant communities in space (patterns) and in time (dynamics, succession).

9.8.2 Competition and succession along productivity gradients

Competitive response (and effect) is not a species-specific trait; it depends on the other (neighbouring) species and the biotic and abiotic circumstances. Several hypotheses have been formulated to indicate the relative importance of competition along productivity gradients, summarized in Fig. 9.4 (after van der Veen 2000). The latter author has shown in a primary succession on a coastal beach plain, that competition can be important from the beginning of succession onwards. Apart from nutrient availability flooding and salinity could also affect the competitive balance between species. Competition also remained important at the intensively grazed sites of the dune succession. Heavy folivory excluded the competition for light but below-ground competition remained high. A change in intensity of interference experienced by a species along a gradient was found in many other studies; see review by Goldberg & Barton (1992). Interestingly, Goldberg *et al.* (1999) did not find a significant relationship between competition intensity and standing crop. Similarly, van der Veen (2000) showed that root competitive intensity experienced by seedlings of seven different species in a productivity gradient was negatively related with standing crop, whereas shoot competitive intensity was positively related. There was a negative

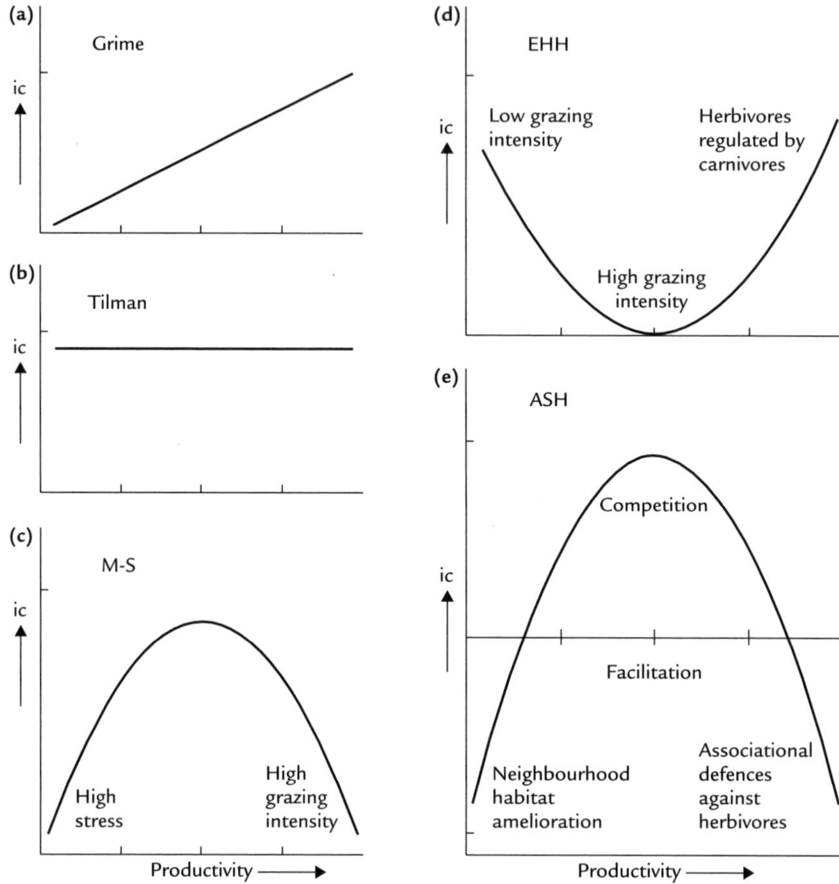

Fig. 9.4 The importance of competition (ic) along a productivity gradient according to (**a**) Grime (1979), (**b**) Tilman (1985), (**c**) Menge & Sutherland (1987), (**d**) the exploitation ecosystem hypothesis EHH (Oksanen *et al.* 1981), and (**e**) the abiotic stress hypothesis ASH (Callaway & Walker 1997), as presented by van der Veen (2000).

relationship between shoot and root competition, in agreement with Tilman's (1985) resource ratio hypothesis.

Competition theory, focusing on mechanisms of resource competition such as maximum resource capture (Grime 1979) and minimum resource requirement (Tilman 1982) or resource ratios (Tilman 1985), has largely been developed independent of succession theory, though a lot of evidence has been presented to suggest competition to be the main mechanism of succession. Tilman (1990) attempted to predict both competition and succession from a similar set of assumptions. The resource-ratio hypothesis assumes that each plant species is a superior competitor for a particular proportion of the limiting resources and predicts that community composition should change whenever the relative ability of two or more limiting resources changes (Tilman 1985). The theory predicts a change from mainly nutrient competitors to mainly light competitors during succession from bare soil. Competition could be equally intense along productivity gradients, although the resource for which competition occurs may change. Gleeson & Tilman (1990), however, showed an increase in proportional root biomass with successional age, in case of secondary succession on poor soil. Similar results were obtained from an additional study of 46 different successional species. Although absolute leaf biomass increased with successional age, the proportion of leaf biomass decreased almost twofold, because absolute root biomass increased almost twice as much as absolute leaf biomass. Their data suggested that the first 40–60 yr of succession are a period of strong competition for soil nitrogen, which they considered a period of transient dynamics of competitive displacement, with a pattern that is, at least in part, caused by a trade-off between maximal growth rate and competitive ability for nitrogen and, in part, a trade-off between colonization ability (seed production) and competitive ability for nitrogen.

9.8.3 Simultaneous or intermittent positive and negative interactions

Callaway & Walker (1997) provided many examples illustrating that species interactions may involve a complex balance of competition and facilitation. *Quercus douglasii* trees had the potential to facilitate understorey herbs by adding considerable amounts of nutrients to the soil beneath their canopies. However, experimental tree root exclusion increased understorey biomass under trees with high shallow-root biomass, but this had no effect on understorey biomass beneath trees with low shallow-root biomass. Thus, the overall effect of an overstorey tree on its herbaceous understorey was determined by the balance of both facilitation and competition.

Shifts in facilitation and competition among aerenchymous wetland plants occur as temperatures change in anaerobic substrates. *Myosotis laxa*, a small herb common in wetlands of the northern Rocky Mountains, benefited from soil oxygenation when grown with *Typha latifolia* at low soil temperatures in greenhouse experiments. At higher soil temperatures, the significant effects of *Typha* on soil oxygen disappeared (presumably because of increased microbial and root respiration) and the interaction between *Myosotis* and *Typha* became competitive. In the field, the overall effect of *Typha* on *Myosotis* was positive, as *Myosotis* plants growing next to transplanted *Typha* were larger and produced more fruits than those isolated from *Typha*. In a recent study in a Tanzanian semi-arid savanna ecosystem, Ludwig (2001) showed that the beneficiary effect of hydraulic lift from *Acacia tortilis* trees to grasses is overruled by competition for water that the grasses experience from the same trees. This would imply that, while the phenomenon of lift has been proven to exist, the ecological effects may be of only little importance, if facilitation is overruled by competition. Something similar holds for the positive effects of nutrient enrichment of the soil by tree litter fall, versus the negative effects from competition for nutrients, but in this case the beneficiary effects may prevail. A conceptual model of these complex interactions is presented in Fig. 9.5.

9.8.4 Mediators of species interactions: third parties

Miller (1994) argued that the success of species in a community is affected not only by direct interactions between species, but also by indirect interactions among

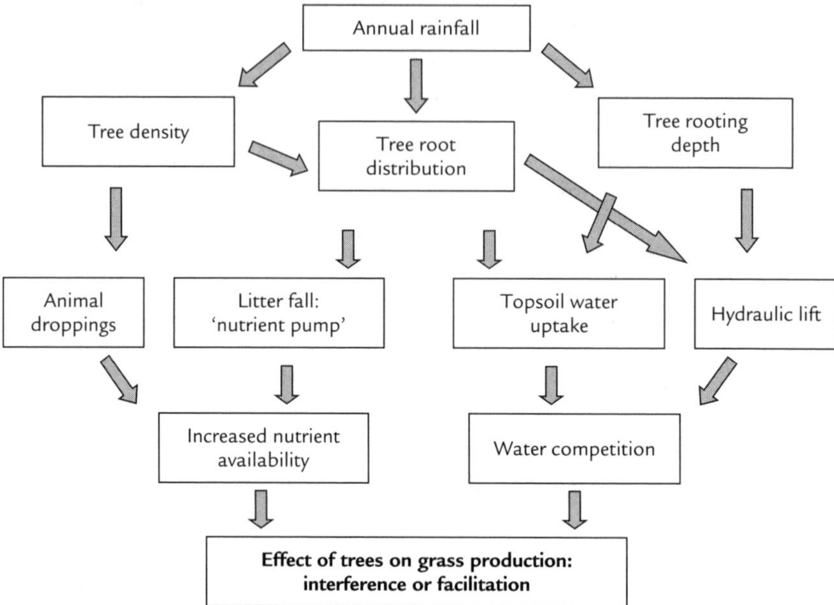

Fig. 9.5 Conceptual model showing the determinants of facilitation and competition in a semi-arid tree-grass savanna ecosystem in Tanzania. After Ludwig (2001). Reproduced by permission of the publisher.

groups of species. In several cases, the magnitude of the indirect positive effect was greater than that of the direct negative effect, resulting in a facilitative overall effect. Indirect interactions occur if a third species modifies the interaction between two other species. This phenomenon, third parties affecting the competitive abilities of two other species, was called mediation of competition by Allen & Allen (1990), similar to Price *et al.*'s (1986) proposal in their review on mediation by parasites, where parasites act as a third party affecting the interference competition or resource competition between two other species (cf. Pennings & Callaway 1996). In the 1990s this was termed 'apparent competition', but the more clear-cut notion of mediation is preferred. Mediation by parasites is very common in nature and must be regarded as one of the major types of interaction in ecological systems, comparable in importance to direct competition, predation, parasitism or mutualism (Price *et al.* 1986). The latter authors reviewed: (i) parasites as agents of interference when one species competes with another; (ii) parasites as instruments of defence for a species against its enemies; and (iii) parasites assisting consumers in exploiting the species they eat.

Perry *et al.* (1989) illustrated that inoculation with ECM fungi neutralized the mutual antagonism between *Pseudotsuga menziesii* and *Pinus ponderosa* when growing in inoculated mixtures. The presence of mycorrhizae has been shown to change the outcome of plant competition in many cases, both for VAM plants and for ECM plants (see references in Ozinga *et al.* 1997). Mycorrhizal linkages were shown to transport ^{15}N and ^{32}P within and between plant species (Chiarella *et al.* 1982;

Finlay *et al.* 1988). This mainly regards nutrients from dying roots, thus N and P will be transported along a shorter route to the plant species which are incorporated in the mycorrhizal network. Furthermore, nutrient surplus (luxury consumption) can be transported from certain plant species to other species by the mycorrhizal fungus. Grime *et al.* (1987) found that, in a microcosm experiment, ^{14}C could be transported through a mycorrhizal network from dominant to subordinate species, which lead to an increase in biomass of the inferior competitors. Differences in the nutrient source utilized by the different mycorrhiza types may create positive feedbacks between plant species, litter quality and mycorrhiza type. Moora & Zobel (1996) tried to answer the question whether a symbiotic interaction with VAM fungi can help young plants resist the competition of older ones with which they naturally co-exist. They chose *Prunella vulgaris* as a target species and *Fragaria vesca* as a neighbouring species. Both are common subordinates in the lowest sublayer of species-rich calcareous grasslands. In interspecific competition with old *Fragaria vesca* plants, the shoots of target plants were 22% greater when inoculated with VAM fungi than when non-mycorrhizal fungi. Thus, if a young *Prunella vulgaris* plant has established itself somewhere in a natural gap, the presence of VAM fungi might make intraspecific competition more severe, but may decrease the strength of interspecific competition. Van der Heijden *et al.* (1998) concluded from their investigations that the species composition of VAM fungal communities could potentially affect the way plant species co-exist and therefore be a determinant of plant community structure. They obtained three pieces of evidence: (i) plant species differed in their dependency on VAM fungi; (ii) specific VAM species and a mixture of these VAM species had significantly different effects on several plant growth variables; (iii) the amount of variation in the growth response of a plant species to four VAM species and to a mixture of VAM species differed among the plant species. Similarly, Kiers *et al.* (2000) provided evidence that host–mycorrhizal (VAM) interactions in tropical forests are characterized by greater complexity than had previously been demonstrated, and suggested that tropical mycorrhizal fungal communities have the potential to differentially influence seedling recruitment among host species and thereby affect community composition.

A volume on plant–herbivore–predator interactions, edited by Olff *et al.* (1999), is worth mentioning as a source of information about trophic interactions that affect plant community structure, both in space and time. Van Andel (1999) summarized interactions as follows. Above-ground herbivores, ranging from insects to mammals, can feed on shoots and roots. As far as their above-ground effects are concerned, they are known to retard succession, while optimizing their food supply at a particular successional stage. Below-ground herbivores feed on roots, which process is known to accelerate succession, at least in early stages. Plant pathogens, both above and below ground, may accelerate succession still further, if they kill dominant plants. Effects on the rate and direction of succession apparently differ between above-ground and below-ground herbivores.

Jefferies (1999) reviewed interactions between herbivores and micro-organisms affecting nutrient fluxes in the ecosystem, and discussed special conditions where increased dominance from herbivores overrides the regulatory controls imposed by other organisms, which leads to trophic cascades and discontinuous vegetation states.

Table 9.3 Distribution of cumulative frequency values (44 species) in the 100-cm^2 quadrats (n = 40) of one of the alvar plots, 1986–1991. Frequency classes are Raunkiaer classes I–V (20% steps). After van der Maarel & Sykes (1993). Reproduced by permission of the publisher.

Frequency Class	Interval	1986	1987	1988	1989	1990	1991
I	1-8	9	8	8	9	11	8
II	9-16	7	9	8	7	5	8
III	17-24	5	3	3	3	3	2
IV	25-32	8	6	5	6	5	4
V	33-40	5	10	13	13	16	18

9.8.5 Co-existence

While non-spatial models predict that no more consumer species can co-exist at equilibrium than there are limiting resources, a similar model that includes neighbourhood competition and random dispersal among sites predicts stable co-existence of a potentially unlimited number of species on a single resource (Tilman 1994). Co-existence occurs because species with sufficiently high dispersal rates persist in sites not occupied by superior competitors. Co-existence requires limiting similarity and two-way and three-way interspecific trade-offs among competitive ability, colonization ability and longevity. Fine-scale repeated observations by van der Maarel & Sykes (1993) in species-rich alvar grassland vegetation, revealed a similar pattern and process, which they labelled with the term 'carousel model' and which also reflects the interpretation of grassland patterning surveyed by Grubb *et al.* (1982); species co-existence at a coarse-grained scale may result from a relatively fast turnover of species at a finer scale (Table 9.3). Similarly, Gigon & Leutert (1996) explained co-existence of a large number of plant species by postulating the 'dynamic keyhole-key model', assuming that plant species diversity (the keys) in a plant community is matched by the diversity of microsites (the keyholes), which both change in the course of time. Both these points of view represent non-equilibrium models of species co-existence, emphasizing dynamic dispersal phenomena rather than niche separation to explain species richness.

The 'resource balance hypothesis of plant species diversity', presented by Braakhekke & Hooftman (1999), relates much more to competition and niche differences and suggests a more static equilibrium. On the basis of a model of competition for multiple resources and related experimental tests, they have given evidence for the idea that opportunities for plant species diversity are favoured when the actual resource supply ratios of many resources are balanced according to the optimum supply ratios for the vegetation as a whole. Their theory predicts that diversity will be relatively low when biomass production of the whole vegetation is limited by a single nutrient, while it can be high when there is co-limitation by several nutrients (see also Olff & Pegtel 1994). Recently, Huisman & Weissing (1999 and later work) offered another type of solution to the so-called plankton paradox, based on the dynamics of the competition itself, by showing that: (i) resource competition models can generate oscillations and chaos when species compete for three or more resources; and (ii) these oscillations and chaotic fluctuations in species abundances allow co-existence of many species on a handful of resources.

9.8.6 Assembly rules

According to Wilson (1999), community ecology is essentially a search for repeated community patterns. They proposed to envisage two basic kinds of plant community pattern, with different causes:

1 Environmentally mediated patterns, i.e. correlations between species due to their shared or opposite responses to the physical environment.

2 Assembly rules, i.e. patterns due to interactions between species, such as competition, allelopathy, facilitation, mutualism, and all other biotic interactions that we know about in theory, and actually affect communities in the real world.

At present, it seems that not much progress has been made during a couple of decades, since Diamond (1975) coined the term 'assembly rules', to deterministically explain stable communities, which he considered as integrated entities. The main factor underlying these rules was, according to him, interspecific competition. Assembly rules were thus defined as constraints determining which species can occur in the community and which combinations are irrelevant. As he stated, through diffuse competition, the component species of a community are selected, and co-adapted in their niches and abundances, so as to fit together and to resist invasions. Though this approach implicitly assumes dispersal to be possible, Diamond's paper was heavily criticized by colleagues who rejected the idea of assembly rules and also questioned the importance of competition. Looijen (2000) analysed and reflected upon the ongoing Diamond–Simberloff debate, in the context of holistic and reductionistic approaches, and plead for co-operating research programmes. Indeed, a community is not just the summing up of its individual components, nor should it be considered as an organismal entity (see Chapter 1). In this context, it is worth mentioning the 'unified neutral theory of biodiversity and biogeography' (Hubbell 2001), which is a step forward in achieving – as he called it – a reconciliation and unification of niche-assembly and dispersal-assembly perspectives in community ecology.

Recognizing that assembly rules are a tool to predict which community will establish where and when, assembly rules can be considered as algorithms that can be used to calculate the estimated density of a 'species' in a community under certain environmental conditions. In this way, the rules can be considered a challenge to explicitly formalize our knowledge of decisions that are implicitly taken by plants in response to their environment during the process of plant community development (Fig. 9.6). There is no difference in assembly rules that concern plant dispersal, plant responses to abiotic factors, and plant–plant responses in the community (Belyea & Lancaster 1999). Indeed, plant–plant responses frequently act in an indirect way, by changing the abiotic environment. Used in this way, assembly rules make ecological knowledge explicit, rather than imply a philosophy on the identity of a plant community.

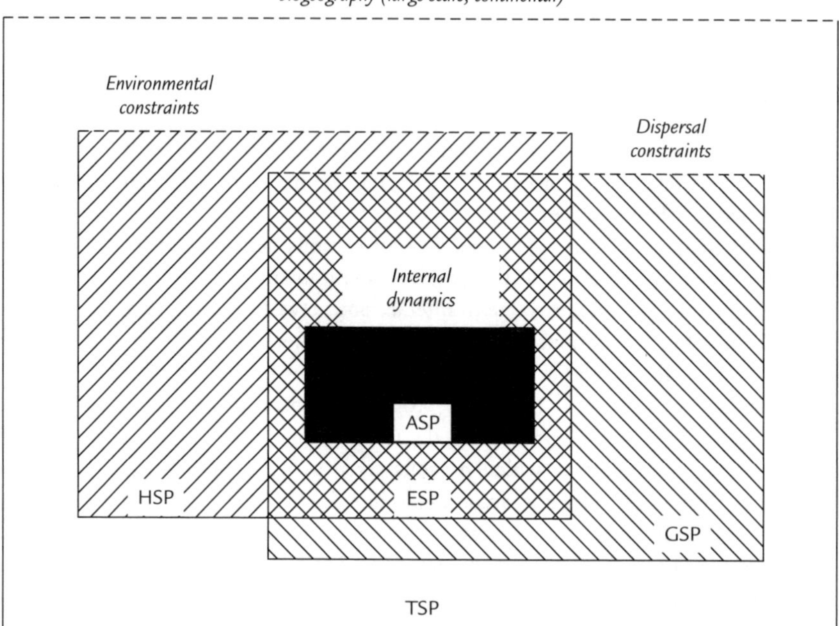

Fig. 9.6 Illustration of the relationships among species pools, and the processes that determine membership of each pool. After Belyea & Lancaster (1999). TSP = total species pool; GSP = geographical species pool; HSP = habitat species pool; ESP = ecological species pool; ASP = actual species pool. Reproduced by permission of the publisher.

References

Aerts, R. (1999) Interspecific competition in natural plant communities: mechanisms, trade-offs and plant-soil feedbacks. *Journal of Experimental Botany* **50**, 29–37.

Allen, E.B. & Allen, M.F. (1990) The mediation of competition by mycorrhizae in successional and patchy environments. In: *Perspectives on Plant Competition* (eds J.B. Grace & D. Tilman), pp. 367–389. Academic Press, London.

Bakker, J.P. (1989) *Nature Management by Grazing and Cutting*. Kluwer Academic Publishers, Dordrecht.

Begon, M., Harper, J.L. & Townsend, C.R. (1996) *Ecology*. 3rd ed. Blackwell Science, Oxford.

Belyea, L.R. & Lancaster, J. (1999) Assembly rules within a contingent ecology. *Oikos* **86**, 402–416.

Berendse, F. & Elberse, W.Th. (1990) Competition and nutrient availability in heathland and grassland ecosystems. In: *Perspectives on Plant Competition* (eds J.B. Grace & D. Tilman), pp. 93–65. Academic Press, London.

Bobbink, R., den Dubbelden, J. & Willems, J.H. (1989) Seasonal dynamics of phytomass and nutrients in chalk grassland. *Oikos* **55**, 216–224.

Braakhekke, W.G. & Hooftman, D.A.P. (1999) The resource balance hypothesis of plant species diversity in grassland. *Journal of Vegetation Science* **10**, 187–200.

Burgess, S.S.O., Adams, M.A., Turner, N.C. & Onk, C.K. (1998) The redistribution of soil water by tree root system. *Oecologia* **115**, 306–311.

Caldwell, M.M. & Richards, J.H. (1989) Hydraulic lift: water efflux from upper roots improves effectiveness of water uptake by deep roots. *Oecologia* **79**, 1–5.

Callaway, R.M. & Walker, L.R. (1997) Competition and facilitation: a synthetic approach to interactions in plant communities. *Ecology* **78**, 1958–1965.

Calow, P. (ed.) (1998) *The Encyclopedia of Ecology & Environmental Management.* Blackwell Science, Oxford.

Chiarella, N., Hickman, J.C. & Mooney, H.A. (1982) Endomycorrhizal role for interspecific transfer of phosphorus in a community of annual plants. *Science* **217**, 941–943.

Connell, J.H. & Slatyer, R.O. (1977) Mechanisms of succession in natural communities and their role in community stability and organization. *American Naturalist* **111**, 1119–1144.

Cui, M. & Caldwell, M.M. (1996) Facilitation of plant phosphate acquisition by arbuscular mycorrhizas from enriched soil patches. *New Phytologist* **133**, 453–460, 461–467.

Dawson, T.E. (1993) Hydraulic lift and water use by plants: implications for water balance, performance and plant-plant interactions. *Oecologia* **95**, 565–574.

de Wit, C.T. (1960) On Competition. *Agricultural Research Report, Wageningen University* **66.8**, 1–82.

Diamond, J.M. (1975) Assembly of species communities. In: *Ecology and Evolution of Communities* (eds M.L. Cody & J.M. Diamond), pp. 332–445. Harvard University Press, Cambridge (Mass.).

Dijk, E., Willems, J.H. & van Andel, J. (1997) Nutrient responses as a key factor to the ecology of orchid species. *Acta Botanica Neerlandica* **46**, 339–363.

Dobson, A. & Crawley, M. (1994) Pathogens and the structure of plant communities. *Trends in Ecology and Evolution* **9**, 393–398.

Finlay, R.D., Ek, H., Odham, G. & Söderström, B. (1988) Mycelial uptake, translocation and assimilation of nitrogen from ^{15}N-labelled ammonium by *Pinus sylvestris* plants infected with four different ectomycorrhizal fungi. *New Phytologist* **110**, 59–66.

Francis, R. & Read, D.J. (1994) The contribution of mycorrhizal fungi to the determination of plant community structure. *Plant and Soil* **159**, 11–25.

Gaudet, C.L. & Keddy, P.A. (1988) A comparative approach to predicting competitive ability from plant traits. *Nature* **334**, 242–243.

Gibson, D.J., Connolly, J., Hartnett, D.C. & Weidenhamer, J.D. (1999) Designs for greenhouse studies of interactions between plants (essay review). *Journal of Ecology* **87**, 1–16.

Gigon, A. & Leutert, A. (1996) The dynamic keyhole-key model of co-existence to explain diversity of plants in limestone and other grasslands. *Journal of Vegetation Science* **7**, 29–40.

Gleeson, S.K. & Tilman, D. (1990) Allocation and the transient dynamics of succession on poor soils. *Ecology* **71**, 1144–1155.

Glenn-Lewin, D.C. & van der Maarel, E. (1992) Patterns and processes of vegetation dynamics. In: *Plant Succession* (eds D.C. Glenn-Lewin, R.K. Peet & Th.T. Veblen), pp. 11–59. Chapman & Hall, London.

Goldberg, D.E. (1990) Components of resource competition in plant communities. In: *Perspectives on Plant Competition* (eds J.B.Grace & D. Tilman), pp. 27–49. Academic Press, London.

Goldberg, D.E. & Barton, A.M. (1992) Patterns and consequences of interspecific competition in natural communities: a review of field experiments with plants. *American Naturalist* **139**, 771–801.

Goldberg, D.E., Rajaniemi, T., Gurevitch, J. & Stewart, O.A. (1999) Emperical approaches to quantifying interaction intensity: competition and facilitation along productivity gradients. *Ecology* **80**, 1118–1131.

Grime, J.P. (1979) *Plant Strategies and Vegetation Processes.* John Wiley & Sons, Chichester.

Grime, J.P., Mackey, J.M., Hillier, S.H. & Read, D.J. (1987) Floristic diversity in a model system using experimental microcosms. *Nature* **328**, 420–422.

Grootjans, A.P., Ernst, W.H.O. & Stuyfzand, P.J. (1998) European dune slacks: strong interactions between vegetation, pedogenesis and hydrology. *Trends in Ecology and Evolution* **13**, 96–100.

Grubb, P.J., Kelly, D. & Mitchley, J. (1982) In: *The Plant Community as a Working Mechanism*, (ed. E.I. Newman) pp. 79–97. Blackwell Scientific Publications, Oxford.

Guevara, S., Meave, J., Moreno-Casasola, P. & Laborde, J. (1992) Floristic composition and structure of vegetation under isolated trees in neotropical pastures. *Journal of Vegetation Science* **3**, 655–664.

Horton, J.L. & Hart, S.C. (1998) Hydraulic lift: a potentially important ecosystem process. *Trends in Ecology and Evolution* **13**, 232–235.

Hubbell, S.P. (2001) *The Unified Neutral Theory of Biodiversity and Biogeography.* Princeton University Press, Princeton and Oxford.

Huisman, J. & Weissing, F.J. (1999) Biodiversity of plankton by species oscillations and chaos. *Nature* **402**, 407–410.

Huisman, J., Jonker, R.R., Zonneveld, C. & Weissing, F.J. (1999) Competition for light between phytoplankton species: experimental tests of mechanistic theory. *Ecology* **80**, 211–222.

Jefferies, R.L. (1999) Herbivores, nutrients and trophic cascades in terrestrial environments. In: *Herbivores: Between Plants and Predators* (eds H. Olff, V.K. Brown & R.H. Drent), pp. 301–330. Blackwell Science, Oxford.

Johnson, N.C., Graham, J.H. & Smith, F.A. (1997) Functioning of mycorrhizal associations along the mutualism-parasitism continuum. *New Phytologist* **135**, 575–585.

Kearns, C.A., Inouye, D.W. & Waser, N.M. (1998) Endangered mutualisms: the conservation of plant-pollinator interactions. *Annual Review of Ecology and Systematics* **29**, 83–112.

Keddy, P.A. (1990) Competitive hierarchies and centrifugal organization in plant communities. In: *Perspectives on Plant Competition*, (eds J.B. Grace & D. Tilman), pp. 265–290. Academic Press, London.

Kellman, M. & Kading, M. (1992) Facilitation of tree seedling establishment in a sand dune succession. *Journal of Vegetation Science* **3**, 679–688.

Kiers, E.T., Lovelock, C.E., Krueger, E.L. & Herre, E.A. (2000) Differential effects of tropical arbuscular mycorrhizal fungal inocula on root colonization and tree seedling growth: implications for tropical forest diversity. *Ecology Letters* **3**, 106–113.

Krebs, C.J. (2001) *Ecology: The Experimental Analysis of Distribution and Abundance* (5th ed). Benjamin Cummings, San Francisco.

Kuijt, J. (1969) *The Biology of Parasitic Flowering Plants.* University of California Press, Berkeley and Los Angeles.

Kuiters, A.T. (1990) Role of phenolic substances from decomposing forest litter in plant-soil interactions. *Acta Botanica Neerlandica* **39**, 329–348.

Kwak, M.M., Velterop, O. & van Andel, J. (1998) Pollen and gene flow in fragmented habitats. *Applied Vegetation Science* **1**, 37–54.

Looijen, R.C. (2000) *Holism and Reductionism in Biology and Ecology.* Kluwer Academic Publishers, Dordrecht.

Looijen, R.C. & van Andel, J. (1999) Ecological communities: conceptual problems and definitions. *Perspectives in Plant Ecology, Evolution and Systematics* **2**, 210–222.

Ludwig, F. (2001) *Tree-grass interactions on an East African savanna: the effects of competition, facilitation and hydraulic lift.* Tropical Resource Management Papers 39, Wageningen University, Wageningen.

MacArthur, R. & Wilson, E.O. (1967) *The Theory of Island Biogeography*. Princeton University Press, Princeton.

Maillette, L. (1988) Apparent commensalism among three *Vaccinium* species on a climatic gradient. *Journal of Ecology* **76**, 877–888.

McCook, L.J. (1994) Understanding ecological community succession: causal models and theories, a review. *Vegetatio* **110**, 115–147.

Menge, B.A. & Sutherland, J.P. (1987) Community regulation: variation in disturbance, competition, and predation in relation to environmental stress and recruitment. *American Naturalist* **130**, 730–757.

Miles, J. (1987) Vegetation succession: past and present perceptions. In: *Colonization, Succession and Stability* (eds A.J. Gray, M.J. Crawley & P.J. Edwards), pp. 1–29. Blackwell Scientific Publishers, Oxford.

Miller, T.E. (1994) Direct and indirect species interactions in an early old-field plant community. *American Naturalist* **143**, 1007–1025.

Moora, M. & Zobel, M. (1996) Effect of arbuscular mycorrhiza on inter- and intraspecific competition of two grassland species. *Oecologia* **108**, 79–84.

Nilsson, M.-C., Gallet, C. & Wallstedt, A. (1998) Temporal variability of phenolics and batatasin-III in *Empetrum hermaphroditum* leaves over an eight-year period: interpretations of ecological function. *Oikos* **81**, 6–16.

Oksanen, L., Fretwell, S.D., Arruda, J. & Niemalä, P. (1981) Exploitation ecosystems in gradients of primary productivity. *American Naturalist* **118**, 240–261.

Olff, H. & Pegtel, D.M. (1994) Characterization of the type and extent of nutrient limitation in grassland vegetation using a bioassay with intact sods. *Plant and Soil* **163**, 217–224.

Olff, H., Brown, V.K. & Drent, R.H. (eds) (1999) *Herbivores: Between Plants and Predators*. Blackwell Science, Oxford.

Ozinga, W.A., van Andel, J. & McDonnell-Alexander, M.P. (1997) Nutritional soil heterogeneity and mycorrhiza as determinants of plant species diversity. *Acta Botanica Neerlandica* **46**, 237–254.

Pennings, S.C. & Callaway, R.M. (1996) Impact of a parasitic plant on the structure and dynamics of salt marsh vegetation. *Ecology* **77**, 1410–1419.

Pennings, S.C. & Callaway, R.M. (2002) Parasitic plants: parallels and contrasts with herbivores. *Oecologia* **131**, 479–489.

Perry, D.A., Margolis, H., Choquette, C., Molina, R. & Trappe, J.M. (1989) Ectomycorrhizal mediation of competition between coniferous tree species. *New Phytologist* **112**, 501–511.

Pieterse, A.H., Verkleij, J.A.C. & ter Borg, S.J. (1994) Biology and management of *Orobanche*. Royal Tropical Institute, Amsterdam.

Price, P.W., Westoby, M., Rice, B., Atsatt, P.R., Fritz, R.S., Thompson, J.N. & Mobley, K. (1986) Parasite mediation in ecological interactions. *Annual Review of Ecology and Systematics* **17**, 487–505.

Pugnaire, F.I., Haase, P. & Puigdefábregas, J. (1996) Facilitation between higher plant species in a semiarid environment. *Ecology* **77**, 1420–1426.

Raffaele, E. & Veblen, T.T. (1998) Facilitation by nurse shrubs of resprouting behavior in a post-fire shrubland in northern Patagonia, Argentina. *Journal of Vegetation Science* **9**, 693–698.

Rice, E.L. (1974) *Allelopathy*. Academic Press, New York.

Smith, D.M., Jackson, N.A., Roberts, J.M. & Ong, C.K. (1999) Reverse flow of sap in tree roots and downward siphoning of water by *Grevillea robusta*. *Functional Ecology* **13**, 256–264.

ter Borg, S.J. (1985) Population biology and habitat relations of some hemiparasitic Scrophulariaceae. In: *The Population Structure of Vegetation* (ed. J. White), pp. 463–487. Dr W. Junk Publishers, Dordrecht.

Tilman, D. (1982) *Resource Competition and Community Structure.* Princeton University Press, Princeton.

Tilman, D. (1985) The resource ratio hypothesis of plant succession. *American Naturalist* **125**, 827–852.

Tilman, D. (1990) Constraints and tradeoffs: toward a predictive theory of competition and succession. *Oikos* **58**, 3–15.

Tilman, D. (1994) Competition and biodiversity in spatially structured habitats. *Ecology* **75**, 2–16.

van Andel, J. (1999) Introductory remarks. In: *Herbivores: Between Plants and Predators* (eds H. Olff, V.K. Brown & R.H. Drent), pp. 169–173. Blackwell Science, Oxford.

van Andel, J. & Dueck, T. (1982) The importance of the physical pattern of plant species in replacement series. *Oikos* **39**, 59–62.

van Andel, J. & Nelissen, H.J.M. (1981) An experimental approach to the study of species interference in a patchy vegetation. *Vegetatio* **45**, 155–163.

van Andel, J., Bakker, J.P. & Grootjans, A.P. (1993) Mechanisms of vegetation succession: a review of concepts and perspectives. *Acta Botanica Neerlandica* **42**, 413–433.

van der Heijden, M.G.A., Boller, T., Wiemken, A. & Sanders, I.R. (1998) Different arbuscular mycorrhizal fungal species are potential determinants of plant community structure. *Ecology* **79**, 2082–2091.

van der Maarel, E. & Sykes, M.T. (1993) Small-scale plant species turnover in a limestone grassland: the carousel model and some comments on the niche concept. *Journal of Vegetation Science* **4**, 179–188.

van der Putten, W.H., van Dijk, C. & Peters, B.A.M. (1993) Plant-specific soil-borne diseases contribute to succession in foredune vegetation. *Nature* **362**, 53–56.

van der Veen, A. (2000) *Competition in coastal sand dune succession.* Ph.D. Thesis, University of Groningen, Groningen.

Wilson, J.B. (1999) Assembly rules in plant communities. In: *Ecological Assembly Rules* (eds E. Weiher & P. Keddy), pp. 130–164. Cambridge University Press, Cambridge.

Wilson, J.B. & Agnew, A.D.Q. (1992) Positive-feedback switches in plant communities. *Advances in Ecological Research* **23**, 263–337.

Yoder, C.K. & Nowak, R.S. (1999) Hydraulic lift among native plant species in the Mojave Desert. *Plant and Soil* **215**, 93–102.

Zackrisson, O., Nilsson, M.-C. & Wardle, D.A. (1996) Key ecological function of charcoal from wildfire in boreal forest. *Oikos* **77**, 10–19.

Zadoks, J.C. (1987) The function of plant pathogenic fungi in natural communities. In: *Disturbance in Grasslands* (eds J. van Andel, J.P. Bakker & R.W. Snaydon), pp. 201–207. Junk Publishers, Dordrecht.

Zamfir, M., Dai, X. & van der Maarel, E. (1998) Bryophytes, lichens and phanerogams in an alvar grassland: relationships at different scales and contribution to plant community pattern. *Ecography* **22**, 40–52.

10

Terrestrial plant–herbivore interactions: integrating across multiple determinants and trophic levels

Mahesh Sankaran and S.J. McNaughton

10.1 Herbivory: pattern and process

Carbon fixed by Earth's primary producers supports life at all other trophic levels. It follows one of three trophic fates in ecological time: it may accumulate in plant tissue, be consumed by herbivores or be channelled into the decomposer pathway as litter. In most ecosystems, the bulk of primary production enters the decomposer pathway (Cebrian 1999). In others however, herbivores can consume as much as 83% of the above-ground foliage production (McNaughton *et al.* 1989).

What factors determine the fate of fixed carbon? Across ecosystems, herbivory levels have been linked to ecosystem productivity. More-productive systems on average support greater herbivore biomass (Fig. 10.1a). Larger herbivore loads in these systems mean that greater absolute amounts of plant biomass are consumed (Fig. 10.1b), resulting in greater secondary productivity (production of herbivore tissue: Fig. 10.1c). A direct positive correlation between ecosystem productivity and herbivore biomass, as suggested by McNaughton *et al.* (1989), is consistent with theories of bottom-up control of trophic structure. Here, organisms at each trophic level are assumed to be food-limited and increases in resource availability to plants therefore translate to increased biomass of organisms at higher trophic levels. However, for a given level of primary production, herbivore biomass and consumption can vary almost 1000-fold between ecosystems (Fig. 10.1a,b), indicating that ecosystem production is only one of many factors regulating herbivory patterns (McNaughton *et al.* 1989; Cebrian 1999).

While bottom-up forces, i.e. resource availability, ultimately constrain both the number and productivity of different trophic levels in an ecosystem, influences imposed by organisms at higher trophic levels (top-down forces) are also believed to be important in regulating herbivore biomass patterns in ecosystems. Formalized as the hypothesis of exploitation ecosystems (Oksanen *et al.* 1981), this viewpoint contends that when ecosystems are productive enough to support carnivores, predators, rather

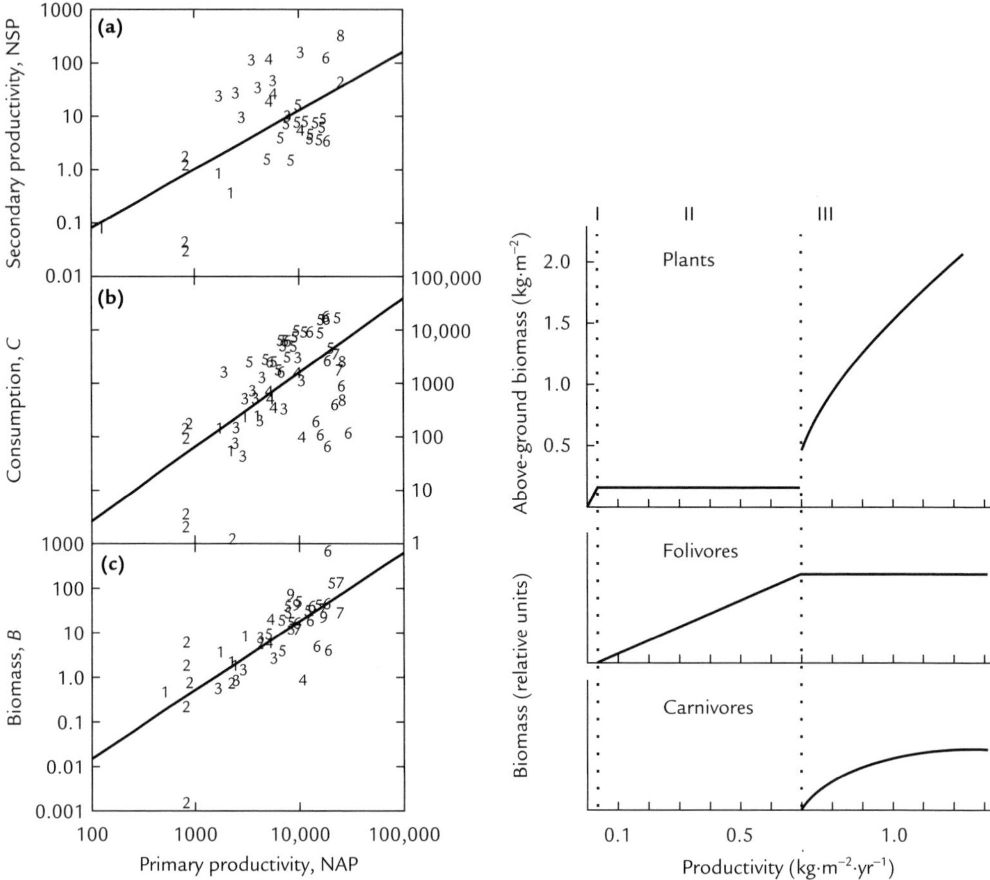

Fig. 10.1 Relationship between net above-ground primary productivity (*x*-axis) and (**a**) herbivore biomass, (**b**) consumption and (**c**) net secondary productivity. Units are in $kJ \cdot m^{-2} \cdot yr^{-1}$ except for biomass which is $kJ \cdot m^{-2}$. Key to ecosystems: 1. Desert; 2. Tundra; 3. Temperate grassland; 4. Temperate successional old field; 5. Tropical grassland; 6. Temperate forest; 7. Tropical forest; 8. Salt marsh; 9. Agricultural tropical grassland. From McNaughton *et al.* (1989). Reproduced by permission of Macmillam Publishers Ltd.

Fig. 10.2 Patterns in plant, herbivore and carnivore biomass across gradients of primary productivity as predicted by the Oksanen hypothesis. Roman numerals indicate the predicted number of trophic links for a given range of plant productivity. From Oksanen & Oksanen (2000). Reproduced by permission of the University of Chicago Press.

than plant production, control herbivore populations. Consequently, herbivore biomass should not be correlated with primary productivity in these systems (Moen & Oksanen 1991; Oksanen & Oksanen 2000). In unproductive areas incapable of supporting viable herbivore populations, plant biomass should increase with increasing productivity (Fig. 10.2, zone I). When productivity increases above the threshold required to support herbivores, herbivory should maintain plant biomass at a constant level with all increases in productivity going towards supporting greater herbivore loads (Fig. 10.2, zone II). Where production is sufficient to support carnivores, predation on herbivores should free plants from the constraints of herbivory such that plant and carnivore biomass increase with productivity, while herbivore biomass remains constant (Fig. 10.2, zone III).

Paucity of data from systems free of human impact and difficulties with experimentally manipulating whole predator communities in terrestrial ecosystems has hampered critical corroboration of these ideas. Further, available data both support and contradict different predictions, precluding a consensus. For example, Moen & Oksanen (1991) re-examined the data of McNaughton *et al.* (1989) and found differences between barren and productive systems in the strengths of correlations between herbivore biomass and plant productivity, suggestive of top-down control in productive systems. At low productivity, herbivore biomass and plant production were strongly correlated, but the relationship was weaker for more productive ecosystems, presumably due to predator control of herbivores in these systems. Similarly, distribution patterns of deer biomass across North America also appear to endorse predictions of exploitation ecosystems (Crete 1999). From the high-arctic to the transition zone between the tundra and forest, where resource availability constrains the number of trophic levels to two, cervid biomass increases with productivity. Within the wolf range in the boreal zone, deer biomass remains relatively constant, whereas south of the wolf range and in wolf-free areas, cervid biomass increases with plant productivity. In contrast, in a two-link system in the Norwegian Arctic, Wegener & Odasz (1998) found no evidence to indicate that reindeer (*Rangifer tarandus platyrhynchus*), in the absence of predators, regulated plant standing crop to a constant low level independent of productivity. Contrary to Oksanen *et al.* (Fig. 10.2, zone II), plant standing crop differed almost threefold between different grazed vegetation types and grazer exclusion had no discernible effect on plant biomass.

Ultimately, neither simple donor-controlled nor consumer-controlled models, by themselves, are likely to fully explain herbivory patterns across productivity gradients since they ignore the inherent heterogeneity among species that characterizes trophic levels in natural systems (Chase *et al.* 2000). Differences within trophic levels in plant (tissue chemistry, nutritional quality and compensatory ability), herbivore (body size, foraging behaviour, interference, anti-predator strategies) and predator characteristics (self-regulation, competition for resources other than food, intra-guild predation) all interact to influence herbivore consumption patterns (e.g. McNaughton *et al.* 1989; Cebrian 1999; Oksanen & Oksanen 2000). More complex models that explicitly incorporate such heterogeneity in their formulations better explain observed patterns of plant and herbivore biomass, as well as herbivore effects on vegetation dynamics and composition, across productivity gradients (Chase *et al.* 2000). Furthermore, most simple models of trophic dynamics essentially treat herbivores as passive conduits of energy flow through ecosystems. However, herbivores are more than just inert components of ecosystems; herbivory constitutes an integral control of plant production.

In reality, natural communities are likely to be characterized by concurrent bottom-up and top-down control, the relative strengths of which depend on the interplay between characteristics of organisms in different trophic levels (see Power 1992). For a plant–herbivore system, the absolute flux of production consumed by herbivores is indicative of the strength of bottom-up control, i.e. the extent to which plant productivity limits herbivore abundance. In contrast, the fraction of primary productivity consumed reflects the importance of herbivores as controls of plant biomass in ecosystems (top-down control; Cebrian 1999). Plant nutritional quality has been implicated as an important determinant of the latter, acting to regulate the relative amounts of carbon that flow through the herbivore versus decomposer pathway (Cebrian 1999). Nutritional quality is often positively correlated with plant relative growth rates or turnover rates. Communities comprised of plants with high relative growth rates tend to lose a greater percentage of primary production to herbivores, and channel a lower percentage as detritus (Fig. 10.3). Presumably, the high tissue nutrient concentrations, specifically nitrogen and phosphorus, required to support fast growth also renders such plants more attractive to herbivores, resulting in greater relative amounts of herbivory in such communities.

Such broad-scale considerations, although invaluable for inferring general trends, obscure within-system specifics of herbivory patterns. Within any community, all plants are not created alike, and herbivores typically face an autotrophic environment that is chemically heterogeneous, both in terms of nutrient quality and feeding deterrents in plant tissue. This heterogeneity is evident at all spatial scales: between tissues within a plant; between genotypes and populations of a species; between species and between communities of different plant species. Plant nutritional quality also varies temporally, both across seasons and over the life cycle of a plant. In addition, herbivores also confront a food base that is nutritionally inadequate. Plant tissues contain a preponderance of low-quality substances such as structural carbohydrates, cellulose and toxins but a dearth of nutrients such as nitrogen and phosphorous. Nutrient concentrations, particularly nitrogen and phosphorous, in herbivore tissue exceed those in plants, sometimes even 5–10 times (Hartley & Jones 1997). The necessity to overcome such stoichiometric imbalances, coupled with the need to avoid plant toxins, has led to a proliferation of feeding strategies in herbivores aimed at maximal exploitation of their food sources.

At its simplest, herbivory is just heterotrophic consumption of plant tissue. Yet, this seemingly straightforward interaction induces suites of responses in eater and eaten alike, and has been a driving force behind the adaptive radiation of both plants and herbivores. From a long-term co-evolutionary perspective, two major groups of present-day terrestrial vascular plants and their affiliated herbivore fauna (McNaughton 1983a) may be recognized: the first, more ancient group, includes non-graminoid plants characterized by diverse and toxic secondary chemistry, and their relatively specialized insect herbivores; the other, more recent group comprises graminoids, by comparison pharmacologically inert, and their allied general-purpose mammalian and orthopteran herbivores. Within these broad evolutionary lines, herbivores vary widely in how they exploit food sources. Most terrestrial herbivores display some measure of feeding selectivity for different plant species. Monophytophagous insects that feed exclusively on a single species occupy one end of the spectrum, and large bulk-feeding mammals that are more catholic in their diets, the other. Herbivores are also fastidious about the plant parts they consume, the degree of selectivity varying with herbivore body size, morphology of mouth parts and digestive system properties (McNaughton 1983a). Feeding mechanisms employed and plant organs consumed

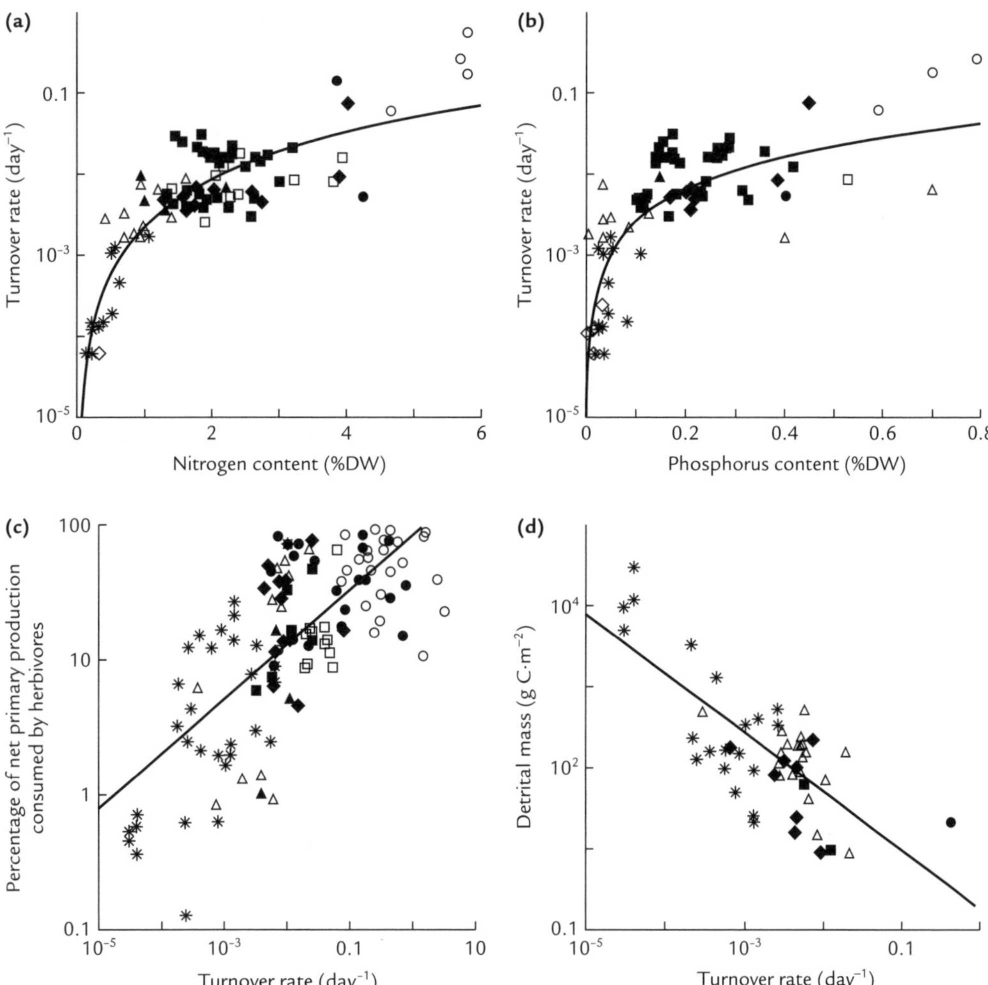

Fig. 10.3 Relationship between plant turnover rates or relative growth rates and (**a**) tissue nitrogen concentrations; (**b**) tissue phosphorous content; (**c**) fraction of production consumed by herbivores; (**d**) amounts of detritus produced across ecosystems. Open circles = phytoplankton; filled circles = benthic microalgae; open squares = macroalgal beds; filled diamonds = freshwater macrophyte meadows; filled squares = sea grass meadows; filled triangles = marshes; open triangles = grasslands; open diamonds = mangroves; asterisks = forests. Adapted from Cebrian (1999). Reproduced by permission of the University of Chicago Press.

not only provide a useful way to functionally classify herbivores (Table 10.1), but also govern plant responses to herbivory.

Over evolutionary time, plants have been selected to reduce the impacts herbivores exert upon them, while herbivores have been selected to maximally exploit their food sources without being overly destructive. These reciprocal effects have led to a proliferation of traits such as physical and chemical defences in plants that operate to reduce or tolerate bouts of herbivory. Herbivores, for their part, have evolved

Table 10.1 A functional classification of herbivores based on feeding modes and feeding targets. Also included are a few representative taxa for each functional class. The reader is directed to McNaughton (1983a) and Mortimer *et al.* (1999) for more details.

Plant organ utilized	Feeding mode	Representative taxa
Foliage	Bulk feeders with grinding and chewing mouth-parts	Mammalian herbivores, some birds, *Orthoptera*, *Hymenoptera*/*Lepidoptera* larvae
	Leaf miners that feed on the mesophyll without destroying the epidermis	*Lepidoptera* and *Hymenoptera*, *Diptera* (family *Agromyzidae*)
	Strip miners that rasp through the epidermis and underlying mesophyll	*Coleoptera*, *Lepidoptera*
Twigs and branches	Stem miners and borers	*Coleoptera*, *Lepidoptera* larvae (family *Cossidae*), *Hymenoptera* larvae (family *Cephidae*)
Xylem, phloem	Xylem and phloem sap feeders	*Homoptera*, *Heteroptera*
Roots	Root bulk feeders	Fossorial vertebrates, particularly rodents
		Vertebrates that feed on roots after disturbing the soil surface
	Young root and root hair feeders	*Collembola*, *Diplura*, *Nematoda* (order *Tylenchida*)
	Internal chewers that feed on roots and storage organs	Insects
	External feeders that consume roots or root epidermal tissue	Insects
	Cell content feeders that either fully or partially enter plants (endo- and semi-endoparasites) or feed from outside plants (ectoparasites)	Nematoda (orders *Tylenchida*, *Dorylaimida* and *Aphelenchida*)
	Sap feeders	Aphids, cicadas
Propagules	Flower, fruit and seed feeders	Mammals, birds, insects (*Bruchidae* and *Megastigmidae*)

elaborate physiological and behavioural mechanisms to breach plant defences such that no plant is totally immune to herbivory at all stages of its life.

10.2 Coping with herbivory

10.2.1 Avoidance or tolerance

Plants deal with herbivory in two basic ways: they try to avoid it or alternatively they tolerate it. Avoidance of herbivore damage can be achieved through investment in mechanical defences, production of secondary compounds, or by escape in space and time. When herbivory is inevitable, plants may instead 'tolerate' herbivory through adaptations that maintain growth and reproduction following damage. Although these alternative strategies are not mutually exclusive, their relative importance varies depending on plant life history, frequency of herbivory and the prevalence of physiological or resource constraints that impede simultaneous investment in both (Rosenthal & Kotanen 1994). From a herbivore's perspective, these alternative strategies have different selective influences as 'tolerance' does not reduce herbivore fitness and so they are under no evolutionary pressure to overcome it (Rosenthal & Kotanen 1994).

10.2.2 Use of secondary chemicals

Among the principal variables affecting a plant's susceptibility to herbivory is the presence of 'secondary' compounds. These are, by definition, not directly involved in the primary metabolism of the plant, i.e. not common to all plants but restricted to select plant groups (Pichersky & Gang 2000). Of the 20,000–60,000 genes estimated to exist in plant genomes, 15–25% may code for products involved in secondary metabolism (Pichersky & Gang 2000). They comprise an exceptionally diverse set of chemicals, many of which are known to have deleterious effects on herbivores. The roles of secondary metabolites are not solely restricted to anti-herbivore defence. Many serve other functions including UV absorption, attraction of pollinators and seed dispersers, drought and salt tolerance (Hartley & Jones 1997; Pichersky & Gang 2000).

Plant secondary chemicals have varied and diverse effects (McNaughton 1983a). They repel herbivores, inhibit their feeding, mask a plant's nutritional suitability, reduce digestibility of plant tissue and are, in some cases, toxic. Some are effective in small doses, while others function in a dosage-dependent manner. They may simultaneously deter several different herbivores, and concurrently serve as attractants for other herbivores, pollinators or seed dispersers. They can stimulate production of secondary compounds in neighbouring plants, or act as allelochemicals to inhibit the growth of neighbours. Their effects overstep trophic boundaries when adapted herbivores successfully appropriate them for their own defence purposes, or when predators and parasitoids use them as cues to locate herbivores. Their presence can also alter the decomposability of plant litter, thereby modifying nutrient recycling rates. As a consequence of these diverse roles, secondary chemicals are important mediators of both herbivore- and decomposer-based food webs (McNaughton 1983a).

10.2.3 Avoidance of herbivory

It is commonly assumed that plants incur a resource cost of defence investment since defence diverts resources away from other potential uses. Selection should therefore favour plants that optimally allocate resources to defence, both in terms of quantity and quality, to maximize their fitness (see Hartley & Jones 1997). Plant investment in defence at any point in time can be simplistically envisioned as a series of 'decisions': whether or not to invest in defences at all, what proportion of resources to allocate to defence, and what kind of defences to invest in.

Several plants maintain background levels of defence compounds at all times (constitutive or passive defences). Others induce production of defence compounds following herbivory or some cue of impending herbivory. High probability of herbivore attack has been implicated as the driving force favouring investment in constitutive defences (Agrawal & Karban 1999). When probability of herbivore attack is low, plants would benefit by inducing defences only when needed, diverting resources to other functions in the mean time. Factors besides saving of allocation costs may also favour induction of defences (Agrawal & Karban 1999). For example, several specialist herbivores that have successfully breached plant defences employ the very same defence compounds to locate host plants. In such cases, constitutive defences make a plant more apparent to herbivores, and induction may be favoured as a means to reduce specialist herbivory. Induced defence may also be favoured over a constitutive strategy if it:

1 simultaneously confers resistance against several different enemies;

2 increases variability in food quality thereby reducing herbivore performance;

3 increases herbivore movement and subsequent predation or parasitism on the herbivore;

4 reduces autotoxicity;

5 is less deleterious to natural enemies of herbivores relative to constitutive defences; or

6 reduces pollinator deterrence (Agrawal & Karban 1999).

Hypotheses to explain the amounts and type of chemical defences deployed by plants invoke a variety of factors such as the probability of herbivore attack, resource availability, kinds of limiting resources and internal physiological constraints and trade-offs between allocation to growth and defence (Hartley & Jones 1997). Suffice it to say, a consensus is still lacking because demonstration of appropriate fitness benefits has largely thwarted ecologists on account of manifold problems with identifying as well as measuring direct and indirect cost-benefit components.

10.2.4 From avoidance to tolerance

Despite the formidable arsenal of defences that plants have erected against herbivores, most plants are not totally immune from herbivory. It stands to reason that plants have evolved ways to deal with or tolerate these bouts of herbivory. The term 'compensation' has often been used synonymously with 'tolerance', particularly with reference to the re-growth capacity of plants following damage. Mechanisms of plant tolerance, albeit complex and interrelated, can be broadly classified as intrinsic and extrinsic mechanisms (McNaughton 1983a,b). Genetically determined responses, specific to species or related sets of species, resulting from physiological or development changes are considered intrinsic mechanisms. These include increased photosynthetic rates following damage, the ability to alter growth form through tillering or branching, reallocation of assimilates from storage organs to meristems, changes in root:shoot ratios, modification of hormonal balance, reductions in rates of tissue senescence and increased nutrient uptake following damage.

In contrast, extrinsic mechanisms of 'tolerance' are not species specific and stem from modification of a plant's immediate environment by herbivory. For example,

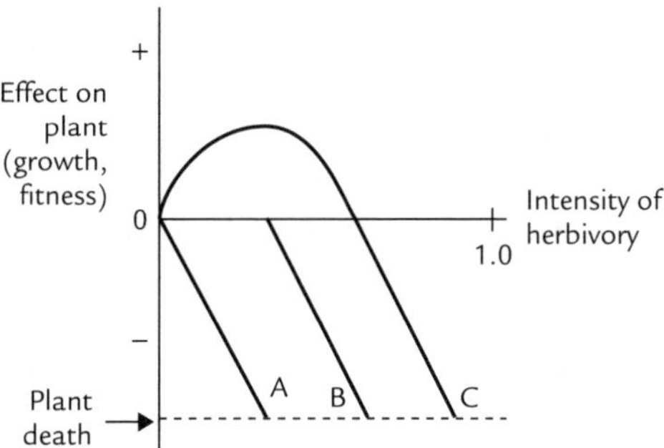

Fig. 10.4 Hypothesized effects of herbivory on plant growth and fitness. From McNaughton (1983a). Reproduced by permission of Blackwell Publishing Ltd.

vertebrate grazing can often result in an increase in light-use efficiency of the remaining ungrazed tissue by reducing mutual leaf shading. Removal of older, less-efficient plant tissue can increase overall photosynthetic rates of plants. Vertebrate grazing can also conserve soil water status by reducing transpiration surface area, which can influence the subsequent ability of plants to compensate for tissue loss. In addition, plant growth following herbivory may also be stimulated by nutrients recycled in more readily available forms such as dung and urine.

10.2.5 A continuum of compensatory responses

Is adequate compensation inevitable in the presence of tolerance mechanisms? The traditional literature distinguishes three contrasting views on tolerance capabilities of plants (Fig. 10.4). The first assumes herbivory is always detrimental (Fig. 10.4, line A); the second that herbivory is only detrimental above a critical threshold (line B); and the third that moderate levels of herbivory can actually result in over-compensation by the plant (line C). Obviously, intensity of herbivory is an important factor determining plant responses, and no plant is likely to tolerate herbivory above a critical threshold. Below this threshold, to assume plant responses are fixed, and plants respond in only one of these three ways, is to treat plants as divorced from all biotic and abiotic components of the ecosystem besides herbivores. Rather than treat a plant's response as deterministic, tolerance to herbivory must be viewed as a continuum of potential responses from under-compensation to over-compensation. Substantial evidence has accumulated in recent years which suggests, at least for vegetative growth, that the level of compensation achieved by plants in nature is contingent on several factors including plant species identity and prevalent environmental conditions (Whitham *et al.* 1991).

Among the factors influencing the ability of plants to tolerate herbivory, intrinsic growth rates are a key determinant (Whitham *et al.* 1991). Slow growth rates make it harder for a plant to replace damaged tissue in a timely manner. The ability to compensate for damage is also contingent on plant phenological status and timing of herbivory. Herbivory during the seedling stage, before root systems and photosynthetic machinery are established, is more likely to have a detrimental effect and result in mortality or under-compensation than herbivory following plant establishment. Similarly, plants in the seed-setting stage are also likely to under-compensate following herbivory. The ability of a plant to compensate generally declines the later herbivory occurs during the growing season, primarily because plants have less time to recover before the end of the growing season. Plant responses are also contingent on stored reserves of carbon and nutrients present at the time of herbivory; the greater the reserves, the higher the probability of compensating for the damage. Similarly, a plant is more likely to compensate when nutrients, water and light are not limiting in the post-herbivory environment, and if it does not have to compete with other plants for these resources.

Besides these factors, type and frequency of herbivore damage, spatial distribution of herbivore damage within the plant, as well as the number of different herbivore species that feed in concert or successively on a plant, all go to determine whether a plant successfully compensates for herbivore damage (Whitham *et al.* 1991). Just as all plants are not created alike, neither are all herbivores. What effects different herbivores have on plants will depend on the type of resource the herbivore consumes, and how damaging the removal of that specific resource is for the plant (Meyer 1993). Furthermore, in certain instances, damage by one species of herbivore can render plants more susceptible to attack by other herbivore species, while in other cases, susceptibility to one pest is associated with resistance to others (Whitham *et al.* 1991). Compensatory responses of plants in such situations will depend on the damage inflicted by each species and whether different species have additive or opposing effects on plant properties. Finally, compensatory ability is likely to be negatively correlated with frequency of herbivory. The greater the recovery period between herbivory bouts, the more likely a plant is to compensate.

Provided conditions are right, plants can over-compensate for tissue loss from herbivory. Over-compensation in response to mammalian herbivory has been demonstrated for plants in the Serengeti ecosystem, where seasonal migratory patterns of herbivores result in conditions that favour stimulation of plant productivity (Fig. 10.5). Ungulate herbivores in this system track pulses of primary productivity associated with rainfall. Herbivory occurs early in the growing season and the migratory nature of herbivores provides plants sufficient time to recover between herbivory bouts. High plant growth rates, coupled with increased nutrient availability from herbivore dung and urine, results in conditions conducive to compensation.

While several studies have shown that plants can equally or over-compensate for tissue loss to herbivores, enhancing vegetative components of fitness, fewer studies have demonstrated increases in terms of the sexual component. Indeed, the majority of studies have documented decreased seed set following herbivore damage. However, *Ipomopsis aggregata* (Paige & Whitham 1987) and *Gentianella campestris* (Lennartsson *et al.* 1998) are examples for over-compensation through seed output following herbivory. Higher seed output following grazing can result if herbivory overcomes a genetic and/or developmental constraint of plants (e.g. removal of apical dominance), or if plants withhold reproductive resources until a herbivory event in situations where there is a high probability of initial attack, but a low probability of secondary attack (Whitham *et al.* 1991). However, the idea that herbivores actually 'benefit' plants

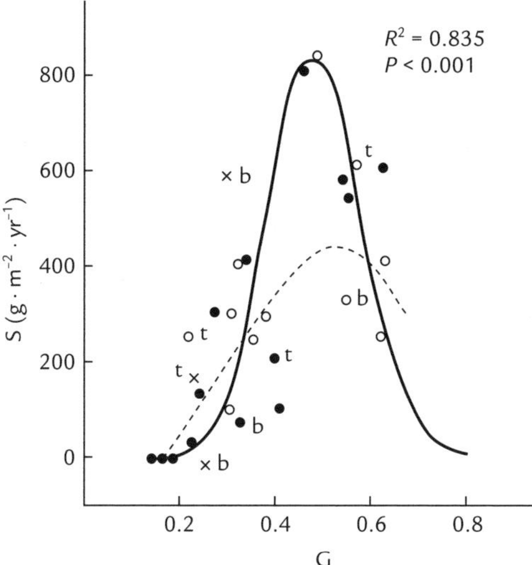

Fig. 10.5 Relationship between grazing intensity (G) and grazer stimulation of above-ground primary productivity (S). o = short grasslands; filled circles = medium-height grasslands; x = tall grasslands; b = lowland; t = hilltop stands at regional study sites. The dotted line fits all sites, while the solid line fits topographically similar sites. From McNaughton (1985). Reproduced by permission of the publisher.

and increase their reproductive fitness by eating them has been strongly contested in the literature (Belsky *et al.* 1993). Part of the controversy stems from differential interpretations of the notion of mutualism (de Mazancourt *et al.* 2001). Mutualism in ecological time (when the performance of each partner is immediately negatively impacted following removal of the other) differs from evolutionary mutualism (over evolutionary time each partner reaches a level of performance not attainable in the absence of the partnership). Agrawal (2000) provides an effective parable to demonstrate the concept. Consider a plant that has the 'ideal' potential to produce 1000 seeds. In a environment damaged predictably by migratory herbivores, consider a genotype that employs herbivory as a cue and phenologically splits its reproductive output 20% pre-herbivory and 80% post-herbivory. In the absence of herbivory, the plant produces 200 seeds, while in the presence of herbivores seed output is 800. An evolutionary consideration (comparison with the 'ideal' plant) suggests a negative impact of herbivores on plant fitness. On the other hand, in ecological time, fitness of plants is higher in the presence of herbivores than in their absence.

Besides directly influencing amounts of resources available for reproductive allocation, herbivory can also indirectly influence plant fitness if either preference or efficiency of pollinators and dispersers is altered following floral or foliar herbivory. Experimental damage in *Oenothera macrocarpa* reduced fruit set by 18% and seed set by 33% (Mothershead & Marquis 2000). Rather than a direct effect through reduced resource availability, herbivory decreased female reproduction by altering floral traits and subsequently changing preference and efficiency of pollinators. Such indirect interactions, although relatively unstudied, are critical to understanding plant fitness consequences of herbivory.

10.3 The continuum from symbiotic to parasitic

10.3.1 Effects of three common symbionts

Terrestrial plants live intimately linked with several micro-organisms, the relationships between which range from mutualistic to parasitic (see Chapter 11). For plants, benefits of mutualistic associations range from an increased ability to acquire limiting nutrients to enhanced capabilities of withstanding abiotic stresses. Such alliances often alter the nutritional status of plants and, in doing so, modulate interactions between plants and herbivores.

10.3.2 Mycorrhizae and plants

Symbiotic associations between plants and mycorrhizal fungi are ubiquitous in nature, with such associations being especially important in nutrient-poor communities. Plants provide mycorrhizae with carbon, and duly obtain several benefits from mycorrhizal infection including increased nutrient uptake, improved water relations and greater tolerance to pathogens (see Chapter 11). By improving plant nutritional status, mycorrhizae can make plants more attractive to herbivores, increasing a plant's susceptibility to attack. Alternatively, mycorrhizal colonization can also potentially reduce a plant's susceptibility to herbivory if enhanced nutrient uptake relative to carbon cost permits greater plant allocation to anti-herbivore defences. Induction of defence compounds that follow infection of plant roots by fungal hyphae, and secondary compounds synthesized by the mycorrhizae themselves, can also act to enhance plant resistance to herbivores. Besides altering plant resistance, mycorrhizal colonization can also improve plant tolerance to herbivory if it enhances a plant's ability to acquire limiting nutrients post-herbivory. Consistent with these potential alternative outcomes, experimental studies of mycorrhizal colonization have demonstrated both increases and decreases in host-plant resistance to herbivory (Gehring & Whitham 1994).

Just as plant–herbivore interactions are influenced by mycorrhizae, herbivores too can affect how a plant interacts with its mycorrhizal symbionts. As much as 10–60% of a plant's photosynthate might be required to support mycorrhizae (Gehring & Whitham 1994). Consequently, when tissue loss to herbivores is high, costs of supporting mycorrhizae can far outweigh benefits, shifting the relationship from mutualistic to parasitic. Many studies have documented reduced mycorrhizal colonization following herbivory, while others show no significant effects or positive effects of herbivory on mycorrhizal colonization (Gehring & Whitham 1994). No significant effects, and possibly increased mycorrhizal colonization, can result if herbivory induces shifts in mycorrhizal communities favouring species or morphotypes with lower carbon requirements (Saikkonen *et al.* 1999). Ultimately, the specific outcome is dependent on how herbivory interacts with prevailing environmental conditions to alter the cost: benefit ratio of association for both involved parties.

10.3.3 The trade-off of N fixation

Nitrogen limits plant growth in many terrestrial ecosystems. Plants have evolved several adaptations to cope with this limitation, including forming symbiotic associations with nitrogen-fixing bacteria. Where nitrogen is limiting, plants involved in symbiotic associations should have competitive advantage over non-fixers. Yet nitrogen-fixing species do not reach widespread dominance. Limitation by nutrients other than nitrogen, inability to quickly colonize early successional sites and high energetic costs of fixing atmospheric nitrogen are potential reasons for the lack of widespread dominance by N fixers. However, herbivores also play critical roles in the observed rarity of N fixers in several ecosystems.

Bentley & Johnson (1991) compared alkaloid content and growth rates of *Lupinus succulentus* plants grown under low-nitrogen concentrations against defoliated and undamaged plants provided with either inorganic nitrogen or with N-fixing bacteria (Fig. 10.6). Unlike plants provided with supplemental inorganic nitrogen, leaf damage in N-fixing plants reduced both alkaloid concentrations and growth rates, suggesting herbivory costs to N fixation. Although N fixers may invest substantially in nitrogenous defence compounds while undamaged (Fig. 10.6), their ability to tolerate herbivory can be compromised once damaged. Presumably, leaf tissue loss reduces photosynthate available to support N-fixing bacteria resulting in N-fixing plants becoming nitrogen stressed under conditions of high herbivory.

Indirect evidence for a herbivory cost of N fixation comes from studies that report increased abundance of N fixers following herbivore removal (Ritchie *et al.* 1998;

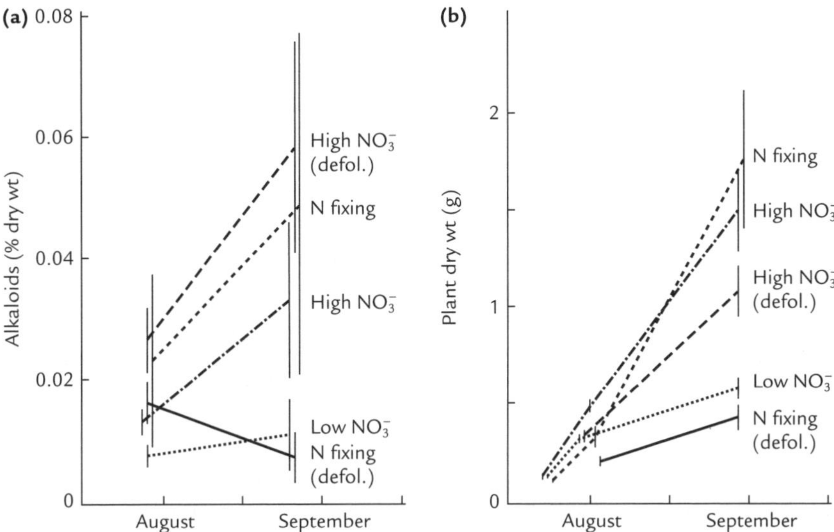

Fig. 10.6 Effects of defoliation on alkaloid concentrations (**a**) and biomass (**b**) of *Lupinus succulentus* plants grown under the indicated nitrogen nutrition treatments. Adapted from Bentley & Johnson (1991). Reproduced by permission of John Wiley & Sons, Inc.

Sirotnak & Huntly 2000). Experimental exclusion of voles (*Microtus* spp.) from a site in Yellowstone National Park, USA, resulted in increased legume abundance within exclosures (Sirotnak & Huntly 2000). However, responses were not consistent across all sites suggesting that herbivore effects on N fixers can vary over space and time and may be contingent on specific site conditions. Since fixation represents a substantial source of nitrogen input into many systems, the interaction between herbivores and N fixers can directly and indirectly affect several aspects of community and ecosystem function.

10.3.4 Fungal endophyte associations

Plants also form mutualistic associations with fungal endophytes that grow intercellularly in leaf and stem cells of plants and infect plants asymptomatically (Omacini *et al.* 2001). Endophytes receive nutrients and protection from plants, and in turn confer plants with increased herbivore, pathogen and drought resistance, enhanced competitive ability and increased germination success. Effects of such associations on plant–herbivore interactions have received relatively little attention in the ecological literature, but recent studies show that fungal endophyte infections can influence herbivory rates as well as the nature of interactions between herbivores and organisms at higher trophic levels. In an experiment involving *Lolium multiflorum* plants, Omacini *et al.* (2001) showed that fungal endophyte infection decreased aphid densities on plants threefold (Fig. 10.7a). Responses, however, differed between aphid species. Endophyte infection also influenced rates of aphid parasitism (Fig. 10.7b). While hatching rates of primary parasatoids (those that attack aphids directly) did not differ

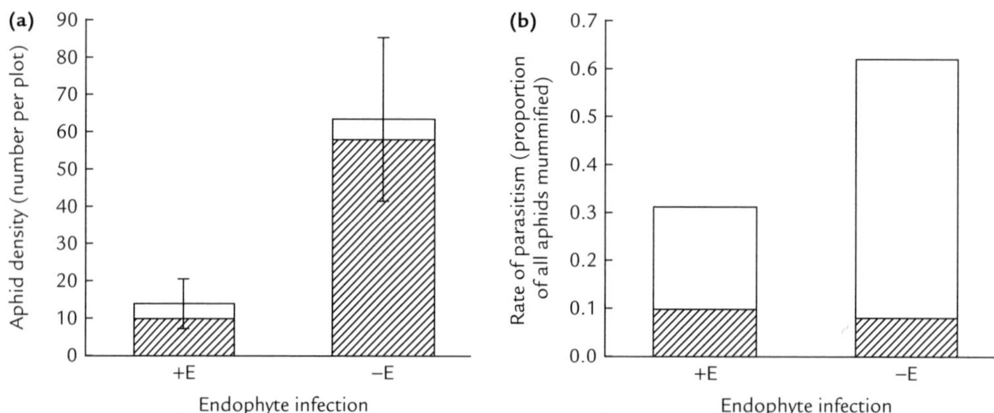

Fig. 10.7 **a.** Differential density responses of two species of aphids, *Rhopalosiphum padi* (shaded bars) and *Metopolophium festucae* (empty bars) to presence (+E) and absence (−E) of fungal endophytes in *Lolium multiflorum* plants. Responses are significant only for *R. padi*. **b.** Endophyte treatment effects on the total aphid parasitism rate. Shaded bars represent proportion of emerged primary parasitoids and open bars, secondary parasitoids. Adapted from Omacini *et al.* (2001). Reproduced by permission of Macmillan Publishers Ltd.

between endophyte-infected and -uninfected treatments, hatching rates of secondary parasatoids (those that attack primary parasatoids) was significantly higher in endophyte-free plants (Fig. 10.7b). Fungal endosymbionts may therefore be important modulators of plant–herbivore interactions and food web structure.

10.4 Community-level effects of herbivory

10.4.1 Herbivores and plant species diversity

Herbivores have varied effects on plant richness of communities, working to either increase, decrease or cause no significant changes in plant richness (see also Chapter 9). Nutrient and water availability, and evolutionary history of grazing are some of the variables hypothesized to have a regulatory effect on herbivore mediation of plant richness (Milchunas *et al.* 1988; Proulx & Mazumder 1998). Comparative studies of a broad array of herbivores and habitat types suggest that herbivore effects on plant species richness may be contingent on nutrient availability. Richness often declines with grazing in nutrient-poor ecosystems, while the outcome is reversed in nutrient-rich ecosystems (Proulx & Mazumder 1998). Presumably, ability to tolerate low-nutrient conditions is the primary factor controlling plant species richness in nutrient-poor ecosystems. When grazed, species intolerant of herbivory are removed from the system. Since few species remain in the pool tolerant to both herbivory and low-nutrient conditions, colonization is low and species diversity declines under grazing. In nutrient-rich systems on the other hand, competitive ability rather than stress tolerance is presumed to be the variable defining plant species richness. In this case, grazing on competitive dominants relaxes competitive interactions, permitting co-existence of inferior competitors. Diversity therefore increases with grazing in such systems. However, increases in grazing intensity beyond a critical threshold, even in nutrient-rich ecosystems, can cause diversity to decline.

Over and above such broad generalizations, herbivore effects are likely to be specific to spatial scales of inquiry (Olff & Ritchie 1998; Stohlgren *et al.* 1999). At small scales, herbivore mediation of competitive interactions may be the dominant process influencing species diversity, while colonization-extinction dynamics may be more important at larger scales. At small scales, herbivores can enhance plant diversity by:

1 selectively consuming competitive dominants, permitting establishment of inferior species;

2 increasing heterogeneity through soil disturbances and permitting species co-existence; and

3 reducing individual plant size and allowing for greater species packing within a given area.

In the absence of grazing, dominants may grow bigger, exclude sub-dominants and lower plant diversity at small scales. However, if overall rates of colonization and extinction are not altered, such differences may not be evident at large scales (Stohlgren *et al.* 1999). Species excluded by grazers at small scales may still persist in 'grazing-safe sites' at larger scales. However, if grazing pressure is strong enough, intolerant species may be weeded out altogether from the regional species pool lowering diversity at larger scales (Stohlgren *et al.* 1999).

10.4.2 Effects of herbivore diversity

Natural communities typically contain several herbivores that vary in body size and differ in their feeding strategies and selectivity. Such a diversity of herbivores can have additive or complementary effects on plant species diversity (Ritchie & Olff 1999). Few studies have experimentally tested effects of herbivore richness or body-size diversity on plant communities. Excluding small herbivores while retaining large ones is rarely feasible in field studies, and most studies progressively exclude larger-bodied herbivores. Results from such studies (reviewed in Ritchie & Olff 1999) indicate widely divergent responses of multiple herbivores on plant diversity. When multiple herbivores feed on the same species, their effects will be additive. Simultaneous feeding on competitive dominants can increase diversity, while that on competitive inferiors decreases it. When herbivores feed on different species, their effects may be complementary, serving to maintain plant species diversity in a quasi-stable state.

The potential for different herbivore species to have additive or complementary effects on plant community diversity is also dependent on whether herbivore species are likely to facultatively diverge or converge in their diets in the presence of other herbivores. Compensatory effects arising from herbivore diet shifts in the presence of competitors has been experimentally demonstrated for grasshoppers feeding on Minnesota old-field plants (see Ritchie & Olff 1999). The proportion of grasses and forbs in the diets of three different grasshopper species changed in the presence of the other species as opposed to when alone. In contrast, additive effects of diverse herbivores can occur if feeding by a particular herbivore makes a plant more attractive to other herbivore species. How such individual effects translate to additive effects at the community level is unclear. Additive effects at the community level can arise if, for example, several different herbivores cue in on a particular plant species following specialist herbivore outbreaks on that species.

Recent theoretical syntheses suggest that herbivore effects should vary predictably across soil fertility and moisture gradients (Olff & Ritchie 1998; Ritchie & Olff 1999). The underlying premise is that tissue nutrient concentrations and palatability of dominant species differ depending on the particular limiting resource, which in turn determines the characteristics of the herbivore community and their consumption patterns. Where dominant plant species tend to be palatable, i.e. have high tissue nutrient concentrations, multiple herbivores can consume the same species in an additive fashion. On the other hand, in communities characterized by abundant low-quality plants and rare high-quality plants, effects of multiple herbivores can lead to compensatory effects. In such cases, large herbivores potentially consume the dominant low-quality plants permitting the co-existence of both high-quality plants as well as the smaller-bodied herbivores that feed on them. Compensatory effects of herbivore diversity in these situations arise from herbivores of different body sizes consuming different plant species. Approaches such as these provide a fruitful avenue of pursuit since they integrate ecosystem-level constraints on plant traits with herbivore feeding selectivity as a function of body size to predict plant community responses to grazing by a diverse herbivore assemblage.

10.4.3 Herbivory and plant succession

Herbivores also influence successional rates and successional trajectories of communities (Ritchie & Olff 1999). Herbivores that feed preferentially on late successional species tend to retard succession. By the same token, selective herbivory on early successional species can hasten establishment of late successional species, thereby accelerating succession. As with species diversity patterns, the presence of a diverse herbivore assemblage can have additive or complementary effects on successional trends. Herbivore assemblages that feed on species characteristic of the same successional state have additive effects on plant species replacement patterns. In such cases, effects of diverse assemblages may be similar to those of individual herbivores, either accelerating or retarding successional rates. On the other hand, when different members of the herbivore assemblage feed on species characteristic of different successional stages, they can 'arrest' plant communities at intermediate stages of succession.

Besides influencing successional rates, herbivores also regulate successional trajectories, thereby defining the qualitative nature of mature plant communities. Seedling herbivory, in particular, is an important pathway through which such herbivore-effects are manifested (Crawley 1997; Hanley 1998). Plants are particularly vulnerable to tissue loss at this stage in their life-cycle and even if herbivory does not result in mortality, it can reduce seedling vigour thereby influencing its competitive ability and chances of long-term survival. The magnitude of such effects can be substantial. For example, in Panamanian forests, mammalian herbivory can cause as much as a sixfold reduction in tree seedling survivorship for certain species (Asquith *et al.* 1997). Several herbivore guilds, from nematodes to large mammals, have been shown to have deleterious effects on seedling survival and establishment; the case of mollusc herbivores on seedling dynamics in temperate systems being particularly well documented (see Crawley 1997; Hanley 1998). Differences between species in herbivore-induced seedling mortality is the primary mechanism through which herbivores influence plant community development and species composition patterns. Differential herbivore-induced mortality can arise from inter-specific variation in seedling palatability, size and morphology, as also from differences in abundance, spatial distribution and timing of seedling emergence (Hanley 1998). However, herbivore effects on seedling establishment need not always be detrimental (Crawley 1997). Besides direct negative effects on vulnerable species that arise from increased mortality or reduced competitive vigour following tissue consumption, herbivores can also have indirect positive effects on seedling establishment of other species. Seedling establishment, particularly for species avoided by herbivores, may be favoured when herbivores enhance microsite suitability through physical disturbances to the environment or when consumption of plant tissue opens up canopies, reduces competition, reduces litter loads and increases light availability at the soil surface, thereby creating opportunities for recruitment. See further Chapter 7.

10.5 Integrating herbivory with ecosystem ecology

The interaction between plants and herbivores has important repercussions for patterns of energy and nutrient flow through the ecosystem because herbivore consumption of plant tissue, plant nutrient uptake and litter decomposition rates are intimately linked. Ecosystem-level studies of energy and nutrient cycling have reported diametrically opposite effects of herbivory on ecosystem processes. Herbivores enhance nutrient cycling in certain systems and retard it in others. Augustine & McNaughton (1998) identified four mechanisms by which herbivores influence energy and nutrient flow in ecosystems:

1 by altering species composition of communities and hence, the quality of litter inputs from uneaten plants;

2 by consuming plant nutrients and returning them to the soil in more readily available forms such as dung and urine;

3 by altering inputs from eaten plants to the soil through changes in the root system, litter quality and other non-detrital inputs such as root exudates; and

4 by altering plant and soil micro-environment.

While the latter three enhance nutrient cycling rates, species compositional changes can either enhance or retard nutrient cycling rates. The eventual outcome depends on whether effects of altered species composition offset effects of the latter three processes.

Species compositional changes influence ecosystem nutrient cycling by modifying the quality of litter inputs to the soil (Fig. 10.8). Herbivory favouring unpalatable, slow-growing species with well-defended or nutrient-poor tissues, results in litter that is of poor quality. Such litter, containing high amounts of structural tissue or secondary chemicals, is broken down more slowly by microbes, reducing rates at which limiting nutrients are recycled between different ecosystem components. When herbivory

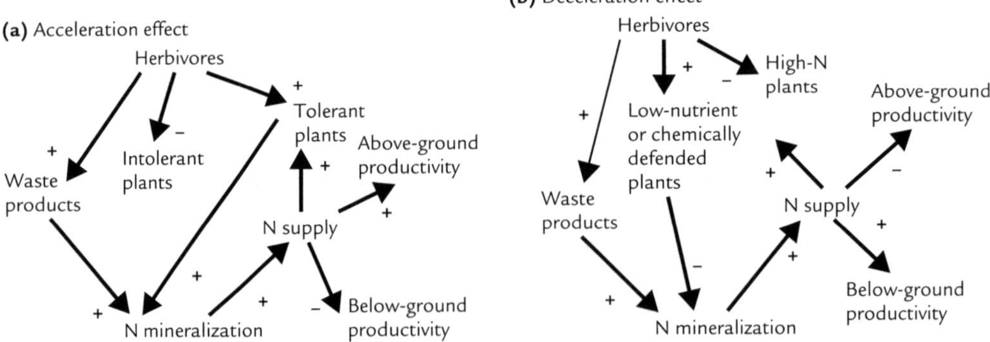

Fig. 10.8 Hypothesized mechanisms by which herbivore feeding preferences can accelerate (**a**) or decelerate (**b**) rates of nutrient cycling through an ecosystem. Arrows indicate the net indirect effect of herbivores on the abundance of plants or the rate of the process. From Ritchie *et al.* (1998). Reproduced by permission of the Ecological Society of America.

favours fast-growing, palatable species with high tissue nutrient concentrations, litter produced is easily broken down by microbes and nutrient cycling rates amplified.

In systems comprising both palatable and unpalatable plant species, herbivores are likely to consume proportionately more tissue from palatable species. Considerations of herbivore intake alone suggests that palatable species must be at a competitive disadvantage in these situations, leading to eventual domination by unpalatable species in these communities. Why then, does herbivory not cause all plant communities to be dominated by unpalatable species? Obviously, greater tissue loss to herbivory is insufficient to tilt the competitive balance in favour of unpalatable species in all systems. Intrinsic and extrinsic mechanisms by which plants tolerate herbivory, intensity and frequency of herbivory, as well as prevalent environmental conditions, all interact to determine the nature of herbivore-induced community change. Conditions favouring persistence of palatable species in a community include high nutrient levels in the system, intermittent herbivory rates such as those resulting from migratory habits of herbivores, early-season and post-fire herbivory, asynchronous phenology of palatable and unpalatable species, herbivore body-size dichotomy, and herding behaviour of herbivores (Whitham *et al.* 1991; Augustine & McNaughton 1998; Ritchie & Olff 1999). Compositional shifts favouring unpalatable species are more likely in systems that are nutrient poor, contain sedentary herbivores that feed selectively on high-quality plants, forage singly or in small groups, and subject plants to chronic levels of herbivory (Augustine & McNaughton 1998).

Herbivore body-size dichotomy can also be important in regulating the balance between palatable and unpalatable species in a community (Olff & Ritchie 1998; Ritchie & Off 1999). In systems where both water and nutrients are non-limiting, plant competition is primarily for light. Plants invest in structural tissues to enhance their light competitive ability, and so dominant species tend to be of low quality, utilized primarily by large-bodied bulk-feeding herbivores. Grazing by large-bodied herbivores on low-quality plants facilitates co-existence of both grazing-tolerant high-quality plants as well as small-bodied herbivores that feed on them. However, reductions in numbers of large-bodied herbivores can lead to low-quality plants dominating the system, causing high-quality plants as well as smaller-bodied herbivores to decline, and reducing overall rates of nutrient cycling. In summary, differential effects of herbivory on plant composition and subsequent ecosystem functioning can arise from differences in:

1 the nature of limiting resources (e.g. water, nitrogen), which in turn defines plant characteristics and herbivore selectivity;

2 herbivory characteristics such as frequency, intensity and timing; and

3 herbivore characteristics including foraging behaviour, herbivore diversity and herbivore body-size dichotomy (Augustine & McNaughton 1998; Olff & Ritchie 1998; Ritchie *et al.* 1998; Ritchie & Olff 1999).

References

Agrawal, A.A. (2000) Overcompensation of plants in response to herbivory and the by-product benefits of mutualism. *Trends in Plant Science* **5**, 309–313.

Agrawal, A.A. & Karban, R. (1999) Why induced defences may be favored over constitutive strategies in plants. In: *The Ecology and Evolution of Inducible Defences* (eds R. Tollrian & C.D. Harvell), pp. 45–61. Princeton University Press, Princeton.

Asquith, N.M., Wright, S.J. & Clauss, M.J. (1997) Does mammal community composition control recruitment in neotropical forests? Evidence from Panama. *Ecology* **78**, 941–946.

Augustine, D.J. & McNaughton, S.J. (1998) Ungulate effects on the functional species composition of plant communities: herbivore selectivity and plant tolerance. *Journal of Wildlife Management* **62**, 1165–1183.

Belsky, A.J., Carson, W.P., Jensen, C.L. & Fox, G.A. (1993) Overcompensation by plants: herbivore optimization or red herring? *Evolutionary Ecology* **7**, 109–121.

Bentley, B.L. & Johnson, N.D. (1991) Plants as food for herbivores: the roles of nitrogen fixation and carbon dioxide enrichment. In: *Plant-Animal Interactions: Evolutionary Ecology in Tropical and Temperate Regions* (eds P.W. Price, T.M. Lewinsohn, G.W. Fernandes & W.W. Benson), pp. 257–272. John Wiley & Sons, New York.

Cebrian, J. (1999) Patterns in the fate of production in plant communities. *American Naturalist* **154**, 449–468.

Chase, J.M., Leibold, M.A., Downing, A.L. & Shurin, J.B. (2000) The effects of productivity, herbivory, and plant species turnover in grassland food webs. *Ecology* **81**, 2485–2497.

Crawley, M.J. (1997) Plant – herbivore dynamics. In: *Plant Ecology* (ed. M.J. Crawley), pp. 401–474. Blackwell Science, Oxford.

Crete, M. (1999) The distribution of deer biomass in North America supports the hypothesis of exploitation ecosystems. *Ecology Letters* **2**, 223–227.

de Mazancourt, C., Loreau, M. & Dieckmann, U. (2001) Can the evolution of plant defence lead to plant-herbivore mutualism? *American Naturalist* **158**, 109–123.

Gehring, C.A. & Whitham, T.G. (1994) Interactions between above-ground herbivores and mycorrhizal mutualists of plants. *Trends in Ecology & Evolution* **9**, 251–255.

Hanley, M.E. (1998) Seedling herbivory, community composition and plant life history traits. *Perspectives in Plant Ecology, Evolution and Systematics* **12**, 191–205.

Hartley, S.E. & Jones, C.G. (1997) Plant chemistry and herbivory, or why the world is green. In: *Plant Ecology* (ed. M.J. Crawley), 2nd ed., pp. 284–324. Blackwell Science, Oxford.

Lennartsson, T., Nilsson, P. & Tuomi, J. (1998) Induction of overcompensation in the field gentian, *Genianella campestris*. *Ecology* 79, 1061–1072.

McNaughton, S.J. (1983a) Compensatory plant growth as a response to herbivory. *Oikos* **40**, 329–336.

McNaughton, S.J. (1983b) Physiological and ecological implications of herbivory. In: *Physiological Plant Ecology III: Responses to the Chemical and Biological Environment*, Vol. 12C (eds O.L. Lange, P.S. Nobel, C.B. Osmond & H. Ziegler), pp. 657–678. Springer-Verlag, Berlin.

McNaughton, S.J. (1985) Ecology of a grazing ecosystem: The Serengeti. *Ecological Monographs* **55**, 259–294.

McNaughton, S.J., Oesterheld, M., Frank, D.A. & Willliams, K.J. (1989) Ecosystem-level patterns of primary productivity and herbivory in terrestrial habitats. *Nature* **341**, 142–144.

Meyer, G.A. (1993) A comparison of the impacts of leaf- and sap-feeding insects on growth and allocation of goldenrod. *Ecology* **74**, 1101–1116.

Milchunas, D.G., Sala, O.E. & Lauenroth, W.K. (1988) A generalized model of the effects of grazing by large herbivores on grassland community structure. *American Naturalist* **132**, 87–106.

Moen, J. & Oksanen, L. (1991) Ecosystem trends. *Nature* **353**, 510.

Mortimer, S.R., van der Putten, W.H. & Brown, V.K. (1999) Insect and nematode herbivory below ground: interactions and role in vegetation succession. In: *Herbivores: Between Plants and Predators* (eds H. Olff, V.K. Brown & R.H. Drent), pp. 205–238. Blackwell Science, Oxford.

Mothershead, K. & Marquis, R.J. (2000) Fitness impacts of herbivory through indirect effects on plant–pollinator interactions in *Oenothera macrocarpa*. *Ecology* **81**, 30–40.

Oksanen, L. & Oksanen, T. (2000) The logic and realism of the hypothesis of exploitation ecosystems. *The American Naturalist* **155**, 703–723.

Oksanen, L., Fretwell, S.D., Arruda, J. & Niemela, P. (1981) Exploitation ecosystems in gradients of primary productivity. *American Naturalist* **118**, 240–261.

Olff, H. & Ritchie, M.E. (1998) Effects of herbivores on grassland plant diversity. *Trends in Ecology & Evolution* **13**, 261–265.

Omacini, M., Chaneton, E.J., Ghersa, C.M. & Muller, C.B. (2001) Symbiotic fungal endophytes control insect host-parasite interaction webs. *Nature* **409**, 78–81.

Paige, K.N. & Whitham, T.G. (1987) Overcompensation in response to mammalian herbivory: the advantage of being eaten. *American Naturalist* **129**, 407–416.

Pichersky, E. & Gang, D.R. (2000) Genetics and biochemistry of secondary metabolites in plants: an evolutionary perspective. *Trends in Plant Science* **5**, 439–445.

Power, M.E. (1992) Top-down and bottom-up forces in food webs: do plants have primacy? *Ecology* **73**, 733–746.

Proulx, M. & Mazumder, A. (1998) Reversal of grazing impact on plant species richness in nutrient-poor vs. nutrient-rich ecosystems. *Ecology* **79**, 2581–2592.

Ritchie, M.E. & Olff, H. (1999) Herbivore diversity and plant dynamics: compensatory and additive effects. In: *Herbivores: Between Plants and Predators* (eds H. Olff, V.K. Brown & R.H. Drent), pp. 175–204. Blackwell Science, Oxford.

Ritchie, M.E., Tilman, D. & Knops, J.M.H. (1998) Herbivore effects on plant and nitrogen dynamics in oak savanna. *Ecology* **79**, 165–177.

Rosenthal, J.P. & Kotanen, P.M. (1994) Terrestrial plant tolerance to herbivory. *Trends in Ecology and Evolution* **9**, 145–48.

Saikkonen, K. *et al.* (1999) Defoliation and mycorrhizal symbiosis: a functional balance between carbon sources and below-ground sinks. *Ecology Letters* **2**, 19–26.

Sirotnak, J.M. & Huntly, N.J. (2000) Direct and indirect effects of herbivores on nitrogen dynamics: voles in riparian areas. *Ecology* **81**, 78–87.

Stohlgren, T.J., Schell, L.D. & Heuvel, B.V. (1999) How grazing and soil quality affect native and exotic plant diversity in Rocky Mountain grasslands. *Ecological Applications* **9**, 45–64.

Wegener, C. & Odasz–Albrigtsen, A.M. (1998) Do Svalbard reindeer regulate standing crop in the absence of predators? A test of the 'exploitation ecosystems' model. *Oecologia* **116**, 202–206.

Whitham, T.G., Maschinski, J., Larson, K.C. & Paige, K.N. (1991) Plant responses to herbivory: The continuum from negative to positive and underlying physiological mechanisms. In: *Plant-Animal Interactions: Evolutionary Ecology in Tropical and Temperate Regions* (eds P.W. Price, T.M. Lewinsohn, G.W. Fernandes & W.W. Benson), pp. 227–256. John Wiley & Sons, New York.

11

Interaction between higher plants and soil-dwelling organisms

Thomas W. Kuyper & Ron G.M. de Goede

11.1 Introduction

Early plant ecologists, notably Schimper, Warming, Clements and Braun-Blanquet, were fast to recognize the importance of soil-dwelling organisms for plant community ecology after the discovery of mycorrhizae and nitrogen-fixing rhizobia in the 1880s. However, interest in these interactions gradually waned as vegetation ecology went increasingly descriptive in the formal recognition of plant community types. The study of these mutualistic and parasitic interactions became part of applied plant ecology (agronomy, forestry), while having a minor impact on the development of vegetation ecological theory. Recently, however, plant ecologists rediscovered the importance of below-ground interactions (e.g. Crawley 1997). In that period it became accepted that studies on interactions between plants without the soil community likely lead to artefacts (e.g. Newsham *et al.* 1995).

Plant roots are a source of carbon and nutrients for many organisms, whose effects range from mutualistic to parasitic. The often major effects on plant production of these organisms can be demonstrated by the use of biocides, which should be highly selective for one taxonomic group without having significant effects on other below-ground groups. However, the claim of selectivity has not always been checked, and one cannot always conclude from such experiments which organisms cause the observed changes in plant productivity. Application of insecticides in pastures resulted in 16–33% yield increases (Clements *et al.* 1987), whereas application of nematicides resulted in increased grassland yields of 25–59% (e.g. Verschoor 2002). Application of benomyl, a supposedly specific fungicide that eliminates arbuscular mycorrhizal fungi, did not reduce plant biomass in an Australian semi-arid herbland, but reduced biomass in experimental microcosms, simulating calcareous grasslands, with 20–40% (van der Heijden *et al.* 1998; O'Connor *et al.* 2002). However, the supposed specificity of benomyl is probably overestimated as it cannot only affect arbuscular mycorrhizal fungi but also the abundance and activity of root pathogenic fungi. Next to effects on plant production, such soil-dwelling organisms may also affect plant community composition.

This chapter treats both mutualistic and parasitic interactions, and looks for similarities and differences in the mechanisms by which these organisms affect plant communities. It is important to realize that:

1 The arena in which the interactions take place includes both mutualistic and antagonistic organisms. Both have direct effects on individual plant species, but moreover their interaction affects plant communities.

2 The plant community is both a determinant of and is determined by mutualistic and antagonistic soil organisms.

3 There is a continuum between mutualistic and antagonistic behaviours between soil-organisms on the one hand and plant (roots) on the other. It is in the context of the environmental conditions how costs and benefits of symbioses affect plants (Johnson *et al.* 1997).

4 The mechanisms by which soil-dwelling organisms affect plant communities can be similar. As to the two main mechanisms, positive feedback and negative feedback, mutualists do not necessarily create positive feedbacks and parasites negative feedbacks.

5 Both for mutualistic and antagonistic soil organisms there is a continuum from associations that are specific for one plant species to those that are shared with other species. Both plants and soil-dwelling organisms show different degrees of specificity or selectivity. Both private and shared associations could amplify or reduce the differences in competitive abilities between plant species.

11.2 The main ecologically important below-ground groups

11.2.1 Introduction

Primary productivity of vegetation is most often limited by nitrogen and phosphorus. Generally speaking, in tropical ecosystems on old weathered soils phosphorus is usually the primary limiting nutrient, whereas in temperate and boreal ecosystems nitrogen limitation is more common (Read 1991). Mutualistic symbioses between micro-organisms and plants contribute to increased possibilities to exploit these scarce resources. Roots of plants of several families are specifically associated with various bacteria that have the capacity to fix atmospheric nitrogen and convert it to mineral nitrogen. The roots of the overwhelming majority of plant species are also associated with beneficial fungi, called mycorrhizal fungi, which enhance the uptake capacity of various plant nutrients, especially those of low mobility such as phosphorus, but also of nitrogen. Roots also attract pathogens and parasites which use carbon and nutrients; this leads to net losses of those resources.

11.2.2 Nitrogen-fixing symbioses

Among the higher plants two major types of symbiotic nitrogen-fixing association are recognized: (i) the legume symbiosis with bacteria collectively referred to as rhizobia; (ii) the actinorrhizal symbiosis of several trees and shrubs with actinomycetes belonging to the genus *Frankia*. The amounts of nitrogen fixed are highly variable. Highly productive, early-successional natural stands of nitrogen-fixing trees, such as *Acacia*,

Robinia, *Alnus*, *Hippophae* or *Myrica*, can fix up to a few hundred kg-N·ha^{-1}·yr^{-1}. Such amounts are largely in excess of the actual demand and uptake capacity of the vegetation. In such cases excess ammonia is converted to nitrate and the nitrate can leach from the system together with a similar quantity of basic cations. During early primary succession in Alaska sites with *Alnus sinuata* are therefore prone to rapid acidification, as a consequence of which *Alnus* disappears and conifers, notably *Picea sitchensis* and *Tsuga heterophylla*, establish. During this development N availability decreases and the subsequent build-up of recalcitrant litter further reduces nitrogen availability (Hobbie *et al.* 1998). The nitrogen-fixing *Alnus* therefore acts as a driving force in primary succession. This process usually lasts less than 100 yr.

Symbiotic nitrogen fixation is often prominent during early successional stages and declines in later successional stages. This pattern seems somewhat paradoxical as many late-successional ecosystems are still N-limited. Why are nitrogen-fixing species virtually lacking from such ecosystems? Most likely the cost of acquisition of resources provides the explanation. Nitrogen-fixing plants can acquire nitrogen from two sources, namely from the soil and through fixation of atmospheric nitrogen. Symbiotic nitrogen fixation is an energy-demanding process, more costly than nitrogen uptake from the soil. Also important for successional interactions is that N-fixers have a lower N-use efficiency (high N demand per a of carbon) by which they become outcompeted by plants with a higher N-use efficiency that acquire nitrogen from the soil. Nitrogen-fixing plants also have a high phosphorus demand, and most N-fixers have specialized organs by which they increase phosphorus uptake: mycorrhizae in case of legumes or cluster roots, bottle-brush-like structures in roots, in case of actinorrhizal plants.

11.2.3 Mycorrhizal associations

The roots of the overwhelming majority of higher plant species are colonized by fungi that live in a mutualistic relationship, called mycorrhizae. In mycorrhizal symbiosis the plants provide carbon to the fungus, whereas the fungus provides essential nutrients especially those of low mobility such as P and several micronutrients, notably Cu and Zn, but also N, K, Mg, and Ca to the plant. Mycorrhizae have also been implicated in a suite of other beneficial effects to plants, e.g. improved water relations; increased protection against acidity, aluminium toxicity and heavy metals; protection against root pathogens. Mycorrhizal symbiosis is therefore multifunctional (Newsham *et al.* 1995).

Regarding the benefits for the mycorrhizal plant, the emphasis is often on the increased biomass as compared to non-mycorrhizal plants. However, the benefit may be expected to be most important in the seedling and establishment stages, and in the reproductive phase, notably on seed quality (Read 1999). Emphasis on above-ground biomass has also lead to the situation where lack of a growth response was interpreted as evidence that mycorrhizal fungi showed parasitic behaviour.

On the basis of the morphology of mycorrhizae, putative functions or beneficial effects, and plant and fungal taxa involved, mycorrhizal associations can best be divided in four broad categories: (i) arbuscular mycorrhizae; (ii) sheathing mycorrhizae, including ectomycorrhizae, ectendomycorrhizae, arbutoid mycorrhizae and monotropoid mycorrhizae; (iii) ericoid mycorrhizae; and (iv) orchid mycorrhizae. Morphological and physiological differences between these mycorrhizal types are treated, amongst others by Smith & Read (1997).

11.2.4 Root-feeding invertebrates

These include earthworms, enchytraeids, molluscs, millipedes, isopods, mites, insects and nematodes (Brown & Gange 1990). The latter two groups are considered the most important root feeders (Mortimer *et al.* 1999). Based on feeding behaviour the insects can be classified as: (i) internal chewers that burrow into large roots or subterranean storage organs; (ii) external chewers that consume whole roots or graze on the root surface; and (iii) sap feeders that feed on the phloem or xylem through specific mouthparts (stylets) (Brown & Gange 1990). Like the sap-feeding insects, all plant-feeding nematodes have a hollow stylet that is used to suck vascular tissue or cytoplasm. Based on feeding behaviour these nematodes can be classified as roothair or epidermal cell feeders, ectoparasites, semi-endoparasites, migratory endoparasites and sedentary endoparasites (Yeates *et al.* 1993). The first two groups live in the rhizosphere and penetrate the root only with their stylet. The semi-endoparasites penetrate the root also with part of their body, whereas the endoparasites live (part of their life cycle) inside plant roots. Migratory endoparasites move freely within the roots and can even exploit several host plants during their life cycle. On the other hand, sedentary endoparasites affect the physiology of plant root cells thereby inducing the development of specific feeding cells that are used by the nematode to feed on. Besides their feeding relationship with plants some species act as disease vectors that contribute to the distribution of plant pathogenic fungi and viruses; these include insects (Brown & Gange 1991) and nematodes (Mortimer *et al.* 1999). Soil invertebrates can also feed on the mycelium of mycorrhizal fungi, thereby lowering the benefits of the mycorrhizal symbiosis (Gange 2000).

11.3 The soil community as cause and consequence of plant community composition

Correlations between plant community composition and the below-ground soil-dwelling community show mutual impacts. However, it is often difficult to determine to what extent the soil community is a cause or a consequence of plant community composition. Hart *et al.* (2001) proposed a qualitative model to separate both mechanisms in the case of mycorrhizal fungi (Fig. 11.1). If mycorrhizal fungi are causes of plant dynamics (driver hypothesis), the presence of specific mycorrhizal fungi is required for the growth of specific plants. Plant species composition would then be a function of (and in principle predictable from!) the presence of these fungi. If soil organisms are merely passive followers of plant species dynamics (passenger hypothesis), specific plants are required to stimulate the growth of specific mycorrhizal fungi. The passenger hypothesis is phytocentric (plants drive succession), the driver hypothesis is mycocentric (soil organisms drive succession). The driver hypothesis has important implications for restoration management, because addition of mycorrhizal fungi would then enhance plant species richness in natural communities. The same

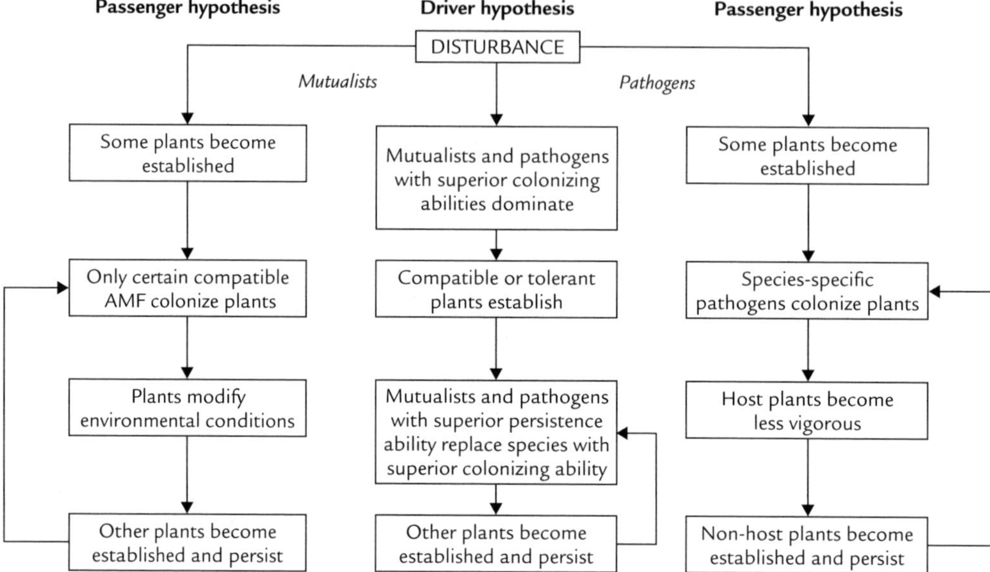

Fig. 11.1 A graphical model of two alternative mechanisms for compositional changes in the mutualistic or antagonistic soil community in interaction with compositional changes of the plant community. In the driver hypothesis changes in the soil community drive vegetation change, whereas in the passenger hypothesis changes in plant community composition result in changes in the soil community. AMF = arbuscular mycorrhizal fungi. Modified after Hart *et al.* (2001).

model is applicable in the case of pathogenic organisms (Fig. 11.1). Verschoor *et al.* (2002) addressed the question to what extent vegetation succession affected the root-parasitic nematode community, and to what extent that community accelerated or decelerated vegetation succession. They concluded that plant species-specific differences in tolerance to generalist root-feeding nematodes, rather than host selectivity of the nematodes, determines plant species replacement and hence plant species composition during succession. This example therefore supports the passenger hypothesis.

11.4 Classification of mutualistic-antagonistic interactions

From an applied ecological perspective it would be obvious to clearly separate between mutualistic and antagonistic soil organisms. However, we should not classify organisms but their behaviours, and the net outcome of interactions between plants and the soil community is context dependent. An interesting (albeit above-ground) example of organisms that cannot be easily classified unambiguously as mutualistic or parasitic are the clavicipitaceous fungi of various grass species. The grass species *Festuca arundinacea* is an introduced plant in North America where it is often infected by *Neotyphodium coenophialum*, which is transmitted only vertically through seed. Endophyte-infected plants have been shown to be less susceptible to insect herbivores and cattle (due to the production of toxins), to be more drought tolerant, to have a higher mineral nutrient uptake, to be less colonized by arbuscular mycorrhizal fungi, and to inhibit soil pathogenic fungi and nematodes compared to uninfected plants. Plots dominated by infected grasses showed lower plant species diversity and increased dominance of *F. arundinacea* (Matthews & Clay 2001). However, while these data suggest the mutualistic nature of the endophyte, experiments by Faeth & Sullivan (2003) showed that *Neotyphodium* endophytes reduced plant performance and reproduction and that the interaction is usually antagonistic rather than mutualistic.

11.5 Specificity and selectivity

11.5.1 Introduction

Interactions between plants and the soil community can also be classified along a continuum from highly specific (private) to non-specific (shared). The gene-for-gene hypothesis, proposed for the co-evolutionary 'arms race' between plants and foliar pathogens, leads to conditions where only certain genotypes of the pathogen can colonize only certain genotypes of a certain plant species. Such highly specific conditions are as yet unknown for soil-dwelling organisms. However, various degrees of ecological selectivity (non-random association between plant and soil organisms under field conditions, or differential effects of soil organisms on different plant species) are common. The argument has been forwarded that parasitic associations show a larger degree of selectivity than mutualistic associations. The argument for this claim is derived from the intuition that symbionts, adapted to rare partners or few species, gain smaller benefits than species that are more promiscuous. However, empirical data (and subsequent theoretical models) strongly suggest that in mutualisms selectivity is also the rule. The issue of specificity has been obscured by the fact that the species level has been considered as the relevant unit. As the number of worldwide species of rhizobia (less than 25) and of arbuscular mycorrhizal fungi (less than 200) is much lower than that of legumes (6000) or arbuscular mycorrhizal plants (presumably 200,000), it is argued that specificity or selectivity must be low. However, genetic variation within both rhizobia and arbuscular mycorrhizal fungi suggests much more scope for selectivity on lower taxonomic levels.

Shared symbioses may or may not equalize competitive abilities among plants, thereby promoting or decreasing floristic diversity (by reducing or enhancing competitive replacement rates), whereas private symbioses may also both increase and decrease differences between plant species. Whether such interactions would increase or decrease species richness seems to depend primarily to the plant's responsiveness to the soil community. If the dominant plants are more responsive to the mutualistic species (or less negatively affected by the antagonistic members) of the soil community, such species will in both cases tend to become more dominant, thereby reducing plant species richness. If the dominant species are less responsive to mutualistic organisms or more affected by antagonistic species than the subordinate plant species, an increase in plant species richness will result.

While selectivity occurs commonly, it is often observed that individual plants harbour more than one rhizobium or mycorrhizal fungus. Simultaneously, individual rhizobia or mycorrhizal fungi can be associated with more than one plant species under field conditions. Multiple colonization may be understood in terms of functional differentiation. Multiple colonization by mycorrhizal fungi would fit with the multifunctional nature of the symbiosis, where next to benefits in terms of increased uptake of mineral nutrients, a suite of other benefits have been reported (Newsham *et al.* 1995). Multiple colonizations by rhizobia, which have not been demonstrated to be multifunctional, are still more difficult to explain. Non-random distribution of mutualistic soil organisms over plants, and multiple colonizations could be a consequence of limited dispersal abilities of plants and soil organisms.

11.5.2 Specificity of rhizobia

Only certain rhizobia are compatible, i.e. have the ability to nodulate and fix atmospheric nitrogen, with specific legumes. In cases of compatibility, association between plant and bacterial genotypes is not a random process. Usually, there is a strong plant genotype × rhizobial strain interaction, which suggests that it is not an intrinsic superiority of rhizobia or plant genotypes that explains its dominance. Next to local adaptation between legume and rhizobia, the interaction is apparently even more complex. Various genotypes of *Trifolium repens* that co-occurred in a meadow were found to be individually associated with (and adapted to) a specific neighbouring grass species (Expert *et al.* 1997). Chanway *et al.* (1989) indicated that the specificity between legume and grass depended on the rhizobia: in the absence of natural *Rhizobium* strains, the legume-grass co-adaptation was lost. Lafay & Burdon (1998) assessed diversity of rhizobia that nodulated on 32 native legume shrubs at 12 sites in southeastern Australia. The occurrence of rhizobia on legume species was non-random, although distribution overlaps were common. True host specificity was not observed, while many legume species were selective. However, part of this selectivity was probably due to the fact that several rhizobia were site-specific. For only three species, *Acacia obliquinervia*, *Goodia lotifolia* and *Phyllota phylicoides*, selectivity was observed, where the dominant rhizobium isolated was different from the most common rhizobium from the site. Specificity or selectivity of the actinorrhizal symbiosis is apparently much lower than that of the rhizobial symbiosis.

11.5.3 Specificity of mycorrhizal fungi

In a microcosm experiment, van der Heijden *et al.* (1998) compared the growth of 11 different plant species when inoculated with four different species of arbuscular mycorrhizal fungi (all belonging to the genus *Glomus*) or with a mixture of these species. They noted that specific fungal species and the species mixture had significantly different effects on plant performance in some plant species but not in others. Plants that were more responsive to mycorrhizal fungi (compared to a non-mycorrhizal control) showed a higher mycorrhizal sensitivity, (variation in effect of the different fungal species). Mycorrhizal sensitivity was largely due to the effects of the fungal species *Glomus geosporum*, which on average was less beneficial than the other fungal species. The non-mycorrhizal *Carex flacca* also showed large effects of mycorrhizal fungus present, also due to the same fungal species. These results suggest that the species composition of arbuscular mycorrhizal fungi can be an important determinant of the composition of the plant community.

Reciprocally, plant species composition can be an important component of the below-ground diversity of arbuscular mycorrhizal fungi, if different fungal species show differences in fitness on different plant hosts. Such differences, as assessed by differential sporulation, have been noted by Bever *et al.* (1996). They noted that two species of *Acaulospora* sporulated most prolifically on *Allium vineale*, while a species of *Glomus* sporulated almost exclusively on *Plantago lanceolata*. Helgason *et al.* (2002) concluded that symbiont compatibility, as expressed by root colonization and plant performance, varies with each fungus–plant combination, even if plants and fungi naturally co-exist. The distribution of mycorrhizal fungi is significantly not random. A clear distinction should be made, however, between selectivity *per se* (non-random distribution of plants and fungi) and functional selectivity (plants associating with a fungal species with which they form the most beneficial combination). In fact, negative feedback between mycorrhizal fungi and plants is possible if the plant does not derive maximum benefit from that mycorrhizal fungus (see section 11.6.2).

11.5.4 Specificity of pathogens and root herbivores

Soil-dwelling pathogens also show a wide variation in selectivity, ranging from host-restricted to species with very wide host ranges. The oomycete *Phytophthora cinnamomi*, a species originally endemic to eastern Australia, has been introduced in many places around the world. Even in its original area, the pathogen can cause rapid vegetation changes; within a period of 5 yr a closed *Eucalyptus* woodland with a dense understorey was transformed into an open woodland with an understorey dominant by the pathogen-resistant sedge *Lepidosperma concavum* (Weste 1981).

Specialized pathogens could have a major impact on vegetation, especially in the case of introduced pathogens. The above-ground pathogenic fungus *Endothia parasitica* almost eradicated its host *Castanea dentata* in North America from many stands (chestnut blight). As a consequence various other trees, such as *Quercus prinus*, *Acer rubrum* and *Liriodendron tulipifera*, increased substantially in importance (Day & Monk 1974). Native, specialized pathogens could also have large impacts on vegetation composition and successional trajectory as in the case of the pathogenic fungus *Phellinus weirii*, a root pathogen exclusive to *Pseudotsuga menziesii*, which creates infection centres that slowly spread outwards and thereby creates gaps by which early successional tree species can survive (Holah *et al.* 1997).

Specific root-herbivore species can be used to control exotic plant species (see Chapter 13). *Lythrum salicaria*, an invasive wetland perennial that was introduced from Europe to North America, can be successfully controlled by the European weevil *Hylobius transversovittatus*. Young larvae feed on the root cortex and older larvae mine the rootstock thereby affecting plant performance. In the absence of root herbivores *L. salicaria* is a superior competitor towards other wetland plant species (possibly because of a trade-off between growth rate and allocation to chemical defence, or because of the absence of species-specific pathogenic rhizosphere fungi, see section 11.7). Under controlled conditions root herbivory in the plant establishment phase affected interspecific competition with the grass *Poa pratensis* resulting in reduced plant height, shoot weight and total dry mass. Root herbivory also negatively affected established plants, although interspecific competition with *P. pratensis* only lead to delayed flowering (Nötzold *et al.* 1998).

Specific root herbivore–plant interactions are not restricted to exotic plants. An example from natural vegetation in California is the caterpillar of the ghost moth *Hepialus californicus* that feeds on roots of *Lupinus arboreus*. These caterpillars are largely monophagous and feed, when young, on the exterior of lupine roots thereafter boring inside the roots. Field surveys showed a strong positive correlation between plant death rates and caterpillar densities inside roots, providing strong evidence that the caterpillar thereby affected vegetation composition and dynamics (Strong 1999).

11.6 Feedback mechanisms

11.6.1 Two mechanisms

Two mechanisms have been proposed by which soil organisms and plants interact (Bever *et al.* 1997):

1 through negative feedbacks; and

2 through positive feedbacks.

Both mechanisms are illustrated in Fig. 11.2.

11.6.2 Negative feedbacks

Negative feedbacks occur under conditions where plants create a rhizosphere (the soil biota in the immediate vicinity of their own roots) that is less beneficial or more detrimental to them than to neighbouring plants of different species. Negative-feedback mechanisms can be easily envisaged in the case of root pathogens. They have been investigated by transplanting plants in soil communities derived from the same ('home') or different ('foreign') plant species. Negative feedbacks in which several below-ground herbivore and/or pathogenic microbe species (pathogen complex) are involved have been described for early successional plant communities such as present in the primary succession of coastal dunes (van der Putten *et al.* 1993). In such vegetation a species-specific pathogen complex will develop in the rhizosphere of early successional plant species which will result in a reduced growth and vigour of these plant species. Later successional species, on the other hand, are not or less affected by this pathogen complex and will replace the early successional species. Subsequently, also in the rhizosphere of the latter plant species such species-specific pathogen complexes will develop which will in their turn decrease the vigour of this plant species thereby favouring the establishment of later successional plant species. Indeed, reciprocal transplant pot experiments with the pioneer grass *Ammophila arenaria* and the later successional graminoids *Festuca rubra* ssp. *arenaria*, *Carex arenaria* and *Elymus*

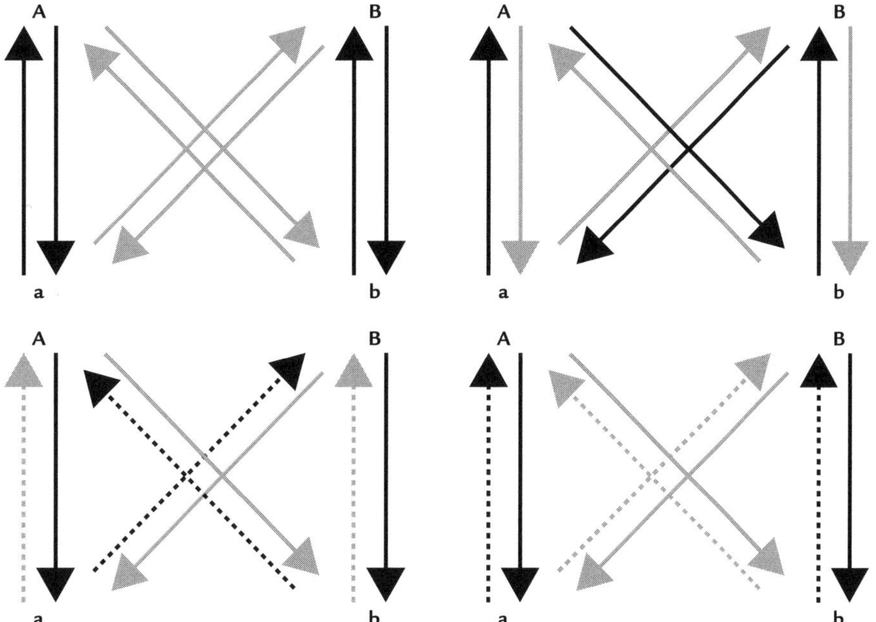

Fig. 11.2 Mechanisms of positive and negative feedbacks between plants (capital A and B) and the soil community (lower-case a and b). Upper left: positive feedback between plants and mutualists; Upper right: negative feedback between plants and mutualists; Lower left: negative feedback between plants and soil pathogens; Lower right: positive feedback (indirect negative feedback) between plants and soil pathogens. Arrows indicate the direction of effects; continuous lines indicate beneficial effects, dashed lines indicate harmful effects. Black lines indicate stronger positive or negative effect than grey lines. Positive feedback can occur if either benefits for plants and mutualists are symmetrical or if damage to plants and benefits for pathogens are asymmetrical. Negative feedback can occur if damage to plants and benefits for pathogens are symmetrical or if benefits for both plants and mutualists are asymmetrical. Modified after Bever (2003).

athericus, showed that each of these species produced more biomass in pre-successional soil than in its own soil, whereas biomass productions in sterilized soils were not significantly different. This suggests that the root zones of these sequential plant species contain pathogens that are specific for their host and pre-successional plant species, whereas effects on the next species in the successional sequence are much less. To survive as a species when being part of and controlled by such biotic relationships, plants need to colonize new sites that are free from their own pathogen complex.

The clonal species *A. arenaria* plays an important role in early soil development in coastal foredunes by the fixation of windblown beach sand. Every year *A. arenaria* extends its root system into the new layer of fresh windblown sand which may be a mechanism to escape from its rhizosphere pathogen complex. The pathogen complex in this primary succession of coastal dunes is thought to comprise mainly plant feeding nematodes and pathogenic fungi. However, so far, laboratory and field experiments in which field densities of ecto- and endoparasitic nematodes were added to sterilized soil, did not result in similar large growth reductions as found for unsterilized soils, suggesting that the organisms involved and their interactions have not yet been completely identified (see section 11.12). Soil invertebrate fauna, especially root-feeding nematodes and larvae of click-beetles, selectively suppressed early successional, dominant plant species and enhanced both late-successional and subordinate early-successional species, thereby increasing plant species diversity and accelerating plant succession (de Deyn *et al.* 2003).

Negative feedbacks between plant and soil communities have been proposed as a potential mechanism to maintain tree species richness in forests, especially in the tropics, the so-called Janzen–Connell hypothesis. Evidence for that mechanism in temperate forests was found by Packer & Clay (2000). In North America, seedlings of *Prunus serotina* had significantly worse survival close to parent trees. The causal agent was identified as a species of the genus *Pythium*. However, in its non-native range in Europe, *P. serotina* readily establishes close to conspecific adults trees and the soil community even enhances seedling growth (Reinhart *et al.* 2003).

There is evidence that negative feedbacks also operate in the case of mutualists. Bever (2002) noted that the arbuscular mycorrhizal fungal species *Scutellospora calospora* grows best together with *Plantago lanceolata* but does not strongly promote the growth of the host plant. In fact, growth of *P. lanceolata* is best promoted by the fungal species *Acaulospora morrowiae* and *Archaeospora trappei* which accumulate with the grass *Panicum sphaerocarpon*. This interaction results in a decline in benefit received by *P. lanceolata*, allowing *P. sphaerocarpon* to increase. An increase in the grass species has an indirect beneficial effect on the growth of *P. lanceolata*, thereby contributing to the co-existence of both competing species. An agronomic example where continuous cropping caused an increased abundance of certain species of arbuscular mycorrhizal fungi leading to decreased yield performance of the crops is consistent with negative feedback within a mutualistic symbiosis (Johnson *et al.* 1992). However, in that case negative feedback through host-specific pathogens cannot be excluded.

11.6.3 Positive feedbacks

Positive feedbacks are linked to conditions where plants create their own rhizosphere environment whereby they locally outperform other species. This occurs if plants are associated with rhizobia or mycorrhizal fungi from which they derive the largest benefits. Direct positive feedbacks with pathogenic soil organisms do not seem to occur. However, indirect positive feedbacks with antagonistic rhizosphere communities have been described by Olff *et al.* (2000); see Fig. 11.2.

How likely is the evidence for positive and negative feedbacks in mutualisms? Bever (1999) tested the occurrence of both feedbacks and, under the assumption that sporulation is a good predictor of arbuscular mycorrhizal fungal fitness, found weak evidence for positive feedback to prevail. Evidence for a negative feedback is also common, however. The question remains to what extent negative and positive feedbacks explain plant species diversity. Positive feedbacks could be sufficient to explain local competitive ability. While this local success could well decrease species richness on small spatial scales, it could on larger scales, for instance the size of a field, still contribute to enhanced plant species diversity if there is dispersal limitation of the below-ground mutualist (Bever *et al.* 2001). Negative feedbacks could allow co-existence of two plant species, but not of species-rich plant communities unless: (i) the soil community exerts effects on plants that previously colonized that site or effects remain notable over various generations of plants; and (ii) individual members are part of different soil communities that exert different feedbacks on other plants. These conditions seem to be met in the primary succession on coastal dunes, but are unlikely valid for rain forests. It is therefore not surprising that feedbacks have been more convincingly demonstrated in relatively species-poor pioneer communities, but not in species-rich, more mature communities. Interestingly, a model proposed by Golubski (2002) indicated that the conditions for negative feedback increase if plants are associated with more than one species of mycorrhizal fungus, and species of mycorrhizal fungi colonize more than one plant species. The model has not yet been tested experimentally, however.

11.7 Soil community feedbacks and the success of invasive plants

The existence of soil community feedbacks could also be an important factor to determine the success of exotic plants (see Chapter 13). Both pathogens and mutualists are involved. In the case of pathogens, the hypothesis has been proposed that exotic plants have been introduced without a complete soil pathogenic community. Consequently, the general negative feedback between two plant species could be broken if the exotic plant outperforms a native plant both in its 'own' soil (due to lack of pathogens as a consequence of low dispersal of the soil community) and in the soil of the native plant (due to the build up of a pathogenic soil community of the indigenous plant). Evidence for this mechanism, where a larger role is assigned to pathogenic than to mutualistic interactions, was presented by Klironomos (2002). Plant species that are considered rare and endangered in Canadian old-fields (meadows) and grassland, were grown from seed in pots that were filled with soil that was collected from a field site that did not contain any of these species. After 20 weeks of plant growth the soil in the pots was defined as 'home' soil for the plant species that was cultured in that soil, or 'foreign' soil for all other plant species. For all rare species tested (i.e. *Agalinis gattingeri*, *Aletris farinosa*, *Gentiana alba*, *Liatris spicata* and *Polygala incarnata*) it was found that their growth in home soil was reduced as compared to their growth in foreign soil (Fig. 11.3). In this experiment strong evidence was found that the growth reduction was at least related to the occurrence of species-specific plant pathogenic fungi that developed in the rhizosphere of the plant species. When grown on sterilized soil that was inoculated with a selection of the most commonly isolated non-mycorrhizal fungi from the rhizosphere of each plant species, the growth of all five rare plant species was significantly reduced when they were grown on soil that was inoculated with fungi isolated from their own rhizosphere and not (with only one exception) when grown on soil with fungi from the rhizosphere of other plants.

Interestingly, the response of these rare species contrasted dramatically with the response of invasive, exotic plant species that were introduced to North America from Eurasia: *Alliaria petiolata*, *Cirsium arvense*, *Euphorbia esula*, *Lythrum salicaria* and

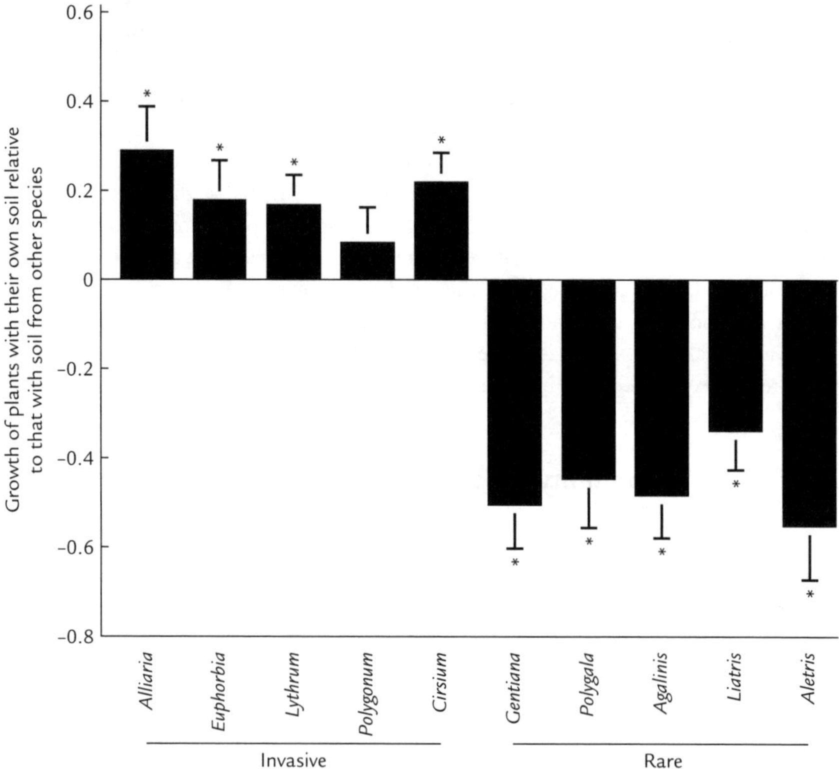

Fig. 11.3 Impacts of arbuscular mycorrhizal fungi and the pathogenic soil community (mainly pathogenic fungi) on the performance of indigenous, rare species and introduced exotics. After Klironomos (2002).

Polygonum cuspidatum. Most of these invasive plant species had better growth on home soil relative to foreign soil. Furthermore, none of the isolated non-mycorrhizal fungi had a significant (neither negative or positive) effect on the growth of the invasive species. Both the rare species and several invasive species (but *A. petiolata* and *P. cuspidatum* are non-mycorrhizal) benefited from interactions with (arbuscular) mycorrhizal fungi. Marler *et al.* (1999) noted that the presence of mycorrhizal fungi enhanced competitiveness of the exotic *Centaurea maculosa* with regard to the native flora. A mycorrhizal role in determining success of exotic plants could both depend on carbon transfer through common mycelial networks (section 11.10) and on increased access to nutrients (Zabinski *et al.* 2002).

For nitrogen-fixing symbioses the possibility that exotic plants could invade existing communities has also been suggested. A good example of this is the success of the exotic plant *Myrica faya* in Hawai'i. Following its introduction soil characteristics changed and indigenous plants were unable to compete with exotics under conditions of higher nitrogen availability. Interestingly, *Myrica faya* also profits from enhanced seed dispersal by introduced birds (Vitousek & Walker 1989).

11.8 The importance of predictability in space and time

This chapter concentrates on below-ground interactions. Above ground, both positive and negative interactions occur. Several fungal endophytes exhibit mutualistic behaviour, while fungal pathogens are also widespread above ground. An important difference between above-ground and below-ground mutualists and antagonists of plants is their mode of dispersal. In most cases of above-ground pathogens, the pathogen is easily dispersed, more easily than the plant. Dispersal through soil as a solid medium is much slower and less effective compared to dispersal by plants. This differential role of dispersal creates a different metapopulation structure, allows for different spatial effects on ecological and evolutionary time scales, ultimately leading to different degrees of specificity of mutualists and evolution of disease-resistance systems. Consistent with this hypothesis is the observation that lack of specificity is more common among annual legumes (and specificity is higher among tree legumes). Equally for arbuscular mycorrhizal fungi the data available suggest that preferential associations occur more commonly with trees than with herbs. Clonal plants could also provide a more predictable habitat for soil-dwelling organisms, and it is therefore not surprising that the most convincing examples of feedbacks have been reported from clonal plants.

11.9 Mycorrhizal symbioses and nutrient partitioning in plant communities

Cornelissen *et al.* (2001) suggested that classifying plant species according to their mycorrhizal association could add a new dimension to classifications of plant functional types. Mycorrhizal functional types could link mycorrhizal type, plant functional type, litter quality, decomposition and mineralization characters, and the relative role of various nutrient sources (organic versus inorganic forms; nitrogen versus phosphorus). Such a model is presented in Fig. 11.4. In general, ectomycorrhizal and ericoid mycorrhizal plants grow in soils and on humus profiles that are rich in organic matter and where decomposition and mineralization are hampered. The enzymatic ability of ectomycorrhizal and ericoid mycorrhizal fungi with regard to organic sources of nitrogen and phosphorus indicates that many of these fungi have the ability to access organic nutrients. On the other hand, arbuscular mycorrhizal fungi and plants are considered to be unable to access organic nitrogen and phosphorus (Read 1991). This difference therefore suggested a mechanism by which different mycorrhizal types could partition resources, and depending on the rate of mineralization through saprotrophic activity – ectomycorrhizal fungi have been hypothesized to compete with saprotrophic fungi and slow down litter decomposition and nutrient mineralization – one or the other group could gain competitive dominance. Such replacements occur with increased atmospheric nitrogen deposition where dominant heathland plants (*Calluna vulgaris*, *Erica tetralix*) with ericoid mycorrhizae are outcompeted by grasses (*Deschampsia flexuosa*, *Molinia caerulea*) with arbuscular mycorrhizae. A similar mechanism could explain successional processes in several tropical forests where species-rich rain forest, composed of arbuscular mycorrhizal trees, is replaced by often

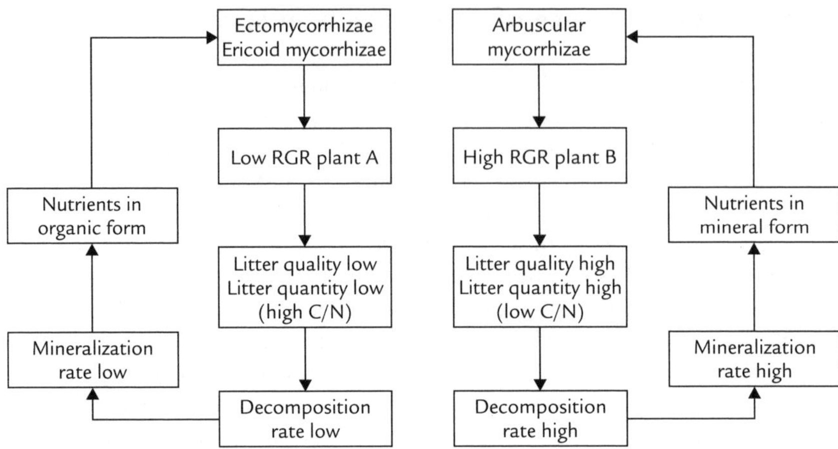

Fig. 11.4 Positive feedbacks between mycorrhizal type, plant functional type, quality of litter produced, and the resulting decomposition and mineralization rate.

species-poor stands composed of ectomycorrhizal *Dipterocarpaceae* or *Caesalpiniaceae* (Connell & Lowman 1989). Although the processes of plant replacement are consistent with the model, one may question whether the mycorrhizal element forms a necessary part of the explanation. Would the explanatory value of the model be different if litter decomposability and the associated trade off between maximizing growth rate and minimizing nutrient loss were the driving force, and mycorrhizal association being only correlated with plant functional type with regard to litter decomposability (Berendse 1994)?

Cornelissen *et al.* (2001) observed, based on data taken from the literature, that arbuscular mycorrhizal plant species had higher relative growth rates (RGR) than ectomycorrhizal and ericoid mycorrhizal plants, and produced litter of higher decomposability. The authors also noted variation within mycorrhizal categories related to life-form characteristics (herbaceous plants show higher RGR than woody plants; litter of graminoids has a lower decomposability than that of dicotyledonous plants), which suggests that the mycorrhizal type may not be the causal factor that determines the variation in carbon-cycling traits. Leaf habits (evergreen versus deciduous) and phylogenetic inertia also affected the patterns (especially of RGR) observed to some extent.

Field studies initially provided support for a mycorrhizal role in differential nitrogen access that drives plant species replacement. Assessment of ^{15}N signatures, widely held to be an indicator for the source of nitrogen during uptake, suggested that plants with different mycorrhizal types had significantly different ^{15}N signals in agreement with the predictions of the theory (Hobbie *et al.* 1999). However, as it became subsequently clear that other factors could equally well explain the different signatures, the use of natural-abundance ^{15}N as an indicator of the nitrogen nutrition of plants is now approached with less certitude. The assumption underlying the use of ^{15}N is that at its natural abundance it acts as a tracer, hence that the isotopic signal from the nitrogen source is preserved in the ^{15}N of the plant. However, plant ^{15}N is not a tracer of the nitrogen source – it integrates isotopic signatures of the source, fractionation events through mycorrhizal associations and internal allocation and recycling of N in the plant (Robinson 2001). The use of organic materials enriched with ^{15}N and ^{13}C has shown that the ability to access organic nitrogen from amino acids is more widespread among non-mycorrhizal and arbuscular mycorrhizal plants than hitherto proposed and not different from that of ectomycorrhizal and ericoid mycorrhizal plants (Persson & Näsholm 2001).

11.10 Mycorrhizal networks counteracting plant competition?

The supposed lack of specificity of mycorrhizal fungi, based on the observation that in laboratory experiments most plants × fungus combinations are functional would make it possible that mycorrhizal plants of different species could be connected below ground through a common mycorrhizal network. The existence of such links has been demonstrated repeatedly and it is clear that carbon and nutrients could move through these networks. However, the issues are still unsettled whether or not carbon and nutrients are exchanged between plants in quantities that are ecologically meaningful, and whether these networks are important for the functioning of plant communities (Robinson & Fitter 1999). The existence of networks, through which carbon is shared among plants, implies that the direction of carbon flow is reversed, i.e. not from plant to fungus but from fungus to plant. Reversed carbon flows are known in the so-called mycoheterotrophic plants. Such plants, achlorophyllous members of the *Monotropoideae* (a subfamily of the *Ericaceae*) are parasitic on both an ectomycorrhizal fungus and an ectomycorrhizal tree at the same time. As there is no direct contact between the monotropoid mycorrhizal plant and the ectomycorrhizal tree, tree defence against cheating (benefiting from the mycorrhizal network without paying carbon costs) is very difficult, as the tree would simultaneously lose its ectomycorrhizal partner.

While there could be negative selection against the parasitized ectomycorrhizal fungus, the 'cheater' apparently manipulates plant and fungus by allowing local ectomycorrhizal root proliferation through which the fungus increases in abundance or biomass (Bidartondo *et al.* 2000).

Reversed carbon flow also occurs with the moss *Cryptothallus mirabilis*, several achlorophyllous and chlorophyllous ectomycorrhizal orchids such as *Neottia*, *Epipactis* and *Cephalanthera* (Gebauer & Meyer 2003) and achlorophyllous arbuscular mycorrhizal *Gentianaceae*. It is not clear whether conditions of reversed carbon flow are a rare exception, or whether the phenomenon could potentially occur under a wider range of conditions. However, the number of cheating plants reported to date is very low. Could it be possible that many plants, for instance many subordinate plants in plant communities, also derive carbon from the dominant neighbouring plants through mycorrhizal fungi, as has been suggested for the plant genus *Gentiana*? Note that this mechanism is different from the mechanism whereby root exudates are transferred between plant species as a consequence of below-ground herbivory (see section 11.11).

Mycorrhizal networks, within which seedlings are integrated, could be essential for the establishment and survival of those plant species which cannot provide sufficient carbon in the seedling stage to maintain the symbiosis. Such plants would be able to establish only in closed vegetation when they could parasitize on the network. This mechanism is a possible example of facilitation sensu Connell & Slatyer (1977), where early occupants modify the biotic environment in a way that makes it more suitable for species with a higher mycorrhizal dependence.

The situation where plants connected in a common network not only benefit in different ways (as a consequence of the multifunctional nature of the mycorrhizal symbiosis), but also pay unequally, could be more widespread. Unequal payment would benefit certain plants if the fungal behaviour to the plant is regulated by the plant demands. It is more likely, however, that fungal behaviour with regard to carbon is primarily regulated by the fungal demands (Fitter 2001). For arbuscular mycorrhizal fungi the evidence to date suggests that carbon transferred between plants remains largely in the fungal mycelium (Robinson & Fitter 1999), and that plants cannot therefore be treated as receivers of carbon donated by neighbouring conspecific or non-conspecific plants. For ectomycorrhizal fungi the evidence is less clear, as data show that the carbon fixed by the donor plant ended up in the foliage of the receiver plant. The opposite situation, where some plant species show a decreased performance in the presence of arbuscular mycorrhizae, is therefore more likely. This has been shown by Francis & Read (1995) for plant species that as a consequence of that mechanism are excluded from closed vegetation dominated by mycorrhizal plants, because the mycorrhizal fungus behaves parasitically on such donor plants in these conditions. They noted that the receiver plant *Plantago lanceolata* had a higher growth rate and higher survivorship in the presence of arbuscular mycorrhizal fungi, whereas the donor plants *Arabis hirsuta*, *Arenaria serpyllifolia*, *Echium vulgare*, *Reseda luteola* and *Verbascum thapsus* showed the opposite pattern, even though they were all colonized by these mycorrhizal fungi.

The existence of mycorrhizal networks also affects plant competition in another way. If different plant species are connected to the same network, the below-ground differences between plants to take up nutrients apparently diminish. If plants are ectomycorrhizal (i.e. all nutrients must be taken up by and pass through the fungus), the condition would imply that the differences between interspecific and intraspecific competition below-ground disappear. As a consequence the balance between competition below ground and above ground could change (Perry *et al.* 1989).

Consequences of this change in balance between both competition modes would depend on the responsiveness of various plants to mycorrhizas. In experimental systems both cases of increase and decrease of diversity have been reported. When the subordinate species are more responsive to mycorrhizae, the presence of mycorrhizae increases diversity, but when the dominant species are more responsive to mycorrhizae, the suppression of mycorrhizae increased diversity (Urcelay & Díaz 2003). Mycorrhizal fungi, due to their better exploration of heterogeneous nutrient resources in the soil, could smooth out the patchy distribution of nutrients – thereby evening out below-ground plant competition. Under such conditions the presence of mycorrhizas could decrease the importance of spatial heterogeneity, and hence have a negative effect on plant species richness.

Mycorrhizal networks could also provide below-ground continuity in cases where the above-ground vegetation shows changes in species composition. This could happen in heathlands where fire can destroy the vegetation, but where certain liverworts, associated with the same ericoid mycorrhizal fungi, maintain below-ground mycorrhizal continuity (Duckett & Read 1995). In a similar way, ectomycorrhizal continuity could be provided for the tree *Pseudotsuga menziesii* in the fire-prone Californian chaparral in the presence of the arbutoid mycorrhizal shrub *Arctostaphylos glandulosa* (Horton *et al.* 1999) or in ectomycorrhizal forests with an understorey of heathland plants with ericoid mycorrhiza (Vrålstad *et al.* 2002).

11.11 Pathogenic soil organisms and nutrient dynamics

Below-ground herbivores can contribute to soil nutrient dynamics which in turn affect nutrient availability and plant productivity of the host, but probably also for companion plant species. Low amounts of root herbivory by, for example, plant-feeding nematodes can result in the leakage of nutrients from the damaged roots. This leads to an increased supply of carbon and nutrients for microbial metabolism in the rhizosphere, affecting nutrient mineralization. A $^{14}CO_2$ pulse-labelling experiment with *Trifolium repens* that was colonized by field densities of clover cyst nematodes (*Heterodera trifolii*) showed a significant increase in the leakage of C from the roots and an increased microbial biomass (Bardgett *et al.* 1999a). Such observations were not restricted to endoparasites like *H. trifolii* but were found also for semi-endoparasites, ectoparasites and epidermal and root-hair cell feeding nematodes. Small migratory ectoparasites, such as species that feed on epidermal and root-hair cells, seemed to increase root exudation relatively more than the other plant feeders. This difference may be explained by the ephemeral feeding behaviour leading to a relatively large number of minor damages in the roots (Bardgett *et al.* 1999b).

At present, the size of this increased root exudation and its stimulating effect on nutrient mineralization relative to the total nutrient uptake by plants is unknown and still needs to be quantified. In a pot experiment where *T. repens* was grown together with *Lolium perenne*, the addition of clover cyst nematodes in field densities below the damage threshold for white clover resulted in a significant increase in root biomass of the host (141%) as well as of the neighbouring *L. perenne* (217%). Furthermore, nitrogen uptake, derived from the clover plants, by *L. perenne* was 3.2 times higher in pots with nematodes than in pots were they were absent. Bardgett *et al.* (1999b) suggested that such increases in nutrient uptake and root growth of *L. perenne* may alter the competitive balance between the two species, most likely to the detriment of the nematode-infested white clover. This may influence plant community structure. Whether this will be true for natural field situations remains to be investigated.

The impact of pathogenic soil organisms could depend on their effect on various plant species. Generalist pathogens that have a broad host choice could increase plant species diversity if they cause more damage on the dominant species, but could also decrease plant species diversity if they cause more damage on the subordinate plants. The importance of this mechanism has been suggested by Olff *et al.* (2000), who described how generalist nematode species thrives under *Festuca rubra* but exerts a larger damage to *Carex arenaria*.

11.12 Community-wide approaches to study the impact of the soil community on vegetation

Two approaches have been applied in investigating the causes and consequences of the above-described feedback processes, namely a whole-community approach (synthesis; black box approach) and an experimental approach (dissection or reconstruction; opening the black box) based on the putatively most important organisms. Basically the first approach involves the use of rhizosphere soil, in which different plants are grown. While such experiments have provided evidence for the operation of biotic feedbacks, they do not indicate the nature of the responsible organisms. In fact, plant responses do not even indicate to what extent pathogens and mutualists are responsible for the negative feedback. The study by Klironomos (2002) – where both mycorrhizal and pathogenic fungi (but not pathogenic nematodes) were included – remains an exception in this regard.

Species-directed approaches have also been regularly applied. In some cases none of the pathogens was by itself powerful enough to provide the full explanation. This was observed in a study of the effect of the soil community on the performance of *Ammophila arenaria* in coastal foredunes. While a comparison of healthy and declining stands of *A. arenaria* showed that the abundance of the nematode *Telotylenchus ventralis* provided the best explanation, experimental tests suggested that the nematode alone, unless when occurring in unrealistically high abundance, was insufficient to cause the decline of the plant. It was also observed that no single fungal species caused a growth reduction of *A. arenaria*, but the combination of all fungi caused a substantial growth reduction in sterile soil, comparable to the growth reduction observed in non-sterile soil. Apparently, other biotic agents, synergistically interacting with the nematode and fungi, are involved too (de Rooij-van der Goes 1995).

Such a multitude of as-yet-unidentified members of the soil community that could equally contribute to patterns of species interactions observed, raises the methodological question about the sufficiency of experimental demonstration of the causal agents of the feedback. To what extent do we need to experimentally replicate the field situation with the addition of various combinations of species groups in order to arrive at a sufficient grasp of the importance of the various mechanisms involved? Such a dissective approach will rapidly run the risk of getting too complicated because of higher-order interactions. Westover & Bever (2001) provided evidence that pathogens can generate the negative feedback between *Anthoxanthum odoratum* and *Panicum virgatum*. Putatively responsible soil organisms for the negative feedback were: (i) accumulation of *Pythium*, which accumulated in *Panicum* soils; and (ii) differentiation of host-selective or host-specific populations of *Bacillus*. The authors also admitted that they most likely had not found all agents that are responsible for the negative feedback between both plant species. They suggested that at least four other components of the soil community could contribute to the negative feedback: (i) rhizosphere bacterial community; (ii) *Fusarium* species; (iii) root-feeding nematodes; and (iv) arbuscular mycorrhizal fungal communities. Not only those components but also their interactions would need to be included in a realistic assessment of the causal agents responsible for the feedback. Westover & Bever (2001) suggested that at least the following interactions would then also need careful attention:

1 Synergistic effects between *Fusarium* and nematodes. Root feeding by nematodes can both directly decrease plant growth, but can also increase the susceptibility of plants to pathogenic fungi and bacteria. In general root-feeding herbivores may be the key species that enhance the sensitivity of plants to other soil organisms. Due to their feeding activity herbivores damage root cells or tissue, which may increase the sensitivity of plants to pathogenic soil fungi, bacteria and viruses.

2 Direct antagonistic effects between arbuscular mycorrhizal fungi and pathogens. Plants such as grasses are characterized by long and thin roots. Consequently, the importance of arbuscular mycorrhiza for phosphorus uptake is less important than for plants with short, thicker roots with few root hairs. An extensive root system is also more susceptible to root pathogens. For the grass *Vulpia ciliata* Newsham *et al.* (1995) demonstrated that arbuscular mycorrhiza had no effect on nutrient uptake, but could effectively protect the grass against the root pathogenic fungus *Fusarium*.

3 Effect of arbuscular mycorrhizal fungi on root exudation and the composition of the rhizosphere community.

4 Interactions between *Bacillus* and *Pythium*.

Finally, soil-dwelling organisms do not act in isolation. Although the task to study the impact of groups of soil organisms and the interactions between the different groups on vegetation processes and plant communities is already Herculean, the picture that could ultimately emerge from such studies is still inevitably partial. The next step therefore should include the interactions between biotrophic (both mutualistic and antagonistic) and saprotrophic soil organisms, as many fungivorous and predatory soil animals prey on both biotrophic and saprotrophic soil organisms. Furthermore, interactions and feedbacks between below-ground herbivores and pathogens and above-ground herbivore grazers and browsers (see Chapter 10) could further complicate the final outcome of these interactions on the plant community. While above-ground grazing by vertebrate herbivores improves nutritional quality of roots and thereby enhances densities of root-feeding insects, above-ground grazing by herbivorous insects generally results in a decreased performance of the below-ground community.

References

Bardgett, R.D., Denton, C.S. & Cook, R. (1999a) Below-ground herbivory promotes soil nutrient transfer and root growth in grassland. *Ecology Letters* **2**, 357–360.

Bardgett, R.D., Cook, R., Yeates, G.W. & Denton, C.S. (1999b) The influence of nematodes on below-ground processes in grassland ecosystems. *Plant and Soil* **212**, 23–33.

Berendse, F. (1994) Litter decomposability – a neglected component of plant fitness. *Journal of Ecology* **82**, 187–190.

Bever, J.D. (1999) Dynamics within mutualism and the maintenance of diversity: inference from a model of interguild frequency dependence. *Ecology Letters* **2**, 52–61.

Bever, J.D. (2002) Negative feedback within a mutualism: host-specific growth of mycorrhizal fungi reduces plant benefit. *Proceedings of the Royal Society of London* **269**, 2595–2601.

Bever, J.D. (2003) Soil community feedback and the coexistence of competitors: conceptual frameworks and empirical tests. *New Phytologist* **157**, 465–473.

Bever, J.D., Morton, J.B., Antonovics, J. & Schultz, P.A. (1996) Host-dependent sporulation and species diversity of arbuscular myorrhizal fungi in a mown grassland. *Journal of Ecology* **84**, 71–82.

Bever, J.D., Westover, K.M. & Antonovics, J. (1997) Incorporating the soil community into plant population dynamics: the utility of the feedback approach. *Journal of Ecology* **85**, 561–573.

Bever, J.D., Schultz, P.A., Pringle, A. & Morton, J.B. (2001) Arbuscular mycorrhizal fungi: more diverse than meets the eye, and the ecological tale of why. *BioScience* **51**, 923–931.

Bidartondo, M.L., Kretzer, A.M., Pine, E.M. & Bruns, T.D. (2000) High root concentration and uneven ectomycorrhizal diversity near *Sarcodes sanguinea* (Ericaceae): a cheater that stimulates its victims? *American Journal of Botany* **87**, 1783–1788.

Brown, V.K. & Gange, A.C. (1990) Insect herbivory below ground. *Advances in Ecological Research* **30**, 1–58.

Brown, V.K. & Gange, A.C. (1991) Effects of root herbivory on vegetation dynamics. In: *Plant root growth. An ecological perspective* (ed. D. Atkinson), pp. 453–470. Blackwell Science, Oxford.

Chanway, C.P., Holl, F.B. & Turkington, R. (1989) Effect of *Rhizobium leguminosarum* biovar *trifolii* genotype on specificity between *Trifolium repens* and *Lolium perenne*. *Journal of Ecology* **77**, 1150–1160.

Clements, R.O., Bentley, B.R. & Nuttall, R.M. (1987) The invertebrate population and response to pesticide treatment of two permanent and two temporary pastures. *Annals of Applied Biology* **111**, 399–407.

Connell, J.H. & Lowman, M.D. (1989) Low-diversity tropical rain forests: some possible mechanisms for their existence. *American Naturalist* **134**, 88–119.

Connell, J.H. & Slatyer, R.O. (1977) Mechanisms of succession in natural communities and their role in community stability and organization. *American Naturalist* **111**, 1119–1144.

Cornelissen, J.H.C., Aerts, R., Cerabolini, B., Werger, M.J.A. & van der Heijden, M.G.A. (2001) Carbon cycling traits of plant species are linked with mycorrhizal strategy. *Oecologia* **129**, 611–619.

Crawley, M.J. (ed.) (1997) *Plant Ecology*. 2nd ed. Blackwell Science, Oxford.

Day, F.P. & Monk, C.D. (1974) Vegetation patterns on a southern Appalachian watershed. *Ecology* **55**, 1064–1074.

de Deyn, G.B., Raaijmakers, C.E., Zomer, H.R., Berg, M.P., de Ruiter, P.C., Verhoef, H.A., Bezemer, T.M. & van der Putten, W.H. (2003) Soil invertebrate fauna enhances grassland succession and diversity. *Nature* **422**, 711–713.

de Rooij-van der Goes, P.C.E.M. (1995) The role of plant-parasitic nematodes and soil-borne fungi in the decline of *Ammophila arenaria* (L.) Link. *New Phytologist* **129**, 661–669.

Duckett, J.G. & Read, D.J. (1995) Ericoid mycorrhizas and rhizoid-ascomycete associations in liverworts share the same mycobiont: isolation of the partners and resynthesis of the associations *in vitro*. *New Phytologist* **129**, 439–447.

Expert, J.M., Jacquard, P., Obaton, M. & Lüscher, A. (1997) Neighbourhood effect of genotypes of *Rhizobium leguminosarum* biovar. *trifolii*, *Trifolium repens* and *Lolium perenne*. *Theoretical and Applied Genetics* **94**, 486–492.

Faeth, S.H. & Sullivan, T.J. (2003) Mutualistic asexual endophytes in a native grass are usually parasitic. *American Naturalist* **161**, 310–325.

Fitter, A.H. (2001) Specificity, links and networks in the control of diversity in plant and microbial communities. In: *Ecology: Achievement and challenge* (eds M.C. Press, N.J. Huntly & S. Levin), pp. 95–114. Blackwell Science, Oxford.

Francis, R. & Read, D.J. (1995) Mutualism and antagonism in the mycorrhizal symbiosis, with special reference to impacts on plant community structure. *Canadian Journal of Botany* **73**, S1301–S1309.

Gange, A. (2000) Arbuscular mycorrhizal fungi, Collembola and plant growth. *Trends in Ecology and Evolution* **15**, 369–372.

Gebauer, G. & Meyer, M. (2003) ^{15}N and ^{13}C natural abundance of autotrophic and myco-heterotrophic orchids provides insight into nitrogen and carbon gain from fungal association. *New Phytologist* **160**, 209–223.

Golubski, A.J. (2002) Potential impacts of multiple partners on mycorrhizal community dynamics. *Theoretical Population Biology* **62**, 47–62.

Hart, M.M., Reader, R.J. & Klironomos, J.N. (2001) Life-history strategies of arbuscular mycorrhizal fungi in relation to their successional dynamics. *Mycologia* **93**, 1186–1194.

Helgason, T., Merryweather, J.W., Denison, J., Wilson, P., Young, J.P.W. & Fitter, A.H. (2002) Selectivity and functional diversity in arbuscular mycorrhizas of co-occurring fungi and plants from a temperate deciduous woodland. *Journal of Ecology* **90**, 371–384.

Hobbie, E.A., Macko, S.A. & Shugart, H.H. (1998) Patterns in N dynamics and N isotopes during primary succession in Glacier Bay, Alaska. *Chemical Geology* **152**, 3–11.

Hobbie, E.A., Macko, S.A. & Shugart, H.H. (1999) Correlations between foliar ^{15}N and nitrogen concentrations may indicate plant – mycorrhizal interactions. *Oecologia* **122**, 273–283.

Holah, J.C., Wilson, M.V. & Hansen, E.M. (1997) Impacts of a native root-rotting pathogen on successional development of old-growth Douglas fir forests. *Oecologia* **111**, 429–433.

Horton, T.R., Bruns, T.D. & Parker, V.T. (1999) Ectomycorrhizal fungi associated with *Arctostaphylos* contribute to *Pseudotsuga menziesii* establishment. *Canadian Journal of Botany* **77**, 93–102.

Johnson, N.C., Copeland, P.J., Crookston, R.K. & Pfleger, F.L. (1992) Mycorrhizae: possible explanation for yield decline with continuous corn and soybean. *Agronomy Journal* **84**, 387–390.

Johnson, N.C., Graham, J.H. & Smith, F.A. (1997) Functioning of mycorrhizal associations along the mutualism – parasitism continuum. *New Phytologist* **135**, 575–585.

Klironomos, J.N. (2002) Feedback with soil biota contributes to plant rarity and invasiveness in communities. *Nature* **417**, 67–70.

Lafay, B. & Burdon, J.J. (1998) Molecular diversity of rhizobia occurring on native shrubby legumes in southeastern Australia. *Applied and Environmental Microbiology* **64**, 3989–3997.

Marler, M.J., Zabinski, C.A. & Callaway, R.M. (1999) Mycorrhizae indirectly enhance competitive effects of an invasive forb on a native bunchgras. *Ecology* **80**, 1180–1186.

Matthews, J.W. & Clay, K. (2001) Influence of fungal endophyte infection on plant-soil feedback and community interactions. *Ecology* **82**, 500–509.

Mortimer, S.R., van der Putten, W.H. & Brown, V.K. (1999) Insect and nematode herbivory below ground: interactions and role in vegetation succession. In: *Herbivores: between plant and predators* (eds H. Olff, V.K. Brown & R.H. Drent), pp. 205–238. Blackwell Science, Oxford.

Newsham, K.K., Fitter, A.H. & Watkinson, A.R. (1995) Multifunctionality and biodiversity in arbuscular mycorrhizas. *Trends in Ecology and Evolution* **10**, 407–411.

Nötzold, R., Blossey, B. & Newton, E. (1998) The influence of below ground herbivory and plant competition on growth and biomass allocation of purple loosestrife. *Oecologia* **113**, 82–93.

O'Connor, P.J., Smith, S.E. & Smith, F.A. (2002) Arbuscular mycorrhizas influence plant diversity and community structure in a semi-arid herbland. *New Phytologist* **154**, 209–218.

Olff, H., Hoorens, B., de Goede, R.G.M., van der Putten, W.H. & Gleichman, J.M. (2000) Small-scale shifting mosaics of two dominant grassland species: the possible role of soil-borne pathogens. *Oecologia* **125**, 45–54.

Packer, A. & Clay, K. (2000) Soil pathogens and spatial patterns of seedling mortality in a temperate tree. *Nature* **404**, 278–281.

Perry, D.A., Margolis, H., Choquette, C., Molina, R. & Trappe, J.M. (1989) Ectomycorrhizal mediation of competition between coniferous tree species. *New Phytologist* **112**, 501–511.

Persson, J. & Näsholm, T. (2001) Amino acid uptake: a widespread ability among boreal forest plants. *Ecology Letters* **4**, 434–438.

Read, D.J. (1991) Mycorrhizas in ecosystems. *Experientia* **47**, 376–391.

Read, D.J. (1999) The ecophysiology of mycorrhizal symbioses with special reference to impacts upon plant fitness. In: *Physiological plant ecology* (eds M.C. Press, J.D. Scholes & M.G. Barker), pp. 133–152. Blackwell Science, Oxford.

Reinhart, K.O., Packer, A., van der Putten, W.H. & Clay, C. (2003) Plant-soil biota interactions and spatial distribution of black cherry in its native and invasive ranges. *Ecology Letters* **6**, 1046–1050.

Robinson, D. (2001) d^{15}N as an integrator of the nitrogen cycle. *Trends in Ecology and Evolution* **16**, 153–162.

Robinson, D. & Fitter, A.H. (1999) The magnitude and control of carbon transfer between plants linked by a common mycelial network. *Journal of Experimental Botany* **50**, 9–13.

Smith, S.E. & Read, D.J. (1997) *Mycorrhizal symbiosis.* 2nd ed. Academic Press, San Diego.

Strong, D.R. (1999) Predator control in terrestrial ecosystems: the underground food chain of bush lupin. In: *Herbivores: between plant and predators* (eds H. Olff, V.K. Brown & R.H. Drent), pp. 577–602. Blackwell Science, Oxford.

Urcelay, C. & Díaz, S. (2003) The mycorrhizal dependence of subordinates determines the effect of arbuscular mycorrhizal fungi on plant diversity. *Ecology Letters* **6**, 388–391.

van der Heijden, M.G.A., Klironomos, J.N., Ursic, M., Moutoglis, P., Streitwolf-Engel, R., Boller, T., Wiemken, A. & Sanders, I.R. (1998) Mycorrhizal fungal diversity determines plant biodiversity, ecosystem variability and productivity. *Nature* **396**, 69–72.

van der Putten, W.H., van Dijk, C. & Peters, B.A.M. (1993) Plant-specific soil-borne diseases contribute to succession in foredune vegetation. *Nature* **362**, 53–56.

Verschoor, B.C. (2002) Carbon and nitrogen budgets of plant-feeding nematodes in grasslands of different productivity. *Applied Soil Ecology* **20**, 15–25.

Verschoor, B.C., Pronk, T.E., de Goede, R.G.M. & Brussaard, L. (2002) Could plant-feeding nematodes affect the competition between grass species during succession in grasslands under restoration management? *Journal of Ecology* **90**, 753–761.

Vitousek, P.M. & Walker, L.R. (1989) Biological invasion by *Myrica faya* in Hawai'i: plant demography, nitrogen fixation, ecosystem effects. *Ecological Monographs* **59**, 247–265.

Vrålstad, T., Schumacher, T. & Taylor, A.F.S. (2002) Mycorrhizal synthesis between fungal strains of the *Hymenoscyphus ericae* aggregate and potential ectomycorrhizal and ericoid hosts. *New Phytologist* **153**, 143–152.

Weste, G. (1981) Changes in the vegetation of sclerophyll shrubby woodland associated with invasion by *Phytophthora cinnamomi. Australian Journal of Botany* **29**, 261–276.

Westover, K.M. & Bever, J.D. (2001) Mechanisms of plant species coexistence: roles of rhizosphere bacteria and root fungal pathogens. *Ecology* **82**, 3285–3294.

Yeates, G.W., Bongers, T., de Goede, R.G.M., Freckman, D.W. & Georgieva, S.S. (1993) Feeding habits in soil nematode families and genera – an outline for soil ecologists. *Journal of Nematology* **25**, 315–331.

Zabinski, C.A., Quin, L. & Callaway, R.M. (2002) Phosphorus uptake, not carbon transfer, explains arbuscular mycorrhizal enhancement of *Centaurea maculosa* in the presence of native grassland species. *Functional Ecology* **16**, 758–765.

12

Vegetation conservation, management and restoration

Jan P. Bakker

12.1 Introduction

In the past few decades the importance of management and, more recently, restoration as tools for nature conservation has increased considerably. This chapter will review the development from agricultural exploitation via maintenance management towards restoration management. The focus will be on north-west Europe.

Conservation is carried out all over the world to maintain existing nature conservation interests (e.g. Pickett *et al.* 1997). Management is mainly carried out in industrialized countries (Westhoff 1983; Spellerberg *et al.* 1991). Restoration is only practised in a few industrialized countries in north-west Europe (e.g. Wheeler *et al.* 1995) and North America (see the many papers in the journals *Restoration Ecology* and *Applied Vegetation Science*).

During the development of north-west Europe we may discern three periods: the natural period, the semi-natural period and the cultural period. The natural period is characterized by the dominance of communities, landscapes and processes without any noticeable human influence (Bakker & Londo 1998). The major patterns in the landscape were largely determined by climatic and geological factors; they were inserted upon the geological matrix. There was grazing and browsing by indigenous herbivores. Hence, the natural landscape can be defined by the species assemblage of the original flora, vegetation and fauna (Westhoff 1983).

The first agricultural immigration took place about 7000 yr BP followed by a second one around 4600 yr BP, both from south-eastern Europe and southern Russia. These people grew arable crops in a shifting cultivation system after the clearance of primeval forest. For the greater part indigenous large herbivores were gradually replaced by livestock. In medieval times degradation and destruction of primeval forest continued and large oligotrophic bogs, mesotrophic fens and eutrophic reedbeds were drained, reclaimed and even completely removed for fuel (Wheeler *et al.* 1995). Not only the natural communities but also certain landscape-building processes disappeared, through the exclusion of the influence of the sea and rivers and through the regulation of hydrological conditions. Although the resulting open landscape was new, many species that invaded grasslands and heathlands were already present as elements in the understorey of open forest, in small glades, fringes along streams, fens and bogs, and in larger open areas along the coast. Hence, the definition of the semi-natural landscape includes the original flora and fauna but also a transformation of the original vegetation by humans (Westhoff 1983). These are the landscapes of the semi-natural period.

The character of human impact also changed. First only the biotic component was influenced by cutting trees and grazing livestock. Abiotic conditions were only influenced indirectly by, for instance, trampling and nutrient transport, and directly by superficial ploughing. Large areas of semi-natural landscapes, such as heathland and grassland on infertile soils used for common grazing, were not enclosed into private fields but belonged to the local community – the 'commons'. The commons were predominantly found on the drier, sandy parts. Here the geological matrix remained more or less intact. As human impact was stable during many centuries, it became superimposed by a historical matrix. In the 'semi-natural period', regulation of hydrological conditions by drains and ditches in wet parts enabled direct influences on the abiotic conditions by reclamation, deep ploughing and soil levelling. These activities, as well as the division of the landscape into private properties, resulted in the enclosed semi-natural landscape, where fields became delimited by ditches and hedgerows. The geological matrix became severely disturbed.

The transition from the semi-natural to the cultural period in north-west Europe was triggered by the introduction of organic manure or waste from large cities and artificial, inorganic fertilizers. The large-scale reclamation and subsequent eutrophication of common grassland and heathland occurred after 1920 when mechanization in agriculture started. It resulted in the development of the cultivated landscape, in which not only the vegetation but also the flora and fauna became heavily influenced by man (Westhoff 1983). Indigenous species were eradicated by herbicides and non-indigenous species were introduced. These landscapes represent the cultural period in which we are living now.

12.2 From agricultural exploitation to nature conservation

Since most semi-natural grasslands and heathlands are marginal from an agricultural point of view, these areas tend to be the first to be neglected or abandoned. This was common practice earlier, but recently it has been enforced by the European Union agricultural policy; this facilitated highly productive farms and led to the closing down of less productive ones, on which so-called low-intensity farming was practised. The total area of 56 million ha of low-intensity farming strongly differs among European countries in the 1990s. Such farming systems feature 82% of the agricultural area in Spain, 61% in Greece, 60% in Portugal, 35% in Ireland, 31% in Italy, 25% in France, 23% in Hungary, 14% in Poland and 11% in the United Kingdom (Bignal & McCracken 1996). Low-intensity farming areas taken out of the agricultural system in, for instance, The Netherlands or Denmark, may still be exploited in, for instance, Spain or France. On the other hand, artificially fertilized grasslands can be taken out of the agricultural system in The Netherlands with the aim of restoring species-rich grassland or heathland – plant communities still widespread, although decreasing, in countries such as Poland.

The degradation of fauna, flora and vegetation in natural and semi-natural landscapes has become a matter of great concern. The problem affects the whole of north-west Europe, and most of all The Netherlands, due to the high population density and the advanced level of agriculture and technical development resulting in expanding urban and industrial areas, connected by a dense network of roads. Although rural areas outside the urbanized areas still have a predominantly agricultural land use, the intensity of agricultural exploitation leaves little room for species diversity.

A backlash against the degradation of flora and vegetation has developed. From the beginning of the 20th century onwards, areas have been acquired by private organizations for landscape and nature conservation purposes, more recently also by the state. Most of these reserves in north-west Europe represent small fragments of areas with conservation interest. Other European countries feature a relatively large proportion of reserves larger than 10,000 ha.

Several forms of management are connected to the aim of nature conservation. In the United Kingdom, Wells (1980) distinguished between 'reclamation management', carried out only once, and regular 'maintenance management'. When grasslands or heathlands are taken out of the agricultural system, i.e. there is cessation of grazing or mowing, coarse grasses, sedges and shrubs take over. When such an abandoned area is destined for amenity use, it first has to be reclaimed. In the United Kingdom and the United States restoration management is referred to as 'biological habitat reconstruction' and 'restoration, reclamation and regeneration of degraded and destroyed ecosystems'.

In The Netherlands two different situations occur. In the first situation conservation interest is great from the beginning, requiring 'management' in a strict sense (Bakker & Londo 1998) to be applied (Fig. 12.1). This implies the continuation of existing management practices – such as coppicing, haymaking, cutting sods and livestock grazing – with a certain regularity in the case of semi-natural landscapes, while nothing should be done in the case of a (near-)natural landscape. Except for sod cutting these practices, whether for agricultural exploitation or for present-day nature conservation, affect the structure of the vegetation by harvesting it.

In the second situation nature conservation interest is low and 'new nature' has to be developed: 'nature development', usually with one or more 'target communities' in mind. Two phases are distinguished (Fig. 12.1). Phase 1, **environmental restoration,** is necessary when the abiotic environment has been degraded – for example, after lowering the groundwater table, levelling the original relief or eutrophication. This may include removal of the eutrophicated topsoil down to 50 cm, restoration of the relief, and raising the groundwater table. The perspectives of environmental restoration will depend on both the quality of the new environmental conditions and species availability.

If the abiotic conditions have been changed after environmental restoration or the abiotic conditions have not been degraded, environmental restoration is not necessary and phase 2, **restoration management**, can be implemented directly (Fig. 12.1). This includes a change in the existing management practices – for example, cessation of fertilizer application, followed by haymaking or grazing at a low stocking rate. In this way existing plant communities are turned into the target communities.

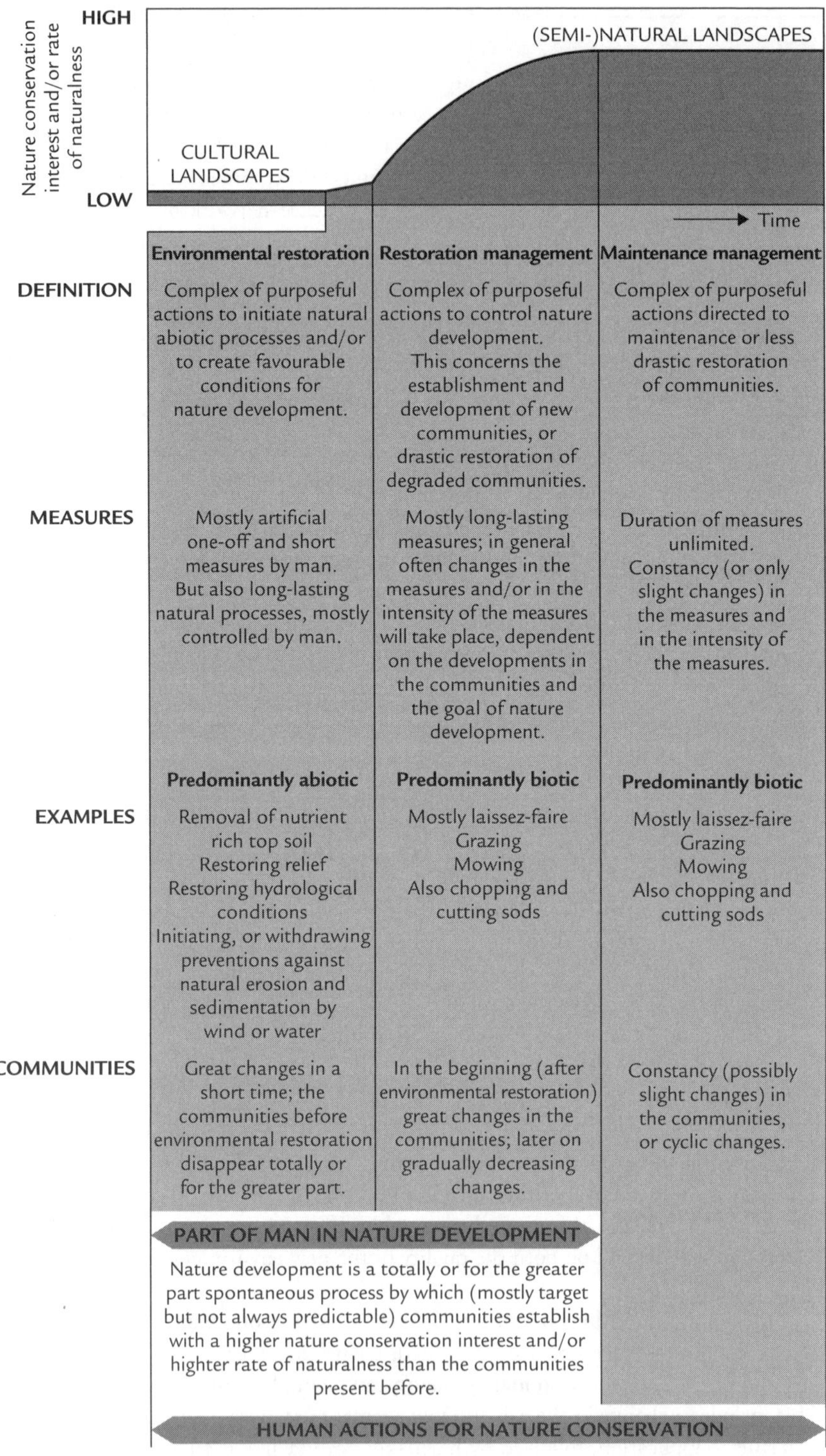

Fig. 12.1 Definitions, measures, examples and communities in relation to human activities for nature conservation in the framework of cultural and (semi-)natural landscapes. After Bakker & Londo (1998). Reproduced by permission of Kluwer Academic Publishers.

12.3 Vegetation management in relation to a hierarchy of environmental processes

Management of plant communities and plant species should take into account the various natural and human-influenced processes. The impacts of these processes can be considered in a hierarchical scheme according to C.G. van Leeuwen (see van der Maarel 1980), ranging – in order of impact – from atmosphere/climate, geology, geomorphology, (ground)water and soil to vegetation and fauna. Londo (1997) elaborated this scheme to indicate the position of environmental restoration and restoration management (Fig. 12.2).

When applying management measures on a lower level one should be aware of the ecological processes occurring at a higher level beyond the influence of the local management. As an example, the mean atmospheric deposition of 40 kg-N·ha^{-1}·yr^{-1} in The Netherlands in the 1990s was about twice that of the critical nitrogen load for plant communities on mesotrophic and oligotrophic soils (Bobbink *et al.* 1998). A lowering of the groundwater table of 60 cm in a *Calthion palustris* fen meadow results in an even more alarming increase of nitrogen availability from 50 to 450 kg-N·ha^{-1}·yr^{-1}. Deep drainage can result in an irreversible desiccation of the soil, the subsequent mineralization of organic matter and acidification because of the replacement of deep calcium-rich seepage water by shallow calcium-poor seepage water. Restoration by simply raising the groundwater table is then insufficient (Grootjans *et al.* 1996).

Vegetation is of course not only influenced by higher-level abiotic processes but also by the fauna (Fig. 12.2). Below-ground herbivores can even act as keystone species in certain ecosystems (Mortimer *et al.* 1999). See further Chapter 11. Geese can destroy the vegetation and even degrade the soil (Jefferies & Rockwell 2002). See also Chapter 10. The establishment of the shrub *Prunus spinosa*, playing a major role in the shifting mosaic of grassland, scrub and forest, is controlled by rabbits rather than by cattle (Olff *et al.* 1999; see Chapter 7).

12.4 Laissez-faire and the wilderness concept

Some managers wish to restore abiotic conditions by removing the topsoil or rewetting, introducing large herbivores and leaving the area as '*new nature*' or wilderness. Here, giving way to natural processes becomes a goal in itself and not a means to reach a well-defined target – a development which cannot be evaluated (Bakker *et al.* 2000).

The idea of wilderness creation was promoted by Vera (2000) who suggested that species-rich open forest can be created or maintained after the introduction of large herbivores related to domestic livestock such as Konik horses, Heck cattle, and wild herbivores such as European bison (*Bison bonasus*), elk (*Alces alces*), red deer (*Cervus elaphus*), roe deer (*Capreolus capreolus*) and beaver (*Castor fiber*). Such a landscape is believed to have occurred before human intervention, and hence is regarded as a natural landscape. However, no indications exist that natural herbivores ever occurred

Natural processes		Human impact	Measures for conservation and restoration	
ATMOSPHERE AND CLIMATE GEOLOGY	ATMOSPHERE AND CLIMATE GEOLOGICAL CONDITIONS	Atmospheric contamination Increased CO_2 level Increased UV level Soil subsidence through mineral mining	Reduction in atmospheric contamination Reduction in CO_2 level Reduction in UV level Reduction in mineral mining	Environmental policy
GEOMORPHOLOGY Running water (sedimentation, erosion) Wind (dune building) Solifluction	RELIEF	Changes in morphodynamics by embankments, sand stabilization, soil levelling, top soil removal, building seawalls, canals	Restoration of geomorphological processes, i.e. water and sand movement, artificial restoration of relief	Environmental restoration
HYDROLOGY Flooding and water withdrawal Running and stagnant water Groundwater dynamics Seepage water Chemistry of groundwater	GROUND- AND SURFACE WATER	Groundwater extraction Desiccation Decrease of seepage Embankments Changes in natural fluctuations Inlet of extraneous water Eutrophication Contamination	Restoration of natural fluctuations Rewetting De-embankment Reduction of extraneous water Reduction of eutrophication Water purification	
SOIL Development of soil profiles Leaching Acidification Accumulation of organic matter	SOIL	Ploughing Digging Manuring Eutrophication Contamination	Cessation of ploughing and digging Removal of nutrients and contaminants by sod cutting or top soil removal	
VEGETATION Succession, e.g. forest development, terrestrialization, patch dynamics, cyclic succession	VEGETATION	Cutting and burning primeval forest Development semi-natural landscapes by chopping, cutting of meadows, grazing of pastures Sowing and planting arable fields	Laissez-faire Cessation of fertilizer application and subsequent haymaking or grazing Transformation of planted forests Re-introduction of locally extinct plant species	Restoration management
ANIMALS Grazing by large herbivores Predation Pollination	ANIMALS	Hunting Fishing Extinction of animals	Cessation/reduction of hunting Cessation of overexploitation Re-introduction of locally extinct animals	

Fig. 12.2 Hierarchical model of levels influencing each other mutually (only influences between adjacent levels are indicated). Thickness of the arrows indicates the strength of the influence. At each level natural processes, human interferences, and measures of nature conservation and restoration are indicated. After Londo (1997). Reproduced by permission of Backhuys Publishers B.V.

in dense populations over large areas and that open forest occurred before agriculture started (Svenning 2002). Only riverine landscapes subjected to natural dynamics may have been open with a shifting mosaic of grassland, tall forb communities, scrub and forest. Such landscapes occur at present along rivers under low stocking densities (Olff *et al.* 1999). Open vegetation may also have occurred on infertile soil.

Human-introduced large herbivores and fire emerged as likely potential key factors in creating open vegetation in north-west Europe. Fire would probably also have been important in the maintenance of light-demanding or short-statured woody species within closed upland forests. Many plant species of calcareous grasslands in West and Central Europe have been found as macrofossils for the first time long after human impact in the landscape started, e.g. *Centaurea scabiosa* and *Primula veris* since the Roman period (Table 12.1) and a currently abundant species such as

Table 12.1 First appearance (+) from archaeological excavations of plant species characteristic of calcareous grasslands (Mesobromion) in the Lower Rhine Valley. Remarkable is that currently occurring characteristic grass species such as *Bromus erectus*, *Koeleria pyramidata*, *K. gracilis*, *Phleum phleoides*, *Festuca cinerea* and *F. rupicola* were not found before the Modern Age. After Poschlod & WallisDeVries (2002).

	Neolithic period	Bronze Age	Iron Age	Roman Empire	Middle Ages
Number of sites for sedges and herbs	66	11	?	> 50	> 80
Number of sites for grasses	34	6	34	27	26
Euphorbia cyparissias	+	+	+	+	+
Potentilla tabernaemontani	+	+	+	+	+
Scabiosa columbaria	+	+	+	+	+
Ajuga genevensis	+	+	+	+	+
Campanula trachelium	+	+	+	+	+
Stachys recta	+	+	+	+	+
Festuca ovina	+	+	+	+	+
Brachypodium pinnatum	+	+	+	+	+
Stipa pennata	+				
Pimpinella saxifraga	+	+	+	+	
Carex caryophyllea	+	+	+		
Medicago lupulina	+	+	+		
Plantago media	+	+	+		
Avenochloa pratensis	+				
Campanula rapunculus	+	+			
Centaurea scabiosa		+	+		
Euphorbia seguierana	+		+		
Hippocrepis comosa			+		+
Peucedanum officinale		+		+	
Primula veris			+		+
Salvia pratensis			+		+
Sanguisorba minor			+		+
Briza media				+	

Bromus erectus only since medieval times (Poschlod & WallisDeVries 2002). This implies that many plant species and communities we wish to preserve depend on human activities in semi-natural landscapes. Hence, the cessation of human activities in semi-natural landscapes might eventually result in closed forest in most lowland biotopes, with almost certainly short-term losses of biodiversity, which may never be restored again. The development of mature natural forests will take centuries, that of fens and bogs even millennia.

12.5 Management and restoration imply setting targets

Management and restoration of vegetation should have targets to follow. The Society for Ecological Restoration (2002) provided nine targets and target-related measures for the restored ecosystem:

1 It contains a characteristic set of the species that occur in the reference system;
2 It consists of indigenous species to the greatest practical extent;
3 All functional groups necessary for its continued development and/or stability are represented or have the potential to colonize;
4 Its physical environment is capable of reproducing populations of the species necessary for its continued stability or development;
5 It functions normally for its ecological stage of development;
6 It is suitably integrated into a larger ecological matrix of landscape;
7 Potential threats to its health and integrity from the surrounding landscape have been eliminated or reduced as much as possible;
8 It is sufficiently resilient to endure the normal periodic stress events in the local environment that serve to maintain the integrity of the ecosystem;
9 It is self-sustaining to the same degree as the reference ecosystem, and has the potential to persist indefinitely under existing environmental conditions.

These targets represent the final situation; it may take a very long time until these are all realized. Examples of how authorities in charge of nature management and restoration deal with targets are discussed below.

Several strategies for the development of targets can be adopted (Bakker & Londo 1998). These are simple for the few natural landscapes left in lowland Europe and America, where human influence has always been modest. When geological processes such as sedimentation and erosion by water and wind are predominant, older successional stages can be eroded locally and young stages can emerge at other places. This may happen in coastal and inland dunes, salt marshes and along rivers.

In certain cultural and semi-natural landscapes a more natural landscape can be restored, first of all through hydrological measures, such as digging side-channels along rivers, building of dams to catch rainwater for bog development or de-embankment for salt-marsh restoration.

Restoration management in communal semi-natural landscapes with grassland, heathland, scrub and/or wooded meadows, can be carried out by the removal or gradual disappearance of former borders in the enclosed landscape. This can be practised by abiotic management such as neglecting drainage systems, giving way to eroding forces of wind and water, sod cutting (up to 10 cm) or top soil removal. Biotic management implies fencing in large areas and grazing by large herbivores of different breeds such as Konik horse, Exmoor ponies, Scottish Highland cattle, Galloways, Schoonebeker or Mergelland sheep, but also heifers of dairy cattle.

Enclosed semi-natural landscapes are typically restored within their field borders including drainage systems. Restoration management deals with oligotrophic or mesotrophic grassland communities by cutting, sometimes with grazing the aftermath with high stocking rates. The process of restoration can be enhanced by topsoil removal, also for small isolated fields with heathland being the target. The effect of restoration management is thought to be enhanced by connecting individual fields by corridors in which species can disperse.

In cultural landscapes species-rich plant communities may be restored in the margin of grasslands and arable fields.

Within semi-natural landscapes more tangible targets are needed. In The Netherlands a system of '*nature target types*' was developed, something between habitat and community types (Bal & Hoogeveen 1995). Each type includes lists of plant and animal species. These types also harbour many Red List species at the national level. It turned out soon that these target types may be useful to strive after eventually, although some Red List species have a regional distribution even in a small country such as The Netherlands. Moreover, the target types are not feasible for the short- and mid-term management practice to be fulfilled. It is recognized now that several ways may lead to certain target types. Landscape matrices show the relationships of the target type with eutrophicated communities at the same substrate and successional relationships of open low canopy, closed low canopy, scrub and forest (Schaminée & Jansen 1998). Such a matrix for the target type of wet heathland, *Ericetum tetralicis*, is shown in Fig. 12.3. Also lists of species to be expected for several time intervals are indicated starting from the dominance of *Molinia caerulea* and other species after sod cutting. Most target species are supposed to emerge immediately, and this expectation must be based on their presence in the established vegetation, or in the soil seed bank for species with a high longevity index (Bekker *et al.* 1998). Species with a low longevity index are supposed to establish later; they need dispersal from elsewhere. Starting from fertilized communities after sod cutting or top soil removal is supposed to show an initial establishment of species of fertilized habitats and a few target species that must have a long-term persistent seed bank; target species with low longevity index are supposed to establish within 10 yr (Fig. 12.3). From 2001 onwards the Dutch government compensates the costs of management based on the fulfilment of targets set during 10 yr. Clearly, authorities in charge of management have to set realistic mid-term targets instead of ideal long-term targets.

The framework of conservation targets should be the plant community system. The recent classification of plant communities in The Netherlands is based on *c.* 350,000 old and more recent relevés (Hennekens & Schaminée 2001). It includes synoptic tables from which lists of target species can be derived belonging to certain plant communities. As the coordinates of the relevés are known, it is also clear whether species have occurred in certain regions of The Netherlands. These data can be accomplished by the 9,000,000 records of occurrence of plant species since the early 20th century in a grid system of 5 km × 5 km. Such data can help to set realistic

(a)

Species	0–1	1–3	3–10	10–25	Longevity index
	Number of years				
Drosera intermedia	—	━	━	━	–
Juncus squarrosus	—	━	━	—	1
Rhynchospora alba	—	━	━		–
Rhynchospora fusca	—	━	━		–
Carex panicea	—	—	━	—	0.36
Calluna vulgaris	—	—	━		0.74
Drosera rotundifolia		—	—	—	–
Lycopodium inundatum		—	━	—	–
Erica tetralix	—	—	━	━	0.42
Molinia caerulea	—	—	—		0.30
Gentiana pneumonanthe		—	━	—	–
Scirpus cespitosus		—	—	━	–
Salix repens		—	—	—	0
Narthecium ossifragum			—		–

(b)

Species	0–1	1–3	3–10	10–25	>25	Longevity index
	Number of years					
Gnaphalium uliginosum	━	━	—	—		0.89
Erigeron canadensis		━	━	—		–
Rorippa sylvestris	━	━	—			–
Cirsium arvense	—	—	—			0.33
Rumex obtusifolius	—	—	—			0.62
Trifolium repens	—	━	━	—	—	0.38
Drosera intermedia	—	━	━	━		–
Ornithopus perpusillus	—	━	—	—		0
Leucanthemum vulgare	—	━	━	—	—	0.36
Juncus squarrosus	—	━	━	—	—	1
Calluna vulgaris	—	━	━	—	—	0.74
Rumex acetosella	—	━	━	━	—	0.71
Holcus lanatus	—	━	━	━	—	0.44
Erodium cicutarium	—	—	━	—		0.14
Carex panicea		—	—	—	—	0.36
Genista anglica		—	—	—	—	–
Filago minima		—	━	—		–
Gentiana pneumonanthe		—	━	—	—	–
Hypochaeris radicata		—	━	━	—	0.39
Erica tetralix	—	—	━	━	━	0.42
Molinia caerulea		—	—	—	━	0.30
Scirpus cespitosus			—	—	━	–
Narthecium ossifragum				—	—	–

Fig. 12.3 Restoration perspectives for wet heathlands on Pleistocene sandy soils in The Netherlands, indicated by the expected occurrence of plant species in various time periods. Thick lines indicate higher abundances. **a.** Starting from heathland overgrown by grasses, and the restoration measurement sod cutting. **b.** Starting from arable fields or grasslands applied with fertilizer, and the restoration measurements sod cutting and re-instalment of the original hydrological conditions. Seed bank data are derived from Thompson *et al.* (1997), the longevity index from Bekker *et al.* (1998). After Schaminée & Jansen (1998).

regional targets for restoration management. Predictions for the establishment of species may be derived from datasets including life-history traits such as seed longevity and dispersal characteristics. Data mining and working with such large data sets is an aspect of '*eco-informatics*'.

The plant community concept assumes that plant species form more or less stable assemblages responding to the local environmental conditions. Using the data set of 350,000 relevés made in The Netherlands in the period 1930–2000, it became clear that in most community types the floristic composition had changed, although the appearance of the vegetation and the presence and abundance of most of the characteristic species has remained the same (Schaminée *et al.* 2002). Likely, eutrophication is responsible for most of the changes. Management authorities have to take into account such internal changes when planning re-introduction of endangered plant species. Although characteristic species may still be present in communities suffering from eutrophication, their populations seem to diminish in size by lack of rejuvenation. It is not sure whether in such circumstances introduction of seeds will result in emerging seedlings (Strykstra 2000). This may become even more questionable when the species has disappeared from the stand.

12.6 Setting targets implies monitoring

The evaluation of targets evokes repeated vegetation monitoring. This can be carried out at different levels of resolution. Repeated aerial photographs can record changes in area and position of bare soil and macro-structural classes such as short vegetation, tall forb communities, scrub and forest, but it reveals no information on nature target types.

Information at the level of plant communities requires repeated vegetation mapping. Aerial photographs or other means of remote sensing, and/or field surveys enable stratified sampling of the elements discerned. The size of vegetation relevés may differ according to the structural class.

Changes in presence/absence or cover of individual species can only be monitored in permanent plots. Long-term permanent plots are important as they can help in separating trends and superimposed fluctuations (Huisman *et al.* 1993), and are needed to test ecological models that are often based on assumptions and not derived from solid field studies (Bakker *et al.* 1996). The study of long-term permanent plots has made it particularly clear that vegetation development in many ecosystems under restoration was different from the final state that was anticipated. This may generate new hypotheses (Klötzli & Grootjans 2001). Because a limited number of permanent plots will not cover all spatio-temporal changes in vegetation, it is better to establish permanent transects with adjacent grid cells varying in size depending on the vegetation structure (Olff *et al.* 1997) or even by large grid cells covering the entire study site (Verhagen *et al.* 2001).

Long-term recordings are needed to validate the effects of management measures. Experimental changes in salt-marsh management (Bos *et al.* 2002) and calcareous grassland management (Kahmen *et al.* 2002) revealed clear changes after 15–20 years, stressing the importance of long-term monitoring.

12.7 Effects of management and restoration practices

12.7.1 Haymaking

Effects of restoration management (Fig. 12.1) and small-scale environmental restoration at the field scale in enclosed previously fertilized or abandoned grasslands and arable fields, will be discussed, especially regarding different mowing regimes. Targets are the decrease in yield and the establishment of more species-rich target communities.

Cessation of fertilizer application and an annual haymaking regime decreased the yield in mesotrophic *Mesobromion erecti* grassland on calcareous soil (Willems 2001) and oligotrophic *Nardo-Galion saxatilis* grassland on sand (Bakker *et al.* 2002a). Moreover, the proportion of species indicating high and low soil fertility decreased and increased, respectively. Annual haymaking decreased the yield, but after 25 yr the standing crop was still about twice the level of the target community, i.e. 200–300 g-dw·m^{-2}. The removal of nitrogen was gradually balanced by the input through atmospheric deposition. Two annual cuts and removal of most nutrients, initially showed the highest species richness, but species number decreased again after 15 yr (Bakker *et al.* 2002a). Annual haymaking regimes, including removal of nutrients, brought the grassland closer to the target *Nardo-Galion* community than regimes where less nutrients were removed, such as mulching or haymaking every second year (Fig. 12.4). Still, the community resulting from the 'best' practice is far from a local reference at *c.* 500 m (Bakker *et al.* 2002a).

Attempts to restore a *Cirsio-Molinietum* wet fen meadow in an agriculturally improved pasture failed because of the very high yield of 1200 g-dw·m^{-2} that did not decrease with annual haymaking. Target species from a nearby (500 m) reference community did not invade. Topsoil removal of 15–20 cm reduced the yield by 50% and the total soil phosphorus amount by 85%, and depleted plant P availability. Target species were planted as seedlings. Where the topsoil had not been removed the vegetation became dominated by a few competitive species and although many of the planted target species were still present after 4 yr, they were found only in trace amounts. Removal of the topsoil created suitable edaphic conditions for all planted target species to remain well established (Tallowin & Smith 2001).

The aforementioned discrepancy between the effects of haymaking and mulching in restoration management was also found in grassland dominated by trivial species such as *Poa trivialis* (Oomes *et al.* 1996). A 25-yr study on calcareous grassland revealed that the swards in the haymaking and mulching regimes resembled each other, but both regimes deviated from the control grazing regime (Kahmen *et al.* 2002). Similar conclusions were reached for an *Arrhenatherion elatioris* community in southern Germany (Moog *et al.* 2002). Apparently cut but not removed biomass decomposed very fast at the high late-summer temperatures in southern Germany.

In the cases of an abandoned calcareous grassland overgrown by the tall perennial grass *Brachypodium pinnatum*, with a subsequent decrease of species richness, and of a mown grassland suffering from atmospheric N deposition, haymaking in August, before the reallocation of nutrients to below-ground storage organs, turned out to be the right management practice to prevent *Brachypodium pinnatum* from becoming

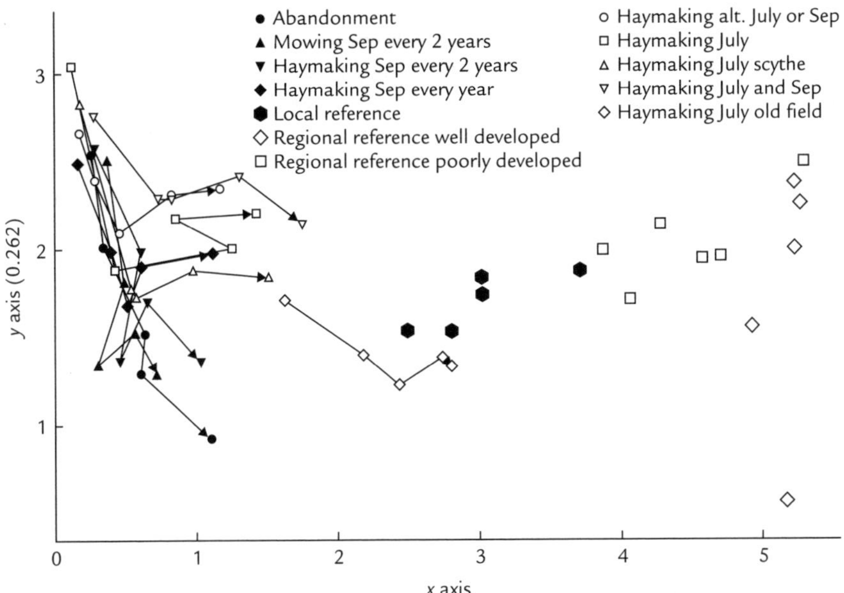

Fig. 12.4 Ordination by Detrended Correspondence Analysis of all species with different management practices in the 'new' and 'old' field, the local reference and the poorly and well-developed regional references of *Nardo-Galion saxatilis* communities. After Bakker *et al.* (2002a). Reproduced by permission of Opulus Press.

dominant (Bobbink & Willems 1991). For moist grasslands it is found that the longer the period of fertilizer application, the more the soil seed bank is depleted (Bekker *et al.* 1997). The problem of lack of diaspores in the soil seed bank can be overcome when fields are 'connected' by haymaking machinery, thus turning mowing machines into sowing machines (Strykstra *et al.* 1996).

12.7.2 Grazing

Grazing as a tool in restoration management and environmental restoration (Fig. 12.1) will be discussed for a communal semi-natural landscape with previously fertilized or abandoned grasslands and arable fields, especially for grazing regimes integrating different units. An area on sandy soil harboured 40 ha of open heathland with grasses and 20 ha of forest. Locally it was dominated by *Deschampsia flexuosa* and *Molinia caerulea* in 1983 when cattle grazing at a stocking rate of 0.2 animal·ha^{-1} started in the entire area, and some additional tree cutting was done in the heathland. Grazing did not reduce the cover of grasses or prevented grass invasion in the heathland. The soil contained a viable *Calluna vulgaris* seed bank. A positive correlation was found between number of seedlings emerged and seed density. Grazing by free-ranging cattle did not prevent encroachment by *Pinus sylvestris* and *Betula pubescens*. It also did not remove the high atmospheric nutrient input. Substantial amounts were redistributed from the grass lawns to the forest (Bokdam & Gleichman 2000).

Livestock grazing is often practised on dry soils of previously fertilized pasture or arable field to restore species-rich grasslands or heathland on oligotrophic soil. Changes from species indicating eutrophic to mesotrophic soil conditions are recorded but then the succession does not proceed beyond stands dominated by *Holcus lanatus* and *Agrostis capillaris*. When wet or moist sites are included, the herbivores tend to avoid these sites and subsequently tall forb stands develop. Livestock grazing is also introduced to control tall-grass dominance in various ecosystems. This often results in patterns of different scale including short grass, tall forbs and scrub (Bakker 1998). Hence, grazing strongly affects the diversity of animals depending on variation in habitats (van Wieren 1998).

12.7.3 Top soil removal

Cutting and/or grazing do not always result in the targets set or only very slowly. Environmental restoration by top soil removal may accerelate the process and render a closer approach of the target (see Fig. 12.1). In order to reduce the amount of nutrients from previously fertilized pastures and arable fields, the top soil was removed to restore heathland and other plant communities characteristic of oligotrophic soil conditions. However, due to lack of money, the top soil is not always removed from the site, but re-allocated. This results in local top soil removal and local accumulation of soil and nutrients with subsequent establishment of non-target species such as *Holcus lanatus* and *Juncus effusus*. Where possible, previous pastures or arable fields are fenced in together with adjacent reference areas which still harbour the target communities; grazing livestock is then supposed to disperse propagules from the reference area into the target area. The majority of viable seeds found in dung includes species of eutrophic soil. Apparently the herbivores feed selectively on species of high forage quality (Mitlacher *et al.* 2002). The similarity between relevés of permanent plots and of reference relevés increased up to 30% for some communities (Verhagen *et al.* 2001). For the sites under study it was possible to collect data on the occurrence of target species in the surroundings of the sites derived from the Dutch 5 km × 5 km grid data. Species occurring in the same grid cell as the study site or in the eight adjacent cells can be regarded as the local species pool (Zobel *et al.* 1998). The majority of the target species were present in the local species pool, but very few target species emerged in most study sites. Part of these must have emerged from the soil seed bank as they are known for their longevity, such as *Erica tetralix*, *Calluna vulgaris* and *Juncus squarrosus*. These species are still found in sites that were reclaimed from heathland more than 70 yr ago (ter Heerdt *et al.* 1997).

12.7.4 Rewetting

Environmental restoration (Fig. 12.1) can also start with changing the hydrological conditions. An increase in the groundwater table of 30 cm in a field on peaty clay overlaying peat resulted in an increase of species indicating wet soil conditions within 5 yr. This was independent of the vegetation management including haymaking or mulching (Oomes *et al.* 1996).

Wet *Cirsio-Molinietum* meadows are annually cut for hay. In The Netherlands, these are all threatened by desiccation, acidification and eutrophication. Restoration in the high Pleistocene part of the country is feasible when the hydrological conditions are only slightly disturbed and dependent on local and regional groundwater systems, or when hydrological measures are carried out in combination with sod cutting. Digging of shallow ditches may promote surface run-off of acid rainwater and promote upward seepage of base-rich groundwater. Restoration prospects in the low Holocene part are small as they depend on very large-scale hydrological systems. Species that re-established seem to have emerged from a long-term persistent seed bank, whereas species with a short-term persistent or a transient seed bank were still locally present in the nature reserves under treatment (Jansen *et al.* 2000). Restoration of *Cirsio-Molinietum* by flooding during winter and spring, and additional sod cutting was not successful. Supposed limiting dispersal capacity of target species was encountered by introduction, but they did not survive after the second year of the experiment. It turned out that the flooding water was poor in base cations (van Duren *et al.* 1998). An overview on possibilities and constraints in the restoration of brook valley systems is given by Grootjans *et al.* (2002a).

Restoration of a cut-over bog on dried and shrunken *Eriophorum-Sphagnum* peat showed that after reflooding after blockage of surrounding ditches, ombrotrophic *Sphagnum* species failed. The very acid and nutrient-poor mire water reaching the surface had to be fertilized to promote any plant growth. After that, a floating mat was formed by minerotrophic *Sphagnum* species which eventually will develop a new bog. As a result of a large precipitation deficit in summer periods, enormous quantities of water have to be stored during winter periods by building high dams. The flooding water contains large amounts of nutrients, hence the mire resumes its functions as a nutrient sink. The vegetation includes very productive reeds and sedges. These stands may eventually transform into low-productive small sedge communities, when a fen acrotelm is formed that can fix nutrients, thus offering Red List species a niche without the need for human intervention (Pfadenhauer & Grootjans 1999).

Restoration of salt marshes, i.e. the building of salt marshes by the interaction of vegetation and sedimentation, recently started deliberately in north-west Europe (Cooper *et al.* 2001). There are also sites where a low summer dyke was breached unintentionally. The de-embanked polder that had a sedimentation deficit was filled fast and many salt-marsh species had established after 10 and 27 yr, respectively (Bakker *et al.* 2002b).

In dune slacks restoration is often carried out by rewetting in combination with top soil removal (Grootjans *et al.* 2002b).

12.7.5 Re-introduction

In order to restore species-rich plant communities, low productivity levels are essential, but these cannot guarantee successful restoration (Berendse *et al.* 1992). Deliberate re-introduction of disappeared species is an issue of restoration management that might be practised more in the future in the present fragmented landscape. Moreover, nowadays the moving ecological infrastructure (i.e. machinery and herds of livestock; cf. Poschlod *et al.* 1996) of the former low-intensity farming system is lacking.

Successful introduction experiments suggest that dispersal of propagules indeed may be a constraint. Diaspore transfer with plant material proved to be a very successful method in restoring species-rich flood meadows. After 4 yr 102 species had established, among them many rare and endangered species. High quality of plant material and suitable site conditions with low competition (after top soil removal) in early stages of the succession seem to be essential prerequisites (Hölzel & Otte 2003). However, other experiments reveal that common species do establish better than rare species (Tiikka *et al.* 2001). Failure of the establishment of target plant species may be due to the lack of accompanying mycorrhizal fungal species (e.g. van der Heijden *et al.* 1998). Experiments including the introduction of soil from reference communities are practised.

12.8 Constraints and strategies in management and restoration

It is clear from the above-mentioned studies that haymaking or grazing alone will not result in restoring target communities on mesotrophic and oligotrophic soils. Because of atmospheric deposition and acidification, succession is unlikely to go beyond communities characteristic of mesotrophic soil. Because of a lack of long-term persistent seeds in the soil seed bank and poor dispersal in the present fragmented landscape many rare and endangered Red List species are unlikely to establish. The combination of reference sites and areas to be restored within one fence in order to be grazed by livestock needs further study with respect to the role of herbivores in plant dispersal. Because of the preference for high-quality forage, dispersal may include transport of diaspores of non-target species. Only experiments including the introduction of target species, can reveal causes of standstill of further succession, be it abiotic conditions or lack of dispersal of propagules from elsewhere. Discussions on the genetic basis of introduced propagules can be overcome by introducing hay from nearby reference communities into the impoverished target area. Ecotypes become established that are adapted to local conditions.

Agricultural intensification has resulted in high nutrient levels in the soil. After the cessation of fertilizer application the levels of nitrogen drop quickly. However, high phosphorus levels may be found even after top soil removal of 50 cm. Restoration and maintenance of soil phosphorus as the primary limiting nutrient is essential where there is a risk of nitrogen becoming non-limited through atmospheric input (Tallowin & Smith 2001). In this respect it is striking that very high numbers of species per 100 m^2 were found in 281 ancient meadows in five European countries when the P content in the soil did not exceed 5 $mg \cdot 100\ g^{-1}$ (Janssens *et al.* 1999).

Drainage has resulted in desiccation of many wetlands. For some plant communities rewetting may be done at the scale of individual fields. However, fragmentation of land ownership often results in failure. Especially, the restoration of communities depending on deep seepage water requires the inclusion of entire catchment areas so that their hydrological conditions are independent from those of neighbouring areas and can therefore be managed separately.

Taking into account the constraints in the restoration of ecological diversity (Bakker & Berendse 1999), it is not unlikely that we have to learn that the results of restoration

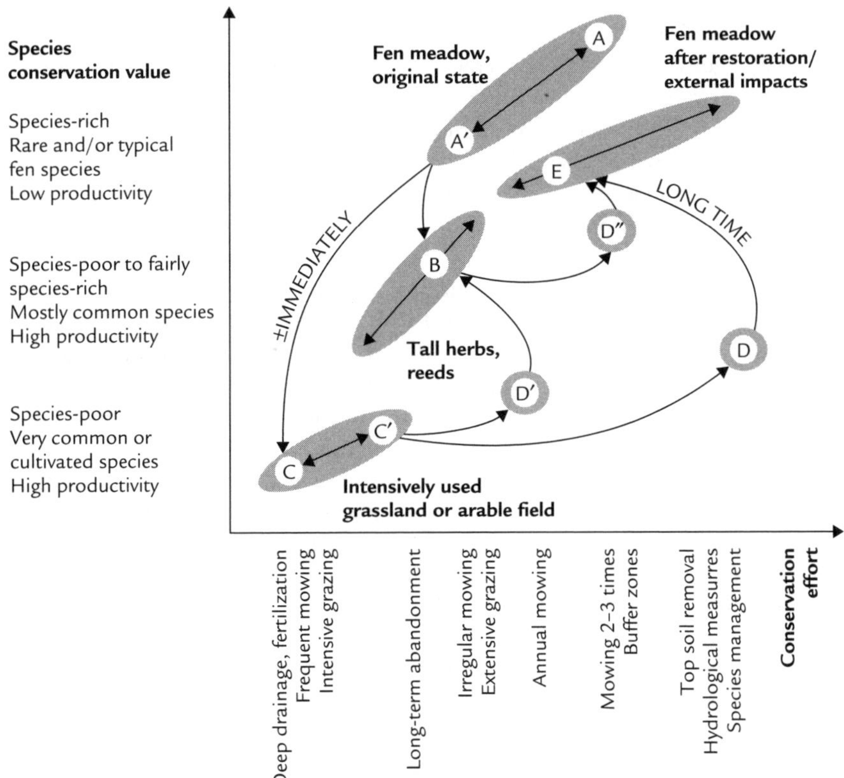

Fig. 12.5 Model relating the species conservation value of a fen meadow site to conservation effort. Each state A–D covers a certain range of effort and of value. This allows reversible changes in management. Beyond a certain threshold, sites can transform into a new state. Restoration to other states requires high conservation efforts and is only partially successful: the original species conservation value may not be reached anymore, and the site to be 'restored' needs more efforts than previously for its maintenance. After Güsewell (1997). Reproduced by permission of the author.

management may differ to some extent from the reference we have in mind (Fig. 12.5) (Güsewell 1997). It is now clear that restoration is difficult, time consuming, expensive and perhaps not always possible in countries with intensive agricultural exploitation. Therefore, one of the best measures in nature conservation at a European level is to maintain the small remnants of (near-)natural and especially the large areas of still existing semi-natural landscapes including the low-intensity farming systems (Bignal & McCracken 1996).

A final point of discussion concerns the targets to be set in relation to existing or desired gradients of intensity of agricultural exploitation within a country, and takes into account the overall differences in agricultural intensity between countries. The highest levels of fertilizer application may be linked to botanically poor grasslands, but can cope with large numbers of winter- and spring-staging geese, for example in

The Netherlands. Common meadow birds can cope with still relatively high levels of fertilizer application, and conventional agricultural practice can be combined with these nature conservation targets. On the other hand, like plants, more '*critical*' meadow birds such as ruff (*Philomachus pugnax*), redshank (*Tringa totanus*) and blacktailed godwit (*Limosa limosa*) can only cope with low levels of fertilizer application and less drainage (van Wieren & Bakker 1998). If farmers are willing to maintain grasslands with suboptimal production in order to achieve nature conservation targets, they need to be economically compensated. The effectiveness of these costly agro-environment schemes is under discussion (Kleijn *et al.* 2001). Botanical conservation interests are only met with to some extent in field margins along ditches outside the area of agricultural exploitation. It should be realized that the role of farmers for many nature conservation goals is very minor in countries with an overall high agricultural production level. Nevertheless they can be important with respect to the scenery, i.e. the open countryside in densely populated countries. In other countries low-intensity farming is very important for many nature conservation goals.

For restoration purposes it is important to realize at which spatial scales different groups of organisms operate or are being affected by environmental conditions. This knowledge makes clear which constraints exist at a certain size of a nature reserve. Cutting regimes act at the field scale. Large herbivores act at the field scale when only used as substitute for haymaking machinery. In small reserves livestock are brought in with the aim of managing species of plants and animals that depend on a short sward (entomofauna, small mammals, birds). Here, the management of the populations of large herbivores is not important: they are covered by veterinary laws and need supplementary food in poor seasons. When management includes the restoration of viable populations of large herbivores, areas of 100–1000 km^2 are necessary (Table 12.2). In such large nature reserves mosaics of grassland, heathland, scrub and forest often occur.

From the above the idea emerges that very large areas beyond the field scale, often practised in cultural and semi-natural landscapes, are necessary for the conservation of viable populations of many plant and animal species, i.e. a high biodiversity. What does this mean for the type of communities and their management? Implications for conservation and restoration can be derived from the pre-agricultural Holocene landscape. Since closed forest, forest glades, pasture-woodland, meadows, grassland, scrub and heathland as well as their associated organisms would have a significant presence in north-west Europe under present-natural conditions, all these habitats should be considered of high conservation interest (Svenning 2002). Large herbivores and natural fires would be key agents creating and maintaining this diversity of habitats. Notably, open spaces within forests are the most important habitats in terms of diversity for many groups of woodland-associated organisms. Free-ranging large herbivores would also be important as dispersers of many herbaceous plants. Furthermore, large herbivores and fires also would provide the special microhabitats needed by a range of dung- and fire-dependent species, many of which have become rare or locally extinct in north-west Europe. Consequently it must be a conservation priority to re-establish native large herbivores and natural fire regimes wherever possible, and mimic their effects by management such as grazing by domestic herbivores and use of prescribed burning where this is not feasible. Finally, it is important to note that

Table 12.2 Relevant spatial dimensions important for the occurence of various groups of organisms (diversity) and the spatial scale at which grazing and haymaking are effective (management). After WallisDeVries (2002).

	dm^2	m^2	ha	km^2	10^2–10^3 km^2	Continent
Management						
Grazing	Bite	Feeding station	Feeding site	Home range	Viable population for wild ungulates	
Cutting			Field size			
Biodiversity						
Plants	Germination, shoot growth	Clonal growth, seed dispersal	Viable population depending on rainwater	Viable population depending on seepage water		
Butterflies	Oviposition	Larval microhabitat	Local population	Viable (meta-)population		
Birds			Individual breeding territory	Viable population, seasonal range of migratory birds		Annual range of migratory birds

the most widespread natural vegetation type would be closed old-growth forest and that scattered old trees and dead wood probably would also be present in the more open vegetation. Many organisms dependent on old-growth forest, old trees, or dead wood have not fully survived in the semi-natural and the cultural landscape. Thus, ancient forest should be protected and the presence of old trees and dead wood promoted in most habitats (Svenning 2002).

References

Bakker, J.P. (1998) The impact of grazing on plant communities. In: *Grazing and Conservation Management* (eds M.F. WallisDeVries, J.P. Bakker & S.E. van Wieren), pp. 137–184. Kluwer Academic Publishers, Dordrecht.

Bakker, J.P. & Berendse, F. (1999) Constraints in the restoration of ecological diversity in grassland and heathland communities. *Trends in Ecology and Evolution* **14**, 63–68.

Bakker, J.P. & Londo, G. (1998) Grazing for conservation management in historical perspective. In: *Grazing and Conservation Management* (eds M.F. WallisDeVries, J.P. Bakker & S.E. van Wieren), pp. 21–54. Kluwer Academic Publishers, Dordrecht.

Bakker, J.P., Olff, H., Willems, J.H. & Zobel, M. (1996) Why do we need permanent plots in the study of long-term vegetation dynamics? *Journal of Vegetation Science* **7**, 147–156.

Bakker, J.P., Grootjans, A.P., Hermy, M. & Poschlod, P. (2000) How to define targets for ecological restoration? *Applied Vegetation Science* **3**, 3–6.

Bakker, J.P., Elzinga, J.A. & de Vries, Y. (2002a) Effects of long-term cutting in a grassland system: possibilities for restoration of plant communities on nutrient-poor soils. *Applied Vegetation Science* **5**, 107–120.

Bakker, J.P., Esselink, P., van Duin, W.E. & de Jong, D.J. (2002b) Restoration of salt marshes in the Netherlands. *Hydrobiologia* **478**, 29–51.

Bal, D. & Hoogeveen, Y. (eds) (1995) *Handboek Natuurdoeltypen in Nederland.* Report 11 IKC-Natuurbeheer, Wageningen. (In Dutch.)

Bekker, R.M., Verweij, G.L., Smith, R.E.N., Reine, R., Bakker, J.P. & Schneider, S. (1997) Soil seed banks in European grasslands: does land use affect regeneration perspectives? *Journal of Applied Ecology* **34**, 1293–1310.

Bekker, R.M., Bakker, J.P., Grandin, U., Kalamees, R., Milberg, P., Poschlod, P., Thompson, K. & Willems, J.H. (1998) Seed shape and vertical distribution in the soil: indicators for seed longevity. *Functional Ecology* **12**, 834–842.

Berendse, F., Oomes, M.J.M., Altena, H.J. & Elberse, W.T. (1992) Experiments on the restoration of species-rich meadows in the Netherlands. *Biological Conservation* **62**, 59–65.

Bignal, E.M. & McCracken, D.I. (1996) Low-intensity farming systems in the conservation of the countryside. *Journal of Applied Ecology* **33**, 413–424.

Bobbink, R. & Willems, J.H. (1991) Impact of different cutting regimes on the performance of *Brachypodium pinnatum* in Dutch chalk grassland. *Biological Conservation* **56**, 1–21.

Bobbink, R., Hornung, M. & Roelofs, J.G.M. (1998) The effects of air-borne nitrogen pollutants on species diversity and semi-natural European vegetation. *Journal of Ecology* **86**, 717–738.

Bokdam, J. & Gleichman, J.M. (2000) Effects of grazing by free-ranging cattle on vegetaion dynamics in a continental north-west European heathland. *Journal of Applied Ecology* **37**, 415–431.

Bos, D., Bakker, J.P., de Vries, Y. & van Lieshout, S. (2002) Vegetation changes in experimentally grazed and ungrazed back-barrier marshes in the Wadden Sea over a 25-year period. *Applied Vegetation Science* **5**, 45–54.

Cooper, N.J., Cooper, T. & Burd, F. (2001) 25 years of salt marsh erosion in Essex: Implementatoin for coastal defence and nature conservation. *Journal of Coastal Conservation* **7**, 31–40.

Grootjans, A.P., van Wirdum, G., Kemmers, R. & van Diggelen, R. (1996) Ecohydrology in The Netherlands: principles of an application-driven interdiscipline. *Acta Botanica Neerlandica* **45**, 491–516.

Grootjans, A.P., Bakker, J.P., Janse, A.J.M. & Kemmers, R.H. (2002a) Restoration of brook valley meadows in the Netherlands. *Hydrobiologia* **478**, 149–170.

Grootjans, A.P., Geelen, H.W.T., Jansen, A.J.M. & Lammerts, E.J. (2002b) Restoration of coastal dune slacks in the Netherlands. *Hydrobiologia* **478**, 181–203.

Güsewell, S. (1997) *Evaluation and management of fen meadows invaded by common reed (Phragmites australis)*. Ph.D. Thesis Swiss Federal Institute of Technology, Zürich.

Hennekens, S.M. & Schaminée, J.H.J. (2001) TURBOVEG, a comprehensive data base management system for vegetation data. *Journal of Vegetation Science* **12**, 589–591.

Hölzel, N. & Otte, A. (2003) Restoration of a species-rich flood meadow by topsoil removal and diaspore transfer with plant material. *Applied Vegetation Science* **6**, 131–140.

Huisman, J., Olff, H. & Fresco, L.F.M. (1993) A hierarchical set of models for species response analysis. *Journal of Vegetation Science* **4**, 37–46.

Jansen, A.J.M., Grootjans, A.P. & Jalink, M.H. (2000) Hydrology of Dutch Cirsio-Molinietum meadows: Prospects for restoration. *Applied Vegetation Science* **3**, 51–64.

Janssens, F., Peeters, A.A., Tallowin, J.R.B., Bakker, J.P., Bekker, R.M., Fillat, F. & Oomes, M.J.M. (1999) Relationship between soil chemical factors and grassland diversity. *Plant and Soil* **202**, 279–298.

Jefferies, R.L. & Rockwell, R.F. (2002) Foraging geese, vegetation loss and soil degradation in an Arctic salt marsh. *Applied Vegetation Science* **5**, 7–16.

Kahmen, S., Poschlod, P. & Schreiber, K.F. (2002) Management practice of calcareous grasslands. Changes in plant species composition and the response of plant functional traits during 24 years. *Biological Conservation* **104**, 319–328.

Kleijn, D., Berendse, F., Smit, R. & Gillissen, N. (2001) Agri-environment schemes do not effectively protect biodiversity in Dutch agricultural landscapes. *Nature* **413**, 723–725.

Klötzli, F. & Grootjans, A.P. (2001) Restoration of natural and semi-natural wetland systems in Central Europe: progress and predictability of developments. *Restoration Ecology* **9**, 209–219.

Londo, G. (1997) *Natuurontwikkeling*. Backhuys Publishers, Leiden. (In Dutch.)

Mitlacher, K., Poschlod, P., Rosén, E. & Bakker, J.P. (2002) Restoration of wooded meadows – comparative analysis along a chronosequence on Öland (Sweden). *Applied Vegetation Science* **5**, 63–74.

Moog, D., Poschlod, P., Kahmen, S. & Schreiber, K.F. (2002) Comparison of species composition between different grassland management treatments after 25 years. *Applied Vegetation Science* **5**, 99–106.

Mortimer, S.R., van der Putten, W.H. & Brown, V.K. (1999) Insect and nematode herbivory below ground: interactions and role in vegetation succession. In: *Herbivores: Between Plants and Predators* (eds H. Olff, V.K. Brown & R.H. Drent), pp. 205–238. Blackwell Science, Oxford.

Olff, H., de Leeuw, J., Bakker, J.P., Platerink, R.J., van Wijnen, H.J. & de Munck, W. (1997) Vegetation succession and herbivory in salt marsh: changes induced by sea level rise and silt deposition along an elevational gradient. *Journal of Ecology* **85**, 799–814.

Olff, H., Vera, F.M., Bokdam, J., Bakker, E.S., Gleichman, J.M., de Maeyer, K. & Smit, R. (1999) Shifting mosaics in grazed woodlands driven by alternation of plant facilitation and competition. *Plant Biology* **1**, 127–137.

Oomes, M.J.M., Olff, H. & Altena, H.J. (1996) Effects of vegetation management and raising the water table on nutrient dynamics and vegetation change in wet grassland. *Journal of Applied Ecology* **33**, 576–588.

Pfadenhauer, J. & Grootjans, A.P. (1999) Wetland restoration in Central Europe: aims and methods. *Applied Vegetation Science* **2**, 95–106.

Pickett, S.T.A., Ostfeld, R.S., Shachak, M. & Likens, G.E. (1997) *The ecological basis of conservation – heterogeneity, ecosystems, and biodiversity*. Chapman and Hall, New York.

Poschlod, P., Bakker, J.P., Bonn, S. & Fischer, S. (1996) Dispersal of plants in fragmented landscapes. In: *Species survival in fragmented landscapes* (eds J. Settele, C. Margules, P. Poschlod & K. Henle), pp. 123–127. Kluwer Academic Publishers, Dordrecht.

Poschlod, P. & WallisDeVries, M.F. (2002) The historical and socioeconomic perspective of calcareous grasslands – lessons from the distant and recent past. *Biological Conservation* **104**, 361–376.

Schaminée, J.H.J. & Jansen, A. (eds) (1998) *Wegen naar natuurdoeltypen*. Report 26 IKC-Natuurbeheer, Wageningen. (In Dutch.)

Schaminée, J.H.J., van Kley, J.E. & Ozinga, W.A. (2002) The analysis of long-term changes in plant communities: case studies from the Netherlands. *Phytocoenologia* **32**, 317–335.

Society for Ecological Restoration Science and Policy Working Group (2002) The SER Primer on Ecological Restoration. www.ser.org/

Spellerberg, I.F., Goldsmith, F.B. & Morris, M.G. (eds) (1991) *The scientific management of temperate communities for conservation*. Blackwell Scientific Publications, Oxford.

Strykstra, R.J. (2000) *Reintroduction of plant species: s(h)ifting settings*. Ph.D. Thesis, University of Groningen, Groningen.

Strykstra, R.J., Verweij, G.L. & Bakker, J.P. (1996) Seed dispersal by mowing machinery in a Dutch brook valley system. *Acta Botanica Neerlandica* **46**, 387–401.

Svenning, J.C. (2002) A review of natural vegetation openness in north-western Europe. *Biological Conservation* **104**, 133–148.

Tallowin, J.R.B. & Smith, R.E.N. (2001) Restoration of a Cirsio-Molinietum fen meadow on an agriculturally improved pasture. *Restoration Ecology* **9**, 167–178.

ter Heerdt, G.N.J., Schutter, A. & Bakker, J.P. (1997) Kiemkrachtig heidezaad in de bodem van ontgonnen heidevelden. *De Levende Natuur* **98**, 142–146. (In Dutch with English summary.)

Thompson, K., Bakker, J.P. & Bekker, R.M. (1997) *The soil seed banks of North West Europe: methodology, density and longevity*. Cambridge University Press, Cambridge.

Tiikka, P.M., Heikkilä, T., Heiskanen, M. & Kuitunen, M. (2001) The role of competition and rarity in the restoration of a dry grassland in Finland. *Applied Vegetation Science* **4**, 139–146.

van der Heijden, M.G.A., Boller, T., Wiemken, A. & Sanders, I.S. (1998) Different arbuscular mycorrhizal fungal species are potential determinants of plant community structure. *Ecology* **79**, 2082–2091.

van der Maarel, E. (1980) Towards an ecological theory of nature management. *Verhandlungen der Gesellschaft für Ökologie* **8**, 13–24.

van Duren, I.C., Strykstra, R.J., Grootjans, A.P., ter Heerdt, G.N.J. & Pegtel, D.M. (1998) A multidisciplinary evaluation of restoration measures in a degraded fen meadow (Cirsio-Molinietum). *Applied Vegetation Science* **1**, 115–130.

van Wieren, S.E. (1998) Effects of large herbivores upon the animal community. In: The impact of grazing on plant communities. In: *Grazing and Conservation Management* (eds M.F. WallisDeVries, J.P. Bakker & S.E. van Wieren), pp. 185–214. Kluwer Academic Publishers, Dordrecht.

van Wieren, S.E. & Bakker, J.P. (1998) Grazing for conservation in the twenty-first century. In: The impact of grazing on plant communities. In: *Grazing and Conservation Management* (eds M.F. WallisDeVries, J.P. Bakker & S.E. van Wieren), pp. 349–363. Kluwer Academic Publishers, Dordrecht.

Vera, F.W.M. (2000) *Grazing ecology and forest history.* CABI International, New York.

Verhagen, R., Klooker, J., Bakker, J.P. & van Diggelen, R. (2001) Restoration success of low-production plant communities on former agricultural soils after top-soil removal. *Applied Vegetation Science* **4**, 75–82.

WallisDeVries, M.F. (2002) Options for the conservation of wet grasslands in relation to spatial scale and habitat quality. In: *Multifunctional Grasslands, Quality Forages, Animal Products and Landscapes* (eds J.L. Durand, J.C. Emile, C. Huyghe & G. Lemaire), pp. 883–892. European Grassland Federation, La Rochelle.

Wells, T.C.E. (1980) Management options for lowland grassland. In: *Amenity Grassland, an Ecological Perspective* (eds I.H. Rorison & R. Hunt), pp. 175–195. Wiley, Chichester.

Westhoff, V. (1983) Man's attitude towards vegetation. In: *Man's impact on Vegetation* (eds W. Holzner, M.J.A. Werger & I. Ikusima), pp. 7–24. Junk, Den Haag.

Wheeler, B.D., Shaw, S.C., Fojt, W.J. & Robertson, R.A. (eds) (1995) *Restoration of temperate wetlands.* Wiley, Chichester.

Willems, J.H. (2001) Problems, approaches, and results in restoration of Dutch calcareous grassland during the last 30 years. *Restoration Ecology* **9**, 147–154.

Zobel, M., van der Maarel, E. & Dupré, C. (1998) Species pool: the concept, its determination and significance for community restoration. *Applied Vegetation Science* **1**, 55–66.

Plant invasions and invasibility of plant communities

Marcel Rejmánek, David M. Richardson and Petr Pyšek

13.1 Introduction

Some 2500 yr ago, Heraclitus of Ephesus said that 'All things change . . . and you cannot step twice into the same stream'. Today, ecologists would not only say the same about streams but also about vegetation. Plant communities change with time due to changes in the environment (Chapter 7), biotic interactions (Chapter 9) and invasions of alien species and genotypes, introduced intentionally or accidentally by humans. Invasions have received detailed attention only recently. There have always been migrating taxa, but now the rate of human-assisted introductions of new taxa is several orders of magnitude higher. In California, for example, more than 1000 alien plant species were intentionally or non-intentionally introduced and established viable populations over the last 250 yr. In the Galápagos Islands, over 3 million yr of their history, only one new plant species arrived with birds or sea currents every 10,000 yr. However, over the last 20 yr the introduction rate has been *c.* 10 species per yr, or some 100,000 times the natural arrival rate (Tye 2001).

Three basic questions arise:

1 What kind of ecosystems are more (or less) likely to be invaded by alien plants?
2 What kind of plants are the most successful invaders and under what circumstances?
3 What is the impact of the plant invaders?

13.2 Definitions and major patterns

Unlike **natives** (taxa that evolved in the region or reached it from another area where they are native without help from humans), **aliens** ('non-native' or 'exotic') owe their presence to direct or indirect activities of humans. Most of them occur only temporarily and are not able to persist for a long time without human-assisted input of diaspores; these are termed **casual**. **Naturalized** taxa form sustainable populations without direct human help but do not necessarily spread; the ability to spread characterizes a subset termed **invasive** taxa. This distinction is critical because not all naturalized taxa reported in floras and checklists are invasive. Not all naturalized plant taxa, and not even all invaders, are harmful invaders – the last-mentioned should

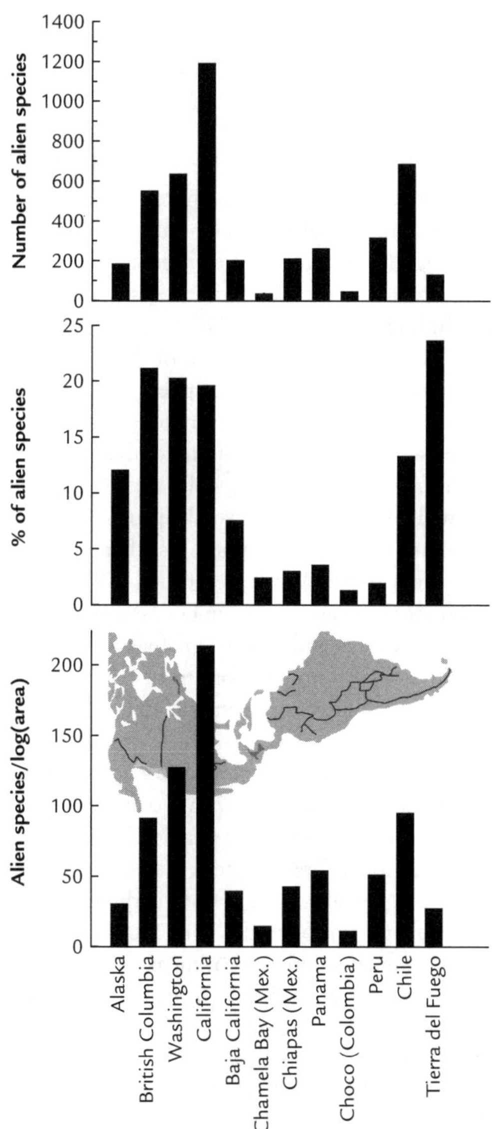

Fig. 13.1 Total number of alien plant species, percentage of alien plant species, and number of alien plant species per log(area) along the Pacific coast of Americas. 'Alien species' here are plants growing in individual areas without cultivation. Not all of them are fully naturalized and even fewer are invasive. Nevertheless, numbers of naturalized and invasive species are proportional to numbers of 'alien species' in this diagram. Primary data or references are in Kartesz & Meacham (1999) and Vitousek *et al.* (1997).

rather be called exotic *weeds* or exotic pest plants (Booth *et al.* 2003; Richardson *et al.* 2000a).

Weeds are both native and alien and the alien element in weed floras varies over the world. Most weedy taxa in southern Europe, Malaya, Mexico or Taiwan are native, whereas most weedy taxa in Australia, the USA, Chile, South Africa, New Zealand, Hawai'i, and many other islands are non-native. There may be inherent differences in invasibility of different parts of the world. Uneven representation of alien, mostly naturalized, plant species in regional floras along the Pacific shore of the Americas illustrates this point (Fig. 13.1). These differences are certainly partly due to the history of human colonization and trade. Nevertheless, similar patterns can also be recognized on other continents (Rejmánek 1996; Lonsdale 1999). For instance, areas with mediterranean climates (with the exception of the Mediterranean Basin itself) seem to be more vulnerable and the tropics appear more resistant to plant invasions. This should not be generalized, however. Savannas and especially disturbed deforested areas in the Neotropics are very often dominated by African grasses such as *Hyparrhenia rufa* and *Melinis minutiflora*, while similar tropical habitats in Africa and Asia are dominated by Neotropical woody plants, e.g. *Lantana camara* and *Opuntia* spp. The absolute number of alien species, therefore, is not necessarily the best indicator of ecosystem invasibility, at least at this scale. Undisturbed tropical forests, however, harbour only a very small number of alien plant species and most of them do not spread beyond trails and gaps (Rejmánek 1996). It is probably not the extraordinary species diversity of tropical forests that is important but simply the presence of fast-growing multilayered vegetation that makes undisturbed tropical forests resistant to invasions.

At the regional scale enormous differences in presence and abundance of invaders among different communities (ecosystems) within one area seem to be the rule. An overview is now available for Central Europe (Table 13.1). Alien species are concentrated mostly in vegetation of deforested mesic habitats with frequent disturbance (Pyšek *et al.* 2002a,b). In general, native forests harbour a low number and proportion of both archaeophytes (introduced before 1500) and neophytes (introduced later); alien species are completely missing from many types of natural vegetation, e.g. bogs, natural *Picea abies* forest, and rare in many natural herbaceous communities. Herbaceous communities of extreme habitats and/or with strong native clonal dominants (*Nanocyperion flavescentis*, *Phragmition*, *Nardion*) seem to be most resistant to invasions of both archaeophytes and neophytes. In general, Californian lowland communities (Fig. 13.2) are more invaded than corresponding communities in Europe. However, there are some important similarities. Open and disturbed communities are more invaded, while undisturbed forests are less invaded.

Data from California (Fig. 13.2) suggest that proportions of alien species numbers are reasonably well correlated with their dominance (cover). This is probably attributable to a simple sampling effect: with increasing proportion of alien species, there is an increasing chance that one or more of them will dominate the community. While there appears to be a general agreement between the proportion of alien species numbers and their actual importance (cover and biomass), some exceptions are quite remarkable. While alien species number in *Chelidonio-Robinion* woodland is certainly not exceptionally high (Table 13.1), the dominant *Robinia pseudoacacia* is an alien

Table 13.1 Numbers of alien species, classified according to the time of introduction into archaeophytes and neophytes, in representative plant communities of the Czech Republic on the phytosociological alliance level. Within each vegetation group, alliances are ranked according to decreasing total number of alien species. Data from Pyšek *et al.* (2002a).

	Number of archaeophytes	Number of neophytes	% invasive among neophytes
Ruderal vegetation			
Sisymbrion officinalis tall-herb comm. of annuals on nitrogen-rich mineral soils	96	106	9.4
Aegopodion podagrariae nitrophilous fringe comm.	16	76	36.8
Arction lappae nitrophilous comm. of dumps and rubbish tips	36	45	31.1
Balloto-Sambucion shrub comm. of ruderal habitats	18	34	41.2
Matricario-Polygonion arenastri comm. of trampled sites	20	20	15.0
Potentillion anserinae comm. of salt-rich ruderal habitats	12	20	10.0
Convolvulo-Agropyrion comm. of field margins and disturbed slopes	24	16	31.3
Onopordion acanthii thermophilous comm. of village dumps and rubbish tips	34	8	12.5
Weed communities of arable land			
Veronico-Euphorbion weed comm. of root crops on basic soils	47	28	21.4
Panico-Setarion weed comm. of root crops on sandy soils	28	15	40.0
Caucalidion lappulae thermophilous weed comm. on base-rich soils	79	11	0.0
Aphanion weed comm. on acid soils	41	8	12.5
Sherardion weed comm. of cereals on medium base-rich soils	47	7	14.3
Grasslands			
Arrhenatherion mesic meadows	15	56	25.0
Festucion valesiacae narrow-leaved dry grasslands	12	12	0.0
Bromion erecti broad-leaved dry grasslands	6	8	0.0
Nardion subalpine grasslands	0	1	0.0
Helianthemo cani-Festucion pallentis rock-outcrop vegetation	2	0	-
Forests			
Alnion incanae ash-alder alluvial forests	4	15	40.0
Carpinion oak-hornbeam forests	6	14	14.3
Chelidonio-Robinion plantations of *Robinia*	5	10	60.0
Genisto germanicae-Quercion dry acidophilous oak forests	1	11	36.4
Tilio-Acerion ravine forests	5	8	37.5
Luzulo-Fagion acidophilous beech forests	0	4	50.0
Quercion pubescenti-petraeae thermophilous oak forests	1	2	0.0
Quercion petraeae acidophilous thermophilous oak forests	0	2	50.0
Salicion albae willow-poplar forests of lowland rivers	0	2	50.0

Table 13.1 (*cont'd*)

	Number of archaeophytes	Number of neophytes	% invasive among neophytes
Alnion glutinosae alder carrs	0	2	0.0
Fagion beech forests	0	1	100.0
Betulion pubescentis birch mire forests	0	0	–
Piceion excelsae spruce forests	0	0	–
Aquatic and wetland vegetation			
Lemnion minoris macrophyte vegetation of eutrophic and mesotrophic still waters	0	3	0.0
Cardamino-Montion forest springs without tufa formation	0	2	50.0
Phragmition reed beds of eutrophic still waters	1	1	0.0
Magnocaricion elatae tall-sedge beds	0	1	0.0
Nanocyperion flavescentis annual vegetation on wet sand	1	0	–

tree from North America. On the other hand, there are many alien species in some grassland communities (*Festucion valesiaceae*, *Bromion erecti*), but dominants are exclusively native and aliens are rarely invasive.

13.3 Invasibility of plant communities

Can we say anything conclusive about differences in invasibility (vulnerability to invasions) of particular ecosystems? Analyses of ecosystem invasibility based just on one-point-in-time observations (*a posteriori*) are usually unsatisfactory (Rejmánek 1989). In most of the cases we do not know anything about the quality, quantity and regime of introduction of alien propagules. Nevertheless, available evidence indicates that only very few non-native species invade successionally advanced plant communities (Rejmánek 1989; Meiners *et al.* 2002). Here, however, the quality of common species pools of introduced alien species – mostly rapidly growing and reproducing *r*-strategists – is probably an important part of the story. These species are mostly not shade-tolerant and many of them are excluded during the first 10 or 20 yr of uninterrupted secondary succession (Fig. 13.3), or over longer periods of primary successions. However, some *r*-strategists are shade-tolerant. e.g. *Alliaria petiolata*, *Microstegium vimineum* and *Sapium sebiferum*. Such species can invade successionally advanced plant communities and, therefore, represent a special challenge to managers of protected areas.

Plant communities in mesic environments seem to be more invasible than communities in extreme terrestrial environments (Rejmánek 1989). Apparently xeric environments are not favourable for germination and seedling survival of many introduced species (abiotic resistance) and wet terrestrial habitats do not provide resources – mainly light – for invaders because of fast growth and high competitiveness of resident

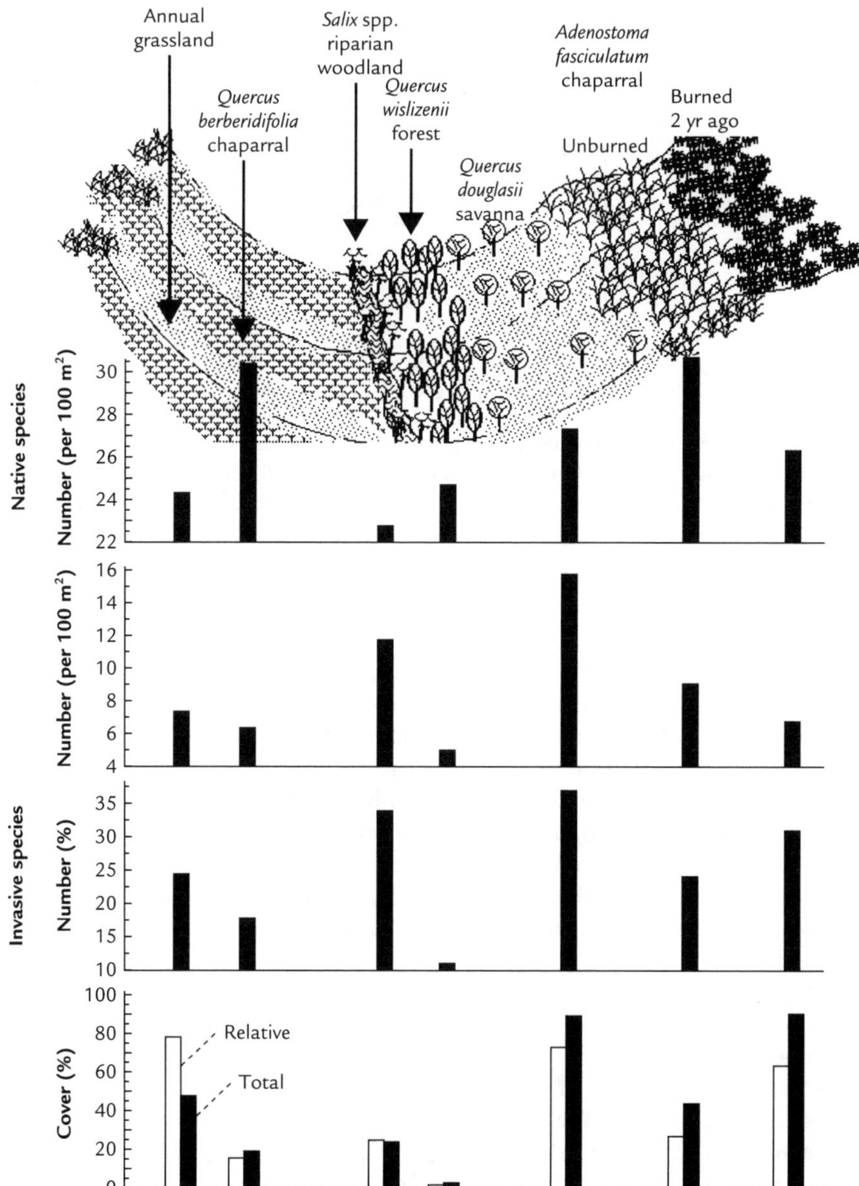

Fig. 13.2 Native and invasive species in seven plant communities of the Stebbins Cold Canyon Reserve, North Coast Ranges, California (150–500 m above sea level). Each column represents a mean from three 100-m^2 plots. 'Relative cover' of invaders is their cover with respect to the cumulative vegetation cover in all strata (herbs, shrubs and trees). Comparing means for individual vegetation types, the only significant correlation is between percentage of invasive species and total cover of invasive species (r = 0.75; n = 7; p = 0.05). Rejmánek (unpublished data).

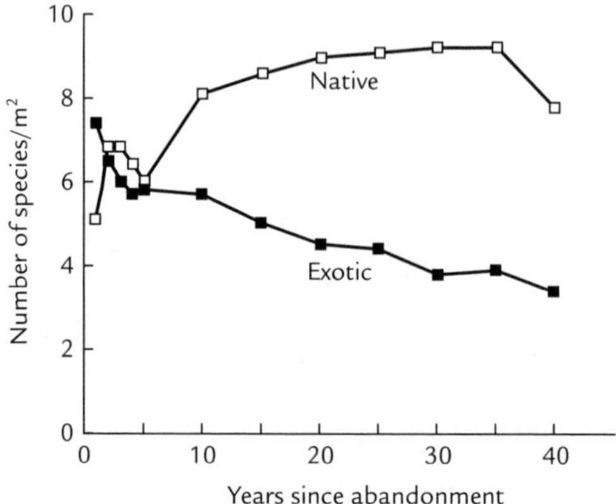

Fig. 13.3 Effect of time since abandonment on the mean species richness of native and exotic species over 40 yr of old-field succession in New Jersey (Meiners *et al.* 2002). Decline of the mean percentage of exotic species is even more dramatic, decreasing from 58 to 28%. Total and relative cover of exotic species declines significantly as well, but there can be a temporary increase during the first 5 or 10 yr of succession; see also Rejmánek (1989).

species (biotic resistance). We have to be cautious, however, in the interpretation of these patterns. When the 'right' species are introduced, even ecosystems that have been viewed as invasion-resistant for a long time may turn out to be susceptible, for instance the Mojave and Sonoran deserts are facing recent invasions by *Brassica tournefortii* and *Pennisetum ciliare.* Open water is notoriously known as open to all kinds of exotic aquatic plants. In general, disturbance, nutrient enrichment, slow recovery rate of resident vegetation, and fragmentation of successionally advanced communities promote plant invasions (Rejmánek 1989; Hobbs & Huenneke 1992; Cadenasso & Pickett 2001). In addition, the increasing CO_2 level will probably accelerate invasions in arid ecosystems (Smith *et al.* 2000).

A general theory of invasibility was put forward recently by Davis *et al.* (2000): intermittent resource enrichment (eutrophication) or release (due to disturbance) increases community susceptibility to invasions. Invasions occur if/when this situation coincides with availability of suitable propagules. The larger the difference between gross resource supply and resource uptake, the more susceptible the community to invasion. This was anticipated by Vitousek & Walker (1987) (Fig. 13.4) and expressed more rigorously by Shea & Chesson (2002). Davis & Pelsor (2001) experimentally manipulated resources and competition in a herbaceous community, and showed that fluctuations in resource availability of as little as one week in duration could greatly enhance plant invasion success (survival and cover of alien plants) up to one year after such events.

Experiments on invasibility of different types of ecosystems have been gaining momentum in recent years (Hector *et al.* 2001; Fargione *et al.* 2003). Crawley *et al.* (1999) and Davis *et al.* (2000) suggested that there is no necessary relationship between invasibility of a plant community and number of species present in that community.

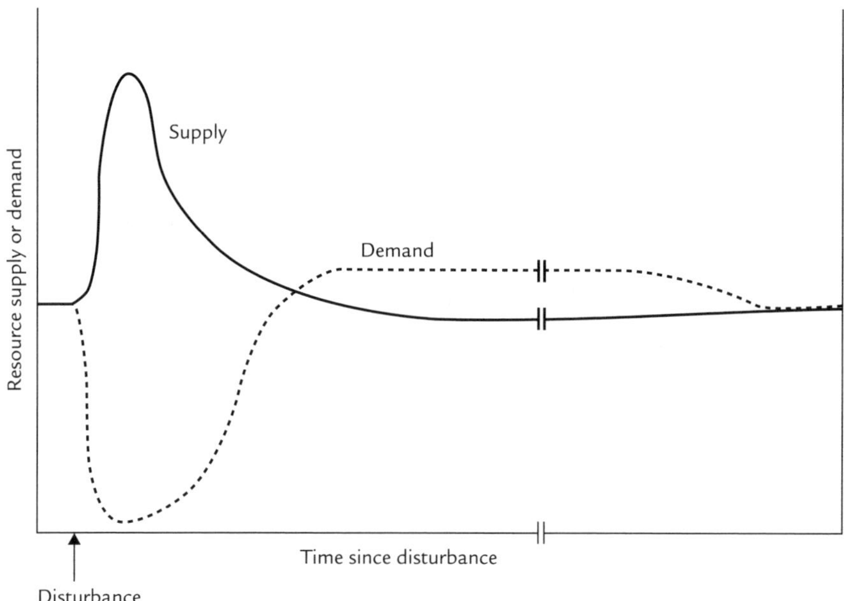

Fig. 13.4 Changes in supply and demand of resources after disturbance in terrestrial ecosystems. Resource availability is generally at its maximum shortly after disturbance, although conditions of bare ground can inhibit seedling establishment in some sites. Modified from Vitousek & Walker (1987).

Other studies show that such a relationship exists: positive at the landscape scale (e.g. Stohlgren *et al.* 1999; Sax 2002) and negative at scales of < 1 m^2 (neighbourhood scales *sensu* Levine 2000). Wardle (2001) provided a valuable methodological criticism of many of these studies. Nevertheless, Kennedy *et al.* (2002) concluded that in herbaceous communities neighbourhood species richness (within 5–15 cm radius) represents 'an important line of defence against the spread of invaders'. Hubbell *et al.* (2001) found that in an undisturbed forest in Panama neighbourhood species richness (within 2.5–50 m radius) had a weak but significantly negative effect on focal tree survival. Is there a generalization emerging from studies on neighbourhood scales? This would not be surprising as vascular plants are sedentary organisms and actual interactions are occurring among neighbouring individuals.

The experimental studies mentioned above relate the number of resident plant species to the number and abundance of alien plant species that establish or become invasive. But, the diversity of organisms at other trophic levels in the receiving environment may well be as important, if not more important, than the number of plant species. We can expect that diverse assemblages of mutualists (pollinators, seed dispersers, microbiota that form symbioses with plant roots) would promote invasibility (Simberloff & Von Holle 1999; Richardson *et al.* 2000b). Recent experiments by Klironomos (2002) on species from Canadian old-fields and grasslands showed that native rare plant species accumulate soil plant pathogens rapidly, while invasive species do not. This result has potentially very important consequences. When introduced

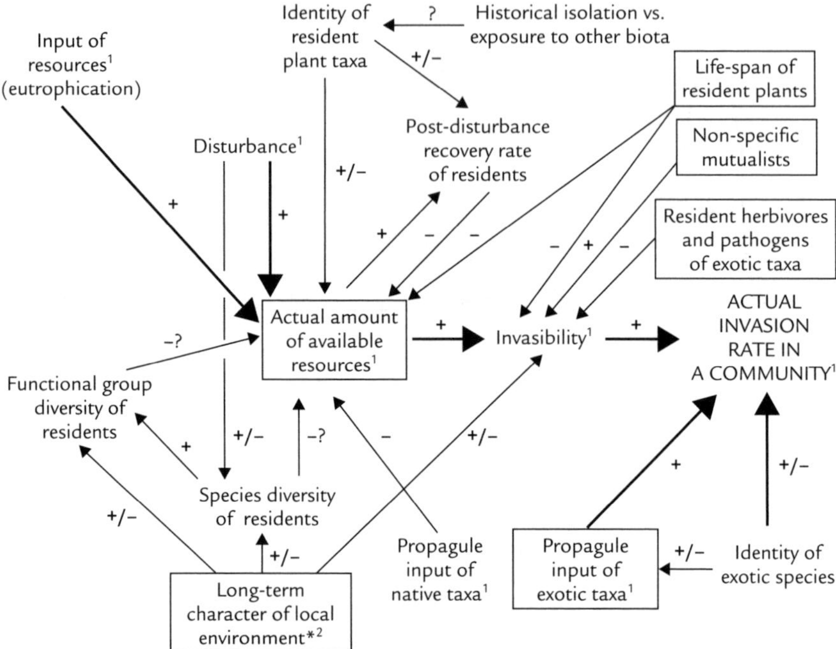

Fig. 13.5 Causal relationships between factors and processes which are assumed responsible for invasions of exotic species into plant communities. The most important relationships are indicated by thick arrows. * = Spatial heterogeneity, (micro)climate and long-term regime of available resources and toxic compounds. 1 = Time scale: days–years. 2 = Time scale: years–centuries. The key components are in boxes.

outside of their native territories, plants are often liberated from their enemies, including soil pathogens. This is a clear advantage that makes natives and aliens, at least temporarily, different (but see Colautti *et al.* 2004). At the same time, however, as many mycorrhizal fungi can associate with a broad range of plant taxa, root beneficial symbionts are likely to be always available to many alien plants.

A conceptual cause-effect diagram (Fig. 13.5) captures all the fundamental aspects of the ongoing debate on the issue of invasibility. The fact that both invasibility and species diversity of residents is regulated in a similar way by the same set of factors – (micro)climate, spatial heterogeneity, long-term regime of available resources – explains why there are so many reports of positive correlation between numbers of native and non-native species when several different communities or areas are compared. Fast post-disturbance recovery of residents may be a key factor making the wet tropics more resistant to plant invasions – measured as number of invading species per log(area) (Rejmánek 1996).

However, there is very likely one extra factor that is currently poorly understood: the historical and prehistoric degree of exposure of resident taxa to other biota (Fig. 13.5). Is this the reason why islands are more vulnerable and Eurasia least vulnerable to invasions? Is instability of so many man-made monocultures a result of the 'lack of any significant history of co-evolution with pests and pathogens' (May 1981)? Actual species richness may not be as important as the complexity of assembly history. In addition to mathematical models and computer simulations (Law 1999) relevant experiments with plant communities will have to be designed. Artificial experimental plant communities that are so often used for invasibility experiments have a clear advantage of homogeneous substrata and microclimates. However, assembly processes here are very short and/or artificially directed via arbitrary species pool selection, weeding, reseeding, etc. The existence of well-established phytosociological associations and the fact that plant species are combined in highly non-random patterns within their natural communities (Gotelli & McCabe 2002) indicate that historical assembly processes cannot be substituted by arbitrary mixtures of species.

Finally, longevity/persistence of resident plants is a distinct component of resistance to invasions (Von Holle *et al.* 2003), especially in forest communities, resulting in 'biological inertia', including allelopathic chemicals produced by living or dead residents.

13.4 Habitat compatibility

Identity of exotic taxa (Fig. 13.5) is important for two reasons. First, they may or may not survive and reproduce in habitats where they are introduced. Second, they may or may be not spread and become invasive. Recipient habitat compatibility is usually treated as a necessary condition for all invasions. The match of primary (native) and secondary (adventive) environments of an invading taxon is not always perfect but usually reasonably close (e.g. Beerling *et al.* 1995; Rejmánek 2000; Widlechner & Iles 2002). In North America, for example, latitudinal ranges of naturalized European plant species from the Poaceae and Asteraceae are on average 15–20° narrower than their native ranges in Eurasia and North Africa. These differences essentially reflect the differences in the position of corresponding isotherms and major biomes in Eurasia and North America. Major discrepancies between primary and secondary ranges have been found for aquatic plants where secondary distributions are often much less restricted than their primary distributions. Vegetative reproduction of many aquatic species seems to be the most important factor. Obviously, secondary ranges, if already known from other invaded continents, should be employed in any prediction of habitat compatibility.

As for plants introduced (or considered for introduction) from Europe, several useful summaries of their 'ecological behaviour' are available. Especially the combination of Ellenberg indicator values (Ellenberg *et al.* 1992) with Grime's functional types (strategies) (Grime *et al.* 1988) can be a powerful tool for predictions of habitat compatibility of European species. The strength of affiliation with phytosociological syntaxa (section 1.4.2) is well known for almost all European taxa. Environmental conditions (climate, soil, disturbance, management) of all syntaxa are available and potential habitat compatibility of taxa can be extracted from the European literature. Knowledge of this 'phytosociological behaviour' of taxa allows predictions about compatibility with analogous (vicarious) vegetation types, even if these will not always be correct.

'Open niches', habitats that can support life forms that are not present in local floras for historical and/or evolutionary reasons, deserve special attention. Dramatic invasions have occurred in such habitats, e.g. *Ammophila arenaria* (a rhizomatous grass) in coastal dunes in California, *Lygodium japonicum* (a climbing fern) in bottomland hardwoods from Louisiana to Florida, *Pinus* spp. and *Acacia* spp. in South African shrubby fynbos, *Opuntia* spp. (Cactaceae) in East African savannas, *Rhizophora mangle* (mangrove) in tree-less coastal marshes of Hawai'i, and the tree *Cinchona pubescens* (Rubiaceae) in mountain shrub communities on Santa Cruz Island, Galápagos. The explanation of such invasions is confirmed by experiments showing that the competitive inhibition of invaders increases with their functional similarity to resident abundant species (Fargione *et al.* 2003).

13.5 Propagule pressure and residence time

The notion of habitat compatibility includes all factors embraced in the concept of habitat. Most effort in assessments of habitat compatibility has been devoted to climatic and substrate compatibility, although it is well known that many other factors influence range limits. Invasions result from an interplay between habitat compatibility and propagule pressure (Fig. 13.5). This is illustrated by the invasion dynamics of the New Zealand tree *Metrosideros excelsa* (Myrtaceae) in South African fynbos (details in Richardson & Rejmánek 1998). Multiple regression of the number of *Metrosideros* saplings on a potential seed rain index (PSRI) and soil moisture revealed that, in this case, both factors are about equally important (Fig. 13.6). This example clearly shows that classification of habitats or communities into 'invasible' and 'non-invasible' cannot be absolute in many situations. Habitats that are currently unaffected (or only slightly affected) by plant invasions may be deemed resistant to invasion. However, as populations of alien plants build up and propagule pressure (Foster 2001) increases outside or within such areas, invasions could well start or increase. Another aspect is the propagule pressure of native species: if propagules of natives are not available, as for instance on abandoned fields in California, the 'repairing' function of ecological succession (Fig. 13.3) does not work.

Residence time – the time since the introduction of a taxon to a new area – represents another dimension of propagule pressure. As we usually do not know exactly when a taxon was introduced, we use a 'minimum residence time' based on herbarium specimens or reliable records. Nevertheless, the number of discrete localities of naturalized species is significantly positively correlated with minimum residence time (Fig. 13.7). One trivial but important conclusion is that the earlier an exotic pest plant taxon is discovered, the better is the chance of its eradication.

13.6 What are the attributes of successful invaders?

The identity of introduced species certainly matters (Fig. 13.5). Extrapolations based on previously documented invasions are fundamental for predictions in invasion ecology. With the development of relevant databases, this approach should lead to

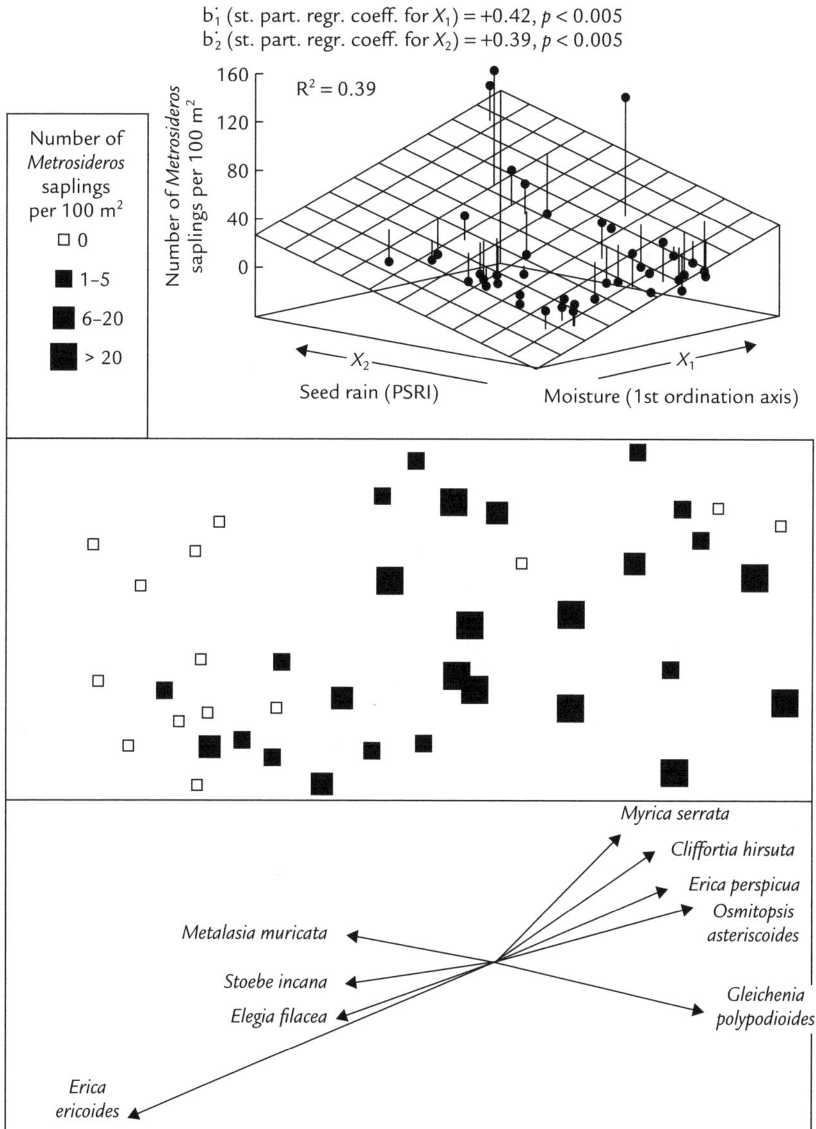

Fig. 13.6 The dependence of the sapling density of *Metrosideros excelsa* on potential seed rain index (PSRI) and moisture in fynbos of the Western Cape, South Africa. PSRI = SUM($1/d_i$), where d_i is distance to the i-th mature tree in metres within the radius 300 m. The first ordination axis (below) serves as a surrogate for moisture gradient. Standardized partial regression coefficients (st. part. regr. coeff.) of the multiple regression are almost identical. Therefore, both independent variables – environment and propagule pressure – are equally important in this case. M. Rejmánek & D.M. Richardson (unpublished data).

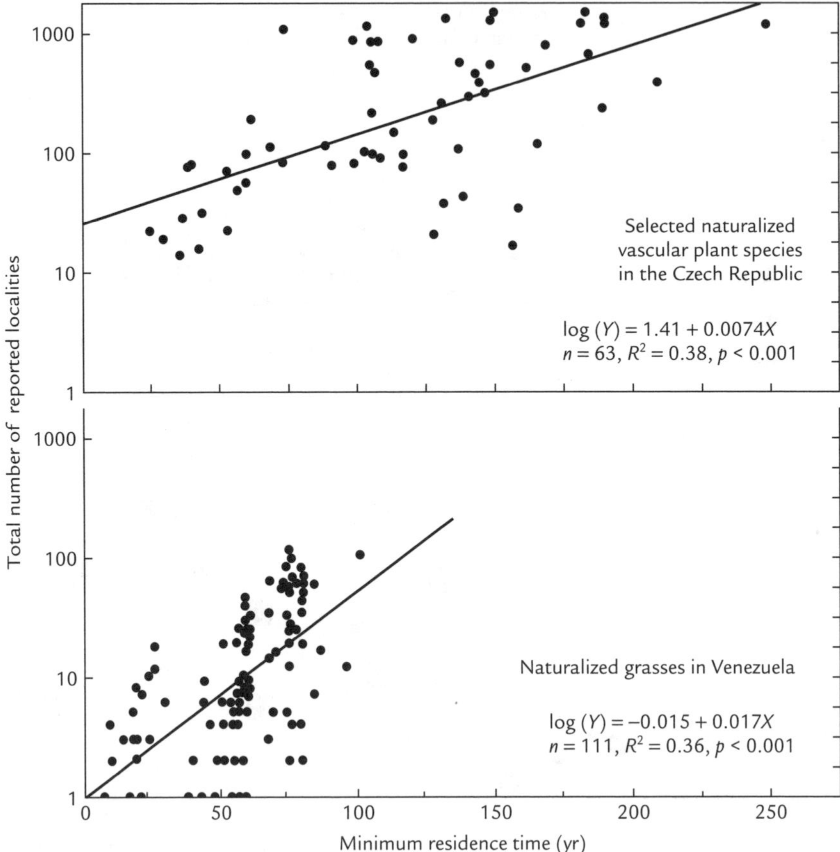

Fig. 13.7 The dependence of the total number of reported localities on the minimum residence time (yr since the first record) of selected naturalized species in the Czech Republic and Venezuela. P. Pyšek & M. Rejmánek (unpublished data).

immediate rejection of imports of many invasive taxa (prevention) and prioritized control of those that have already been established. Such transregional, taxon-specific extrapolations are very useful in many situations, but our lack of understanding makes them intellectually unsatisfactory. Understanding how and why certain biological characters promote invasiveness is very important since even a global database will not cover all potentially invasive taxa. In New Zealand, for example, Williams *et al.* (2001) reported that 20% of the alien weedy species collected for the first time in the second half of the 20th century had never been reported as invasive outside New Zealand. For these reasons, several attempts have been made to find differences in biological characteristics of non-invasive and invasive taxa or, at least, between native taxa and non-native invasive taxa in particular floras. Major predictions made by an emerging theory of plant invasiveness were summarized recently by Booth *et al.* (2003), Myers & Bazely (2003) and Rejmánek *et al.* (2004):

1 Fitness homoeostasis (the ability of an individual or population to maintain relatively constant fitness over a range of environments) promotes invasiveness.
2 Genetic change can facilitate invasions, but many species have sufficient phenotypic plasticity to exploit new environments.
3 Several characters linked to reproduction and dispersal are key indicators of invasiveness.
4 Seed dispersal by vertebrates is implicated in many plant invasions.
5 Low relative growth rate of seedlings and low specific leaf area are good indicators of low plant invasiveness in many environments.
6 Large native range is an indicator of potential invasiveness. However, several important exceptions are known (*Impatiens parviflora*, *Pinus radiata*).
7 Vegetative reproduction is responsible for many plant invasions, especially in aquatic and wetland environments.
8 Alien taxa are more likely to invade a continental area if native members of the same genus (and family) are absent, partly because many herbivores and pathogens cannot switch to phylogenetically distant taxa (Agrawal & Kotanen 2003; Mitchell & Power 2003). However, invaders on islands seem to exhibit the opposite tendency (Duncan & Williams 2002).
9 The ability to utilize generalist mutualists greatly improves an alien taxon's chances of becoming invasive.
10 Efficient competitors for limiting resources are likely to be the best invaders in natural and semi-natural ecosystems (Tilman 1999; Shea & Chesson 2002).
11 Characters favouring passive dispersal by humans greatly improve an alien plant taxon's chance of becoming invasive.

Points 3, 4 and 5 are particularly relevant. Reproduction and dispersal are key issues. Consistent seed production in new environments is usually associated with rather simple or flexible breeding systems. For example, rare and endangered taxa in the genus *Amsinckia* (e.g. *A. furcata*, *A. grandiflora*) are heterostylic, while derived invasive taxa (*A. menziesii*, *A. lycopsoides*) are homostylic and self-compatible. Self-pollination has been consistently identified as a mating strategy in colonizing species. Nevertheless, not all sexually reproducing successful invaders are selfers. Pannel & Barrett (1998) examined the benefits of reproductive assurance in selfers versus outcrossers in model metapopulations. Their results suggest that an optimal mating system for a sexually reproducing invader in a heterogeneous landscape should include the ability to modify selfing rates according to local conditions. In early stages of invasions, when populations are small, plants should self to maximize fertility. However, later, when populations are large and pollinators and/or mates are not limiting, outcrossing will be more beneficial, mainly due to increasing genetic polymorphism.

Invasiveness of woody taxa in disturbed landscapes is associated with small seed mass (< 50 mg), a short juvenile period (< 10 yr), and short intervals between large seed crops (1–4 yr) (see Fig. 13.8 and Rejmánek & Richardson 1996, 2003). These three attributes contribute, directly or indirectly, to higher values of three parameters which are critical for population expansion: net reproduction rate, reciprocal of mean age of reproduction, and variance of the marginal dispersal density. For wind-dispersed seeds, the last parameter is negatively related to terminal velocity of seeds which is positively related to $\sqrt{}$(seed mass) (Rejmánek *et al.* 2004). Because

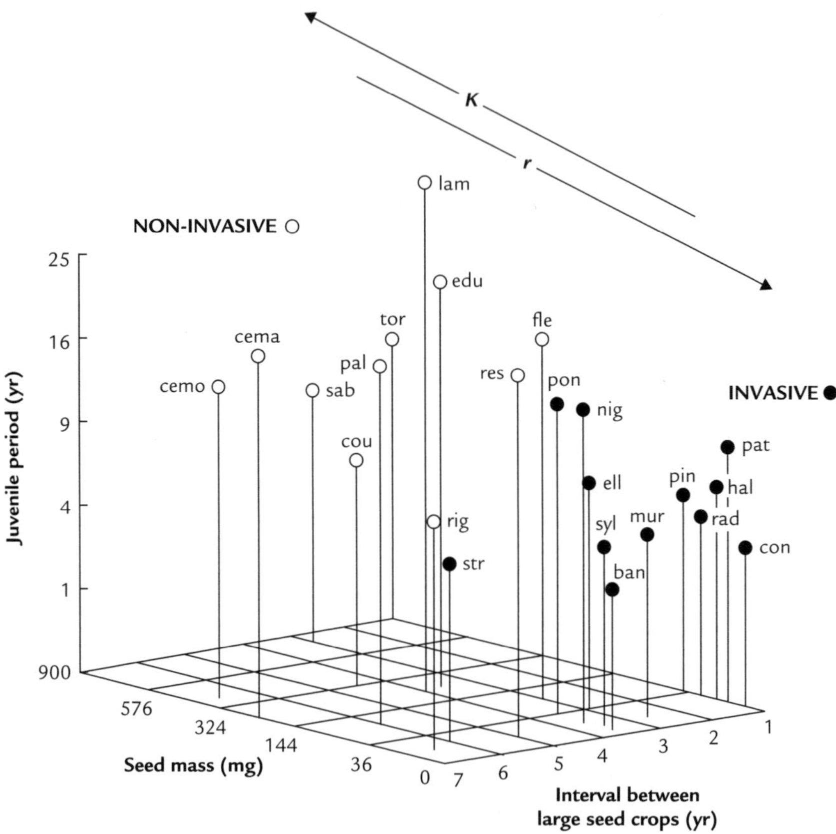

Fig. 13.8 Distribution of 23 frequently cultivated *Pinus* species in a space created by three biological variables critical in separating invasive and non-invasive species. The *K*–*r* selection continuum running from the upper left to the lower right corner of the diagram also represents the direction of the discriminant function (Z) separating non-invasive and invasive *Pinus* species. $Z = 23.39 - 0.63\sqrt{M} - 3.88\sqrt{J} - 1.09S$, where M = mean seed mass (in mg), J = minimum juvenile period (in yr), and S = mean interval between large seed crops (in yr). Pine species with positive Z scores are invasive and species with negative Z scores are non-invasive. Species abbreviations: ban = *banksiana*; cema = *cembra*; cemo = *cembroides*; con = *contorta*; cou = *coulteri*; edu = *edulis*; ell = *elliottii*; eng = *engelmannii*; fle = *flexilis*; hal = *halepensis*; lam = *lambertiana*; mur = *muricata*; nig = *nigra*; pal = *palustris*; pat = *patula*; pin = *pinaster*; pon = *ponderosa*; rad = *radiata*; res = *resinosa*; sab = *sabiniana*; str = *strobus*; syl = *sylvestris*; tor = *torreyana*.

of the trade-off between seed number and mean seed mass, small-seeded taxa usually produce more seeds relative to biomass. Invasions of woody species with very small seeds (< 3 mg), however, are limited to wet and preferably mineral substrata (Rejmánek & Richardson 1996). Based on invasibility experiments with herbaceous species, it seems that somewhat larger seeds (3–10 mg) extend species habitat compatibility (Burke & Grime 1996). As seed mass seems to be positively correlated with habitat shade, large-seeded aliens may be more successful in undisturbed, successionally more mature plant communities.

Seed dispersal by vertebrates is responsible for the success of many invaders in disturbed as well as 'undisturbed' habitats (Binggeli 1996; Rejmánek 1996; Richardson *et al.* 2000b; Widlechner & Iles 2002). Even some very large-seeded alien species like *Mangifera indica* can be dispersed by large mammals. The proportion of naturalized plant species dispersed by vertebrates seems to be particularly high in Australia: over 50% (Tables 2.1 and 7.3 in Specht & Specht 1999). The assessment of whether there is an opportunity for vertebrate dispersal is an important component of the screening procedure for woody plants (Table 13.2). However, vertebrate seed dispersal in relation to invasions is complicated (Richardson *et al.* 2000b).

Table 13.2 General rules for detection of invasiveness of woody seed plants based on values of the discriminant function Z^*, seed mass values, and presence or absence of opportunities for vertebrate dispersal. Modified from Rejmánek & Richardson (1996).

		Opportunities for vertebrate dispersal	
		Absent	***Present***
$Z > 0$	Dry fruits and seed mass > 3 mg	Likely[1]	Very likely[2]
	Dry fruits and seed mass < 3 mg	Likely in wet habitats[3]	Likely in wet habitats[3]
	Fleshy fruits	Unlikely[4]	Very likely[5]
$Z < 0$		Not unless dispersed by water[6]	Possibly[7]

$^*Z = 23.39 - 0.63\sqrt{M} - 3.88\sqrt{J} - 1.09S$; where M = mean seed mass (mg); J = minimum juvenile period (yr); S = mean interval between large seed crops (yr). Z was derived on the basis of *a priori* defined groups of invasive and non-invasive *Pinus* species. The function was later successfully applied on other gymnosperms and, as a component of this table, even on woody angiosperms. Note that parameters in this discriminant function are somewhat different from those in Rejmánek & Richardson (1996). This is due to exclusion of *Pinus caribaea* from the data set used for estimation of the parameters. This species, that is in general non-invasive in many countries, is highly invasive in New Caledonia.

[1] For example *Acer platanoides, Cedrela odorata, Clematis vitalba, Cryptomeria japonica, Cytisus scoparius, Pinus radiata, Pittosporum undulatum, Pseudotsuga menziesii, Robinia pseudoacacia, Senna* spp., *Tecoma stans.*

[2] Species with large arils (*Acacia cyclops*) are dispersed by birds.

[3] For example *Alnus glutinosa* in New Zealand, *Eucalyptus globulus* in California, *Melaleuca quinquenervia* in southern Florida, *Tamarix* spp. in the south-western USA, *Cinchona pubescens* in Galápagos and *Baccharis halimifolia* in Australia.

[4] *Feijoa sellowiana* and *Nandina domestica* are frequently cultivated but non-invasive species in California. The second species, however, is dispersed by birds and water in the south-eastern USA.

[5] *Berberis* spp., *Clidemia hirta, Crataegus monogyna, Lantana camara, Lonicera* spp., *Myrica faya, Passiflora* spp., *Psidium guajava, Rubus* spp., *Schinus terebinthifolius, Solanum mauritianum.*

[6] *Nypa fruticans* is spreading along tidal streams in Nigeria and Panama. *Thevetia peruviana* can be dispersed over short distances by surface run-off in Africa.

[7] Examples of invasive species in this group are *Pinus pinea, Melia azedarach,* and *Maesopsis eminii* in Africa, *Quercus rubra* in Europe, *Mangifera indica* in the Neotropics, and *Persea americana* in Galápagos.

Table 13.3 Differences between means of growth related variables for non-invasive, unclassified and invasive *Pinus* species. Same superscript letters for each variable denote means that are not significantly different ($p > 0.05$, Scheffé test). From Grotkopp *et al.* (2002).

Variable	Non-invasive	Unclassified	Invasive
	$n = 8$	$n = 8$	$n = 8$
Relative growth rate (RGR, $mg{\cdot}g^{-1}{\cdot}d^{-1}$)	23^a	33^b	37^b
Net assimilation rate (NAR, $mg{\cdot}cm^{-2}{\cdot}d^{-1}$)	0.505^a	0.559^a	0.572^a
Leaf area ratio (LAR, $cm^2{\cdot}g^{-1}{}_{plant}$)	50^a	67^b	73^b
Specific leaf area (SLA, $cm^2{\cdot}g_{leaf}$)	79^a	101^b	111^b
Relative leaf production rate ($leaf{\cdot}leaf^{-1}{\cdot}d^{-1}$)	0.014^a	0.022^b	0.024^b

Many ecologists assume that a high relative growth rate should be an important characteristic of invasive plant taxa in disturbed or open areas, especially in resource-rich environments. Only a few studies have demonstrated this experimentally (Pattison *et al.* 1998; Grotkopp *et al.* 2002). An analysis of seedling growth rates for 29 *Pinus* species (Table 13.3) revealed that: (i) relative growth rate (RGR) of invasive species is significantly higher than that of non-invasive species; (ii) differences in RGR are primarily determined by leaf area ratio (LAR; leaf area per plant biomass); and (iii) LAR is primarily determined by specific leaf area (SLA; leaf area per leaf biomass). Consequently, invasive species have significantly higher specific leaf area. Moreover, there is a highly significant ($R^2 = 0.685$; $p < 0.001$) positive relationship between RGR and invasiveness of the *Pinus* species (Fig. 13.8, Table 13.2). Identical results were obtained using phylogenetically independent contrasts (Grotkopp *et al.* 2002). High SLA was implied as an important factor associated with invasiveness of grasses and other plants. In general, SLA < 90 $cm^2{\cdot}g^{-1}$ (dry leaf mass) seems to be a good indicator of non-invasive or, at least, less-invasive evergreen woody plants and SLA < 150 $cm^2{\cdot}g^{-1}$ probably means the same for other vascular plant taxa.

Basic taxonomic units used in plant invasion ecology are usually species or subspecies. However, genera are certainly worth considering as well. Species belonging to genera notorious for their invasiveness or 'weediness', e.g. *Amaranthus*, *Echinochloa*, *Ehrharta*, *Myriophyllum*, should be treated as highly suspicious. On the other hand, a continuum from highly invasive to virtually non-invasive species is also common in many genera, e.g. *Acer*, *Centaurea*, *Pinus*. Recently, some attention has been paid to taxonomic patterns of invasive plants (Daehler 1998; Pyšek 1998; Rejmánek & Richardson 2003). In terms of relative numbers of invasive species, some families seem to be over-represented: Amaranthaceae, Brassicaceae, Chenopodiaceae, Fabaceae, Gramineae, Hydrocharitaceae, Papaveraceae, Pinaceae and Polygonaceae. Among the larger families, the Orchidaceae is the only under-represented one.

13.7 Impact of invasive plants, justification and prospects of eradication projects

Numerous studies have documented the wide range of impacts caused by invasive plants. Many invasive taxa have transformed both the structure and function of ecosystems by, for example, changing disturbance- or nutrient-cycling regimes (D'Antonio *et al.* 1999). In many parts of the world, impacts have clear economic implications for humans, e.g. as a result of reduced stream flow from watersheds in South African fynbos following alien tree invasions (Van Wilgen *et al.* 2001), or through disruption to fishing and navigation after invasion of aquatic plants such as *Eichhornia crassipes*.

It is important to stress, however, that the impact of invasive plants on biodiversity is much less dramatic than impact of exotic pathogens, herbivores or predators. It seems that most of the naturalized/invasive plant species have hardly detectable effects on biotic communities (Williamson & Fitter 1996; Meiners *et al.* 2001). There are at least 2000 naturalized plant species in North America and more than 1000 of them are invasive. However, not a single native plant species is known to have been driven to extinction due to interactions with alien plants alone. Even on islands, where numbers of exotic plant species are often increasing exponentially, extinctions of native plant species cannot be attributed to plant invasions *per se* (Sax *et al.* 2002). Also, the often reported correlation between numbers of native and exotic plant species on the landscape scale can be interpreted as a lack of mechanisms for competitive exclusion of native plants by exotic ones. Nevertheless, we should be careful with conclusions – many invasions are quite recent and extinction takes a long time.

Considerable progress has recently been made in developing methodologies for making biological, ecological and economic assessments. In attempting to quantify the value of ecosystem services of South African fynbos systems and the extent to which these values are reduced by invasions, Higgins *et al.* (1997) showed that the cost of clearing alien plants was very small (< 5%) as compared to the value of services provided by these ecosystems. Their conclusion was that pro-active management could increase the value of these ecosystem services by at least 138%. The most important ecosystem service was water, and much work has been done on developing models for assessing the value (in monetary terms) of allocating management resources to clearing invasive plants from fynbos watersheds. Among the most dangerous invaders in riparian areas within the USA are species of the Old World genus *Tamarix* (salt cedar). An instructive economic evaluation of *Tamarix* impacts is provided by Zavaleta (2000).

As we showed earlier, the most reliable predictions based on biological characters are limited to invasiveness (likelihood of species establishment and spread). Predictions of potential impacts will always be less reliable. Because decline in native species richness is dependent on cover of invaders (Fig. 13.9; Richardson *et al.* 1989; Meiners *et al.* 2001), indices based on a ratio of cover to frequency should be tested as impact predictors for individual taxa. Other obvious impact indicators may be biological characters of plants that are known to have ecosystem consequences (e.g. high transpiration rates or nitrogen fixation).

Invasiveness and impact are not necessarily positively correlated. Some fast-spreading species, like *Aira caryophyllea* or *Cakile edentula*, exhibit little (if any) measurable environmental or economic impact. On the other hand, some relatively slowly spreading species, e.g. *Ammophila arenaria* or *Robinia pseudoacacia*, may have far-reaching environmental effects (stabilization of coastal dunes in the first case and nitrogen soil enrichment in the second).

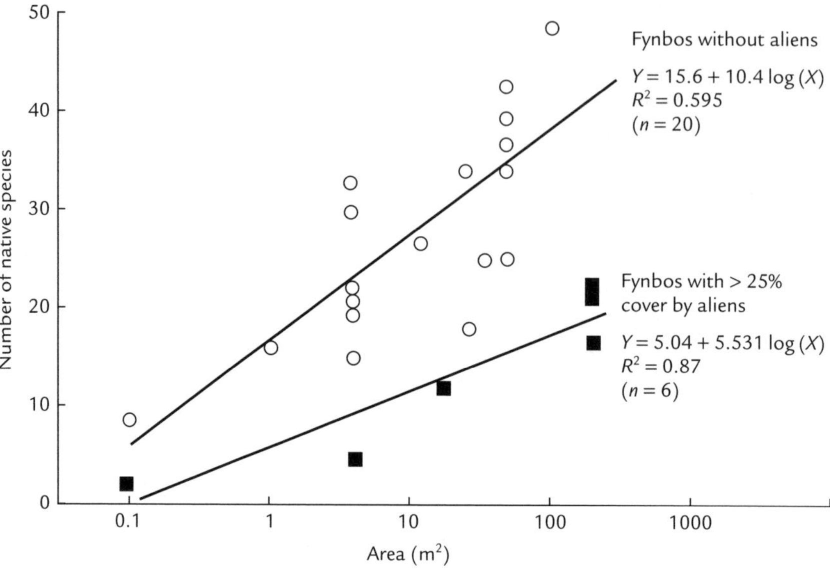

Fig. 13.9 Species-area relationships for native vascular plant species in South African fynbos areas densely infested (squares) by alien woody plants and in uninfested areas (circles). Elevations of the two regression lines are significantly different ($p < 0.001$). Sources of the data are acknowledged in Richardson *et al.* (1989).

There is a need for a universally acceptable, and objectively applicable, term for the most influential invasive plant taxa within given regions, or globally. A potentially useful term to use in this regard is 'transformer species' (Richardson *et al.* 2000a). Such species, comprising perhaps only about 10% of invasive species, have profound effects on biodiversity and clearly demand a major allocation of resources for containment/control/eradication. Several categories of transformers may be distinguished:

1 Excessive users of resources: water – *Tamarix* spp., *Acacia mearnsii*; light – *Pueraria lobata* and many other vines, *Heracleum mantegazzianum*, *Rubus armeniacus*; water and light – *Arundo donax*; light and oxygen – *Salvinia molesta*, *Eichhornia crassipes*; high leaf area ratio, LAR, of many invasive plants (discussed earlier) is an important prerequisite for excessive transpiration;

2 Donors of limiting resources: nitrogen – *Acacia* spp., *Lupinus arboreus*, *Myrica faya*, *Robinia pseudoacacia*, *Salvinia molesta*;

3 Fire promotors/suppressors: promotors – *Bromus tectorum*, *Melaleuca quinquenervia*, *Melinis minutiflora*; suppressors – *Mimosa pigra*;

4 Sand stabilizers: *Ammophila* spp., *Elymus hymus* spp.;

5 Erosion promotors: *Andropogon virginicus* in Hawai'i, *Impatiens glandulifera* in Europe;

6 Colonizers of intertidal mudflats – sediment stabilizers: *Spartina* spp., *Rhizophora* spp.;

7 Litter accumulators: *Centaurea solstitialis, Eucalyptus* spp., *Lepidium latifolium, Pinus strobus, Taeniatherum caput-medusae*;
8 Soil carbon storage modifiers: promotor – *Andropogon gayanus*; suppressor – *Agropyron cristatum*;
9 Salt accumulators/redistributors: *Mesembryanthemum crystallinum, Tamarix* spp.

The potentially most important transformers are taxa that add a new function, such as nitrogen fixation, to the invaded ecosystem (Vitousek & Walker 1989). Many impacts, however, are not so obvious. For example, invasive *Lonicera* and *Rhamnus* change vegetation structure of the forest, affecting nest predation of birds (Schmidt & Whelan 1999), and *Lythrum salicaria* and *Impatiens glandulifera* can have negative impacts on pollination and reproductive success of co-flowering native plants (Grabas & Laverty 1999; Chittka & Schürkens 2001).

It follows from the discussion on impacts of exotic plants that careful prioritization is needed before starting often very expensive and time-consuming eradication projects. Maintenance of biodiversity is dependent on the maintenance of ecological processes. Our priority should be protection of ecological processes. Attempts to eradicate widespread invasive species, especially those that do not have any documented environmental impacts (including suppression of rare native taxa), may be not only useless but also a waste of time and resources. Exotic taxa with large-scale environmental impacts (transformers) are usually obvious targets for control and eradication. But when is complete eradication a realistic goal?

There are numerous examples where small infestations of invasive plant species have been eradicated. There are also several encouraging examples where widespread alien animals have been completely eradicated. Can equally widespread and difficult alien plants also be eradicated? On the basis of a unique data set on eradication attempts by the California Department of Food and Agriculture on 18 species and 53 separate infestations targeted for eradication in the period 1972–2000 (Table 13.4), it is shown that professional eradication of exotic weed infestations smaller than 1 ha is usually possible. In addition, about 1/3 of infestations between 1 and 100 ha and 1/4 of infestations between 101 and 1000 ha have been eradicated. However, costs of eradication projects increase dramatically. With a realistic amount of resources, it is very unlikely that infestations larger than 1000 ha can be eradicated (Table 13.4).

Table 13.4 Areas of initial gross infestations (at the beginning of eradication projects) of exotic weeds in California, numbers of eradicated infestations, numbers of on-going projects, and mean eradication effort for five infestation area categories. The data include 18 noxious weedy species (two aquatic and 16 terrestrial) representing 53 separate infestations. From Rejmánek & Pitcairn (2002).

Initial infestation (ha)	< 0.1	0.1–1	1.1–100	101–1000	> 1000
No. of eradicated infestations	13	3	5	3	0
No. of on-going projects	2	4	9	10	4
Mean eradication effort per infestation (work hours)					
Eradicated	63	180	1496	1845	-
On-going	174	277	1577	17,194	42,751

Early detection of the presence of an invasive harmful taxon can make the difference between being able to employ offensive strategies (eradication) and the necessity of retreating to a defensive strategy that usually means an infinite financial commitment. Nevertheless, depending on the potential impact of individual invaders, even infestations larger than 1000 ha should be targeted for eradication effort or, at least, substantial reduction and containment. If an exotic weed is already widespread, then species-specific biological control may be the only long-term effective method able to suppress its abundance over large areas (Myers & Bazely 2003).

Regardless of their environmental and/or economical effects, plant invasions provide unique chances to understand some basic ecological processes that are otherwise beyond the capacity or ethics of standard ecological experiments. We are just beginning to fully appreciate these opportunities. However, we have a long way to go to achieve a more complete understanding and more rational decision making.

Acknowledgements

We thank Jennifer Erskine and Jennifer Sedra for comments on a draft of this chapter.

References

Agrawal, A.A. & Kotanen, P.M. (2003) Herbivores and the success of exotic plants: a phylogenetically controled experiment. *Ecology Letters* **6**, 712–715.

Beerling, D.J., Huntley, B. & Bailey, J.P. (1995) Climate and the distribution of *Fallopia japonica*: use of an introduced species to test the predictive capacity of response surfaces. *Journal of Vegetation Science* **6**, 269–282.

Binggeli, P. (1996) A taxonomic, biogeographical and ecological overview of invasive woody plants. *Journal of Vegetation Science* **7**, 121–124.

Booth, B.D., Murphy, S.D. & Swanton, C.J. (2003) *Weed Ecology in Natural and Agricultural Systems*. CABI Publishing, Wallingford.

Burke, M.J.W. & Grime, J.P. (1996) An experimental study of plant community invasibility. *Ecology* **77**, 776–90.

Cadenasso, M.L. & Pickett, S.T.A. (2001) Effect of edge structure on the flux of species into forest interiors. *Conservation Biology* **15**, 91–97.

Chittka, L. & Schürkens, S. (2001) Successful invasion of a floral market. *Nature* **411**, 653.

Colautti, R.I., Ricciard, A., Grigorovich, I.A. & MacIsaac, H.J. (2004) Is invasion success explained by the enemy release hypothesis? *Ecology Letters*, in press.

Crawley, M.J., Brown, S.L., Heard, M.S. & Edwards, G.G. (1999) Invasion-resistance in experimental grassland communities: species richness or species identity? *Ecology Letters* **2**, 140–148.

Daehler, C. (1998) The taxonomic distribution of invasive angiosperm plants: ecological insights and comparison to agricultural weeds. *Biological Conservation* **84**, 167–180.

D'Antonio, C.M., Dudley, T.L. & Mack, M. (1999) Disturbance and biological invasions: direct effects and feedbacks. In: *Ecosystems of the World: Ecosystems of Disturbed Ground* (ed. L.R. Walker), pp. 413–452. Elsevier, Amsterdam.

Davis, M.A. & Pelsor, M. (2001) Experimental support for a resource-based mechanistic model of invasibility. *Ecology Letters* **4**, 421–428.

Davis, M.A., Grime, J.P. & Thompson, K. (2000) Fluctuating resources in plant communities: a general theory of invasibility. *Journal of Ecology* **88**, 528–534.

Duncan, R.P. & Williams, P.A. (2002) Darwin's naturalization hypothesis challenged. *Nature* **417**, 608–609.

Ellenberg, H., Weber, H.E., Düll, R., Wirth, V., Werner, W. & Paulißen, D. (1992) Zeigerwerte von Pflanzen in Mitteleuropa. *Scripta Geobotanica* **18**, 1–248.

Fargione, J., Brown, C.S. & Tilman, D. (2003) Community assembly and invasion: an experimental test of neutral versus niche processes. *Proceedings of the National Academy of Sciences of the United States of America* **100**, 8916–8920.

Foster, B.L. (2001) Constrains on colonization and species richness along a grassland productivity gradient: the role of propagule availability. *Ecology Letters* **4**, 530–535.

Gotelli, N.J. & McCabe, D.J. (2002) Species co-occurrence: a meta-analysis of J.M. Diamond's assembly rules model. *Ecology* **83**, 2091–2096.

Grabas, G.P. & Laverty, M. (1999) The effect of purple loosestrife (*Lythrum salicaria* L.; Lythraceae) on the pollination and reproductive success of sympatric co-flowering wetland plants. *EcoScience* **6**, 230–242.

Grime, J.P., Hodgson, J.G. & Hunt, R. (1988) *Comparative Plant Ecology*. Unwin Hyman, London.

Grotkopp, E., Rejmánek, M. & Rost, T.L. (2002) Toward a causal explanation of plant invasiveness: Seedling growth and life-history strategies of 29 pine (*Pinus*) species. *American Naturalist* **159**, 396–419.

Hector, A., Dobson, K., Minns, A., Bazeley-White, E. & Lawton, J.H. (2001) Community diversity and invasion resistance: An experimental test in a grassland ecosystem and a review of comparable studies. *Ecological Research* **16**, 819–831.

Higgins, S.I., Turpie, J.K., Costanza, R., Cowling, R.M., Le Maitre, D.C., Marais, C. & Midgley, G.F. (1997) An ecologically-economic simulation model of mountain fynbos ecosystems: Dynamics, valuation and management. *Ecological Economics* **22**, 155–169.

Hobbs, R.J. & Huenneke, L.F. (1992) Disturbance, diversity and invasion: implications for conservation. *Conservation Biology* **6**, 324–337.

Hubbell, S.P., Ahumada, J.A., Condit, R. & Foster, R. (2001) Local neighborhood effects on long-term survival of individual trees in a Neotropical forest. *Ecological Research* **16**, 859–875.

Kartesz, J.T. & Meacham, C.A. (1999) *Synthesis of the North American Flora*. CD-ROM Version 1.0. North Carolina Botanical Garden, Chapel Hill.

Kennedy, T.A., Naeem, S., Howe, K.M., Knops, J.M.H., Tilman, D. & Reich, P. (2002) Biodiversity as a barrier to ecological invasion. *Nature* **417**, 636–638.

Klironomos, J.N. (2002) Feedback with soil biota contributes to plant rarity and invasiveness in communities. *Nature* **417**, 67–70.

Law, R. (1999) Theoretical aspects of community assembly. In: *Advanced Ecological Theory* (ed. J. McGlade), pp. 143–171. Blackwell Science, Oxford.

Levine, J.M. (2000) Species diversity and biological invasions: relating local process to community pattern. *Science* **288**: 852–854.

Lonsdale, W.M. (1999) Global patterns of plant invasions and the concept of invasibility. *Ecology* **80**, 1522–1536.

May, R.M. (1981) Patterns in multi-species communities. In: *Theoretical Ecology. Principles and Applications* (ed. R.M. May), pp. 197–227. Blackwell Scientific, Oxford.

Meiners, S.J., Pickett, S.T.A. & Cadenasso, M.L. (2001) Effects of plant invasions on the species richness of abandoned agricultural land. *Ecography* **24**, 633–644.

Meiners, S.J., Pickett, S.T.A. & Cadenasso, M.L. (2002) Exotic plant invasions over 40 years of old field succession: community patterns and associations. *Ecography* **25**, 215–223.

Mitchell, C.E. & Power, A.G. (2003) Release of invasive plants from fungal and viral pathogens. *Nature* **421**, 625–627.

Myers, J.H. & Bazely, D.R. (2003) *Ecology and Control of Introduced Plants.* Cambridge University Press, Cambridge.

Pannel, J.R. & Barrett, S.C.H. (1998) Baker's Law revisited: reproductive assurance in a metapopulation. *Evolution* **52**, 657–668.

Pattison, R.R., Goldstein, G. & Ares, A. (1998) Growth, biomass allocation and photosynthesis of invasive and native Hawaiian rain forest species. *Oecologia* **117**, 449–459.

Pyšek, P. (1998) Is there a taxonomic pattern to plant invasions? *Oikos* **92**, 282–294.

Pyšek, P., Jarošík, V. & Kučera, T. (2002b) Patterns of invasion in temperate nature reserves. *Biological Conservation* **104**, 13–24.

Pyšek, P., Sádlo, J. & Mandák, B. (2002a) Catalogue of alien plants of the Czech Republic. *Preslia* **74**, 97–186.

Rejmánek, M. (1989) Invasibility of plant communities. In: *Biological Invasions. A Global Perspective* (eds J.A. Drake, H.A. Mooney, F. di Castri, R.H. Groves, F.J. Kruger, M. Rejmánek & M. Williamson), pp. 369–388. John Wiley & Sons, Chichester.

Rejmánek, M. (1996) Species richness and resistance to invasions. In: *Diversity and Processes in Tropical Forest Ecosystems.* (eds G.H. Orians, R. Dirzo & J.H. Cushman), pp. 153–72. Springer-Verlag, Berlin.

Rejmánek, M. (2000) Invasive plants: approaches and predictions. *Austral Ecology* **25**, 497–506.

Rejmánek, M. & Pitcairn, M.J. (2002) When is eradication of exotic plant pests a realistic goal? In: *Turning the Tide: The Eradication of Invasive Species* (eds C.R. Veitch & M.N. Clout), pp. 249–253. IUCN, Gland, Switzerland and Cambridge, UK.

Rejmánek, M. & Richardson, D.M. (1996) What attributes make some plant species more invasive? *Ecology* **77**, 1655–1661.

Rejmánek, M. & Richardson, D.M. (2004) Invasiveness of conifers: extent and possible mechanism. *Acta Horticulturae* **615**, 375–378.

Rejmánek, M., Richardson, D.M., Higgins, S.I., Pitcairn, M.J. & Grotkopp, E. (2003) Ecology of invasive plants: State of the art. In: *Invasive Alien Species: Searching for Solutions* (eds H.A. Mooney, J.A. McNeelly, L. Neville, P.J. Schei & J. Waage), Island Press, Washington.

Richardson, D.M. & Rejmánek, M. (1998) *Metrosideros excelsa* takes off in the fynbos. *Veld & Flora* **85**, 14–16.

Richardson, D.M., Allsopp, N., D'Antonio, C.M., Milton, S.J. & Rejmánek, M. (2000b) Plant invasions – the role of mutualisms. *Biological Reviews of the Cambridge Philosophical Society* **75**, 65–93.

Richardson, D.M., Macdonald, I.A.W. & Forsyth, G.G. (1989) Reductions in plant species richness under stands of alien trees and shrubs in the fynbos biome. *South African Forestry Journal* **149**, 1–8.

Richardson, D.M., Pyšek, P., Rejmánek, M., Barbour, M.G., Panetta, F.D. & West, C.J. (2000a) Naturalization and invasion of alien plants: concepts and definitions. *Diversity and Distributions* **6**, 93–107.

Sax, D.F. (2002) Native and naturalized plant diversity are positively correlated in scrub communities of California and Chile. *Diversity and Distributions* **8**, 193–210.

Sax, D.F., Brown, J.H. & Gaines, S.D. (2002) Species invasions exceed extinctions on islands world-wide: a comparative study of plants and birds. *American Naturalist* **160**, 766–783.

Schmidt, K.A. & Whelan, C.J. (1999) Effects of exotic *Lonicera* and *Rhamnus* on songbird nest predation. *Conservation Biology* **13**, 1502–1506.

Shea, K. & Chesson, P. (2002) Community ecology theory as a framework for biological invasions. *Trends in Ecology & Evolution* **17**, 170–176.

Simberloff, D. & Von Holle, B. (1999) Positive interactions of nonindigenous species: Invasional meltdown? *Biological Invasions* **1**, 21–32.

Smith, S.D., Huxman, T.E., Zitzer, S.F., Charlet, T.N., Housman, D.C., Coleman, J.S., Fenstermaker, L.K., Seeman, J.R. & Nowak, R.S. (2000) Elevated CO_2 increases productivity and invasive species success in an arid ecosystem. *Nature* **408**, 79–82.

Specht, R.L. & Specht, A. (1999) *Australian Plant Communities.* Oxford University Press, Oxford.

Stohlgren, T.J., Binkley, D., Chong, G.W., Kalkham, M.A., Schell, L.D., Bull, K.A., Otsuki, Y., Newman, G., Bashkin, M. & Son, Y. (1999) Exotic plant species invade hot spots of native plant diversity. *Ecological Monographs* **69**, 25–46.

Tilman, D. (1999) The ecological consequences of changes in biodiversity: a search for general principles. *Ecology* **80**, 1455–1474.

Tye, A. (2001) Invasive plant problems and requirements for weed risk assessment in the Galapagos Islands. In: *Weed Risk Assessment* (eds R.H. Groves, F.D. Panetta & J.G. Virtue), pp. 153–175. CSIRO Publishing, Collingwood.

Van Wilgen, B.W., Richardson, D.M., Le Maitre, D.C., Marais, C. & Magadlela, D. (2001) The economic consequences of alien plant invasions: Examples of impacts and approaches for sustainable management in South Africa. *Environment, Development and Sustainability* **3**, 145–168.

Vitousek, P.M. & Walker, L.R. (1987) Colonization, succession and resource availability: ecosystem-level interactions. In: *Colonization, Succession and Stability* (eds A.J. Gray, M.J. Crawley & P.J. Edwards), pp. 207–223. Blackwell Science, Oxford.

Vitousek, P.M. & Walker, L.R. (1989) Biological invasion by *Myrica faya* in Hawai'i: plant demography, nitrogen fixation, ecosystem effects. *Ecological Monographs* **59**, 247–265.

Vitousek, P.M., D'Antonio, C.M., Loope, L.L., Rejmánek, M. & Westbrooks, R. (1997) Introduced species: A significant component of human-caused global change. *New Zealand Journal of Ecology* **21**, 1–16.

Von Holle, B., Delcourt, H.R. & Simberloff, D. (2003) The importance of biological inertia in plant community resistance to invasion. *Journal of Vegetation Science* **14**, 425–432.

Wardle, D.A. (2001) Experimental demonstration that plant diversity reduces invasibility – evidence of a biological mechanism or a consequence of sampling effect? *Oikos* **95**, 161–170.

Widlechner, M.P. & Iles, J.K. (2002) A geographic assessment of the risk of naturalization of non-native woody plants in Iowa. *Journal of Environmental Horticulture* **20**, 47–56.

Williams, P.A., Nicol, E. & Newfield, M. (2001) Assessing the risk to indigenous biota of plant taxa new to New Zealand. In: *Weed Risk Assessment* (eds R.H. Groves, F.D. Panetta & J.G. Virtue), pp. 100–116. CSIRO Publishing, Collingwood.

Williamson, M. & Fitter, A. (1996) The varying success of invaders. *Ecology* **77**, 1661–1666.

Zavaleta, E. (2000) Valuing ecosystem services lost to *Tamarix* invasion in the United States In: *The Impact of Global Change on Invasive Species* (eds H.A. Mooney & R.J. Hobbs), pp. 261–300. Island Press, Washington.

14

Vegetation ecology and global change

Brian Huntley and Robert Baxter

14.1 Introduction

The term 'global change' refers to the changes that are currently taking place in various aspects of the global environment as a consequence of human activities. These changes can be grouped into two broad categories. In the first category are changes to components of the Earth system that are inherently global in their extent and impact, principally because they are changes to 'well-mixed' Earth system components, notably the atmosphere. Amongst these are included:

1 Climate change;

2 Increasing atmospheric concentrations of CO_2;

3 Increased fluxes of UV-B as a consequence of 'thinning' of the stratospheric O_3 layer;

4 Increasing rates of deposition of N compounds (NO_x and NH_3) from the atmosphere; and

5 Increasing tropospheric concentrations of various pollutants, notably SO_2, NO_x and O_3.

In the second category are changes that individually are of local to regional extent, but that, because they recur world wide, have a global impact upon biodiversity and/or upon Earth system processes. The latter will especially arise where the changes taking place alter land surface qualities and hence impact upon transfers of energy and materials between the land surface and the overlying atmosphere. In this category are included:

6 Land-cover changes as a consequence of human land use, often resulting in habitat loss and fragmentation, as well as in changes of land surface qualities;

7 Selective pressures upon ecosystem components as a result of human activities (selective felling, hunting, persecution of carnivores) that, in addition to their biodiversity impacts, may result in changes in ecosystem structure and/or function with consequent impacts upon the participation of ecosystems in global geochemical cycles; and

8 Both deliberate and accidental introductions of 'alien' species that, as a result of the consequent changes in ecosystem composition, also may result in changes in ecosystem structure and/or function, with similar ultimate impacts.

This chapter will mainly deal with the first five of these changes that are truly global in their extent, and more particularly with climate change, because this has a qualitatively different and more far-reaching impact than the other four. First the impacts of climate change upon vegetation will be discussed, and then the confounding impacts of increases in CO_2 concentration, UV-B flux, nitrogen deposition and concentrations of atmospheric pollutants in the troposphere. The focus will be upon the general principles underlying the observed or expected responses. The chapter will also discuss some of the further factors and phenomena that will confound and limit our efforts to predict the consequences of global changes that might be expected over the next 100 yr or so.

14.2 Vegetation and climate change

14.2.1 Individualistic species responses

Fundamental to the response of vegetation to climate change is the individualism of species' responses to their environment (see Chapter 2). Historically, vegetation scientists debated the extent to which plant communities could be considered 'organismal' in character, responding in their entirety to environment, as opposed to being comprised of component species that were independently and 'individualistically' responding to their environment. Today the individualistic view is widely accepted; this hypothesis receives support not only from the observations of continuous spatial variation in present plant communities, but also from the Quaternary palaeo-ecological record that shows plant species assemblages changing continuously through time (Huntley 1990, 1991).

The principal responses of vegetation to climate changes can be considered to be:

1 Quantitative changes of composition and/or structure;

2 Qualitative changes in composition;

3 Adaptive genetic responses of the component species.

Although the discussion will focus upon responses of vegetation to climate change, it is important to realize that the vegetation cover of land surfaces has fundamental influences upon their interaction with the surrounding atmosphere and hence upon climate. Vegetation should be considered as part of the global climate system, actively participating in changes in the system rather than passively responding to such changes. This active participation is primarily through important feedback mechanisms and has been demonstrated by climate modelling experiments (Street-Perrott *et al.* 1990; Mylne & Rowntree 1992; Foley *et al.* 1994). Amongst these feedbacks, the best documented relate to the effects of vegetation cover on albedo (e.g. boreal forest versus tundra, especially during spring months when trees mask snow cover) and the 'recycling' of moisture to the atmosphere by transpiration and evaporation (e.g. large-scale clearance of rain forest). The 'greening' of much of the Sahara during the early Holocene is likely to have altered regional climate through both mechanisms.

Of great importance when we consider the interactions of vegetation and climate change are considerations of both spatial and temporal scale. The responses differ according to the scale that we consider (Fig. 14.1).

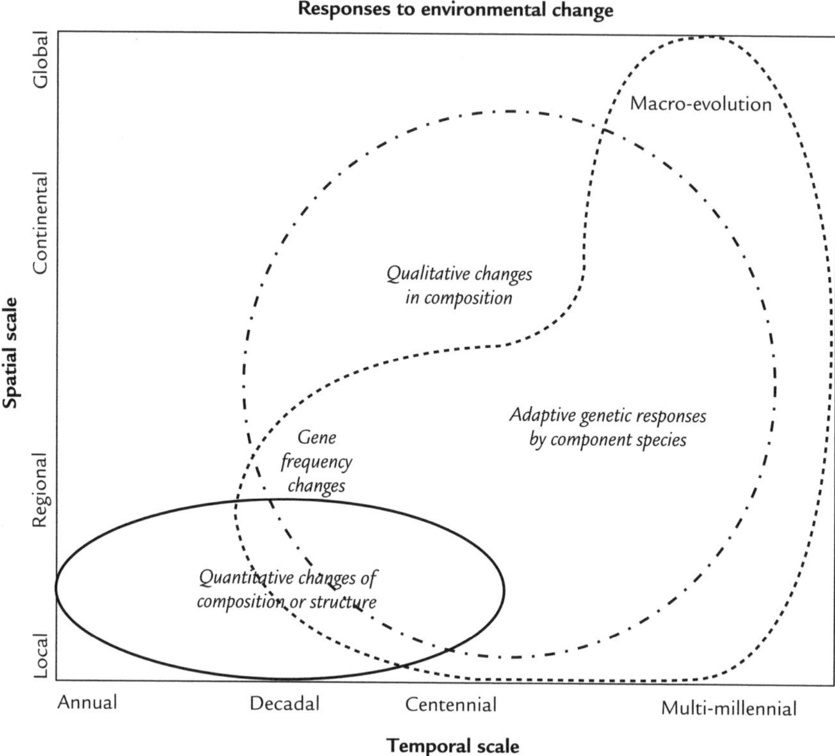

Fig. 14.1 Schematic illustration of the relationships between spatial and temporal scale and the principal responses of vegetation to environmental change.

14.2.2 Quantitative changes

These are changes that involve only alterations in the abundance, whether measured as number of individuals (plant units), biomass, etc., of the component species of a plant community. Such changes may result in changes in the structure of the vegetation, although these are unlikely to be major changes; a climate change that results in a substantial change in vegetation structure, e.g. savanna to closed woodland, is likely also to lead to qualitative changes in species composition, and thus belongs in the next category. Quantitative changes occur within vegetation stands and hence are local in spatial scale. They are also potentially rapid in terms of temporal scale, being limited in their response time only by the inherent growth and life-cycle characteristics of the species comprising the plant community. Thus a community dominated by annual plant species may exhibit very marked inter-annual changes in the relative abundance of the component species in response to inter-annual changes in climatic conditions differentially affecting the germination, establishment and survival of different species.

Communities of long-lived perennial herbaceous species can also exhibit inter-annual variations in biomass or productivity amongst species in response to inter-annual climate variability (Willis *et al.* 1995; see Fig. 14.2). Quantitative changes in

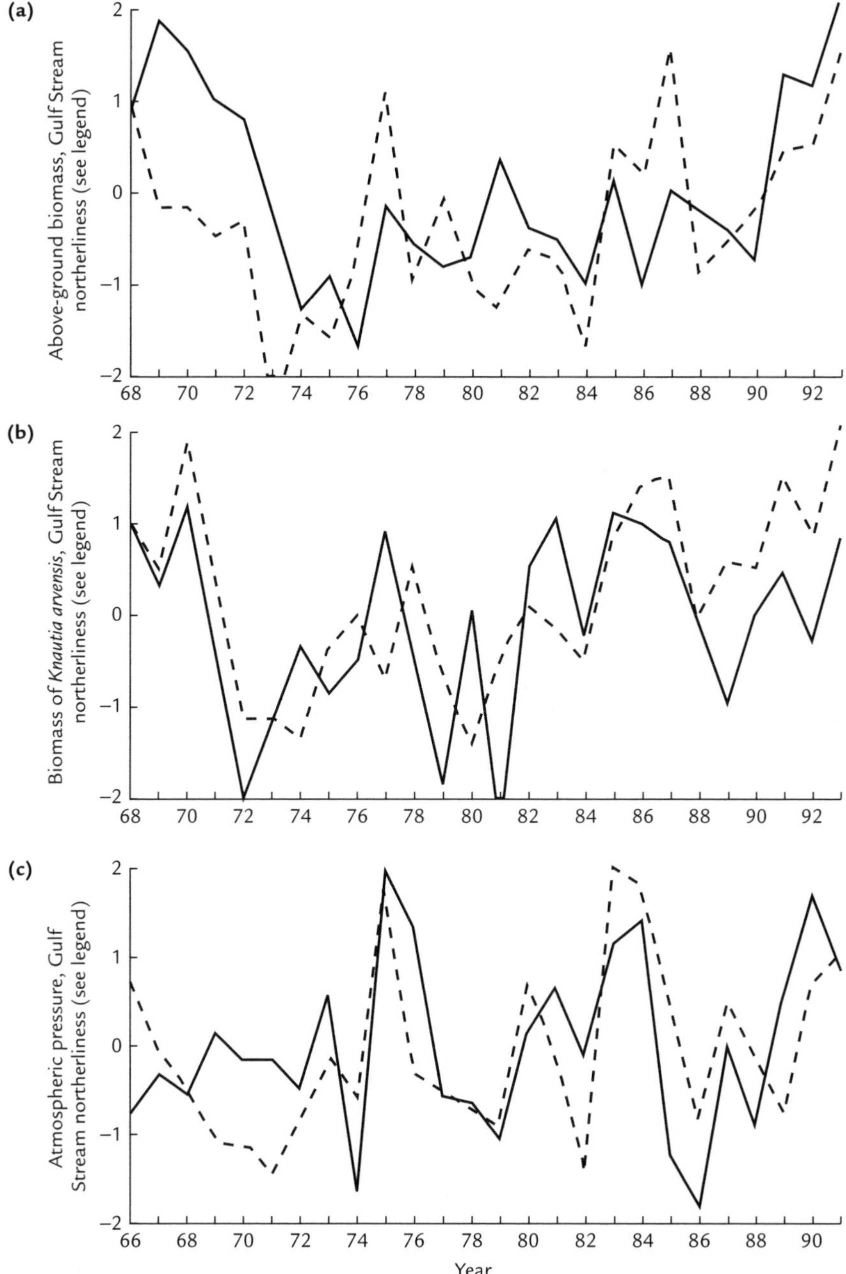

Fig. 14.2 Inter-annual quantitative changes in a community of herbaceous perennials on road verges at Bibury, Gloucestershire (Willis *et al.* 1995). Standardized time series (zero mean, unit variance). **a.** Mean above-ground biomass ($g \cdot m^{-2}$) of vegetation in the Bibury plots (solid line) and Gulf Stream northerliness index in the next-to-previous spring and summer months (March to August, broken line), $r = 0.489$. **b.** Mean July biomass ($g \cdot m^{-2}$) of *Knautia arvensis* in the Bibury plots (solid line) and Gulf Stream northerliness index in the previous autumn and winter months (September to February, broken line), $r = 0.496$. **c.** Log mean atmospheric pressure at sea level (mbar, solid line) and Gulf Stream northerliness index in the same August (broken line), $r = 0.551$.

the numbers of individuals, however, will occur more slowly, in response to decadal scale shifts in mean climatic conditions (Watt 1981). At centennial scales even communities of woody species will show quantitative changes in composition as fluctuations in long-term mean climatic conditions differentially alter both survival and recruitment rates of species.

Quantitative responses thus can provide resilience to the climatic variability that occurs on inter-annual to centennial time scales but which characteristically does not involve persistent trends of change in mean conditions that are maintained over centennial time scales. These responses do not enable vegetation to respond either to rapid, large-magnitude or less rapid, but persistent changes in mean climate conditions.

14.2.3 Qualitative changes

These changes involve losses and gains of species from the plant community and can result not only in changes in composition but also, potentially, in profound changes in the structure of the plant community. In contrast to quantitative changes, qualitative changes may occur across a wide range of spatial scales, from local to continental, and across a correspondingly wide range of temporal scales, from decadal to multi-millennial.

At the scale of the individual stand the time scale for such changes is often dependent upon the frequency of disturbance and/or the characteristic longevity of component species of the present vegetation. In gap-regenerating communities of perennials the death of individuals provides the opportunity for species favoured by the changed climate conditions to become established. In such systems the incoming species take many years to completely replace species that are no longer favoured by the prevailing climate conditions. As a consequence, a transient effect of the climate change may be an increase in within-stand species diversity comparable to the enhanced diversity often observed spatially in relation to ecotones between adjacent stands of different communities. The transient plant community can perhaps be considered the temporal equivalent of such an ecotonal community.

In communities that regenerate episodically following landscape-scale disturbance events (e.g. forest fires, windstorms, defoliating insect outbreaks) only the characteristic spatial and temporal scales of the process differ from those in gap regenerating communities. Spatially, entire stands of such communities may be transformed following the disturbance event, with prevailing climate conditions favouring the establishment and growth of a qualitatively different community to that which was present before the disturbance (Bradshaw & Zackrisson 1990). Temporally, the replacement of the previous community occurs within the time scale for stand regeneration, typically decades to a century. However, the inertia of the pre-disturbance community appears to enable it to persist in a progressively greater degree of disequilibrium with the changing climate until it is disturbed, so that the rate at which the qualitative change occurs across the landscape as a whole is determined by the characteristic return time for disturbance events. Thus, in boreal forest systems, where fire return times may typically be 200–500 yr (Foster 1983; Segerström *et al.* 1996), it may require many centuries to pass before the entire landscape is transformed. In consequence, the period of transition is again characterized by a transient peak in diversity, in this case at the landscape scale and among stands. Once again, this temporal phenomenon can be considered equivalent to the spatial heterogeneity and associated biodiversity peak typically found when ecotonal zones are viewed at a landscape scale.

At larger spatial scales, and over longer time scales, qualitative changes in vegetation composition are the resultant of the 'migration' of species, i.e. their shifts in geographical distribution, as a response to persistent large-scale changes in climate conditions. Such migrations of plant species, including long-lived woody taxa, are well documented by Quaternary palaeo-ecological data, and are the predominant response of species to persistent shifts in climatic conditions (Huntley & Birks 1983; Huntley & Webb 1989). Thus, as climate shifted from glacial to interglacial conditions around 11,400 yr ago, temperate forest trees in Europe, North America and eastern Asia 'migrated' at rates of 0.2–2 $km{\cdot}yr^{-1}$, shifting their geographical ranges in response to the changing climate (Fig. 14.3). Such range changes have continued throughout the Holocene: in Europe species such as *Fagus sylvatica* and *Picea abies* continued to exhibit changes of range and of prevalence in forests at the landscape scale during recent millennia (Björse & Bradshaw 1998). The qualitative changes seen at local and landscape scales are indeed generally an expression of the larger scale migrations of species; conversely, these large-scale migrations are achieved as a result of, or at least facilitated by, the smaller-spatial-scale qualitative changes. Perhaps all such changes should best be viewed as part of one overall process of spatial response of species to climate change.

At the geological time scales of glacial-interglacial cycles and beyond a further factor can contribute to qualitative changes in the composition of vegetation, namely extinction. Trivially, the qualitative change in composition of a stand that leads to the loss of a species can be considered to be a local extinction, and the same can be argued at landscape or regional scales. However, the persistent impact of an extinction only comes to be felt when the taxon becomes regionally extinct, for example, *Tsuga* in Europe during the late Pleistocene, or globally extinct, for example *Picea critchfieldii* at the last glacial-interglacial transition around 11,000 to 12,000 yr ago (Jackson & Weng 1999). Such extinctions most probably reflect the loss from the region where the taxon occurs of any area of the climatic conditions favouring that taxon (Huntley 1999). Subsequently conditions may once again be favourable for the taxon, but its extinction results in qualitatively different vegetation now developing under those climate conditions.

14.2.4 Adaptive genetic responses

Plant species may also exhibit adaptive genetic responses when climate changes, enabling them to persist in communities that may, as a result, exhibit neither qualitative nor quantitative changes. Such adaptive responses are in principle possible at spatial scales ranging from local to continental and across time scales from decadal to the multi-millennial scales of geological time. In practice, such responses are primarily polarized to the two extremes of this range of time scales because they occur as a result of two processes with very different temporal characteristics.

At local to regional scales, and over relatively short time scales, the process is predominantly one of selection amongst the many genotypes that arise each generation

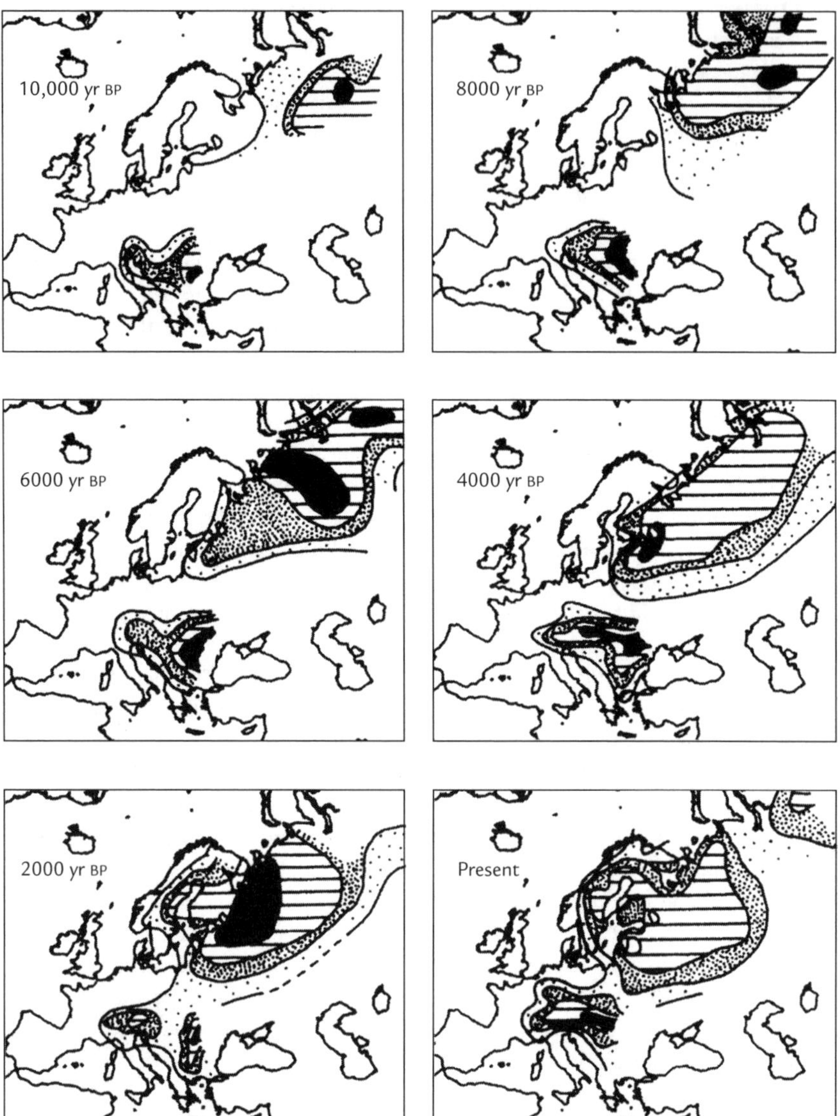

Fig. 14.3 Isopoll map sequence for *Picea* in Europe during the Holocene (Huntley 1988) showing how its pattern of distribution and abundance changed as climate changed during this period. Maps are contoured at 2, 5, 10 and 20%. Reproduced by permission of Kluwer Academic Publishers.

as a result of the recombination of alleles during sexual reproduction. Such 'micro-evolution' does not alter the species overall climatic range, in terms of its tolerances or requirements, but does enable a local or regional population of the species to adapt to changing climate conditions. The capability of such an adaptive response, however, is limited by the species' overall climatic adaptability; in many cases it may be even more limited in scope if elements of the overall genetic variability exhibited by the species in relation to climate are limited in their distribution to only a fraction of the species' overall geographical distribution.

At somewhat longer time scales the latter limitation may be overcome by the 'migration' of genotypes as a result of pollen/spore as well as seed/fruit dispersal. This will be relatively more rapid and more extensive in the case of anemophilous species, in which pollen transport may facilitate the rapid and widespread migration of newly favoured alleles. At the longest time scales, species may evolve new capabilities with respect to climate as the result of the chance mutation of a gene to produce an allele that provides an extension of the range of climate conditions under which the species may persist. Such mutations will arise infrequently, however, and, as a result, such 'macro-evolution' is only able to play any significant role at longer geological time scales.

14.3 Confounding effects of other aspects of global change

14.3.1 Increasing atmospheric concentrations of CO_2

Increased atmospheric CO_2 concentration not only leads to warming of the lower atmosphere, but also increases availability of the primary substrate for photosynthesis. In principle, higher atmospheric concentrations of CO_2 should stimulate photosynthesis and lead to faster growth. This generally is the case for species using the C_3 photosynthetic pathway (*c.* 95% of the world's higher plants). In these species, increased atmospheric partial pressure of CO_2 leads to reduced photorespiration and hence a net gain in carbon fixation, by an average of *c.* 30% (range –10 to +80%) for a doubling of CO_2 concentration. In addition, these species also benefit from physiological gains, notably a reduction in stomatal opening, as well as a potential reduction in stomatal density, that leads to increased water-use efficiency as a result of reduced transpirational water loss. In contrast, species using the C_4 photosynthetic pathway (only *c.* 5% of the world's higher plants, but including *c.* 50% of grass species) experience only a very modest net gain in C fixation, estimated at *c.* 7%, for the same doubling of atmospheric CO_2 concentration. This substantial difference in photosynthetic stimulation reflects evolutionary adaptations of physiology and morphology in C_4 species, including saturation of the key CO_2 fixation enzyme of C_4 photosynthesis (PEP carboxylase – phosphoenolpyruvate carboxylase) at much lower CO_2 concentrations than are required to saturate its C_3 counterpart (RuBP carboxylase – ribulose bisphosphate carboxylase), and the so-called Kranz anatomy, in which initial CO_2 fixation is spatially separated from the remainder of the photosynthetic pathway, C_4 acids formed when CO_2 is fixed in the mesophyll being shunted to bundle sheath cells where CO_2 is released to enter the 'normal' C_3 pathway, but at higher concentrations than in the surrounding atmosphere. The benefits of this CO_2-concentrating mechanism are seen when the CO_2:O_2 ratio is low; they include: (i) minimal photorespiration; and (ii) an inherently higher water-use efficiency. The latter is a consequence of a significantly lower stomatal conductance, C_4 species typically having lower stomatal densities and opening their stomata for shorter periods of the day. When the CO_2:O_2 ratio increases, however, C_4 species are at a disadvantage relative to their C_3 counterparts because they expend greater amounts of energy concentrating the CO_2 in the bundle sheath cells.

The range in CO_2 response exhibited by C_3 species results, at least in part, from the phenomenon of 'acclimation'. When plants are exposed to elevated concentrations of atmospheric CO_2 there is, through time, a strong tendency for them to adapt physiologically and morphologically to their new environment. As a consequence initial increases in C-fixation rates are, in many cases, not maintained through time. Predictions of future vegetation changes resulting directly from increasing concentrations of atmospheric CO_2 thus must be made with care. This is especially true when considering communities that may be impacted by the differential responses of slow-growing, long-lived perennial species, as opposed to fast-growing perennials. Whereas the former, often characteristic of more diverse but less productive communities, are relatively unresponsive, the latter, often dominant in less diverse communities of productive habitats, show a strong response to elevated CO_2 (Hunt *et al.* 1993).

C_4 species have significantly higher temperature optima for photosynthesis than C_3 species. This is associated with a strong latitudinal trend in the distribution and relative abundance of the former. C_4 species are increasingly predominant in the hotter and drier environments at lower latitudes where their high water-use efficiency also contributes to their ability to out-compete C_3 species. Given the important differences in their physiological and morphological adaptations, some key questions remain with respect to the direct effects of increased atmospheric CO_2 concentration upon vegetation. In particular, will the increasing atmospheric CO_2 concentration, through its direct effects, be the primary determinant of changes in the composition of plant communities, especially where C_3 and C_4 species co-occur, or will the altered climate, itself an indirect effect of increased atmospheric CO_2 concentration, have a greater impact, masking or even overriding any direct effects? Studies of Quaternary palae-ovegetation and palaeo-environments can help answer this question. The partial pressure of atmospheric CO_2 was markedly lower during the last glacial stage, at around 190 ppmv than during either the preceding interglacial stage or the pre-industrial post-glacial, during both of which it was around 280–300 ppmv. Studies of glacial-interglacial variations in relative abundance of C_3 versus C_4 plants, however, suggest that regional climates exerted the strongest influence upon their relative abundance, and that in the absence of favourable moisture and temperature conditions, a low partial pressure of atmospheric CO_2 alone was insufficient to result in an increased proportion of C_4 plants in the vegetation (Huang *et al.* 2001). Thus, climate seems likely to be more important than the direct effects of CO_2 concentration in determining the presence or absence of C_4 species in a given environment, and thus is likely to have the greater role in determining future changes, quantitative and qualitative, in the relative contributions of C_3 and C_4 species to plant communities.

Many succulent plants employ a third mode of CO_2 carboxylation, CAM, crassulacean acid metabolism. CAM plants comprise *c.* 10% of the world's higher-plant flora. Like C_4 species, they have the PEP carboxylase enzyme in addition to the enzymes of the C_3 Calvin cycle. In contrast to C_4 plants, however, the carboxylation process is temporally rather than spatially separated from the remainder of the photosynthetic pathway. CAM plants minimize transpirational water losses by opening their stomata during the cooler conditions at night, rather than during the heat of the day. CO_2 fixation by PEP carboxylase can occur in the dark, when cooler conditions reduce the amount of water lost for a given amount of CO_2 uptake, and the CO_2 is then released from the C_4 acids during daylight hours and utilized by the 'normal' photosynthetic pathway.

In addition to those species that use only CAM (e.g. most Cactaceae, many Bromeliaceae and Orchidaceae) there are facultative CAM species that employ CAM under conditions of stress resulting from drought or salinity (e.g. members of the Aizoaceae). This ability of facultative CAM species (some of which employ diverse variants of the pathway) to 'switch' metabolism gives them a plasticity that enables them to occupy otherwise sparsely inhabited ecological niches. The extent of any response by such species to increased atmospheric CO_2 concentration will very much depend upon the impacts of both prevailing temperature and water availability in the particular environments that they occupy. It must also be noted that productive investment of carbon fixed by photosynthesis requires adequate availability of nutrients, especially nitrogen. In part that N-requirement may be met by reallocation within the plant, for example through increased nitrogen-use efficiency as a result of reduced investment of N in enzymes of the photosynthetic pathway. Resultant changes in tissue chemistry, however, may translate into altered litter quality that in turn might impact upon decomposition rates of such material and hence upon biogeochemical cycling. When grown in elevated CO_2 atmospheres, a wide range of species exhibits reduced leaf tissue nitrogen content, along with increases in lignin and cellulose, compared to the same species grown at present day CO_2 concentrations. A higher lignin content renders leaves more recalcitrant in terms of their decomposability. Altered litter chemistry, including higher C:N ratios, thus is likely to slow down nutrient-cycling rates in ecosystems, in turn reducing their responsiveness to elevated CO_2 concentrations (Bazzaz 1996; Fig. 14.4). Leaf tissue chemistry also has profound implications with respect to herbivory. Plant survival of herbivory may be altered as a result of changes in allocation of C and N to secondary metabolites that act as anti-herbivore defences, whilst increased C:N ratios will reduce the quality of foliage as food for herbivores. Lower nitrogen content per unit mass of foliar tissue also requires the herbivore to consume more tissue in order to gain the same amount of nitrogen, increasing the impact of herbivory upon the plant. Such an increase in tissue predation levels is likely to impact upon competitive interactions amongst plant species, with resulting changes in plant community structure and composition. Evidence from studies of experimental plant communities exposed to herbivory suggests that species that do relatively well in these circumstances are characterized by their general competitive ability rather than by the extent of their CO_2 responsiveness (Bazzaz *et al.* 1995).

The potential consequences of increased atmospheric CO_2 concentration and increased temperatures, outlined above, ultimately may interact to alter the flux of carbon between the atmosphere and the soil–plant continuum. Any potential resulting shift in the balance between the carbon in the atmosphere and that sequestered in biomass, litter and soil is highly relevant to the discussion of future potential climates.

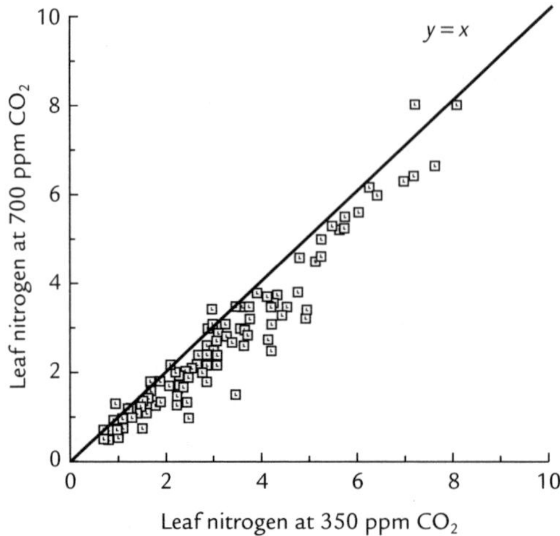

Fig. 14.4 Decline in leaf nitrogen concentration when plants are grown in a high CO_2 environment. From Bazzaz (1996).

The possibility of a net flux of carbon to the atmosphere as a result of increased rates of decomposition of biotic materials (as CO_2 from aerobic and as CH_4 from anaerobic, e.g. wetland, environments) generating a positive feedback through the consequent increased radiative forcing, as well as the direct effects upon vegetation of increased CO_2 concentration, is a key issue. Recent studies have demonstrated that carbon cycling in Boreal and Arctic wetlands strongly influences the global climate (Panikov 1999). Plant productivity was found to exert very important biological controls on CH_4 flux both through stimulation of methanogenesis, by increasing C-substrate availability (input of organic substances to soil through root exudation and litter production), and through bypassing of potential sites of CH_4 oxidation in the upper layers of the soil (Christensen *et al.* 2003), as a result of enhanced gas transport from the soil to the atmosphere *via* root aerenchyma.

Many of the ecosystems on earth are currently net sinks for carbon, fixing more than they release. A shift in this balance could have significant impacts through positive feedbacks, resulting in faster and more extensive climate changes than have hitherto been recorded.

14.3.2 Increased fluxes of UV-B as a consequence of 'thinning' of the stratospheric ozone layer

Anthropogenic alterations to the trace-gas composition of the atmosphere have not been restricted to increases in naturally occurring greenhouse gases – CO_2, CH_4 and N_2O – but have included, especially during the latter half of the 20th century, introducing increasing amounts of chlorofluorocarbons (CFCs) to the atmosphere. CFCs may reside in the atmosphere for many decades, accumulating in the stratosphere where, as a result of photolysis reactions, they release reactive halogens and halogen compounds, including chlorine and chlorine oxide, implicated in the breakdown of stratospheric O_3. Decreases in total-column ozone are now observed over large parts of the globe, permitting increased penetration of solar UV-B radiation to the Earth's surface. Depletion of O_3 in the atmospheric column is not uniform around the globe, but is more intense at higher latitudes and especially in polar regions. Since the 1970s, UV radiation reaching the surface during winter and spring has increased by *c.* 4–7% in northern and southern hemisphere mid-latitudes; over the same period UV reaching the surface during spring in the Antarctic and Arctic has increased by 130% and 22% respectively. Although it is at present difficult to predict longer-term future UV-B levels, current best estimates indicate a slow return to pre-ozone depletion levels may occur within *c.* 50 yr. However, confounding influences that remain poorly understood at the present time, especially the interactions of CFCs with other greenhouse gases and the atmospheric chemistry of CFC substitutes, render these predictions subject to considerable uncertainty.

Most of the UV-B radiation penetrating plant cells is absorbed, potentially causing acute injuries as a result of its high quantum energy. In addition to its photo-oxidative action, UV radiation causes photolesions, particularly in biomembranes. Although absorption of UV radiation by epicuticular waxes and by flavonoids dissolved in the cell sap provides higher-plant cells with considerable protection from radiation injuries, some damage does occur to DNA, membranes, photosystem II of photosynthesis and photosynthetic pigments. Recent research has shown the importance of the dynamic balance between damage and protection/repair mechanisms (e.g. DNA excision repair, scavenging of radicals formed by absorption of UV-B photons), and of the great variation between species with respect to this balance. Whereas, for example, some species have a high capacity for repair of DNA damaged by UV-B irradiation, others have a much weaker capacity. A growing body of evidence suggests that the effects of UV-B irradiation may be exerted primarily through altered patterns of gene activity rather than through physical damage (e.g. alteration of life-cycle timing, altered morphology, altered production of secondary metabolites leading to changes in palatability and in plant–herbivore interactions). Evidence to date strongly indicates that the primary responses of vegetation to increased UV-B levels result principally from shifts in the balance of competition between individual higher plant species in a community rather than from negative impacts upon the performance of individual species. However, it is currently difficult to predict the sign of such UV-mediated changes in species' interactions. It should also be noted that the responses of plants exposed to UV-B are modulated strongly by other environmental factors, such as atmospheric concentration of CO_2, water availability, temperature and nutrient availability, all of which, as we have already seen, are also changing.

14.3.3 Increasing tropospheric concentrations and deposition of various pollutants – SO_2, NO_x, NH_3 and O_3

Across much of the developed world, emission of the gas sulphur dioxide and its subsequent deposition from the atmosphere has historically been of great significance. This gas modifies plant growth responses through either acute, or more often chronic, toxic action, although with few, if any, visible symptoms. The phytotoxic effects of SO_2 gas and of its solution products have been studied extensively, particularly in relation to those taxa most susceptible to gaseous pollutants, such as the bryophytes and lichens that lack the protection of a cuticle.

Over recent decades, the shift from coal-burning to gas- and oil-fired power stations, coupled with large increases in road traffic volumes, has led to decreasing SO_2 emissions but increasing emissions of oxides of nitrogen (NO_x). In addition, there has been a substantial increase in emission of reduced nitrogen compounds, predominantly arising from intensive agricultural activities (e.g. ammonia (NH_3) from intensive animal rearing). A further complexity arising in the case of NO_2 emissions to the atmosphere is the sequence of complex photochemical interactions that ensue leading to the tropospheric formation of O_3, itself a highly phytotoxic gas. Compared to the greenhouse gases, the residence times of SO_2, NO_x and O_3 in the atmosphere are short because of their highly reactive nature and relatively rapid deposition back on to the Earth's surface. Nevertheless, atmospheric monitoring networks have provided clear evidence of measurable deposition of N-containing and acidic compounds remote from their sources. The high latitudes of the arctic regions are an excellent example, where local sources are negligible.

The duration of the passage of pollutant species through the atmosphere, following emission, has an important bearing upon the state in which they are deposited upon, or interact with, plants. They may interact with plant tissues whilst still in the gaseous state, or following deposition in aqueous droplets derived either from rain or from the fine droplets of mist or cloud – so-called dry, wet and 'occult' deposition respectively. In addition, aqueous deposition may be in either un-dissociated (e.g. $H_2SO_{4(aq)}$, $HNO_{3(aq)}$) or dissociated (e.g. H^+, SO_4^{2-}, NO_3^-, NH_4^+) state. Furthermore, in many situations concurrent deposition will occur of two or more of these pollutants. The impacts of deposition of these pollutants must therefore be considered not only in terms of acidity, toxicity and nutrient ion content, but also in terms of potential antagonistic or synergistic effects arising from combined deposition.

Of present-day atmospheric pollutants, increased deposition rates of nitrogen compounds and significant tropospheric concentrations of ozone represent serious threats to vegetation in a wide range of terrestrial ecosystems, although their impacts vary greatly between species. For example, the varied impacts upon plant growth of atmospheric deposition of N-compounds reflect great differences between species with respect to their N-requirements. Many plant species in natural and semi-natural ecosystems are adapted to grow in oligotrophic (low-nutrient) environments. Such species often compete successfully with other species only in such environments. An increase in nitrogen supply may lead to quantitative or even qualitative changes in the composition of the vegetation as a result of competitive displacement of species adapted to grow under low nutrient conditions. This is the case, for example, in Dutch heathlands, where increased inputs of nitrogen (mainly as NH_3) and increased soil acidity have been associated with declines of a number of heathland species, increased abundance of non-native species and accelerated replacement of dwarf-shrub heath by grass heath as part of a succession progressing towards woodland (Bobbink 1991; Houdijk *et al.* 1993). The overall result is a reduction in both species richness and species diversity. Such qualitative and quantitative changes in plant community composition and structure may have positive feedbacks to rates of N-cycling as a result of, for example, altered rates of soil nitrogen mineralization and nitrification (Table 14.1). However,

Table 14.1 Changes in ecosystem properties associated with vegetation change in Dutch heathlands. After vanVuuren *et al.* (1992). As species composition has changed, the potential for nitrogen loss, as nitrate, has increased. Nitrogen mineralization is a measure of the rate at which nitrogen is made available to plants; the percent nitrified is the percentage of the mineralized nitrogen that is converted to nitrate, which has the potential for being leached from the soil.

Vegetation	Net primary production ($g{\cdot}m^{-2}{\cdot}yr^{-1}$)	Net N mineralization ($g\text{-}N{\cdot}m^{-2}{\cdot}yr^{-1}$)	% nitrified
Calluna (native)	730	6.2	4.8
Molinia (invader)	2050	10.9	33.0
Deschampsia (invader)	430	12.6	42.9

high rates of deposition of N-compounds will not always promote vegetation growth and development; in many plant communities other nutrient species, especially P, will be limiting. Increasing N-deposition in such cases is more likely to result in increased N-saturation of the soil and consequent increased leaching of N into drainage waters (Wilson *et al.* 1995).

14.3.4 Interactive effects of pollutants, their deposition products and human land use

No one tropospheric pollutant is found in isolation, nor operates in isolation. As mentioned above, antagonistic and/or synergistic interactions are often seen, interactions between pollutants leading either to less damage when present together or, conversely, to greater damage when present in combination than when either is present alone. The character of the soil, including moisture status, temperature, nutrient status, etc., are also all of key importance in determining the impact of pollutants. These soil characteristics in turn may be a function of other pressures operating upon the ecosystem, such as management practices associated with particular land use. For example, in a recent modelling study of the carbon dynamics of northern hardwood forests, Ollinger *et al.* (2002) have shown that historical increases of atmospheric concentration of CO_2 and of N-deposition (over the period 1700–2000) have stimulated forest growth and carbon uptake. However, the degree of stimulation differs depending upon the intensity of human land management because this alters soil C- and N-pools and hence also alters the degree of growth limitation by C versus N. When other components of atmospheric pollution (e.g. tropospheric ozone) are factored into the model, this substantially offsets the increases in growth and C uptake resulting from CO_2 and N-deposition. Thus, for modelled intact temperate forests, at least, there is little evidence of altered growth since before the Industrial Revolution, despite substantial changes in the chemical environment experienced by these forests. Whether this result will be borne out by field measurement and manipulation studies under way in temperate forest stands remains to be seen. Findings to date, however, from both experiments and models, highlight the potential importance of interactions between confounding factors that are operating at the present time. They also highlight the need to understand these interactions if we are to be able accurately to predict their likely future impacts upon vegetation. It should finally be noted that there also remain many uncertainties associated with the range of experimental protocols used and hence with the conclusions reached.

14.4 Conclusions

Amongst the various aspects of global environmental change, the impacts of climate change are the most important. Climate change elicits large-scale spatial responses by species that in turn lead to qualitative changes in the composition, as well as structural changes, in many cases, of vegetation. The palaeo-ecological record shows the potential rates and magnitudes of these spatial responses, and also reveals their individualistic character. This individualism leads to the development of plant communities without a present-day analogue; such no-analogue communities result primarily from the response of species to combinations of environmental conditions, including climate, that lack a present analogue. One of the few certainties about the new century is that the combinations of carbon dioxide concentration, UV-B flux and climate that will develop are without current analogues. We can thus predict with reasonable confidence that plant communities without any present-day analogue, in terms of their composition and/or structure, will arise during the next century in many parts of the world.

In contrast to climate change, the other aspects of global change – CO_2, UV-B, N deposition and tropospheric pollutants – are secondary in the magnitude of their impacts, many of which are likely to be quantitative. Nonetheless, the impacts of these secondary factors are likely in many cases to become visible more quickly than will the impacts of climate change. This rapid visibility of their impacts must not be allowed to distract vegetation scientists' attention from the longer-term but very profound impacts of climate change.

The emergence of no-analogue environments will result in the emergence of new plant communities in the future, just as it did in the past. This effect is likely to be amplified by two further factors. Firstly, as noted above, transient communities and landscapes of greater species diversity may emerge in some cases during the period of transition from present to future communities. Secondly, because of the relatively rapid rate at which climate is expected to change over the next century, compared to past changes, species' migration capabilities may be exceeded. As a result, species' migrations may lag behind the changing climate. This is likely to affect the relatively slower growing and longer-lived species (S strategists *sensu* Grime 2001) much more than it will affect the rapidly growing species with short life cycles (R and C–R strategists). Because many of the latter species are also characterized by the production of numerous, small, widely-dispersed propagules, they are likely to experience a short-term benefit, because they will be able to migrate rapidly to exploit newly available areas. Many of them also will benefit from increased availability of N in areas with high rates of deposition of N-compounds from the atmosphere, and some too are likely to benefit from the increased atmospheric concentration of CO_2. As a result, many plant communities beyond present climatic ecotones may be replaced initially by communities dominated by shorter-lived, opportunist species of a relatively 'weedy' or ruderal character. Such communities may be analogous to early-successional communities within that present climatic boundary, although it is more likely that they will be without such analogues, because they will persist and develop without the influence of the late-successional species that would today come to dominate such stands in place of the early-successional taxa.

The potential for micro-evolution to select 'cryptic' genotypes with climatic tolerances or requirements not exhibited by a species' present population will also arise. This will result from selection of genotypes favoured by newly available no-analogue conditions, but that are destined always to die under present conditions. The extent to which this may occur is extremely difficult to assess; it is clear, however, that species were able to adapt to no-analogue conditions during the late-Quaternary. If it does occur to any significant extent, then selection of such 'cryptic' genotypes will render the prediction of species, and hence vegetation, responses to climate change fraught with even greater difficulties. Such potentially unpredictable adaptive changes may lead to 'surprise' responses of species and vegetation to climatic changes, especially when they occur in tandem with the many other ongoing changes to the global environment.

References

Bazzaz, F.A. (1996) *Plants in changing environments. Linking physiological, population and community ecology.* Cambridge University Press, Cambridge.

Bazzaz, F.A., Miao, S.L. & Wayne, P.M. (1995) Microevolutionary responses in experimental populations of plants to CO_2-enriched environments: parallel results from two model systems. *Proceedings of the National Academy of Science of the United States of America* **92**, 8161–8165.

Björse, G. & Bradshaw, R. (1998) 2000 years of forest dynamics in southern Sweden: suggestions for forest management. *Forest Ecology and Management* **104**, 15–26.

Bobbink, R. (1991) Effects of nutrient enrichment in Dutch chalk grassland. *Journal of Applied Ecology* **28**, 28–41.

Bradshaw, R.H.W. & Zackrisson, O. (1990) A two thousand year record of a northern Swedish boreal forest stand. *Journal of Vegetation Science* **1**, 519–528.

Christensen, T.R., Panikov, N., Mastepanov, M., Joabsson, A., Stewart, A., Öquist, M., Sommerkorn, M., Reynaud, S. & Svensson, B. (2003) Biotic controls on CO_2 and CH_4 exchange in wetlands – a closed environment study. *Biogeochemistry* **64**, 337–354.

Foley, J.A., Kutzbach, J.E., Coe, M.T. & Levis, S. (1994) Feedbacks between climate and boreal forests during the Holocene epoch. *Nature* **371**, 52–54.

Foster, D.R. (1983) The history and pattern of fire in the boreal forest of southeastern Labrador. *Canadian Journal of Botany* **61**, 2459–2471.

Grime, J.P. (2001) Plant Strategies, Vegetation Processes, and Ecosystem Properties. 2nd ed. John Wiley & Sons, Chichester.

Houdijk, A., Verbeek, P., van Dijk, H. & Roelofs, J. (1993) Distribution and decline of endangered herbaceous heathland species in relation to the chemical composition of the soil. *Plant and Soil* **148**, 137–143.

Huang, Y., Street-Perrott, F.A., Metcalfe, S.E., Brenner, M., Moreland, M. & Freeman, K.H. (2001) Climate Change as the Dominant Control on Glacial-Interglacial Variations in C3 and C4 Plant Abundance. *Science* **293**, 1647–1651.

Hunt, R., Hand, D.W., Hannah, M.A. & Neal, A.M. (1993) Further responses to CO_2 enrichment in British herbaceous species. *Functional Ecology* **7**, 661–668.

Huntley, B. (1988) Glacial and Holocene vegetation history: Europe. In: *Vegetation History* (ed. B. Huntley & T. Webb III), pp. 341–383. Kluwer Academic Publishers, Dordrecht.

Huntley, B. (1990) European post-glacial forests: compositional changes in response to climatic change. *Journal of Vegetation Science* **1**, 507–518.

Huntley, B. (1991) How plants respond to climate change: migration rates, individualism and the consequences for plant communities. *Annals of Botany* **67**, 15–22.

Huntley, B. (1999) The dynamic response of plants to environmental change and the resulting risks of extinction. In: *Conservation in a changing world* (eds G.M. Mace, A. Balmford & J.R. Ginsberg). Cambridge University Press, Cambridge. *Symposia of the Zoological Society of London* **72**, 69–85.

Huntley, B. & Birks, H.J.B. (1983) *An atlas of past and present pollen maps for Europe: 0-13000 B.P.* Cambridge University Press, Cambridge.

Huntley, B. & Webb III, T. (1989) Migration: species' response to climatic variations caused by changes in the earth's orbit. *Journal of Biogeography* **16**, 5–19.

Jackson, S.T. & Weng, C. (1999) Late Quaternary extinction of a tree species in eastern North America. *Proceedings of the National Academy of Science of the United States of America* **96**, 13847–13852.

Mylne, M.F. & Rowntree, P.R. (1992) Modeling the Effects of Albedo Change Associated with Tropical Deforestation. *Climatic Change* **21**, 317–343.

Ollinger, S.V., Aber, J.D., Reich, P.B. & Freuder, R.J. (2002) Interactive effects of nitrogen deposition, tropospheric ozone, elevated CO_2 and land use history on the carbon dynamics of northern hardwood forests. *Global Change Biology* **8**, 545–562.

Panikov, N.S. (1999) Fluxes of CO_2 and CH_4 in high latitude wetlands: measuring, modelling and predicting response to climate change. *Polar Research* **18**, 237–244.

Segerström, U., Hörnberg, G. & Bradshaw, R. (1996) The 9000-year history of vegetation development and disturbance patterns of a swamp-forest in Dalarna, northern Sweden. *Holocene* **6**, 37–48.

Street-Perrott, F.A., Mitchell, J.F.B., Marchand, D.S. & Brunner, J.S. (1990) Milankovitch and albedo forcing of the tropical monsoons: a comparison of geological evidence and numerical simulations for 9000 y BP. *Transactions of the Royal Society of Edinburgh* **81**, 407–427.

vanVuuren, M.M.I., Aerts, R., Berendse, F. & de Visser, W. (1992) Nitrogen mineralisation in heathland ecosystems dominated by different plant species. *Biogeochemistry* **16**, 151–166.

Watt, A.S. (1981) Further observations on the effects of excluding rabbits from grassland A in East Anglian Breckland: the pattern of change and factors affecting it (1936–73). *Journal of Ecology* **69**, 509–536.

Willis, A.J., Dunnett, N.P., Hunt, R. & Grime, J.P. (1995) Does Gulf Stream position affect vegetation dynamics in Western Europe? *Oikos* **73**, 408–410.

Wilson, E.J., Wells, T.C.E. & Sparks, T.H. (1995) Are calcareous grasslands in the UK under threat from nitrogen deposition? – an experimental determination of a critical load. *Journal of Ecology* **83**, 823–832.

Index

Page references to figures are in *italic*, to tables are in **bold**